Dielectric Materials for Electrical Engineering

Dielectric Materials for Electrical Engineering

Edited by
Juan Martinez-Vega

iSTE

WILEY

First published 2007 in France by Hermes Science/Lavoisier in two volumes entitled: *Matériaux diélectriques pour le génie électrique* © LAVOISIER 2007
First published 2010 in Great Britain and the United States by ISTE Ltd and John Wiley & Sons, Inc.

ISTE Ltd
27-37 St George's Road
London SW19 4EU
UK

www.iste.co.uk

John Wiley & Sons, Inc.
111 River Street
Hoboken, NJ 07030
USA

www.wiley.com

Library of Congress Cataloging-in-Publication Data

Matériaux diélectriques pour le génie électrique. English
Dielectric materials for electrical engineering / edited by Juan Martinez-Vega.
p. cm.
Includes bibliographical references and index.
ISBN 978-1-84821-165-0
1. Dielectric devices. I. Martinez-Vega, Juan. II. Title.
TK7872.D53M3813 2010
621.3--dc22

2009041391

British Library Cataloguing-in-Publication Data
A CIP record for this book is available from the British Library
ISBN 978-1-84821-165-0

Printed and bound in Great Britain by CPI Antony Rowe, Chippenham and Eastbourne

FSC
Mixed Sources
Product group from well-managed forests and other controlled sources
Cert no. SGS-COC-002953
www.fsc.org
© 1996 Forest Stewardship Council

Table of Contents

PART 1

General Physics Phenomena

Chapter 1

Physics of Dielectrics

1.1. Definitions

A dielectric material is a more or less insulating material (with high resistivity and with a band gap of a few eV), that is polarizable, i.e. in which electrostatic dipoles exist or form under the influence of an electric field.

Like any material, it is an assembly of ions with positive and negative charges which balance, for a supposedly perfect solid, so as to ensure electrical neutrality. This neutrality is observed at the scale of the elementary structural motifs which constitute solids with ionocovalent bonding (ceramics, for example) and on the molecular scale in molecular solids (polymers and organic solids).

The action of an electric field at the level of these element constituent of solids manifests itself by dielectric polarization effects. Let us remember that the dipole moment of a charge q with respect to a fixed system of reference centered in O is:

$$\vec{m}_t\,(0) = q\,\vec{r} \qquad [1.1]$$

where $\vec{r}$ is the vector which connects the point O to the charge's position.

If due to a force (caused, for example, by a magnetic field), the charge moves $\delta\vec{r}$, then the variation of the moment will be:

$$\delta\,\vec{m}_t(0) = q\,\delta\vec{r} \qquad [1.2]$$

Chapter written by Guy BLAISE and Daniel TREHEUX.

$\delta \vec{m}_t$ represents *the polarization effect* of the field on the charge. The generalization of expressions [1.1] and [1.2] to a collection of charges occurs by vectorial summation of the moments of each charge. An important case is that of a set of two charges $\pm q$, whose positions are defined by $\vec{r}_1$and $\vec{r}_2$ (see Figure 1.1). The application of [1.1] to the two charges gives:

$$\vec{m}_t(0) = q\,\vec{r}_1 - q\,\vec{r}_2$$

Setting $\vec{r}_1 = \vec{r}_2 + \vec{\ell}$, we get:

$$\vec{m}_t(0) = q\,\vec{\ell} = \vec{p} \quad [1.3]$$

$\vec{p}$ is called the dipole moment formed by the two charges, oriented from the negative charge to the positive charge (see Figure1.1).

The dipole moment appearing in a solid, during the application of a field $\vec{E}$, is (to a first approximation) proportional to it. We can then write:

$$\vec{p} = \varepsilon_0 \alpha \vec{E} \quad [1.4]$$

In this equation, α characterizes the polarisability of the species which gave the dipole and ε_0 the vacuum permittivity.

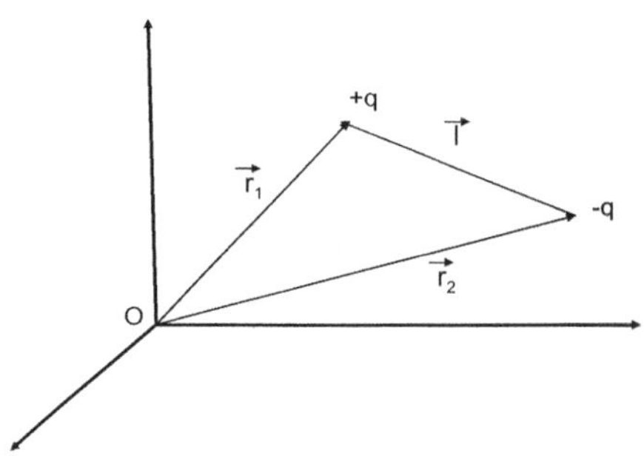

Figure 1.1. *Calculation of the dipole moment formed by 2 charges +q and –q*

1.2. Different types of polarization

To study dielectrics, it is necessary to first of all describe the different types of polarization. In order to do so, we must distinguish two types of solids: polar solids and non-polar solids.

1.2.1. *Non-polar solids*

In the case of non-polar solids, the centers of gravity of positive and negative charges coincide, and the dipole moment is therefore null (in the absence of a field). This is the case for solids with metallic bonding, or of numerous ionocovalent solids (ceramic Al_2O_3, ZrO_2, ZnO, SiO_2, etc.). Thus, the tetrahedron SiO_4 which constitutes the motif of quartz has a null dipole moment. It is the distortion of this tetrahedron, under the effect of a mechanical stress, which will make a polarization and the piezoelectric effect appear (see Figure 1.2).

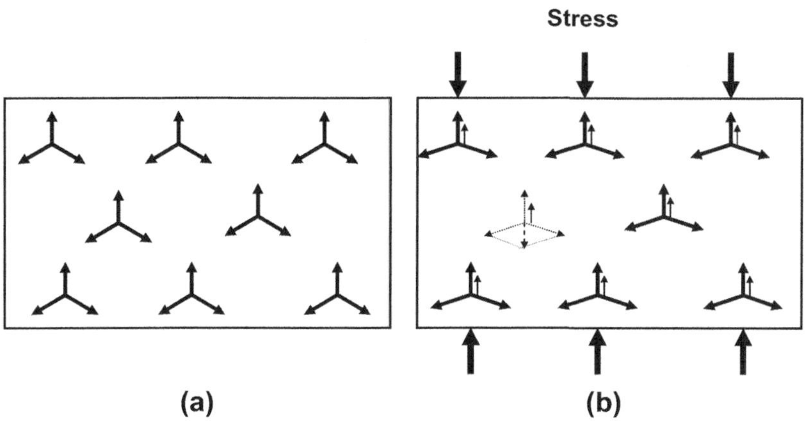

Figure 1.2. *(a) Quartz cristal at rest; it posesses a symmetry axis of order 3. The arrows represent the dipole moments whose resultant is null. (b) Subject to a mechanical stress according to the direction indicated, the network is distorted and the resulting dipole moment (small vertical arrow), is no longer null: an electric field appears*

1.2.2. *Polar solids*

Polar solids are composed of polar molecules for which the centers of gravity of the positive and negative charges do not coincide (for example a water molecule); this is molecular polarization. This is the case for most molecular solids and ferroelectric solids, which present a spontaneous polarization. Figure 1.3 gives, for example, the structure of barium titanate, a typical case of a ferroelectric body (and therefore also piezoelectric).

1.2.3. *Electronic polarization*

Let us consider the spherical orbital of an electron. Under the influence of an external electric field $\vec{E}$, the electrons are subject to a force $-e\vec{E}$ and the orbital gets distorted (see Figure 1.4). Consequently, the centers of gravity of the positive and negative charges which were initially merged, no longer are: this is electric polarization, and this leads to the formation of an electrostatic dipole; therefore, a dipole moment internal to the atom is characterized by:

$$\vec{p}_{\text{elect}} = \varepsilon_0 \alpha_{\text{elect}} \vec{E} \qquad [1.5]$$

which opposes itself to the field $\vec{E}$. α_{elect} is called the electronic polarisability. The polarization disappears if the field is removed.

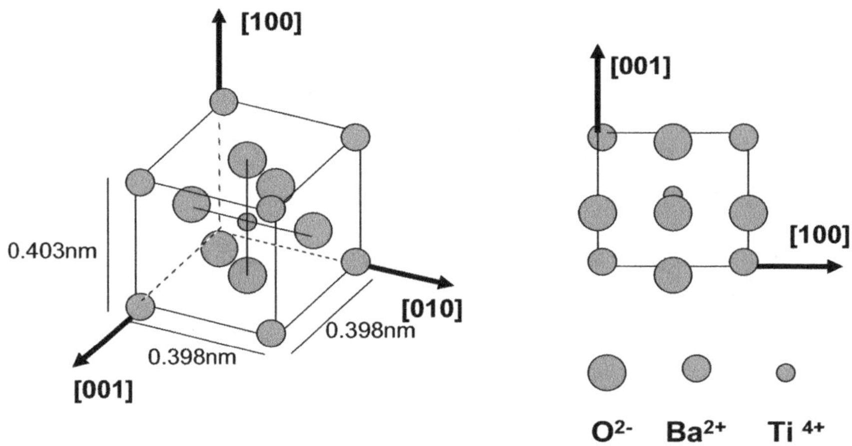

Figure 1.3. *Non-centrosymmetric crystalline structure of barium titanate $BaTiO_3$*

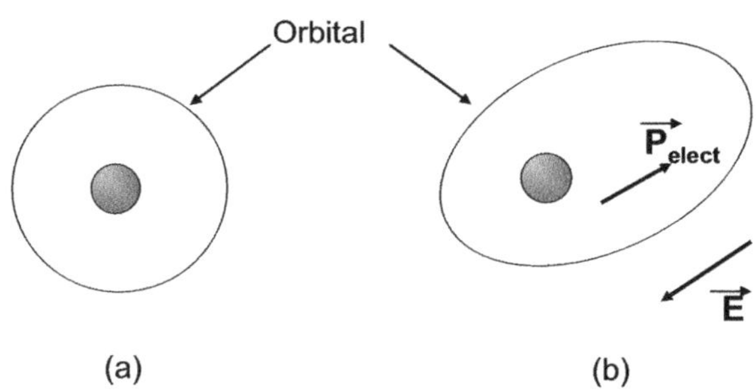

Figure 1.4. *(a) Orbital in the absence of electric field; (b) distortion of the orbital and appearance of electronic polarization $\vec{p}_{elect}$ in the presence of a field $\vec{E}$*

1.2.4. *Ionic polarization*

In the case of ionic crystals, the average position of positive and negative ions changes under the influence of a field $\vec{E}$. Suppose the ion is perfectly rigid from every angle. The action of the field will be to move it a quantity $\vec{\ell}$ with respect to a fixed mark centered in O; hence a variation of the polar moment:

$$\vec{p}_{ion} = q\,\vec{\ell} = \varepsilon_0 \alpha_{ion} \vec{E} \qquad [1.6]$$

This is the induced ionic polarization, proportional to the field (elastic distortions); where α_{ion} is the ionic polarisability.

The total dipole moment attached to the displacement of the ion and to the distortion of the electronic orbitals is, to a first approximation, the sum of [1.5] and [1.6], that is to say:

$$\vec{p}_t = \vec{p}_{ion} + \vec{p}_{elect} \qquad [1.7]$$

1.2.5. *Orientation polarization*

When we subject a polar molecule, carrier of a permanent dipole moment $\vec{p}_0$, to an electric field $\vec{E}$, its dipole tends to turn towards the direction of the field, which leads to a distortion of the molecule related to a torque: this is orientation polarization. This distortion is not instantaneous. There is the appearance of a hysteresis, on the one hand because the molecular forces tend to block its motion and, on the other hand, the thermal agitation will tend to disorient the molecules with respect to one another.

If $\vec{p}_0$ makes an angle θ with the direction of the field, the torque is:

$$\vec{\Gamma} = \vec{p}_0 \wedge \vec{E}$$

The application of a field will have the effect on each molecule of producing a polar component in the direction of the field, whose first-order expression is:

$$\vec{p}_{or} = \varepsilon_0 \alpha_{or} \vec{E} \qquad [1.8]$$

α_{or} is called orientational polarisability. In general, $\vec{p}_{or} << \vec{p}_o$.

1.2.6. *Interfacial or space-charge polarization*

This type of polarization plays a part when the material possesses different phases or permittivity zones. Subject to a low-frequency electric field (from 10^{-1} to 10^{2} Hz), this material will behave as though it contains electric charges with interfaces separating the zones. However, these charges are not real charges, but known as "polarization" charges (see section 1.3.3).

1.2.7. *Comments*

Units: a dipole moment is the product of a charge by a distance; it is therefore measured in C.m. A commonly used unit is the Debye:

$$1 \text{ Debye} = \frac{1}{3} 10^{-29} \text{ C.m} = e \times 2.08\ 10^{-11} \text{ m (e charge of the electron).}$$

A dipole moment is measurable, unlike q and l. The dipole moment must be taken as an entity, in the same way as an electric charge.

Polarization vector: this is the dipole moment per unit volume. By analogy with a capacitor, we can write:

$$\vec{P} = \varepsilon_0 (\varepsilon_r - 1)\ \vec{E} = \varepsilon_0\ \chi\ \vec{E} \qquad [1.9]$$

where ε_0 is the vacuum permittivity, ε_r is the dielectric constant of the material and $\chi = (\varepsilon_r - 1)$ is its dielectric *susceptibility*.

1.3. Macroscopic aspects of the polarization

1.3.1. *Polarization of solids with metallic bonding*

Rather than address non-conductive materials, it is interesting first of all to describe the polarization phenomena appearing in a metal. In this case, each ion of the solid is neutralized on the scale of the atomic volume by "free" electrons. The ions do not move under the action of the field and subsequently do not introduce any dipole moment. On the other hand, the conduction electrons go up the field (polarisability $\alpha \rightarrow \infty$) until they reach the limit of the solid: the electrons accumulate on the surface of the solid by which the field enters, leaving an excess of positive charges on the surface by which it comes out (see Figure 1.5). This giant dipole creates an internal field $\vec{E}_{ch}$ within the solid which opposes itself to the applied field $\vec{E}_a$. The motion of charges takes place until the total field is null:

$$\vec{E}_{internal} = \vec{E}_{ch} + \vec{E}_{a} = \vec{0} \qquad [1.10]$$

We say that the free charges come to screen the applied field. The total charge of the solid is null but its surface is positively charged on one side and negatively on the other.

1.3.2. *Polarization of iono-covalent solids*

Unlike metals, there are no free charges in a perfect iono-covalent solid: there is therefore no screen with the applied field. Each ion of the elementary structural motif is subject to a polarization, such that the solid presents a dipolar structure at the atomic or molecular scale (see Figure 1.6). The solid being neutral, the internal field at a point is the sum of the applied field and the field created by all of the dipoles.

$$\vec{E}_{internal} = \vec{E}_{dip} + \vec{E}_{a} \neq \vec{0} \qquad [1.11]$$

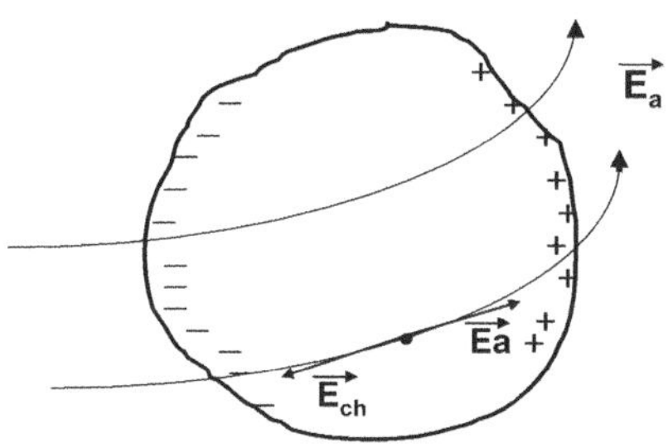

Figure 1.5. *Polarization of a solid with metallic bonding*

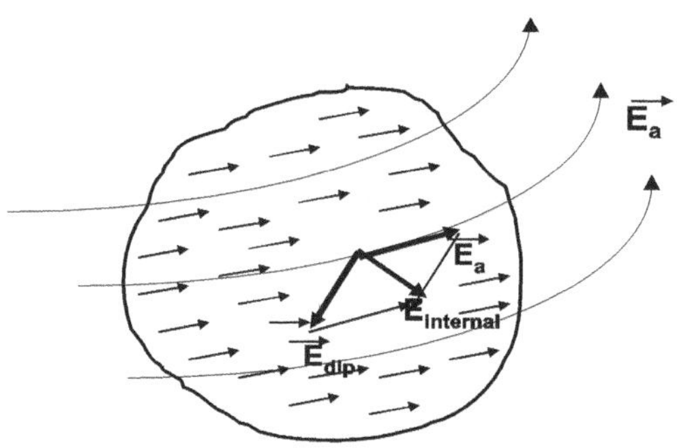

Figure 1.6. *Polarization of an iono-covalent solid*

This internal field is called the local field ($\vec{E}_{internal} = \vec{E}_{loc}$). It is this field which is responsible for the polarization of the medium whose description was given in section 1.2.

If each atom, i, of a solid with a cubic lattice of parameter, a, carries a dipolar moment, $\vec{p}_i$, the polarization vector is defined by:

$$\vec{P}_i = \frac{Na}{a^3} \vec{p}_i \qquad [1.12]$$

where N is the number of atoms per cell.

If $\vec{P}(\vec{r})\, d^3r$ is the dipole moment at point $\vec{r}$ of an element with continuous volume d^3r, the potential $dV(\vec{R})$ created by this dipole moment at a point $\vec{R}$ (see Figure 1.7) is:

$$dV(\vec{R}) = \frac{1}{4\pi\varepsilon_0} \vec{P}(\vec{r})\, \vec{grad} \frac{1}{|\vec{r} - \vec{R}|} d^3r \qquad [1.13]$$

Let us consider a continuous solid of volume $\mathcal{V}$ and surface S, totally neutral (with no excess charges in the medium), subject to an external field $\vec{E}_a$ (see Figure 1.7). It presents a polarization $\vec{P}(\vec{r})$. The potential $V(\vec{R})$ created at point $\vec{R}$ by the dipole moment $\vec{P}(\vec{r})\, d^3r$ is (from [1.13]):

$$V(\vec{R}) = \frac{1}{4\pi\varepsilon_0} \iiint_V \vec{P}(\vec{r})\, grad \frac{1}{|\vec{R} - \vec{r}|} d^3r \qquad [1.14]$$

and the created field has a value of:

$$\vec{E}(\vec{R}) = -\vec{grad}\, V(\vec{R}) \qquad [1.15]$$

Using mathematical operations (Ostrogradski and Green), we get:

$$V(\vec{R}) = \frac{1}{4\pi\varepsilon_0} \iiint_V \frac{-\mathrm{div}\vec{P}}{|\vec{R} - \vec{r}|} d^3r + \frac{1}{4\pi\varepsilon_0} \iint_S \frac{\vec{P}d\vec{s}}{|\vec{R} - \vec{r}|} \qquad [1.16]$$

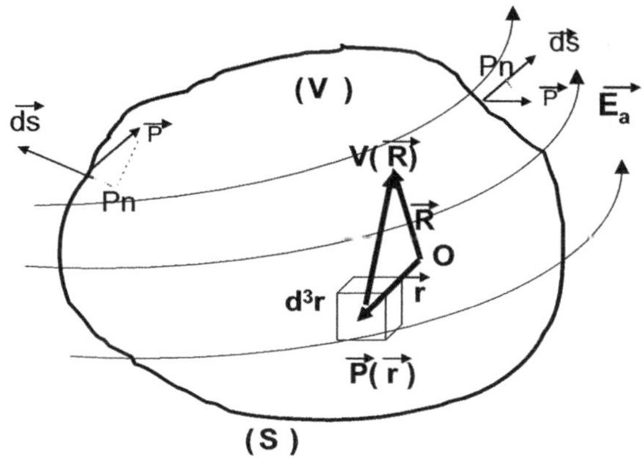

Figure 1.7. *Representation of a solid (V), of surface (S) subject to an external field* $\vec{E}_a$

1.3.3. *Notion of polarization charges*

The two integrals of [1.16] are interpreted as being Coulomb integrals. Indeed, we can write the second integral of [1.16] as: $\vec{P}d\vec{s} = P_n ds$, with P_n being the projection of $\vec{P}$ following the direction of the surface element, directed towards the outside (see Figure 1.7). P_n has the dimension of a surface charge density σ_p. Similarly for the first integral $\rho_p = -\text{div}\vec{P}$ represents a density per unit volume of charges. The densities σ_p and ρ_p are known as *polarization charge* densities.

These polarization charges, in a neutral medium (without excess charges) are not real electric charges; it is a convenient equivalence.

1.3.4. *Average field in a neutral medium*

To the potential $V(\vec{R})$ given by [1.16], due to the polarization, we must add the potential $V_a(\vec{R})$ due to the applied electric field. The potential in $\vec{R}$ is therefore:

$$V_M(\vec{R}) = V(\vec{R}) + V_a(\vec{R}) \quad [1.17]$$

And the field in $\vec{R}$ is given, by using the polarization charges, by:

$$\vec{E}_M(\vec{R}) = \vec{E}_a(\vec{R}) + \vec{E}_{dep}(\vec{R}) + \vec{E}_{\rho_p}(\vec{R}) \quad [1.18]$$

$\vec{E}_{dep}(\vec{R})$ is the field due to surface polarization charges whose effect within the dielectric is to oppose itself to the applied field $\vec{E}_a$. As a result of this, it is called the depolarization field. $\vec{E}_{\rho p}(\vec{R})$ is the field created by the volume polarization charges.

$\vec{E}_M(\vec{R})$ is the average field. It is the usual macroscopic field defined at all points of the medium and the one that we measure (for capacities, for example). By analogy with [1.9] we can write:

$$\vec{P} = \varepsilon_0 (\varepsilon_r - 1)\, \vec{E}_M = \varepsilon_0 \chi \vec{E}_M \qquad [1.19]$$

where χ is the dielectric susceptibility.

The volume polarization charges ensuing from [1.19] have a density of:

$$\rho_p = -\text{div}\vec{P} = -\text{div}(\varepsilon_0 \chi \vec{E}_M) = -\varepsilon_0 \chi\, \text{div}\vec{E}_M - \varepsilon_0 \vec{E}_M\, \text{grad}\, \chi \qquad [1.20]$$

with, from [1.18], $\text{div}\vec{E}_M = \text{div}\,\vec{E}_a + \text{div}\vec{E}_{dep} + \text{div}\vec{E}_{\rho p}$

The sources of the fields $\vec{E}_a$ and $\vec{E}_{dep}$ are either outside, or at the periphery of the medium, subsequently: $\text{div}\,\vec{E}_a = \text{div}\vec{E}_{dep} = 0$.

As for the field, $\vec{E}_{\rho p}$, due to the polarization charges, its divergence has the expression: $\text{div}\vec{E}_{\rho p} = \rho_p / \varepsilon_0$ and, subsequently:

$$\text{div}\vec{E}_M = \rho_p/\varepsilon_0 \qquad [1.21]$$

Plugging [1.21] into [1.20], the polarization charge is written:

$$\rho_p = -\,\varepsilon_0 \frac{E_M}{1+\chi}\, \vec{grad}\, \chi \qquad [1.22]$$

Subsequently, in a neutral medium, the polarization charges are due to the gradient of dielectric susceptibility. In other words, a medium in which susceptibility varies presents volume polarization charges.

The electric induction is defined by:

$$\vec{D} = \varepsilon_0 \vec{E}_M + \vec{P} \qquad [1.23]$$

So, taking into account [1.19]: $\vec{D} = \varepsilon_0 (1+\chi)\vec{E}_M$, and setting $\varepsilon = \varepsilon_0 (1+\chi)$:

$$\vec{D} = \varepsilon \vec{E}_M \quad [1.24]$$

where ε is the dielectric permittivity of the material, sometimes called the dielectric constant.

From [1.21] and [1.23], it follows:

$$\mathrm{div}\vec{D} = \varepsilon_0\, \mathrm{div}\vec{E}_M + \mathrm{div}\vec{P} = \rho_p - \rho_p = 0$$

In a neutral medium the divergence of the induction is null ($\vec{D}$ is at conservative flux). Equally, we can theorize the proportionality between $\vec{P}$ and the depolarization field:

$$\vec{P} = -\varepsilon_0/\lambda\ \vec{E}_{dep} \quad [1.25]$$

The minus sign indicates that $\vec{E}_{dep}$ has an opposite effect to that of $\vec{E}_a$ on the polarization, i.e. it has a depolarizing effect. The fact that $\vec{E}_{dep}$ reduces the polarization produced by $\vec{E}_a$ implies that $|\vec{E}|_M < |\vec{E}|_a$.

1.3.5. *Medium containing excess charges*

Let ρ_a be the charge density of charges and $\vec{E}_{\rho a}$, the field they produce. This field must be added to the expression [1.18], so:

$$\vec{E}_M(\vec{R}) = \vec{E}_a(\vec{R}) + \vec{E}_{dep}(\vec{R}) + \vec{E}_{\rho p}(\vec{R}) + \vec{E}_{\rho a}(\vec{R}) \quad [1.26]$$

The presence of charges in the medium imposes $\mathrm{div}\vec{D} \neq 0$.

The problem can be tackled in two ways:

– we can treat these excess charges as charges external to the dielectric and associate their field with the applied field $\vec{E}_a$ whose sources are outside the medium. We will therefore set: $\vec{E}'_a = \vec{E}_a + \vec{E}_{\rho a}$ and subsequently $\mathrm{div}\vec{E}'_a = \rho_a/\varepsilon_0$,

– we can also integrate these charges to the medium because, as we will see, these excess charges are generally *trapped charges*, which affect the physical characteristics of the medium considerably. We then set:

$$\vec{E}'_{ch} = \vec{E}_{\rho_a} + \vec{E}_{\rho_p} \text{ et } \operatorname{div}\vec{E}'_{ch} = (\rho_p+\rho_a)/\varepsilon_0$$

The final result is identical according to both approaches with:

$$\operatorname{div}\vec{E}'_M(R) = (\rho_p+\rho_a)+/\varepsilon_0 = \rho/\varepsilon_0$$

where ρ is the total density of charges (polarization charges plus excess charges).

The application of [1.23] gives the expression for the induction:

$$\operatorname{div}\vec{D} = \varepsilon_0 \operatorname{div}\vec{E}_M + \operatorname{div}\vec{P} = \rho_p+\rho_a-\rho_p = \rho_a$$

Thus, in a charged medium, the divergence of the induction is equal to the density of excess charges.

1.3.6. *Local field*

We have seen (in equation [1.11]) that, within a dielectric subject to an applied field $\vec{E}_a$, an internal field (known as a local field) prevails, such as:

$$\vec{E}_{loc} = \vec{E}_{dip} + \vec{E}_a \neq \vec{0}$$

where $\vec{E}_{dip}$ is the field created by all of the dipoles.

Each ion of a solid is therefore solicited by this local field different to the applied field. The dipole moment which appears on a site j of the lattice is therefore expressed by:

$$\vec{P}_j = \varepsilon_0\alpha_j \vec{E}_{loc}(j)$$

where α_j is the induced or orientational polarisability, depending on the material. The field created by all of the dipoles in j is the vectorial sum of the fields of each dipole, that is to say:

$$\vec{E}_{loc}(j) = \sum_{i \neq j} \vec{E}_i(j)$$

Each dipole i will create a potential at the point j whose expression is similar to equation [1.13].

1.3.7. *Frequency response of a dielectric*

When the applied field varies over time (harmonic field E(ω,t)), this field induces a polarization P(ω,t), defined from the expression [1.19]:

$$P(\omega,t) = \varepsilon_0 \chi(\omega) \vec{E}(\omega,t).$$

This polarization is the sum of each type of polarization. But the reaction of a material to a type of polarization is not instantaneous. Thus, there is a phase difference δ between an alternative electric field ($E= E_0 \sin\omega t$) and the polarization $P=P_0 \sin(\omega t-\delta)$. In complex notation, we can write $P^*= P_0 \exp(i(\omega t-\delta))$ and $E^*= E_0 \exp(i\omega t)$. The values of the polarization and of the dielectric constant depend on the ease with which the dipole moments reorient themselves when the direction of the field varies. The time required for this reorientation to take place is called relaxation time, τ, and its inverse the relaxation frequency, f.

Given that the relaxations are related to thermal agitation, the frequency, f, of the material is a function of the temperature (f increases with T). When the frequency electric field is much stronger than the relaxation frequency of a type of polarization, this polarization cannot be produced. Conversely, if the frequency of a field is much less than the relaxation frequency, the polarization is produced instantaneously and the phase difference between P and E is null. But if the frequency of the field and the relaxation frequency are close, the phase difference, δ, increases to reach a maximum value. In this case, the curves D=f(E) or P=f(E) form a hysteresis buckle. The area of this buckle represents the energy loss per cycle and per unit volume of the material.

Let a solicitation S= S'+ iS'', in complex notation. The relationship S''/S' = tgδ allows us to find δ, the "loss angle" or "dissipation coefficient". This coefficient corresponds to all of the energy dissipations in the material during its solicitation.

The most classical solicitations are mechanical or electrical. In the first case, we find the anelasticity phenomenon encountered for the mechanical properties of the materials. We note G' and G'' the real and complex modules which lead to *the mechanical loss angle* δ_m. For a dielectric, we consider the real permittivity ε' and the complex permittivity ε'' and *the dielectric loss angle* δ_e.

Electronic and ionic polarizations, which bring about short-distance rearrangements, persist in a large range of frequencies. On the other hand, for molecular materials, the orientation polarization imposes reorientations of the dipoles at the molecular scale. It can't take place above a certain frequency, determined by the size of the molecules and by the molecular dipole moments. We must then take into account the variations of the dielectric constant and the loss factor according to the frequency.

From an experimental point of view, the mechanical solicitations have frequencies in the range 10^{-6} and 10^{7} hertz, which permits them to act especially on the molecular chains in polymers. The electrical solicitations can have larger ranges of frequencies, up to 10^{15} hertz, which allows ionic and electronic vibrations to be analyzed.

1.4. Bibliography

[JON 83] JONSCHER A.K., *Dielectric Relaxation in Solids*, Chelsea Dielectric Press, London, 1983.

[KEL 89] KELDYSH L.V., KIRZHNITZ D.A., MARADUDIN A.A. (Eds), *The Dielectric Function of Condensed Systems: Modern Problems in Condensed Matter Sciences*, North Holland, Elsevier Science Ltd, 1989.

[LAN 69] LANDAU L., LIFCHITZ E., *Electrodynamique des milieux continus*, éditions MIR, 1969.

[SCA 89] SCAIFE B.K.P., *Principles of Dielectrics*, Clarendon Press, 1989.

Chapter 2

Physics of Charged Dielectrics: Mobility and Charge Trapping

2.1. Introduction

For the lay person, an insulating material is a material which does not conduct electricity. In fact, however, it is necessary to analyse this assertion more closely.

The conductivity of a material is defined by:

$$\sigma = ne\mu \qquad [2.1]$$

where n is the density of the charge carriers and μ the mobility.

Table 2.1 gives the values, at room temperature, of the conductivity and mobility of the electrons in high purity copper and alumina (aluminum oxide). We note that there are 23 regions between the two conductivities, but the mobility is twice as high in alumina as in copper. These values, which could be surprising, are due to the fact that, from [2.1], alumina contains much less than an electron per cm^3 which is used for conduction while copper has about 4.5×10^{22} electrons per cm^3 used for conduction (Table 2.1).

Alumina has, therefore, a very weak conductivity because it has practically no free charges to conduct (and not because it is inept at conducting charges). For a

---

Chapter written by Guy BLAISE and Daniel TREHEUX.

pure material with a large band gap, we have, then, an insulating state, through lack of charge carriers.

For conduction to be observed in an insulator, we need to inject charges into it. Unlike a metal, the bulk of which remains neutral during conduction, an insulating material is thus charged. The injection of charges can be done in many different ways, as will be developed in later chapters of this book. Here, we mention, for example, irradiation (electronic, radiative, etc.), corona discharge, the application of an electrical constraint, and also mechanical or tribological ones.

When we manage to inject charges in a conduction band, the strong mobility of these charges (Table 2.1) is explained by the fact that the electron-electron interactions are reduced to nothing, as a result of the weak density of the free charges. In a perfect insulator, only the electron-phonon interaction remains to limit mobility. But, if the solid contains impurities, they will also reduce the mobility. This explains, as we shall see, the very important role which impurities take in the properties of insulators.

	Copper	**Alumina**
σ (Ω^{-1}cm^{-1})	$5.8.10^{5}$	10^{-18}
μ (cm^2.V^{-1}.s^{-1})	80	200
n (e^{-1} .cm^{-3})	$4.5.10^{22}$	$3.1.10^{-2}$

Table 2.1. *Conductivity σ and mobility μ of pure monocrystalline copper and alumina*

2.2. Localization of a charge in an "ideally perfect" and pure polarizable medium

2.2.1. *Consideration of the polarization*

Polarization must be taken into account to describe the conduction properties of the dielectrics. The electric field which polarizes the medium is the coulomb field of the charge in motion in the conduction band.

Let us consider an electron introduced into the polarizable medium. It will interact with one of the electrons of the medium situated at a distance r. The potential energy of interaction for two electrons in the permittivity medium ε is equal to $e^2/4\pi\varepsilon r$. Without polarization of the medium, this same interaction would be that produced in a vacuum, which is $e^2/4\pi\varepsilon_0 r$. The fact of polarizing the medium then lowers the potential energy by a quantity:

$$\Delta W = e^2/4\pi\varepsilon r - e^2/4\pi\varepsilon_0 r \qquad [2.2]$$

We notice that ΔW is negative and that, consequently, the electron is trapped in a potential well. This observation led Landau [LAN 69] to introduce the *polaron concept* in 1933, which will be further analyzed here from an electrostatic as well as a quantum point of view.

Let an electron of kinetic energy w_e interact with an electron connected to the medium whose orbital frequency is $\omega_0 = \hbar / mr_0^2$, where m is the mass of the electron and r_0 its orbital extension. The time t taken by the electron to cover the distance r is equal to:

$$r = \sqrt{\frac{2w_e}{m}}\, t \qquad [2.3]$$

For an electron connected to the medium to be able to follow the displacement of the electron over the distance r, it must describe, numerous times, its trajectory during time t, that is to say: $t \geq 2\pi/\omega_0$. So only the electrons situated at a distance greater than r_e, defined by the condition $\omega_0 t = 2\pi$, could follow the motion of the electron. Expressing this condition and replacing ω_0 by its value ($\omega_0 = \hbar / mr_0^2$), we get, from equation [2.3]:

$$r_e = 2\pi \frac{r_0^2}{\hbar} \sqrt{2m w_e} \qquad [2.4]$$

Only the electrons situated outside the sphere with radius r_e centered on the moving electron could fit its motion by an electronic polarization.

Numerically, if the energy w_e gets very big, r_e tends to infinity and there is no electronic polarization, and the electron then moves like in vacuum (with permittivity ε_0). Conversely, if the energy of the electron tends to kT, at room temperature, r_e tends to values lower than interatomic distances (r_e less than 0.15 nm for frequencies ω_0 of the order of 10^{15} to 10^{16} s^{-1} in the UV): the electron then moves in a polarized medium of permittivity $\varepsilon(\omega_0)$.

The ionic polarization around the moving charge can be analyzed in the same way, but as the highest oscillation frequencies of ions ($\omega \cong 10^{12}$ s^{-1}) are situated in the IR, the radius associated with this polarization is much greater than r_e.

The variation of the polarization with distance is shown in Figure 2.1. Very near the moving charge ($r < r_e$) there is very little polarization and then the permittivity is that of vacuum ε_0. For $r_e < r < r_i$ the electronic polarization is the only one which plays a part. The ionic polarization will come into action for $r > r_i$ and, at lower and lower frequencies, we reach static permittivity $\varepsilon(0)$.

This variation of permittivity leads then, as seen in Chapter 1, to the formation of polarization charges with an opposite sign to the charge in motion. The result is negative interaction energy between the two types of charges, which leads to a confinement of the moving charge.

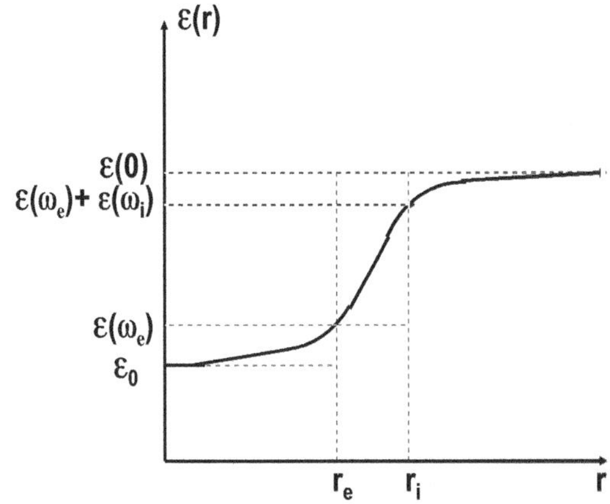

Figure 2.1. *Variation of the polarization as a function of the distance r*

2.2.2. *Coupling of a charge with a polarizable medium: electrostatic approach*

For simplicity, let us take a fixed charge q, surrounded by a permittivity medium $\varepsilon(r)$ varying radially between $\varepsilon_1 = \varepsilon(\infty)$ for $r < r_1$ and a maximum value ε_2 for $r > r_2$ (see Figure 2.2). Polarization charges appear whose density is given by equation [1.20]:

$$\rho_p(r) = -\text{div}\vec{P}(r) = -\text{div}(\varepsilon_0 \chi \vec{E}_M) = \varepsilon_0\, \varepsilon(r)\vec{E}(r)\ \text{grad}\,[1/\varepsilon(r)] \qquad [2.5]$$

With $\vec{P}(r) = [\varepsilon(r) - \varepsilon_0]\vec{E}(r)$, polarization at distance r and $\vec{E}(r) = \dfrac{1}{4\pi\varepsilon(r)}\dfrac{q}{r^2}$, average field in r.

The polarization charges contained between two spheres of radii r and r+dr are:

$dQ_p(r) = 4\pi r^2 \rho_p(r)dr = 4\pi\, \varepsilon_0\varepsilon(r)\, r^2\vec{E}(r)\ d\,[1/\varepsilon(r)]$. It follows after integration:

$$Q_p(r) = \left[\frac{1}{\varepsilon(r)} - \frac{1}{\varepsilon 1}\right] q \qquad [2.6]$$

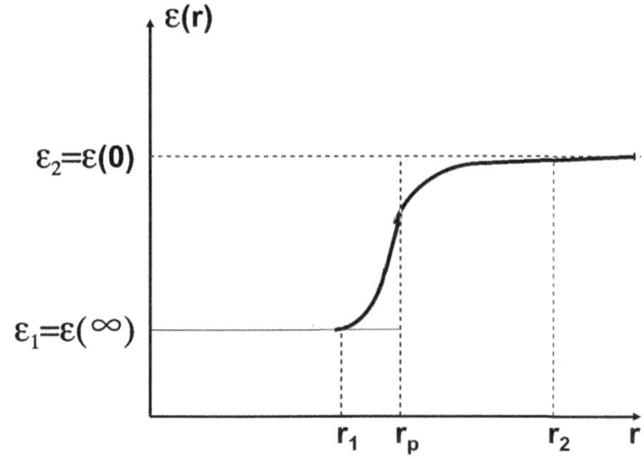

Figure 2.2. *Variation of the permittivity of the medium as a function of distance*

The potential energy resulting from the interaction between the charge and the polarization charges is then written as:

$$W= \frac{q^2}{4\pi\varepsilon_0}\int_{r1}^{r2}\frac{\rho_p(r)}{r}4\pi r^2 dr = \frac{q^2}{4\pi}\int_{r1}^{r2}\frac{d[1/\varepsilon(r)]}{r} \quad [2.7]$$

Integration by parts leads to:

$$W = \frac{q^2}{4\pi}\left[\frac{1}{\varepsilon_2 r_2}-\frac{1}{\varepsilon_1 r_1}\right]+\frac{q^2}{4\pi}\int_{r1}^{r2}\frac{1}{\varepsilon(r)r^2}dr \quad [2.8]$$

If r_1 and r_2 tend to a common value r_p, the second term tends to 0 and consequently:

$$W = -\frac{q^2}{4\pi r_p}\left[\frac{1}{\varepsilon_1}-\frac{1}{\varepsilon_2}\right] \quad [2.9]$$

This expression corresponds to the binding energy of the polaron defined by Landau, where r_p is the radius of the polaron. We note that, the smaller r_p is, the more the charge will be localized and the higher the binding energy.

2.2.3. *Coupling of a charge with a polarizable medium: quantum approach*

This approach is based on the polaronic concept introduced by Landau in the 1930s [LAN 69], and resumed later by Austin and Mott [AUS 69]. The idea stemmed from a study which expressed the interaction between a given charge q and the longitudinal optical vibrations of a network. In this quantum frame, the localization of an electronic charge (e) in a dielectric medium ensues from its coupling with different polarization fields, which develop:

– a field created by the electronic polarization in the domains of UV frequencies;

– a field stemming from the atomic vibrations (phonons) in the infrared;

– a static field created by the displacement of ions.

2.2.3.1. *Coupling with the electronic polarization field*

Valence electrons are the quickest to react to the action of a field. The time constant, in the order of 10^{-15}– 10^{-16} s, corresponds to the opposite of the orbital frequencies ω_{UV}^{-1} situated in the ultraviolet range. If the time taken by an electron to traverse the basic cell of the material is greater than ω_{UV}^{-1}, we can consider that all the electrons of the medium can fit the motion of the electron perfectly. It implies that the electron has a kinetic energy less than 10eV. The electron then moves in a medium containing electronic dipoles which produce an electronic polarization field $\vec{P}_e$. Being in motion, the electron can then couple itself with this field, hence the confinement effect according to Landau. It moves at an average speed V, in an electronic permittivity medium $\varepsilon(\omega_e)$ where ω_e represents the electronic pulsation: it is the electronic polaron which is a very fluid pseudo-particle.

2.2.3.2. *Coupling with the ionic polarization field*

The atomic vibration modes (phonons) of an ionic component are sources of a field. Taking the case of ions vibrating in phase opposition, we see that each couple of ions forms a dipole. When the dipoles are parallel to each other, the field is very intense and the coupling with the charge is very strong. This situation corresponds to the domain of atomic vibrations in IR frequencies $\left(\omega_{IR} \approx 10^{12} - 10^{13}\ s^{-1}\right)$. The result of this coupling is a lattice polaron: it is the usual quantum polaron described by Fröhlich [FRÖ 54], [FRÖ 62] which describes the coupling of a charge with a longitudinal optical phonon.

By definition, the quasi-particle formed by the charge in motion and the polarization charges which surround it in a radius rp form the polaron. The radius rp depends on the force of coupling. We can then understand that the polaron is affected by an effective mass m*, clearly greater than the mass m of the electron, and that its motion is very dependent on its radius rp.

The longitudinal waves respond at best to the solicitations of the field and consequently maintain a better coupling. Fröhlich uses the approximation of a coupling with only one optical mode of vibration, the dispersion being neglected. The expression for the infrared polarization is written:

$$\vec{P}_{IR} = \varepsilon_0 \left[\frac{1}{\varepsilon(\infty)} - \frac{1}{\varepsilon(0)} \right] \vec{D} \qquad [2.10]$$

where $\vec{D}$ is the electric induction, $\varepsilon(\infty)$ the optical (electronic) permittivity and $\varepsilon(0)$ the static permittivity.

The total potential energy (potential energy of the charge carrier q surrounded by the polarization charge whilst taking into account the polarization energy of the surrounding medium) which ensues is, then:

$$W_p = - \frac{1}{8\pi} \left[\frac{1}{\varepsilon(\infty)} - \frac{1}{\varepsilon(0)} \right] \frac{q^2}{r_p} \qquad [2.11]$$

We note that equation [2.11] has the same form as equation [2.9] deduced from the electrostatic approach.

The result is that, since $\varepsilon(\infty) \ll \varepsilon(0)$, the bonding effect is controlled by electronic permittivity, i.e. the permittivity of the medium very near the charge. Besides, the potential deduced from [2.11] does not preexist on all the sites likely to receive a polaron. It is only present at the site where the polaron is, and, in this way, we say that the charge "digs its own potential well".

We distinguish, in fact, two types of situations:

– *a large polaron*: if the perturbation region around an electron is of an order of size greater than the lattice parameter, we talk about a large polaron. In this case, the electron-phonon interaction and the effective mass are both weak. The polaron conserves, in its potential well, a notable kinetic energy W_c which must be added to the potential energy W_p: $W_t = W_c + W_p$. The radius of the large polaron is:

$$r_p = 2\pi^2 \frac{\varepsilon_p m}{\varepsilon_0 m^*} r_B \qquad [2.12]$$

with: $\frac{1}{\varepsilon_p} = \frac{1}{\varepsilon(\infty)} - \frac{1}{\varepsilon(0)}$ and r_B, Bohr radius.

For reasonable values, we find r_p of the order of 20 r_B, and thus clearly greater than the interatomic distances. The large polaron especially affects the semi-conductors.

– *a small polaron*: this is the case where the electron-phonon interaction is strong. The effective mass of the polaron is then very big. The kinetic energy of the electron localized in its potential well is weak and negligible to a first approximation ($W_t \approx W_p$), so r_p becomes smaller than the interatomic distance. Its radius is estimated from a simple approximation of the optical phonon spectrum:

$$r_p \cong \frac{1}{2}\left(\frac{\pi}{6N}\right)^{1/3} \qquad [2.13]$$

where N is the number of centers per unit volume on which the polaron can form.

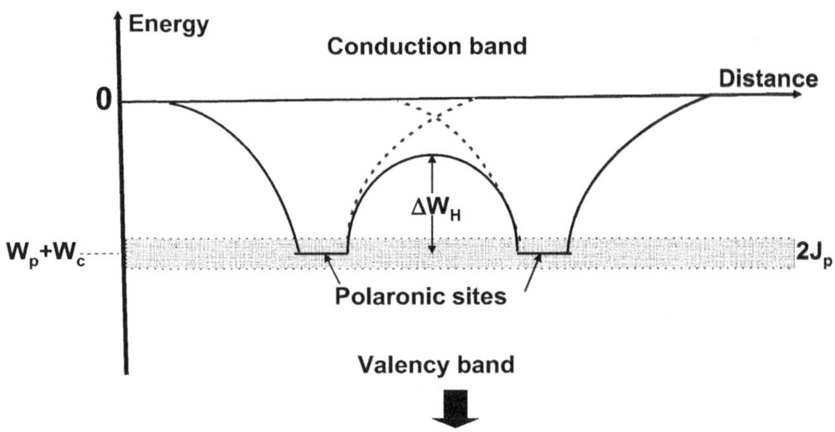

Figure 2.3. *Potential wells associated with the polaronic sites in a medium with uniform polarisability forming a polaronic band, shaded, of width* $2\ J_P$

In the case of simple orderly insulators (MgO, Al_2O_3), we are dealing with small polarons. The density of polaronic sites is equal to the density of cationic sites, which is approximately half of the atomic density. This strong density leads to a certain recovery between two adjacent sites (see Figure 2.3): this recovery leads to the formation of a polaronic band, situated at $W_t = W_c + W_p$ below the conduction band and due to a recovery of the orbital wave functions of the different adjacent polaronic sites. The polaronic sites possess the periodicity of the cristalline structure, and the recovery of the wave functions form a narrow band of width $2\,J_P$ such that $\frac{m^*}{m} \propto \frac{1}{2} J_P$, which results in values of the order of 10^{-3} eV for J_P.

2.2.4. *Conduction mechanisms*

In the process of conduction described by Austin and Mott, the moving charge "carries away" with it the polarization cloud as though it were a heavy particle of given effective mass. Generally, we distinguish two conduction mechanisms: band conduction and hop conduction. The transition between these two mechanisms occurs at a temperature $\left(T_t = {}^{T_D}\!/_2\right)$; ($T_D$ represents the Debye temperature).

2.2.4.1. *Low-temperature conduction* $\left(T \prec {}^{T_D}\!/_2\right)$

At low temperature, the charge moves by band conduction mechanism, which does not bring into play the activation energy, analogous to the motion in a conduction band with a mobility $\mu = e\tau/m^*$, where τ is the average time between two collisions. All states of the polaronic band are equally populated, the band being narrow and less than kT. Consequently the mobility takes the form:

$$\mu \cong \frac{e\,a^2}{\hbar^2}\tau\frac{J_P}{k_B T} \qquad [2.14]$$

where a represents the lattice parameter, and τ represents average time between two collisions (diffusion by the phonons).

Consequently, at very low temperature, the mobility is inversely proportional to the absolute temperature.

2.2.4.2. *Conduction at high temperature* $\left(T \succ {}^{T_D}\!/_2\right)$

At higher temperature, we observe a conduction mechanism by hops (*hopping*) from one polaronic site to the other. This second process requires the overstepping of a barrier of energy W_H (see Figure 2.3) which can be supplied by the phonons (thermal activation).

The mobility is expressed by:

$$\mu = \mu_0 \exp\left(-\frac{W_H}{k_B T}\right) \qquad [2.15]$$

where W_H is the *hopping* energy between the different adjacent polaronic sites.

The detailed analysis of the mobility given by [2.15] is complicated because we must distinguish the adiabatic displacement process, by quantum tunnelling, from the non-adiabatic process:

– if the process is adiabatic by the tunnel effect, we have: $\mu_0 = \dfrac{ea^2\omega_0}{kT}$,

– if the process is non-adiabatic, the electron is directly excited by a phonon, and we then have: $\mu_0 = \dfrac{\sqrt{\pi}}{2}.\dfrac{e\,a^2}{\hbar}.\dfrac{J_p{}^2}{\sqrt{W_H}}.\dfrac{1}{(k\,T)^{3/2}}$.

Whatever the process brought into play, there is always a possibility of displacement for the polaron: a polaron (free or free polaron) is thus a moving charge, contrary to the bonded polaron, described later, trapped on an impurity.

2.2.4.3. *Comments*

In a polarizable medium, the total energy of the carrier is lowered by Wp. The binding energy of a polaron being negative, the polaronic band is situated in the forbidden band. This energy is quite weak and of the order of 0.1eV for oxides.

2.3. Localization and trapping of carriers in a real material

2.3.1. *Localization and trapping of the small polaron*

For a small polaron, let us remember that the kinetic energy is negligible, and only the potential energy W_p remains, to a first approximation. In expression [2.11], which gives the potential energy, $\varepsilon(\infty)$ is systematically at least 2 to 4 times smaller than $\varepsilon(0)$; consequently the bonding effect is controlled by the electronic permittivity, i.e. by the polarisability of the atoms of the cell occupied by the polaron. We can understand, then, how a polaron can get trapped. Indeed, let us consider the electronic polarisability $\alpha_{elec}(\vec{r},\infty)$ (Chapter 1) at the point $(\vec{r})$ and the corresponding permittivity $\varepsilon(\vec{r},\infty)$: if the electronic polarisability on a given site $(\vec{r}_i)$ (impurity, for example) is $\alpha_{elec}(\vec{r}_i,\infty) < \alpha_{elec}(\vec{r},\infty)$, with $(\vec{r}_i) \neq (\vec{r})$, that is to say if $\varepsilon(\vec{r}_i,\infty) < \varepsilon(\vec{r},\infty)$, we see from [2.11] that the binding energy on the site $(\vec{r}_i)$ is reinforced (see Figure 2.4), so: $|W_P(\vec{r}_i)| > |W_P(\vec{r},\infty)|$, and Wp is then lowered by ΔWp.

Differentiating equation [2.11], and neglecting $1/\varepsilon(0)$ in front of $1/\varepsilon(\infty)$ gives:

$$\Delta Wp = \frac{q^2}{8\pi r_p}\left[\frac{\Delta\varepsilon(\infty)}{\varepsilon^2(\infty)}\right] \qquad [2.16]$$

where, for an electron of charge e:

$$\Delta Wp = E_R\left[\frac{\Delta\varepsilon(\infty)}{\varepsilon^2(\infty)}\right]\frac{r_B}{r_p} \qquad [2.17]$$

r_B: *Bohr radius* of the order of 0.05 nm

E_R: *Rydberg Energy*: $E_R = e^2/ 8\pi\varepsilon_0 r_B = 13.6$ eV

In other words, if $\Delta\varepsilon(\infty) = \varepsilon(\vec{r}_i,\infty) - \varepsilon(\vec{r},\infty) < 0$, a polaron will stabilize on the site $(\vec{r}_i)$ because the charge is more bonded, so its hopping frequency is reduced and its maintenance time on this site is longer. We can then speak of carrier localization, associated with a reduction of its mobility. But, if the carrier is thus localized during a sufficient time so the ions can move, the ionic polarization is to be added to the electronic polarization and consequently the energy of the carrier is reduced again by ΔW_m (see Figure 2.4). The electron-lattice binding energy therefore becomes more and more negative, which reinforces the bonding between the electron and the polarization charges thus contributing to the stable trapping of the carrier.

2.3.2. *Localization and intrinsic trapping of the carriers*

In an ideal structure where all sites are equivalent, there is no possible carrier localization: this is the case with pure monocrystalline annealed MgO [VAL 99]. In pure corindon (monocrystalline Al_2O_3) we observe a strong localization of low-temperature charges (<-20°C) which can be likened to trapping on unoccupied octaedric sites of the hexagonal structure of aluminum. This has been confirmed by a molecular dynamic calculation [RAM 97]. Thus, Figure 2.5 shows that at 200K an electron is *confined* on a site of the hexagonal structure. On the other hand, at 300K the electron can migrate from one site to another. Thus even in a very pure and orderly material, there can be trapping, but with an energy variation of the polaron of less than 1eV, which allows easy thermal detrapping.

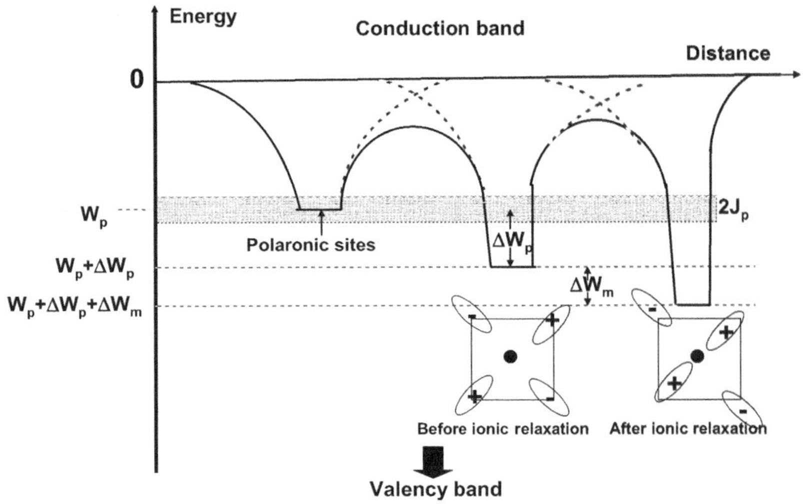

Figure 2.4. *Illustration of the localization and trapping of a carrier before and after relaxation of the ions around the charge*

2.3.3. *Trapping on structure defects and impurities*

The intrinsic point defects (vacancies, interstitials, etc.) or extrinsic (impurities) will then also be potential trapping sites. In oxides (MgO, Al_2O_3, ZrO_2, etc.), studies show that vacancies are particularly important to consider and in particular the oxygen vacancies whose formation energy is weaker than that of cationic vacancies (25 eV for oxygen vacancies against 55 eV for aluminum vacancies in alumina). An oxygen vacancy in alumina is doubly charged, by the absence of an O^{2-} ion. This center, known as F^{2+}, is then likely to trap 1 (center F^+) or 2 electrons (center F). The trap is very stable, with a binding energy of 3.8 eV in the first case, and 3eV in the second.

Impurities are equally responsible for the apparition of deep trap levels for the charges. They can be of a positively charged empty donor character or an empty neutral acceptor character. It appears, here again, that the traps are very deep and the detrapping possibility by thermal activation is weak.

When the defects density is fairly high, an impurity band can appear localized in the forbidden band. We can then envisage conduction in this impurity band.

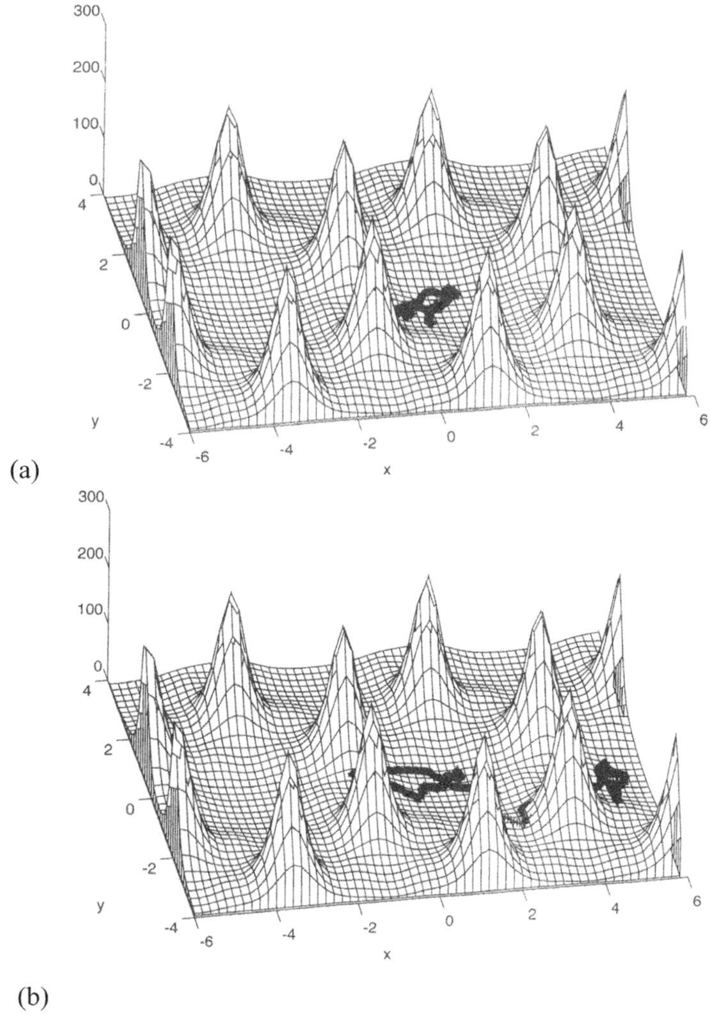

Figure 2.5. *Modeling by molecular dynamics of the motion of an electron in ideally pure alumina α, (a) at 200K, and (b) at 300K (from [RAM 97])*

Extended defects, such as the interfaces between phases, are places where there is a variation of dielectric susceptibility and so trapping possibility, but the deepness of this type of trap remains weak (<0.1 eV), so the detrapping is at first sight easy.

We note, however, that these extended defects can themselves be the segregation place of impurities or intrinsic defects, which can lead to a large range of trap deepness.

2.3.4. *Localization related to disorder*

Electron theory in the lattice has been established for a crystalline material. However, in non-crystalline materials we find conducting materials and insulating materials. We must conclude that the existence of bands and forbidden energy gaps is not something specific to crystals. Anderson [AND 58] has modeled disorderly matter by processing the problem of an electron located in a "random potential" (see Figure 2.6).

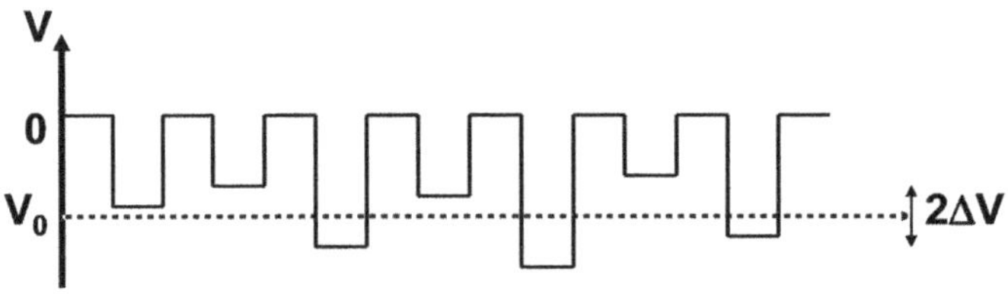

Figure 2.6. *Anderson's random potential*

Figure 2.6 is made up of a collection of square wells, centered at the geometric points of the crystal but non-identical: their deepness V varies randomly around an average value V_0, with a distribution width ΔV. The disorder is characterized by the parameter ΔV. Once ΔV is not null, *localized states* appear at the bottom of the bands (valence and conduction bands). These are "Anderson states" which exist from the apparition of a disorder, whether the material is crystalline, or amorphous.

In these conditions, the true electronic configuration of a real material will be characterized (see Figure 2.7):

– by delocalized or extended states (valence and conduction bands);

– by Anderson's localized states at the bottom of the bands, whose expanse depends on the disorder;

– by states known as impurity states (from impurities but also structure defects in general) for electrons or holes, localized in the forbidden band.

We then see that the classic description of *conduction band–band gap–valence band* is very simplistic and the reality is more complex. In these conditions, it is the boundaries between the forbidden band and the Anderson state bands which become badly defined. Mott has introduced critical energies E_C and $E_{C'}$ [MOT 74] which delimit the extensive states from the states localized at the bottom of the bands (see Figure 2.7).

Blaise studied the mobility in disorderly materials by using Mott's work. Figure 2.8 summarizes this work and shows the evolution of the effective mobility $\mu^* = \mu^*_{ext} + \mu^*_{hop}$ where μ^*_{ext} and μ^*_{hop} are respectively the mobilities in extended and in localized states. The disorder is expressed with respect to N_C/N where N_C is the density of the localized states at the bottom of the conduction band and N the atomic density of the material. We can conclude that the mobility in extended states decreases with disorder, whereas the mobility in localized states increases with disorder.

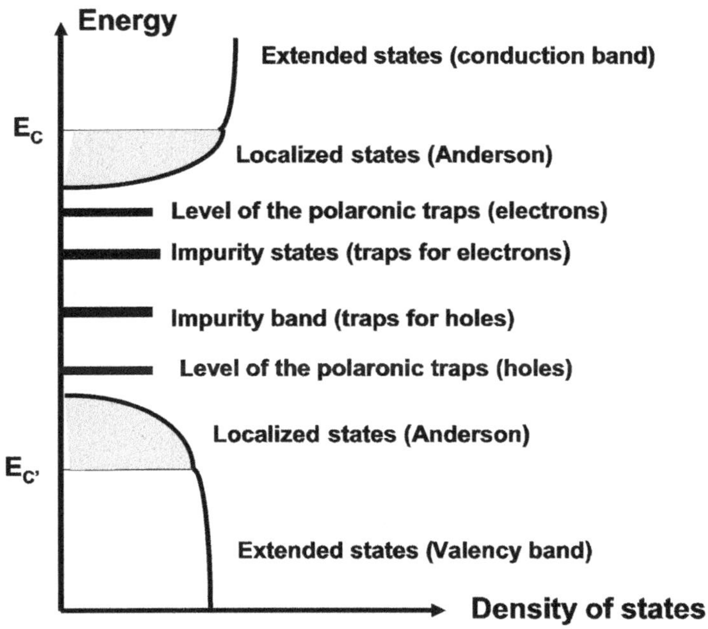

Figure 2.7. *Sketch of the energy bands in an insulating material. The position and the amplitude relative to the levels are given for information only*

Thus for a pure material without disorder, such as pure monocrystalline alumina, the insulator can assimilate and evacuate charges towards the mass at room temperature. It is clear that it would be useless to carry out metallization to observe the material using a Scanning Electron Microscope; on the other hand, if we lower the temperature, trapping of intrinsic sites of the material could be seen.

Conversely, for a high disorder ($N_C/N > 5.10^{-2}$), the mobility which prevails is that of Anderson's localized states. It can lead to a current within the sample and an expansion and diffusion of the injected charges.

For intermediate disorders ($10^{-2} < N_C/N < 5.10^{-2}$), the effective mobility passes through a minimum, and we see a strong accumulation of charges. The relaxation of these charges will be extremely slow; the secondary electronic emission could be a way of regulating this charge excess.

The same reasoning can be applied to the holes, but as they are much less movable than electrons, the accumulated positive charges will release more slowly.

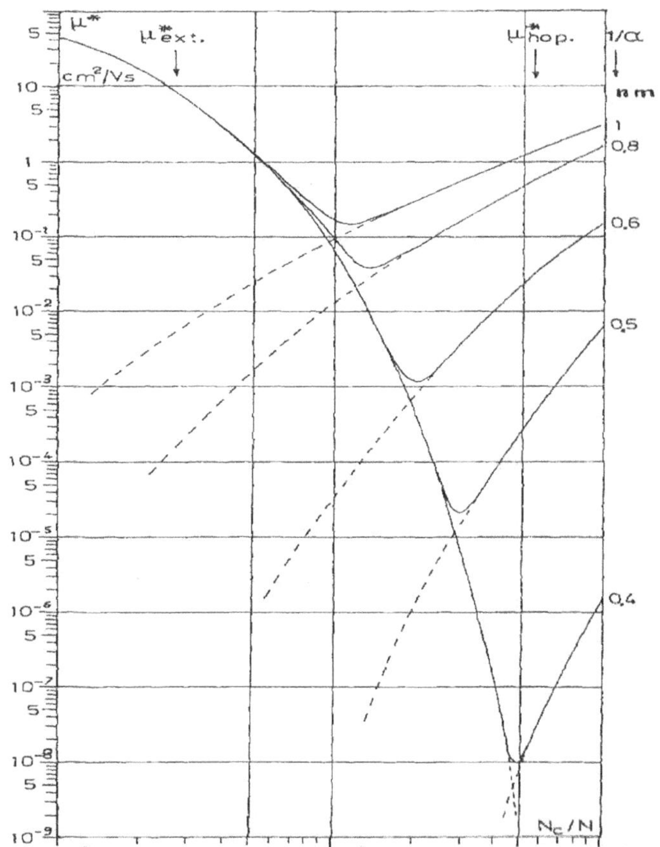

Figure 2.8. *Variation of the effective mobility of the electrons as a function of disorder (from [BLA 01])*

2.3.5. *Mechanical energy related to the trapping of one charge*

Trapping imposes a restructuring of the network around the localized polaron, creating a local distortion. Different calculations [RAM 97], [BLA 91] have shown that a trapped charge can result in the localization of a mechanical energy of 5eV or more, which is considerable, of the order of the energy of a defect formation. This energy localization leads to a metastable state which "weakens" the material and can trigger catastrophic effects (strain, fracture, wear, etc.) or, to say the least, results in an ageing of the material.

2.4. Detrapping

The trapped charges can be detrapped if a sufficient energy is brought to the system. Of particular mention are thermal detrapping and detrapping under the influence of an electric field, which lowers the trapped energy (the Poole-Frankel effect).

2.4.1. *Thermal detrapping*

In order to be detrapped, an electron trapped at a negative energy level $-E_a$ in the forbidden band, must reach the bottom of the conduction band (level 0). By applying Boltzman statistics, the detrapping probability is proportional to $\exp(-E_a/kT)$. We can then see that for shallow traps (< 0.1 eV) the detrapping probability is high at room temperature (300 K); conversely for deep traps (3eV), thermal detrapping at room temperature is not likely to occur.

The phenomenon is, however, more complex than a simple mono-electronic transition. Indeed, calorimetric measurements show that trapping is endothermic whereas detrapping is exothermic. This shows that in a trapping situation, the lattice is constrained and forms a metastable state which releases its energy at the moment of detrapping; hence the observed exothermic effect and, in other experiments, a difference between the trapping temperature and that of detrapping. It appears then that trapping-detrapping phenomena are related to collective lattice distortion processes.

2.4.2. *Detrapping under an electric field by the Poole-Frankel effect*

This phenomenon describes the lowering of a trapped charge's potential barrier to cross in a potential well, in the presence of an electric field F: the energy to supply in order to detrap the charge will then be weaker. The activation energy is then $E_a = E_T - \Delta E_{PF}$ where E_T is the activation energy in the absence of field (see Figure 2.9). To determine ΔE_{PF}, we seek the local maximum of the potential in r_0. We find $\Delta E_{PF} = \beta_{PF} F^{1/2}$ where F is the applied electric field and β_{PF} the Poole-Frankel constant:

$$\beta_{PF} = 2\frac{e^{3/2}}{(4\pi\varepsilon_0\varepsilon_r)^{1/2}} \qquad [2.18]$$

where ε_r is the static dielectric constant of the material.

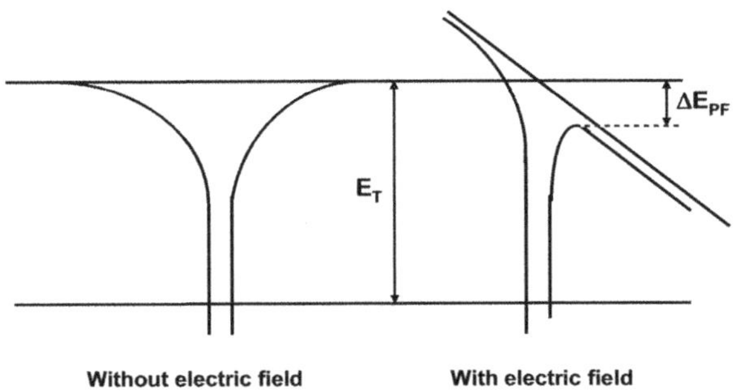

Figure 2.9. *Potential well model for a coulomb trapping site, with and without its applied electric field on the trap. In the presence of the field, the energy to supply is lowered by* ΔE_{PF}

This effect, well known for gases, is open to criticism for solids on account of their strong atomic density. However, transport simulations by hops [MAD 02] in a medium with coulomb traps bring out behaviour of Poole-Frenkel type with a coefficient β very slightly less than the coefficient β_{PF} calculated from expression [2.18].

The drop in the potential barrier is even stronger since the permittivity is weak and, if the activation energy of the trap is less than the drop in the potential barrier, the charge can immediately be trapped. Thus, for MgO, the calculations show that, at room temperature, the detrapping probability for an activation energy of 0.5 eV is about 10^{-9}, and it becomes greater than 0.65 for a field of 4MV/cm because such a field lowers the barrier by 0.49eV. Consequently, the detrapping process of a charge distribution will occur preferentially in the zones where the field is strong, i.e. alongside the distribution of the trapped charges. Knowing that a trapped charge corresponds to a stored internal mechanical energy of over 5eV, when the density of trapped charges reaches a critical value causing detrapping, we see an energy liberation which, according to the ability of the material to evacuate it, will be transformed into thermal agitation which itself might lead (as we have already indicated) to an ageing of the material, i.e. the formation of shockwaves which are the cause of catastrophic phenomena (dielectric strain, mechanical rupture, or wear, etc.).

2.5. Bibliography

[AND 58] ANDERSON P.W., "Absence of diffusion in certain randol lattice", *Physics Review,* vol. 109, p. 1492–1505, 1958.

[AUS 69] AUSTIN I.G., MOTT N.F., "Polarons in crystalline and non crystalline materials", *Advances in Physics*, vol. 18, no. 71, p.41–102, 1969.

[BLA 91] BLAISE G., LE GRESSUS C., "Charging and flashover induced by surface polarization relaxation process", *Journal of Applied Physics,* vol. 69, p. 6334–6339, 1991.

[BLA 01] BLAISE G., "Charge localization and transport in disordered dielectric materials", *Journal of Electrostatics*, 50, p. 69–89, 2001.

[FRÖ 54] FRÖLICH H., *Adv. Phys.* 3, p. 325, 1954.

[FRÖ 62] FRÖLICH H., *Polarons and Excitons*, in C.G. Kuper, G.D. Whitfield, (eds), Oliver & Boyd, Edinburgh & London, p. 1–32, 1962.

[LAN 69] LANDAU L., LIFCHITZ E., *Electrodynamique des milieux continus*, éditions MIR, 1969.

[MAD 02] MADY F., Modélisation des phénomènes de transport et des effets de charge dans les matériaux isolants, Doctoral thesis, Nantes University (France), 2002.

[MOT 74] MOTT N.F., *Metal Insulator Transition*, Taylor & Francis, London, 1974.

[RAM 97] RAMBAUT C., OH K.H., JAFFREZIC H., KPHANNOF J., FAYEULLE S., "Molecular dynamics simulation of electron trapping in the sapphire lattice", *International Atomic Energy Agency. International Centre for Theoretical Physics*, IC/95/330, *Journal of Applied Physics,* 81, p. 3263, 1997.

[VAL 99] VALLAYER B., BLAISE G., TREHEUX D., "Space charge measurement in a dielectric material after irradiation with a 30 kV electron beam: Application to single-crystal's oxide trapping properties", *Review of Scientific Instruments*, vol. 70 no. 7, p. 3102–3112, 1999.

Chapter 3

Conduction Mechanisms and Numerical Modeling of Transport in Organic Insulators: Trends and Perspectives

3.1. Introduction

Most of the physical concepts currently used to describe charge transport and breakdown in solid dielectrics have been known for over 20 years. With regard to polymers which are disordered materials, these concepts essentially stem from amorphous semiconductor physics [DIS 92] with the fundamental notions of *hopping* conduction, space-charge controlled current, interface states, etc. The dielectric nature of the media under consideration explains the references to the concept of the polaron [see Chapters 1 and 2; BLA 01] which describe the coupling between an electrical charge and its environment, or that of molecular charge states in polymers [DUK 78]. Some excellent review articles, among which some very recent, describe the basis of our current knowledge in this domain [LEW 02], [LEW 90], [WIN 90], [BLA 98], [WIN 99], [LEW 98], [MIZ 04], [BOG 05].

Despite this background in physical mechanisms, work on charge transport modeling has been rare over the course of the past 20 years. Most of the time, they have concerned analytical models in simple case studies or under very restrictive conditions when compared with real conditions of materials in electrical systems.

Chapter written by Fulbert BAUDOIN, Christian LAURENT, Séverine LE ROY and Gilbert TEYSSEDRE.

The difficulty for the development of such models has been the lack of basic data, such as the nature of carriers and traps, charge mobility and their local density, etc.

Furthermore, information on the microscopic process controlling charge transport was unavailable, simply because all the experimental measurements were based on techniques said to be "integral" which brought the information in time and/or in space (for example: surface potential measurement of a dielectric resulting from an internal distribution of charges).

Two developments have contributed to change the situation: the first is related to the tremendous increase of the calculating power of computers and to the development of numerical techniques [WIN 90]; the second concerns the development of experimental techniques which today allow the measurement of the internal distribution of the space charge as a function of time. The birth and the development of these measurement techniques in the course of the past 20 years unquestionably marks an important step in the dielectric domain. Several methods to measure the space charge have emerged, and have provided new information, principally on polymers [FLE 05], [SES 97], [TAK 99a,b], [BAU 03], [LEW 05]. Nowadays, these methods permit a range of information to be obtained, such as: the polarity of the carriers (although only the net charge is detected and we also measure the polarization gradients); the charge carriers mobility or the depth of the traps in different field and temperature situations. By combining the measurement techniques of the space charge with different spectroscopic methods such as luminescence experiments, more information can be obtained on the recombination processes which can indicate, for example, whether the transport is bipolar or unipolar [TEY 01a,b].

Another, cultural, factor has considerably changed the situation: research in the engineering sciences has been made in closer collaboration with manufacturers. Thus, this research has been made with a view to supplying a simple but realistic representation of physical, electrical and chemical phenomena, at the microscopic scale, so that it can be exploited in macroscopic models. Among the reasons which have provided an incentive to the development of models capable of describing the charge transport phenomena within insulating polymers under thermoelectrical stress, are:

– the evolution towards more and more compact systems in power electronics and, more generally, in all domains of electrical engineering, leading to an increase in power density (increasing voltage and current), with two consequences for polymeric insulations: first, these materials are continuously subjected to strong stresses, often close to the limits they can sustain; second, manufacturers have to develop new tools capable of making best estimates of the intrinsic limits of the insulator used, in order to guarantee the security of their installation;

– the demand for more and more reliable electrical systems, principally due to their use in critical applications (large-scale electrical networks, embedded networks, complex systems, etc.). This is why new predictive models capable of anticipating the consequences of chemical and/or physical defects need to be elaborated;

– the research of new materials for electrical insulation such as, for example, the composites with micro- or nano-metric size inorganic charges [TAN 05], with chemical functionalization (by addition or grafting) of the matrix or of the charges. The models must be sufficiently elaborated to fit these data.

Our objectives in this chapter are neither to inspect the published works on modeling nor to introduce new dielectric physical concepts. We aim to show that, considering the accumulated knowledge on the physics of transport, there is a real window of opportunity to further develop simulation activities, in order to better control the behavior of polymer insulators under electrical stress. The term "modeling" here is meant in the largest sense: some approaches can be based on microscopic concepts while others, according to the nature of the problem, will be based on macroscopic models. These approaches lead to different information and can be used for various reasons, contributing either to the understanding of the physics of dielectrics, or to new technological and/or industrial applications. We think that all the conditions are met – the computer tools, the understanding of physical phenomena, the development of experimental techniques – to develop transport and degradation models which integrate different levels of physical approaches. The objective of this chapter is to underline the important steps made over these past 20 years in transport simulation and to anticipate the trends for the following years.

The first part of the chapter concerns atomistic modeling which starts from atomic or molecular properties to build collective properties for combinations of such elementary components On the basis of these models, band gap properties, trap levels, and the mobility of carriers have been defined. In the second part, a macroscopic approach is established so as to describe and analyse the different existing transport models. These models permit the simulation of the electrical response of a dielectric under stress. The second part also deals with macroscopic approaches: the electrical response is modeled using "effective" parameters which encompass microscopic phenomena without explicitly describing them (nor really controlling them). The last part of the chapter (section 3.4) presents trends and perspectives in this research field. Many microscopic processes are still poorly defined (interfaces, the impact of the structure on trapping), but the limiting step for progress is no longer of an experimental order: numerous experimental methods today allow space charges to be probed with a continuously improved resolution in time [FLE 05] and in space [STE 05].

Amongst insulating materials with a large band gap, polyethylene (PE) stands out by its simple chemical structure (repetitive unit CH_2), its chemical inertness, and its numerous applications in electrical engineering. Although this chapter is not exclusively devoted to polyethylene, we are forced to note that this material has been most often considered as a model for the fundamental studies (molecular descriptions, macroscopic modeling) which makes it by far the most studied material.

3.2. Molecular modeling applied to polymers

Molecular simulation techniques are used to predict materials properties resulting from their physical or chemical structure. Indeed, the individual properties of atoms or molecules allow collective phenomena, which develop by the physical or chemical bonding of these elements, to be understood. In the case of organic polymers, which have strong covalent bonds along the chain and weak inter-chain bonds (these being essentially caused by electrostatic forces, of the Van der Waals type, or of hydrogen in specific cases), different interaction levels must be considered. With regard to the properties of a unique isolated macromolecule, the elementary entities to consider are the monomer units. On the other hand, the properties of amorphous or crystalline domains are described exclusively if many segments of adjacent chains are considered. The techniques referenced in this section have been used from 1997 onwards to foresee the trapping properties of polyethylene (PE) and to relate molecular-scale defects to the formation of space charges [MEU 97], [MEU 98]. Molecular simulations involve both the ab-initio quantum method (Density Functional Theory (DFT)) and classical molecular dynamics techniques. DFT has really developed since 1964 by becoming an alternative to the usual quantum mechanics representations of the electronic state and by supplying more direct information of interest for chemists (electronegativity, rigidity, reactivity for example). It was used to solve the Schrödinger equation for systems comprising a high number of atoms, and to derive diverse electronic properties (such electron affinity) to which the trapping properties for a given molecule ensues. As for molecular dynamics, it has been used to generate states of metastable molecular conformation of *n*-alkanes chains (C_nH_{2n+2}), considered as models for polyethylene. Further, it allows the generation of a glassy structure in which molecules are fixed. However, a limit in the simulations quickly appears when the number of atoms reaches 100 or 120 (i.e. $n > 35$–40 for the *n*-alkanes series). Starting from an isolated chain, a realistic representation of the material must take into account, including (as well as the idealized structural aspects) a certain number of defects. A first category of these defects is formed by physical defects, in which we find conformational disorder (which is the rule in the case of an amorphous structure), the free volume or the presence of voids. The second category refers to the defects of a chemical nature: inter-chain bonding, chain terminations, chemical irregularities, additives and residues, for example.

In the next section, we recall how the representation in terms of energy diagrams, stemming from inorganic semiconductor physics, can be extended to the case of polymer insulators. This is a necessary step, as molecular modeling calls on band gap concepts, extended energy states compared to localized states, etc. Subsequently, we present a state-of-the-art in this domain referring to recently published works.

3.2.1. *Energy diagram: from the n-alkanes to polyethylene*

Even the very concept of an energy diagram is not totally agreed in the domain of polymers, insulators in particular. Criticisms have been expressed, regarding the very applicability of this concept to insulating polymers, or its usefulness to understand the electrical insulation properties of these media, even if it is admitted that there are few alternatives. But why this reluctance? First, because the communities have different approaches (organic chemistry on one side, solid state physics on the other, to be very schematic), and hence a certain mutual incomprehension. Second, because polymers are complex media. "Defects" in a material with crystalline structure are generally localized and can be identified through a chemical, physical or energetic representation. In disordered media, defects are no longer the exception, but the rule so that the representation of such complex systems is delicate.

The objective here is not to argue or identify the *pros* and the *cons*, but to briefly recall the history and bring out the input of physical and chemical quantum theories. The results of molecular simulation from the last decade to describe the transport and trapping properties constitute real advancements and rely on the concept of the energy diagram. It is, then, important to demystify solid state theories so as to establish a language base common to chemists, physicists and engineers.

Numerous theoretical studies, at different levels of complexity, have been developed around polyethylene, PE. The interest for this material is double: it presents a simple chemical structure and it is the subject of numerous applications in the world of insulation. In order to understand and interpret the experimental results on the band gap of the polymers derived from classical techniques, such as XPS (X-ray photoelectron spectroscopy) or EELS (electron energy loss spectroscopy), band structure calculations have been carried out. These calculations have been made easier by the small size of the elementary cell of the material under consideration. We direct readers, at this point, to the excellent overview by R. Hoffmann *et al.*

[HOF 91], written from an educational perspective and intended to develop a common language for physicists and chemists.

The band structure construction of PE from a series of related molecules, from methane to *n*-alkanes, is represented in Figure 3.1. In this series, from methane to PE, the number of orbitals increases with the size of the molecule, which leads to a narrower and narrower separation between the electronic orbitals. The extrapolation of a continuous band is therefore an idealized representation of a high number of very close electronic levels. The molecular orbitals merge with the energy domains covered by the bands, which proves the close relationship (and even the equivalence) between the states of molecular energy and those of the crystal passing from molecules to extended chains for these hydrocarbons.

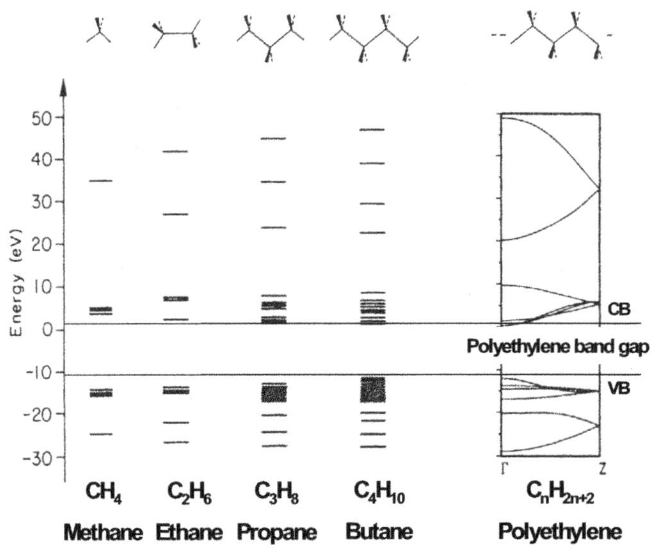

Figure 3.1. *Distribution of energy levels of the molecular orbitals from methane to PE in all-trans conformation. The reference for the energies is the vacuum level. The x-axis in the polyethylene band diagram is the wave number (k) scale, spanning from the center (k=0, term Γ) to the edge (k=π/a, term Z) of the Brillouin zone. CB and VB refer to the conduction and valence bands. Adapted from [HOF 91]*

From Figure 3.1, we can see that PE is a material with a large band gap, which signifies that there are no energy levels accessible to the carriers between the top of the valence band and the bottom of the conduction band. The width of the forbidden band, whether it is deduced from calculation or experimentally determined, is around 8.8 eV [LES 73], which means that we would have a perfect insulator with an intrinsic conductivity of the order of 10^{-45} S/m [COE 93]. Of course, we know that such a level of conductivity has never been measured because the reality is far more complex: other levels and processes assist the transport. However, some of these complex features have been studied and form the subject of this section.

Modifications of the "idealized" band structure of PE have been proposed; see for example [DIS 92] as well as the excellent overview by G. Blaise [BLA 01]. Coming back to the fundamentals, two factors must be considered:

– the first is related to the fact that polymer molecules can adopt different conformations, while the linear arrangement, *all-trans*, of lower energy, has been considered until now. These conformations introduce disorder in the organization. Here as well, according to the author's culture, we talk about physical, conformational or topological disorder. This has consequences for the energy diagram, with the introduction of localized states (accessible to carriers) in the forbidden band. Their density is high (of the order of 10^{22} cm^{-3}) and they are situated in a range of 0.1 to 1 eV with respect to the extended states of the bands (see Figure 3.2). These states are denoted differently according to the author, for example shallow traps, conduction states, or Anderson states. The residence time of the carriers on these sites varies from 10^{-13} s for a level to 0.1 eV to 500 s for 1 eV [MEU 00a,b]. These different terminologies refer to a common physical process: the transport, according to two possible mechanisms. Localized states can be in thermal equilibrium with the extended states of the valence or conduction band, the transport occurring by thermal activation of the carriers toward the bands. This mechanism leads to the transport phenomenon by successive trapping/detrapping, with passage toward the extended band at each step. The term "shallow trap" stems from this approach where the carriers move from one trap to another with a passage by extended levels. The second transport mechanism refers to the possible interaction between localized levels considering their large density (the wave functions of the traps are superimposed). Carriers can then migrate from one trap to another by quantum tunelling without passing through the extended states defining a tunnel mechanism assisted by phonons (*hopping*). The energy E_c separating extended and localized states is a critical parameter since the mobility of the carriers increases abruptly beyond this value by passing from the effective mobility associated to the transport in the localized states (controlled by the residence time of the carriers in the traps), to the band mobility (free electrons in the conduction band, free holes in the valence band). Some descriptions of the electronic carriers' behavior under very high electric field consider that the mobility of the carriers is null for energies $< E_c$ and infinite above it, the hypothesis on which is based the field limited space charge model (FLSC) initially proposed by H.R. Zeller [HIB 86], and subsequently refined by S. Boggs [BOG 05];

– the second factor concerns chemical disorder. It can be related to the presence of one or several atoms or bonds non-representative of the monomeric unit, or the isolated molecules. In PE and in polymers in general, it can be a matter of intra-chain defects (unsaturation, lateral groups, bonding, chain ends), additives (antioxidants, catalysts), reaction residues (cross-linking, for example), and impurities (pollution) which introduce the concept of impurity states [BLA 01]. Chemical disorder introduces physical disorder in turn in its environment, owing to the difference in size and position of the atoms constituting the defect. The electronic properties of the defect are different from that of the host medium, so new energy levels can be introduced in the band gap, generally as deep levels (see Figure 3.2) and with a lower density than the traps known as shallow traps (as we shall see later). It is generally admitted that these energy levels control the space charge phenomenon: we talk about "deep traps" or "real traps" because their contribution to the conduction processes is weak, unlike physical traps which assist transport.

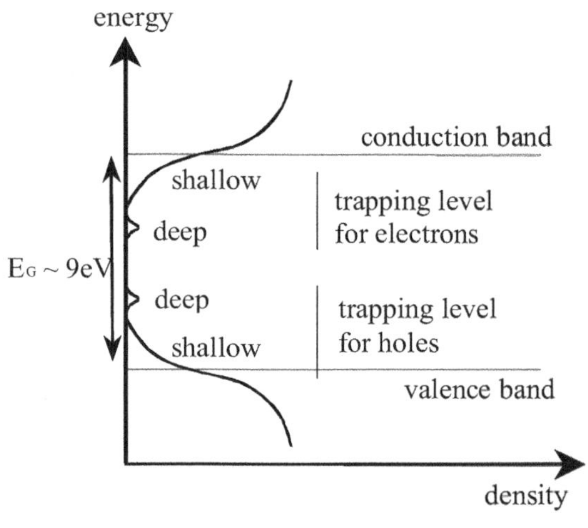

Figure 3.2. *Schematic representation of the density of states in a disordered dielectric. The deep and shallow traps are respectively related to the chemical and physical disorder*

Having said that, other processes can have a large influence, or even control the electric behavior of insulators: the presence of ions, electrochemical phenomena, for example. From another point of view, the formulation of technical polymers can be so complex that it becomes delicate to distinguish additives from the matrix... All these parameters must be taken into account to determine if the band model that we conceive can adjust to the structure that we process. It is nonetheless true that the band diagram constitutes a fundamental support for the understanding and the development of models.

3.2.2. *Results of modeling*

3.2.2.1. *N-alkanes models of polyethylene*

We have already seen (Figure 3.1) that the *n*-alkanes series represent a reasonable model of polyethylene with regard to the characteristics of the band gap. Further, the behavior of an excess electron in *n*-alkanes has been thoroughly studied and remains a paradigm for the behavior of excess carriers in dielectrics. Numerous experimental and simulation work has been done on low field mobility, the optical absorption associated to excess electrons or the properties of the forbidden band. They show that the properties of PE become very close to those of the *n*-alkanes with n of the order of 15 to 20 at the minimum [PIR 76], even though this assertion is to be adjusted considering recent results showing a difference of 0.2eV in the position of the minimum of the conduction band of PE and $C_{27}H_{56}$ [CUB 03a,b]. Previous experimental measurements on the hexatriacontane (solid) $C_{36}H_{74}$ and on other paraffins have been made to extrapolate the electronic and electrical properties of PE [REI 90], [RAC 87], [VAN 62]. Recent molecular simulations have relied on the *n*-alkanes to approach the properties of PE: molecules such as C_9H_{20} [MEU 97], $C_{13}H_{28}$ [MEU 98], [MEU 00a,b], $C_{15}H_{32}$ [MEU 97] and $C_{27}H_{56}$ [CUB 03a,b], [CUB 02] have been considered.

3.2.2.2. *Volume properties: amorphous and crystalline phases*

Most calculations on the electronic structure of polymers use methods ranging from semi-empirical or *ab initio* approaches to DFT [MIA 96], the latter being less widespread. While the states of valence bands have been well characterized, those of conduction bands, which are, however, of capital importance in transport, have not been so characterized. It is only very recently that studies on the electronic and structural properties of valence and conduction bands have been made. The first report on this subject, concerning purely crystalline PE in its orthorhombic crystallographic form, dates back to 1998 [SER 98].

Using DFT and similar techniques, it was possible to describe the nature of non-occupied electronic states (the conduction band) and occupied states (the valence band). It was then shown that conduction states have an interchain character (the wavefunction is maximum at the half-distance between chains, so at about 5 Å from the chains) while the valence band states have an intrachain character [SER 00]. In a solid, the electronic affinity (E_A) is the energy required to carry an electron from the vacuum level to the conduction band (see Figure 3.3).

It has long been known that PE has a negative electronic affinity, which signifies that the lowest state of the conduction band is situated above the vacuum level [LES 73], [RIT 82]. The interchain character of the conduction states provides a physical interpretation of this phenomenon. A "free" electron would thus be directed toward the regions of weaker density, such as the surface of the crystalline lamellae, or to an internal cavity. However, although the crystalline PE and alkane crystals present a similar structure as well as a negative electronic affinity, the conduction processes in both materials are different: it is interchain in PE and interlayer in the alkanes (the alkane molecules are organized in layers, which structure the interlayer zones with a weak density). This defines the limit of similarity between PE and the *n*-alkane series [CUB 03a,b]. Another important conclusion can be drawn concerning the conduction by holes in the valence band. If the holes are confined in the chain while the electrons are excluded from it, a distinction between conduction paths for electrons and holes appears probable [LEW 02]. Furthermore, if we take into account a heterogenous density of PE at the local scale, the electrons, which contribute to the space charge phenomenon, have to be uniformly spread out at a nanometric scale. Other publications by different groups [CUB 03a,b], [CUB 02] confirm the interchain character of the wavefunction of the excess electrons in the conduction band.

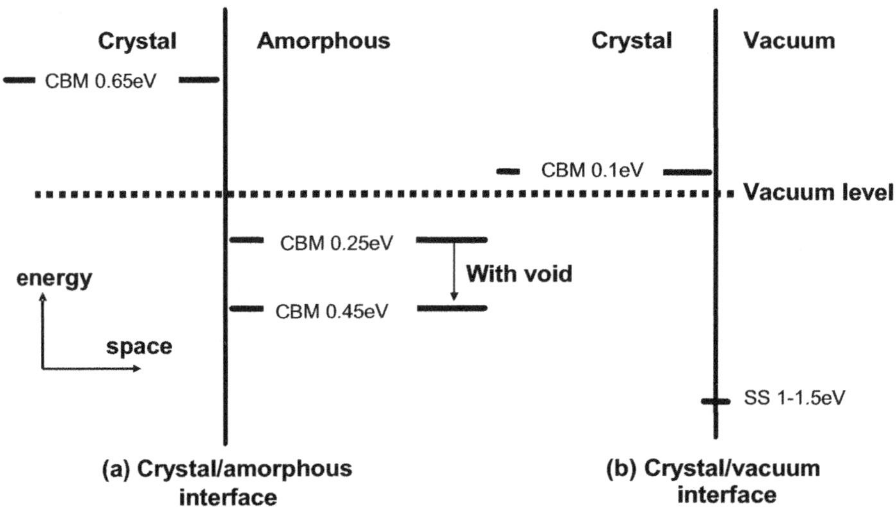

Figure 3.3. *Schematic representation of the position of the conduction band minimum (CBM) with respect to the vacuum level of PE. (a) amorphous-crystal internal interface; (b) crystal-vacuum external interface; E_A is the electronic affinity, defined as the energy difference between CBM and vacuum level. It is negative (by convention) for crystalline PE in volume (-0.65eV) and on surface (-0.1eV), and positive for amorphous PE in volume. A surface state (SS) able to assist the injection is represented. A void in the amorphous phase behaves like a trap. Traps introduced by physical or chemical defects are not represented*

A fundamental energy state of 0.65 eV has been identified as the bottom of the conduction band, which brings the electronic affinity of polyethylene to approximately this value; however, all the factors have not been taken into account. At a temperature of 300K, disorder is introduced in the crystalline regions by thermal activation, in the form of a structural parameter distribution (with rotation around the bonds) in respect to their value at 0K. This introduces localized states which can play the role of electron traps (see section 2.1) with depths of the order of 0.1 to 0.2 eV.

The amorphous regions have been modeled by considering a unique chain composed of 360 atoms [CUB 02]. The fundamental energy state (conduction) has been situated at -0.25 eV, i.e. it is localized below the vacuum level, which brings about significant differences between the amorphous and crystalline phase properties. The amorphous regions behave like trapping sites (of a depth of the order of 1 eV) with respect to the crystalline phase. The interfacial regions between amorphous and crystalline phases would also act as traps. For materials with a weak crystallinity rate, the conduction is probably dominated by trapping/detrapping phenomena in the amorphous phase of PE.

3.2.2.3. *Surface and nano-void properties*

The electronic properties of the surface of a PE crystal (at the interface with a vacuum) have been studied considering different orientations of crystals (and so of the molecules) with respect to the surface [RIG 01]. The surface has a negative electronic affinity, like the volume of PE, with energies of -0.17 and -0.10 eV for chains respectively parallel and perpendicular to the surface. The two types of surface have surface states with energies of -1.2±0.5 eV with respect to the conduction band minimum in the volume. This signifies that an electron situated near a cavity in the volume of the PE crystal will be spontaneously emitted in a vacuum (or in a cavity) and the surface represents a deep trapping state with respect to the volume of the crystalline PE. The creation of nano-voids in the amorphous PE brings the fundamental energy state to -0.45eV. The wave function of an electron is then centered on the cavity [CUB 02], which then behaves like a trap.

3.2.2.4. *Trapping sites*

The objectives here are to calculate the energy levels associated with physical or chemical defects in a given environment, in order to determine if these sites constitute potential trapping sites. This principally concerns electron traps, even though some calculations have been made for the holes. The trap depth is defined with respect to the electronic affinity of the system, estimated with and without the defect as illustrated in Figure 3.4. A positive value of E_t represents a potential well for the electron, since in this case the electron has more electronic affinity for the trap than for the material without defect, and similarly for the holes.

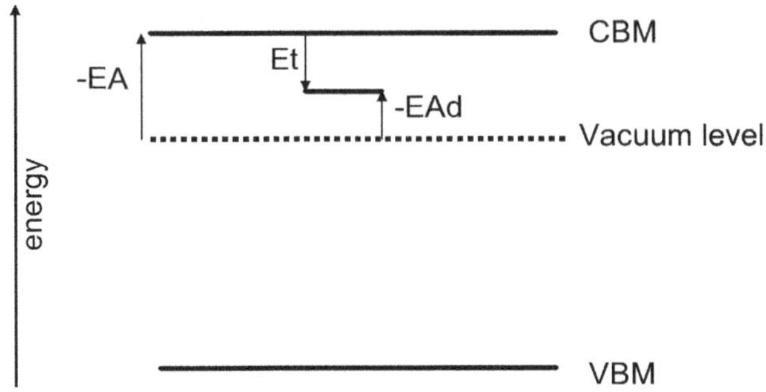

Figure 3.4. *Definition of the trap depth for the electrons (E_t=E_{Ad}-E_A), where E_{Ad} is the electron affinity of the defect and E_A that of the matrix without defect. VBM and CBM are the minima of the valence and conduction bands*

3.2.2.4.1. Physical defects

The first calculations [MEU 97], [MEU 98] were made based on tridecane ($C_{13}H_{28}$) in its minimum energy configuration (*all-trans*), thus giving the reference for electronic affinity. The conformational disorder of PE has been simulated from the crystalline phase of the tridecane by molecular dynamics. The conformational defects thus produce shallow traps (<0.3 eV) [MEU 00a,b], with density of the order of 10^{+20} cm^{-3}, for an average intertrap distance of 15 Å. The residence time in the traps is of the order of 10^{-11} to 10^{-13} s, which appears consistent with the fact that these sites assist the transport but do not permit the stabilization of the charges which we detect in the long term by classical techniques (hence the term “localized conduction states” rather than “traps”). This first approach was limited to the effect of conformational defects with respect to an *all-trans* structure and did not take into account the density fluctuations of the amorphous phase nor the energy states for excess electrons in the crystalline phase of PE.

A finer approach of the amorphous phase properties has subsequently been developed, relying on the pseudo-potential technique [CUB 03a,b]. The electronic states and the mobility of excess carriers have been deduced for the amorphous PE. The minimal energy states have thus been situated below the vacuum level, at E_c = -0.32 eV (compared with -0.25 eV in Figure 3.3) and the mobility limit (E_o) is very close to the vacuum level, which leads to trap depths related to local disorder of the order of 0.3eV (very close to previous estimations). This results in an expression for the mobility of the form:

$$\mu = \mu_o.\exp-(\frac{E_c - E_o}{kT}) \qquad [3.1]$$

where μ_o has been estimated at 7.5×10^{-2} m^2/V.s, giving a mobility of 2×10^{-7} m^2/V.s with an error bar of a factor of 10. Equation [3.1] is valid in the limit $E_o - E_c >> kT$ and comes from an approximation of the Kubo–Greenwood relation [CUB 03a,b].

3.2.2.4.2. Chemical defects

Two groups of chemical defects have been considered with $C_{13}H_{28}$ as a reference molecule in *all-trans* conformation for the electronic affinity. The first group concerns chain imperfections (hydroxyl and ketone functions, double or triple bonding, branching, etc.) and has been simulated by bringing a chemical defect on to the alkane chain [MEU 00a,b], [MEU 01]. The modification of the minimal energy of conformation due to the integration of the defect has been taken into account here. We can see in Table 3.1 that the defects containing carbonyle groups or double bonds form the deepest electron traps. The second group concerns crosslinking by-products of PE [MEU 97]. Among those, acetophenone and α–methylstyrene have been identified as deep electrons traps. The affinity of the holes has also been considered for some of the species in order to estimate their propensity to stabilize positive charges [TEY 01a,b]. A summary of trap depths thus estimated is proposed in Table 3.1, where most values have been deduced without taking into account the polarization that the neighboring molecules can generate on the model chain.

To estimate this polarization effect, calculations have been made on a *cluster* model where the defect is surrounded by alkane chains [MEU 01]. The modifications made have been minimal in this case. However, polarization effects can be significant in the general case, in particular for polar polymers [EKS 68]. Finally, the trap depth of chemicals present in the semi-conducting screens of cables, and likely to diffuse in a PE insulator, have recently been published [CAM 02].

3.2.2.5. *Self-trapping and polaron concept*

The molecular modeling presented so far has neglected the local changes of conformation associated with the presence of trapped charges. Now, it has long been known that an excess charge in a dielectric polarizes the medium [BLA 01], [BLA 98], see Chapter 2, section 2.3.

In practice, an electron with kinetic energy greater than 10 eV can be considered as mobile in a medium with vacuum permittivity, i.e. there is no perturbation of molecules due to the displacement of these charges. Conversely, a "thermal" electron moves in a polarized medium, whose permittivity $\varepsilon(\omega)$ varies radially around the charge starting from the vacuum permittivity near the charge to the static permittivity $\varepsilon(0)$ at long distance [BAI 89], [TAY 92]. The radial variation of the polarization around the charge can be taken into account by introducing a polarization charge density, which leads to the *polaron* concept originally introduced by Landau [FRÖ 63].

This phenemenon leads to the gradual confinement of the charge in a potential well whose depth increases when kinetic energy is reduced. This has direct consequences on the energy stored in a material in the presence of a space charge, and on transport. The polarization energy due to a trapped charge can be of the order of 1 to 10 eV for usual inorganic insulators, depending on their dielectric susceptibility [BLA 98]. A polaron formed by capture of an electron by a shallow trap will behave like a particle with high effective mass, the charge "transporting" its own polarization field.

<table>
<tr><th colspan="2" rowspan="2">Nature of the defect</th><th rowspan="2">Molecule with:</th><th colspan="2">Trap depth (eV)</th></tr>
<tr><th>electrons</th><th>holes</th></tr>
<tr><td colspan="2">6-tridecene* ($C_{13}H_{26}$)</td><td>In-chain C=C</td><td>0.16</td><td>0.57</td></tr>
<tr><td colspan="2">5,7 tridecene* ($C_{13}H_{24}$)</td><td>In-chain conjugated C=C</td><td>0.51</td><td>1.35</td></tr>
<tr><td colspan="2">6-tridecanone* ($C_{13}H_{26}O$)</td><td>Side chain C=O</td><td>0.49</td><td>0.43</td></tr>
<tr><td colspan="2">5-decanone** ($C_{10}H_{20}O$)</td><td>Side chain C=O</td><td>0.453</td><td></td></tr>
<tr><td colspan="2">5-decene** ($C_{10}H_{20}$)</td><td>In-chain non-conjugated C=C</td><td>0.122</td><td></td></tr>
<tr><td colspan="2">4, 6-decene** ($C_{10}H_{18}$)</td><td>In-chain conjugated C=C</td><td>0.443</td><td></td></tr>
<tr><td colspan="2">5-decyne** ($C_{10}H_{18}$)</td><td>In-chain non-conjugated C≡C</td><td>0.041</td><td></td></tr>
<tr><td colspan="2">5-vinyl nonane** ($C_{10}H_{20}$)</td><td>Non-conjugated C=C</td><td>0.157</td><td></td></tr>
<tr><td colspan="2">5-decanol** ($C_{10}H_{21}O$)</td><td>Hydroxyl</td><td>0.186</td><td></td></tr>
<tr><td colspan="2">5-decanal** ($C_{10}H_{20}O$)</td><td>Carbonyl group</td><td>0.445</td><td></td></tr>
<tr><td colspan="2">4-propyl heptane** ($C_{10}H_{22}$)</td><td>Saturated</td><td>0.121</td><td></td></tr>
<tr><td rowspan="4">Crosslinking by-products with dicumyl-peroxide as crosslinking agent</td><td>α-methylstyrene* (C_9H_{10})</td><td>Aromatic</td><td>1.53</td><td>0.79</td></tr>
<tr><td>Cumylic alcohol * ($C_9H_{12}O$)</td><td>Aromatic + Hydroxyl group</td><td>0.28</td><td>0.36</td></tr>
<tr><td>Acetophenone* (C_8H_8O)</td><td>Aromatic + Carbonyl group</td><td>0.9</td><td>0.04</td></tr>
<tr><td>Cumene** (C_9H_{12})</td><td>Aromatic</td><td>0.04</td><td></td></tr>
</table>

Table 3.1. *A summary of trap depth values estimated by* ab initio *methods for chemical traps of PE. *from [TEY 01a,b], **from [MEU 01]*

Let us consider now how this effect has been evaluated in organic materials. Some modifications of the trap depth by molecular reorganization have been brought to the fore in fluid alkanes [SHI 72], [WIL 73], [ABR 92]. The stabilization of excess electrons in the hydrocarbons has been particularly studied by means of optical absorption spectroscopy. The stabilized electrons form absorption bands in the near infra-red, which move toward the shorter wavelength over time, this being clearer in polar matrices than in hydrocarbons. This phenomenon has been attributed to a strengthening of the trap by reorientation of the surrounding molecules [KLA 72]. Experimental evidence of a trap deepening by rotation of the molecular dipoles has been reported on the basis of recombination-induced luminescence experiments on a model molecule frozen in a glass [HO 77]. In solid polymers, the polarization energy will lower the energy state of an electronic trap (in reference to the vacuum level) [LEW 02]. For certain polymers, notably those composed of lateral groups, the energies at work can be of 3 eV [DUK 78]. As for PE, the calculations made so far neglect the molecular reorientation associated with the excess electrons. If the approximation is acceptable for treating extended states (the environment not being very perturbed by the moving electrons), this becomes more debatable for trapping sites. Preliminary calculations taking into account these effects in the case of extended states show that, after a certain time, the electron "traps itself" with an energy corresponding to the bottom of the conduction band. The variation of energy at work is weak here (0.1eV, being in the error bar of the simulation method), as expected for a non-polar material [CUB 03a,b]. For a chemical trap, the problem has not been treated but should be, in particular for polar polymers where the polarization energy can be very significant.

3.3. Macroscopic models

The layman might wonder why the models for describing transport in polymer insulators are so numerous and so disparate in their hypotheses, whereas transport, Poisson and continuity equations are very general since they are valid for semi-conductors, dielectrics, but also for solid, liquid and gaseous states. A first answer can be found in the fact that, in electrical engineering, the research strategies, from where scientific developments ensue, are led more by system-oriented rather than materials-oriented approaches. Further, for organic semi-conductors, the experimental behavior is generally more satisfactorily described by available physical models than for insulators. To dwell on this comparison, a certain number of specificities of polymer insulators need to be underlined:

– the complexity of the very structure of the materials. Polymers have a complex physical structure (amorphous and crystalline regions, chain conformation, etc.) and chemical structure (branching, in-chain defects, residues, additives, etc.). These issues have been broached in the previous section;

– response time. Transient phenomena in semi-conductors are fast, such that, quite often, the physical models are only solved in stationary conditions [WAL 02], [TOR 84], [SHE 98] by taking into consideration processes such as injection, transport and recombination. Conversely, transient responses in the insulators are very slow, such that we never know if a stationary regime is reached. This is why the models are defined and established in a way that respects the time scale of these devices. For years, studies on transient phenomena in insulators have been made by macroscopic observations, such as charge or discharge currents or surface potential measurements. The more recent measurements of internal charge densities have had a considerable impact on the evolution of models dedicated to simulation, by considering new processes to insert in analytical and/or numerical modeling. Another problem concerning the behavior of insulators is the superimposing of two physical phenomena: polarization and space charge [WIN 88], [SHO 71]. The very representation of the dipolar response of polymers is debated, with diverse approaches ranging from the "universal" relaxation law [JON 96] to mode-coupling theory [RAJ 91], and to distributed processes [BOT 78]. The minimum we need here is to have a good approximation (by independent measurements and phenomenological simulations) of the response in polarization alone. However, it is not obvious to obtain this kind of information when the polar species are molecules in weak concentration, like in PE and in materials which are not generally very polar;

– the variety of phenomena potentially involved in transport. This concerns the hypotheses taken concerning the origin and nature of carriers and traps, as well as the transport and trapping mechanisms. While the previous models for describing the transient currents and voltage/current characteristics are based on a unique species of carriers, the space charge measurements show that transport is, most often, bipolar. In the same manner, it is not conceivable to imagine a generation of charges only by intrinsic processes, since the space charge measurements show the accumulation of homo-charges near the electrodes.

Next, we consider the different physical phenomena which have been inserted in the macroscopic models. The problems related to the choice of numerical methods, algorithms and meshing [WIN 03a], [WIN 03b], will not be discussed here. Whichever physical model of transport is considered, and neglecting the polarization, the mathematical problem is reduced to equations [3.2], [3.3] and [3.4] below, considering a 1D problem along the spatial coordinate x.

$$j(x,t) = \mu.E(x,t).\rho(x,t) - D\frac{\partial \rho(x,t)}{\partial x} \qquad [3.2]$$

$$\frac{\partial \rho(x,t)}{\partial t} + \frac{\partial j(x,t)}{\partial x} - D\frac{\partial^2 \rho(x,t)}{\partial x^2} = s(x,t) \qquad [3.3]$$

$$\frac{\partial E(x,t)}{\partial x} = \frac{\rho(x,t)}{\varepsilon} \qquad [3.4]$$

where j is the current density for each type of species, e the elementary charge, μ is the mobility, D is the diffusion coefficient, E is the applied field, ρ is the net charge density. Equations [3.2] and [3.3] must be written for each type of carrier defined in the model. The term s is the source term which represents the local density changes due to processes other than the transport, such as the internal generation, the recombination, etc. These equations can have a specific form for the interfaces and are completed by the different boundary conditions like applied voltage.

Whatever the system considered, there are at least two processes to define the construction of the transport model: how the carriers are generated and how they are transported. Most often, the models are defined for two types of charged species (which are not necessarily mobile) and therefore an additional set of physical phenomena must be considered defining the interactions or exchanges between carriers of differents types. In the sections below, a description of these two processes (generation and charge transport) is first made. Then, an analysis is made of how these elementary mechanisms have been combined to supply a set of results which enable the experimental behavior of insulating polymers under electrical stress to be approached.

3.3.1. *Elementary processes*

3.3.1.1. *Generation of charges*

If we set aside the generation of charges due to discharges in the insulation environment, and all other sources of irradiation (photo-generation, X-rays, electronic irradiation, etc.), several mechanisms still remain for the generation of carriers, namely: electrode injection, ionic species and donor states.

3.3.1.1.1. Electrode injection

The most discussed mechanism to describe charge injection in organic semi-conductors is the Schottky effect, also known as field assisted thermo-ionic emission [DIS 92], [DWY 73] where the dependence of the current density, with field, is in the form of $exp(E^{1/2})$. This law is also used in insulators, with (however) some inconsistencies on the parameter values: the experimental value of the injection barrier (around 1 eV) is much smaller than the presumed value for a dielectric-metal contact (about 4 eV, depending on the material). Furthermore, the pre-exponential factor is much smaller than expected [DIS 92], [HUG 80]. One of the reasons for such a low injection barrier could be that the conditions for which the work functions have been measured (clean surfaces in the ultra vacuum) are no longer respected when we work in normal conditions: the interfaces must present far more imperfections. Another injection process currently touched on is injection by field effect, also known as the Fowler–Nordheim effect [DWY 73], generally when the electric field is very high (beyond 100 kV/mm).

Apart from these two models, we find some alternative descriptions for charge injection, notably by thermally assisted quantum tunneling from metal to localized states [ABK 95], quantum tunneling toward polaron levels in the polymers [CON 97], injection limited by interfacial states induced by the dipoles [BAL 01], thermally-assisted injection in a random *hopping* system, [ARK 98], Schottky emission limited by diffusion [EMT 66], [SCO 99], or hot carrier injection followed by a thermalization and a carrier motion, according to the Onsager theory [BLO 74]. Further, a certain number of models suppose a constant charge density or an infinite reservoir of charges on the electrodes (ohmic contact). The current is then entirely controlled by the volume transport.

3.3.1.1.2. Internal generation

A weak concentration of ions can be present in polymer materials, associated for example with chemical reactions implying sulphur present in the antioxidants or organic ions resulting from the oxidation of the polymer [GIL 92]. An alternative to ion generation is the dissociation of impurities or water molecules under the influence of an electric field – referred to as the Onsager mechanism [PAI 75]. Finally, ionic species can be generated at the metallic or semi-conductor electrode level and diffused within the dielectric [TAK 99a,b], [HO 01]. As an example of transient current simulation, let us cite Kahn and Maycock [KAH 67] who have considered a bipolar ionic model with blocking electrodes. The transport is controlled by diffusion and effective mobilities are also defined. Iwamoto [IWA 96] has considered other mechanisms, a *hopping* conduction and blocking electrodes in the case of ionic impurities blocked at the interfaces.

Carriers in the volume (for example, the electrons) can originate from donor states, and the hole thus generated could be mobile or not. Extraction from donor states can be produced from processes such as the Poole–Frenkel effect [RAS 02].

3.3.1.1.3. Carrier extraction

According to their nature, the carriers can be extracted or not by the opposite electrodes. In the case of ions, it is difficult to envisage mass transport through electrodes. This is why ions can be considered blocked at the interfaces [IWA 96]. However, they can interact with carriers with opposite polarity injected at the electrode, thus re-establishing electrical neutrality. In the case of electronic carriers, extraction can be made with or without an extraction barrier.

3.3.1.2. *Charge transport*

The charge transport processes first depends on the nature of the carriers. The first classification is a distinction between ionic and electronic processes. In electronic processes, a second classification can be established depending on whether the transport is made via the conduction band (Poole–Frenkel mechanism) or not (*hopping*). The Space Charge-Limited Current (SCLC) descriptions are based at the same time on an "ohmic" type of conduction (with a band mobility or limited by traps) and on a distorsion of the local electric field due to the accumulation of space charges. This is why this model is not considered as an elementary mechanism. Figure 3.5 shows some basic mechanisms concerning electronic transport.

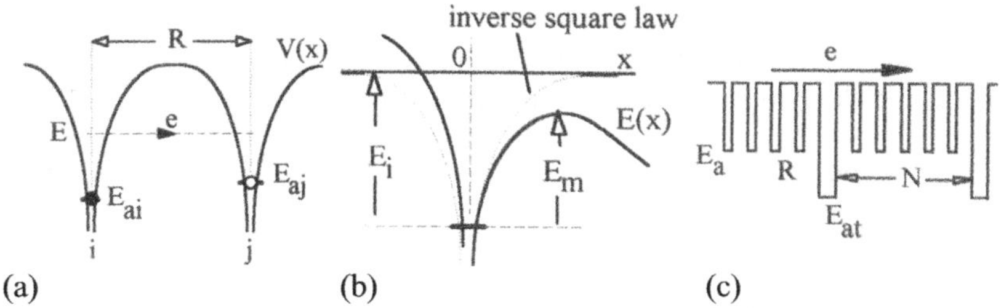

Figure 3.5. *Some basic mechanisms of the transport [LEW 98]. (a) Tunnel resonating between two states. Here, i and j are acceptor states. (b) Poole–Frenkel. (c) Transport controlled by traps: the mobility depends on the average residence time of the carriers in the traps*

3.3.1.2.1. Hopping conduction

The process by which an electron can move from one place to another is made by thermal excitation above a potential barrier, by quantum tunneling through the barrier, or by a combination of both phenomena. The *hopping* conduction corresponds to an intermediate situation in which an electron is brought by thermal activation to a level of similar energy to that of an empty site towards which it moves by quantum tunneling [DIS 92]. The hopping probability depends on the field. In a more general manner, the term refers to the mechanisms which do not involve the conduction band (or the valence band for holes).

In applications such as photocopiers, where charges are generated in a similar manner to time-of-flight type experiments, the transit of a charge occurs with a time-dependent mobility. This phenomenon, called *dispersive transport*, has been interpreted by a hopping frequency distribution, associated with an inter-site distance. It has been described in the frame of random walk theory [SCH 75]. The models based on this approach have attempted to explain the Poole–Frenkel law for mobility [EKS 68], [DUN 95] – see the bottom two rows of Table 3.2.

3.3.1.2.2. Poole–Frenkel Effect

The Poole–Frenkel mechanism in the volume is analogous to the Schottky injection at the metal–insulator interface. This mechanism is based on the decrease of the potential barrier height needed to get over by application of an electric field. It applies to deep donor (or acceptor) states in insulators with a large gap. This process is subject to strong controversies [WIN 99], [PAI 75], [WIN 98]. Indeed, the barrier maximum is at a further distance from the site than the diffusion length, such that the excited electron has a high probability to be thermalized before it gets over the barrier. The observed slope on the curve $j(E^2)$ is often smaller than expected, and the observed low field saturation is not explained. Finally, another problem is the lack of evidence for the presence of donor states in the polymers, even though this hypothesis can be found in recent models. An alternative process, judged more "physical" and based on the Onsager dissociation theory, has often been proposed as an alternative to the Poole–Frenkel mechanism [BLO 74], [PAI 75].

3.3.1.2.3. SCLC models

The main objective of models based on the Space Charge-Limited Current (SCLC) is to determine the external current under steady state regime in conditions where the carrier concentration is sufficiently high to produce a significant variation of the electric field and, therefore, a variation in the carriers drift velocity. These models enable the external current of electronic carriers to be estimated, in a medium without traps, with only one level of trapping, or with an exponential distribution of trapping levels [ROS 55], [MAR 62]. In the simplest version, the free carriers have a band mobility and the contacts are assumed to be ohmic. Besides, the non-linear character of the current/voltage characteristics under a stationary regime has led to resolutions in a non-stationary regime [MAN 62], [ZAH 74], [HAR 75]. Finally, improvements have been made by taking bipolar models, mobility functions of the electric field and the distributions of traps into consideration [LAM 70], [NAT 02].

3.3.1.2.4. Ionic conduction

The carriers are considered thermally activated above a potential well. When a field is applied, the potential barrier in the direction of this field decreases (and increases for the opposite direction), thus increasing the probability that the carrier will be directed in the direction of the field. A field-dependent mobility derives from this simple outline, the first reference to a *hopping* type mobility (a variation of *sh(E)*) for ions dating from 1948 [MOT 64].

3.3.1.2.5. Diffusion

Diffusion is a natural motion of species – atoms, molecules, ions [MUL 94] or electronic carriers [TOR 84] – which is due to a gradient of chemical potential. At a first approximation [MUL 94], transport by diffusion can be seen as the first derivative of species concentration in the three directions of space, see equation [3.2]. For a highly diluted ion solution, mobility is related to the diffusion coefficient by the Nernst–Einstein equation:

$$\mu = \frac{qD}{kT} \qquad [3.5]$$

This relationship supposes that ion mobility is closely related to the molecular motion within the polymer [BAM 03]. In polymers, diffusion is often neglected in the transport equations. This does not signify that the elementary transport mechanism is different, but that equation [3.2] can be approximated by solely considering the drift current:

$$j = \mu.e\left(E.n - \frac{kT}{e}\frac{\partial n}{\partial x}\right) \cong \mu.e.E.n \qquad [3.6]$$

The hypothesis of negligible diffusion comes from the fact that the concentration gradients are generally not very significant, but also that the nature of ions is uncertain and therefore identical for the coefficient values. Indeed, even when the simplest case of neutral species is considered, the diffusion coefficients can change in a very significant manner, according to the size of the molecules which diffuse. In the case of ions, other phenomena add up, such as the electrostatic and polarization interactions with the matrix and inter-ion interactions [MUL 94].

3.3.1.2.6. Mobility

With regard to electronic transport in disorderly materials, the mechanism generally admitted for the transport is *hopping* conduction [DYR 88], implying a diagonal disorder (see Figure 3.5), or off-diagonal; see [BAS 93] for a detailed description. Thus, mobility would be more related to one or the other of the elementary mechanisms described above than to conduction in the extended states. This is why, for each of these mechanisms (SCLC, *hopping*, Poole–Frenkel, etc.), the dependence in field and temperature of the mobility is either explicitly defined, or described by hypothesis; see Table 3.2. We should point out here that we do not refer to ohmic conduction as a basic mechanism for charge transport; here, it can be a band mobility, improbable in polymeric insulators, or a trap limited ohmic conductivity. In the case of shallow traps where the charges, near the conduction band, are in thermal equilibrium, the apparent mobility is weaker than band mobility, in a ratio corresponding to the ratio between the time spent by the carriers in these traps and the time spent in the conduction band [DIS 92]. The resulting expression is a mobility independent to the electric field. This model is commonly used to describe charge transport.

So far, only one energy level has been considered. When the trap depths are distributed, the situation becomes much more complex. Indeed, interactions between the different trap levels must be considered. In the same manner, a distribution of inter-trap distances can considerably change the results from a numerical simulation. Such a disorder is expected in polymer materials, but supplying a detailed description of these traps (in terms of depths, distances, etc.) is a much more difficult problem.

Process	Expression for the effective mobility	References
ionic conductivity: low field	$\mu = \frac{\alpha}{kT}\exp(\frac{-\beta}{kT})$	[DWY 73]
high field	$\mu = \frac{\alpha}{E}\exp(\frac{-\beta}{kT})sh(\frac{\gamma E}{kT})$	[DWY 73]
SCLC – space charge limited current	$\mu = \alpha\exp(\frac{-\beta}{kT})$	[DIS 92]
Hopping: above a potential barrier	$\mu = \frac{\alpha}{E}\exp(\frac{-\beta}{kT})sh(\frac{\gamma E}{kT})$	[MOT 64]
through the barrier	$\mu \propto \exp(\frac{-2\alpha}{3kT})^2\exp\left(\frac{\beta\sqrt{E}}{(kT)^2}\right)\exp(-\gamma\sqrt{E})$	[WIN 03]
Poole–Frenkel	$\mu = \alpha\exp(\frac{-\beta}{kT})\exp(\frac{\gamma E^{1/2}}{kT})$	[JON 67]

Table 3.2. *Some expressions for mobility as a function of electric field and temperature. α, β and γ are constants*

3.3.2. *Some models characterizing the experimental behavior*

In the literature, most of the articles dedicated to the modeling of charge transport are models established to recover current/voltage characteristics under a stationary regime. Only a few articles allow estimates of the external current under a non-stationary regime: they are subject to the first part of this section. Among these articles, only a few models are capable of reproducing both the external current and the space charge distribution as a function of time. In the second part of the section, a synthesis of these different models is proposed. For each model, a detailed presentation of the hypotheses, considered physical phenomena and results obtained is given. Finally, we consider high field models, whose principal difficulty consists of simulating charge packet phenomena.

3.3.2.1. *Transient current models*

Most of the models were established in the 1970s, to address problems related to semi-conductors and insulators used in reprography systems [SCH 75], [HAR 75], [LAM 70]. Transient currents have been simulated within the framework of the SCLC model [MAN 62], [MOR 62], [HEL 62]. In the Many and Rakavy model [MAN 62], a single carrier type was considered and only one trapping level is defined. The contacts are assumed to be ohmic.

Figure 3.6 shows the shape of the transient current. When there is no trap (a) or when the traps are shallow (b), a current peak is observed at the moment when the principal charge front reaches the opposite electrode. When the traps are shallow (c), the trapped charges accumulate near the electrode where the carriers are injected; this phenomenon induces a decrease, then a vanishing of the current density. It is obvious that this decrease will be even faster, as the trapping probability will be high. Iwamoto [IWA 96] has simulated transient currents produced by ion transport with extraction blocking. Only one carrier type is considered, with a *hopping* type of transport and including diffusion. The electric field distortions induced by the space charge are integrated. The general form of the current (see Figure 3.6d), is composed of a slight short-term reduction, followed by an increase up to a maximum, and finally a monotonic decrease. Despite being completely different hypotheses, these two models give similar results: the curves obtained look practically the same.

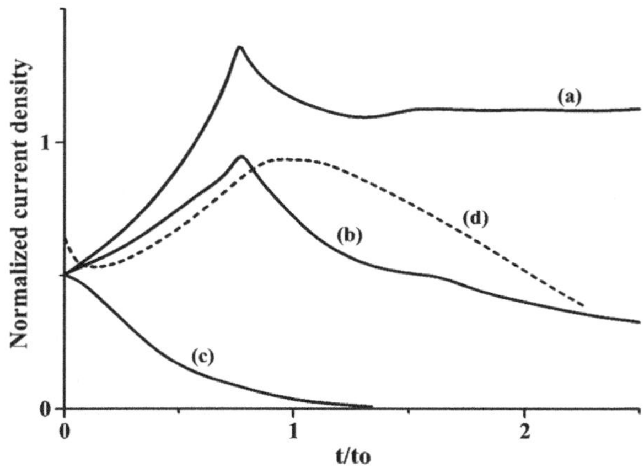

Figure 3.6. *(a–c): Transient current in a space charge limited model for a step-like applied voltage (a) no trapping; (b) moderate trapping; (c) strong trapping. t_o is the carriers transit time in the absence of space charge effect. Adapted from [MAN 62]. (d): transient current for an ionic transport and blocking electrodes. The time is in arbitrary units. Adapted from [IWA 96]*

3.3.2.2. *Models associated with the space charge measurements*

The physical processes which potentially contribute to the transport and generation of charges, presented at the beginning of this section, and the complexity of the materials studied show the difficulty, for modelers, of identifying simplifying hypotheses. Until now, experiments and simulations, permitting the dynamics of space charges in the organic insulators to be characterized, have been combined only for polyethylene-type materials. The different data obtained by space charge measurements on these materials indicate the necessity of considering the deep trapping notion at the same time for positive and negative charges [TEY 01a,b], [ALI 94]. In the same manner, the recombination of charges must be considered to explain electroluminescence phenomena [TEY 01a,b]. Hence, there are a number of observations, going from space charge profiles, to conduction current and to electroluminescence measurements, which force us to consider injection, bipolar charging processes, deep trapping and recombination phenomena. The goal is to establish models capable of explaining a maximum of experimentally observed physical phenomena, which inevitably leads to a parameter vector to adjust, generally of high dimension, in a way such as to approach experimental measurements [WIN 03]. The obtaining of a set of parameters giving acceptable results (by comparison with the experiment) does not necessarily signify that the physical processes inserted in the model are exact. The solution, to reinforce the physical aspects of the established model, is probably to confront results from the numerical simulation with experimental data obtained by using different measurement parameters, different protocols and a large variation interval in temperature, electric field, etc. Unfortunately, the optimization of parameters becomes even more delicate.

The common difficulty to all approaches is how to take into consideration all phenomena related to trapping whilst dismissing those related to conduction. In other words, all of the phenomena must be considered and pondered according to their energy level. As we have already specified, the most commonly used mechanism to describe charge transport in polymer insulators is to define a trap-controlled mobility. The carriers move in the conduction band, for the electrons, (or in the valence band, for holes) and can be trapped in shallow traps by remaining in thermal equilibrium. However, this approach can hardly explain the persistent charge in depolarization. For that, a deep trapping level is introduced. The principal conduction models in the polymer insulators are presented in Table 3.3. All of these models consider bipolar transport and, for some of them, the trapping and recombination phenomena are taken into consideration.

Alison and Hill [ALI 94] have proposed a bipolar model in which the available charge density for the injection is the difference between a constant source density and the trapped charge density near the electrode. Internal generation and diffusion are neglected. The model is intended to describe the behavior of cross-linked PE. Charge transport is modeled using an effective mobility. The trapping for electrons and holes occurs on only one level, without detrapping. Recombination is defined by coefficients, for each combination of species. The results from these simulations are essentially space charge profiles as a function of time. They have been favorably compared to space charge measurements obtained by Li and Takada [LI 92].

Fukuma *et al.* [FUK 94] have developed a more complete model than that previously set out. The generation of charges is made by two-electrode Schottky injection. They also considered the presence of charges in the dielectric before application of a voltage. The transport is described by *hopping*, the two mobile species being able to be trapped for a short time on only one shallow trapping level.

A potential barrier is defined at both interfaces for the extraction of carriers. The recombination between the mobile carriers is taken into consideration. All the barriers (*hopping*, injection, extraction) are symetrical for both species. As for the previous model, the results are space charge profiles which are compared to the experimental data of Li and Takada [LI 91]. The model has been applied for a material containing an internal interface between two layers of cross-linked PE [FUK 95].

Kaneko *et al.* [KAN 99] have presented a macroscopic model, whose hypotheses are similar to those of Alison and Hill's, the simulated material being low density PE. Bipolar charges are generated by a Schottky type mechanism and the transport is achieved through *hopping*. The recombination between the holes and the electrons is taken into account; there is no deep trapping and the extraction barrier is assumed to be null (by continuous extraction). The results, which are compared to experimental measurements, are space charge and current density profiles. The hypotheses did not permit them to obtain results consistent with experimental data, underlining the necessity to take into account deep trapping effects.

Le Roy *et al.* [LER 04], [LER 05] have recently proposed a model to characterize charge transport, trapping and recombination phenomena, experimentally observed by charge and discharge current measurements, space charge and electroluminescence measurements, for low density PE. The current and charge measurements on "intrinsic" PE (without additives) indicate that electronic processes are dominant [KAN 99], [LER 04], the charges most likely being generated by injection at the electrodes. Furthermore, the space charge measurements do not show any accumulation of hetero-charges during the polarization time, so no extraction barrier has been applied. The model follows the other hypotheses put forward by Hill and Alison [ALI 94], but detrapping is introduced. The approach is justified by the existence of two very distinct zones in the trap energy distribution in polyethylene. The first one, very close to the conduction band, is characterized by trap depths which do not exceed 0.3 eV. The second, due to chemical disorder, ranges from 0.5 to 1.5 eV. The model appears able to reproduce the essential features observed under DC stress: the space charge profiles and their dynamics under polarization and depolarization, the charge and discharge currents and electroluminescence [LER 05]. Regarding the electroluminescence data, a relation has been established for the minimum recombination rate that provides a signal above the noise level of the detection system [LER 04].

Finally, a model has been developed by Boufayed *et al.* [BOU 04], [BOU 05] in which an exponential trap distribution truncated by a maximum depth is considered, the material studied being cross-linked PE. The distribution parameters rely on the distribution of the physical and chemical traps estimated by Quirke *et al.* [MAR 02], [ANT 02] by molecular modeling. The discrete distribution of these traps is converted into a continuous distribution. Transport is made by a *hopping* type mechanism. Quirke *et al.* suppose that these traps are filled from the lower energy level upwards. The recombination is not taken into consideration here. The results from the numerical simulation are not in total agreement with the experiment [BOU 06], which is explained on the one hand by very interdependent processes (transport, trapping, mobility variation), and on the other hand by the very complexity of the material considered.

The main hypotheses of these different models are summarized in Table 3.3.

Reference:	Alison & Hill [ALI 94]	Fukuma *et al.* [FUK 94]	Kaneko *et al.* [KAN 99]	LeRoy *et al.* [LER 04]
Generation of charges	Constant charge source	Schottky Injection	Schottky Injection	Schottky Injection
Extraction of charges	No extraction barrier	Extraction barrier	No extraction barrier	No extraction barrier
Transport	Conduction by free carrier	Conduction by *hopping*	Conduction by *hopping*	Conduction by free carrier
Trapping	Single trap level	Single trap level	Two trap levels	Single trap level with detrapping
Recombination	Between free charges and trapped charges	Free charges	Free charges	Between free charges and trapped charges
Others		Joule effect Charges initially presented		Density of initial charges

Table 3.3. *Characteristics of the physical models for bipolar transport*

3.3.2.3. *High field models*

So far, the different models presented have been established for relatively weak fields. In the case of polyethylene materials and for applied fields greater than 70 kV.mm^{-1}, the space charge measurements – by pulsed electro-acoustic technique (PEA) or by laser induced pressure pulse (LIPP) [KAN 99], [HOZ 94], [HOZ 98] – show repetitive charge waves propagating in the bulk of the insulation, usually referred to as charge packets. These charge packets have been observed for dielectrics of 0.1 mm to a few mm of thickness. The observed oscillations in the current density have clearly been associated with the propagation of charge packets [HOZ 94], [ALI 98]. Indeed, studies have shown that these oscillations, which depend on the field and the material under consideration, have an oscillation period similar to that of the charge fronts. Further, the current peak is associated with the disappearance of the charge packet when it reaches the opposite electrode [ALI 98], [SUZ 98], coinciding with the formation of a new charge front. Finally, the oscillations observed in the electroluminescence signal and the current are synchronous [TEY 01a,b], which signifies that the rate of recombination is maximal at the moment when a charge front disappears and/or the next one forms.

Numerical models have been made to reproduce these phenomena [LI 91], [HOZ 94], [HOZ 98], the constraints on the predictions being that: (i) charge fronts must form; (ii) they must move; (iii) they must maintain their form during the transit; (iv) the process must be repetitive. Usually, these charge packets appear near an electrode, such that a massive charge injection or an ionization induced by the field have been proposed as mechanism initiators of the packets. A necessary hypothesis for these models is the presence of charges of both polarities, which is in agreement with the observed electroluminescence in these field regimes (associated with the radiative (i.e. giving rise to luminescence) recombination of charges).

Generally, the models consider a two-electrode injection. The principal difficulties encountered in simulation are the attenuation and dispersal of the charge fronts during their transit, and the lack of repetitivity of the phenomenon. The attenuation of the fronts leads to a displacement of a short distance and a limited number of successive fronts. In fact, the form of the charge packet is maintained during the transit only if charges are generated in the volume (as opposed to injection at the electrodes) during the front transit. This generation has been simulated by considering donor centers, or by ionization in the region where the local field is maximal, generally before the charge front [HOZ 98], [FOT 00]. The apparent charge motion would then result from the difference in mobility of positive and negative carriers.

3.4. Trends and perspectives

3.4.1. *Unification of atomistic and macroscopic approaches*

A first step towards a unification of macroscopic and microscopic approaches has been attempted by Quirke *et al.* [MAR 02], [ANT 02] where a prediction of the current-field characteristics of PE has been proposed relying on *ab initio* calculations. The method has three steps:

1. Estimation of energies, density and residence time of the conformational traps of PE, using DFT. A similar approach has been made for chemical traps.

2. Calculation of local electronic mobilities for a transport model by multiple hops including the influence of the traps filling. An important result here is that, even though the chemical traps are largely a minority, electronic mobilities are controlled by the deep traps for low carrier densities.

3. Switch to the macroscopic scale by resolution of the Poisson's and transport equations. The boundary conditions are a constant electrode charge source.

In its current form, the model is limited, in the sense that the transport is supposed to be unipolar (electrons only), where the behavior at the interfaces is simplified and where the agreement with the experimental behavior is approximate, particularly at high field. However, the model is unified in the sense that it gets to a macroscopic description of the transport, starting from the traps' characteristics simply deduced from the physical and chemical structure of the material. To our knowledge, it is the one and only attempt at an integrated model in the insulating polymers domain.

What are the difficulties of using this method? Certainly, the atomistic models are very greedy in terms of computational power. However, progress in this domain has been made, and it does not seem that calculation times are always a limitating factor; in any case they will not remain so. Elementary simulations still suffer from an incomplete description of what these materials really are. Even for the simplest of them (PE), the chemical structure is complex: not all chemical defects are identified and quantified, and the antioxidant chemistry is particularly disconcerting. Certainly, a concentration which is extremely weak in defects or impurities is sufficient to explain the net space charge densities which are measured: by considering the order of 1 C/m^3 for these latter, a charged center density of $6x10^{12}$ cm^{-3} is obtained. The repetitive unit density of PE (-C_2H_4-) being $6x10^{22}$ cm^{-3}, a density of (charged) defects of 1 in 10^{10} repetitive units is sufficient to explain what is measured [TEY 02]. Such a concentration is beyond the scope of chemical analysis techniques for these materials. It is also immensely weaker than the concentration in any known residue. Thus, occupation of a tiny fraction of potential chemical traps described above is sufficient to describe the experimental behavior, unless the measurement only represents a slight imbalance between positive and negative trapped charges. The same problem is expected for transport sites considering the very high density of estimated physical traps.

The way to establish the continuity between microscopic and macroscopic descriptions is probably to bring quantum calculations to simple (or simplified) outputs, as is the case in the approaches of Quirke *et al.* [ANT 02]. In this example, energy distributions of traps have been produced, indeed by considering a limited number of defects. This type of information can be taken as the starting point of macroscopic models. We expect new developments of these *ab initio* methods, for example to obtain information on hole trapping and transport at the same level as what was done for the electrons, to better define the interactions between electronic carriers (charge recombination), to tackle the internal generation phenomena by dissociation assisted by the field, or to evaluate reactivity of charged centers.

Coming back to macroscopic models, one of the difficulties is that the physical models must be relatively complex to reproduce a continually growing set of experimental observations. This necessarily leads to an increasing number of parameters, most of them being hardly accessible by independent experiments. The numerical techniques can henceforth bring solutions to the physical models in reasonable time, as long as we introduce the right parameters. The problem is that the adjustment becomes tedious when the number of parameters increases, especially if it is empirical (see section 3.3.2). The contribution of inverse methods to adjust these parameters becomes urgent.

3.4.2. *Interface behavior*

The metal/dielectric interfaces processes are very critical in the sense that, in numerous cases, they control the carrier generation. We have previously shown that a limited number of models are available to account for interface effects. What is more, their relevance is questionable. The Schottky mechanism does not work with theoretical barrier heights. Another problem: the barrier maximum is situated at such a long distance from the electrode that an electron has a very strong probability of being thermalized by collisions before getting over the barrier [WIN 98]. Such a mechanism might explain why apparent barrier heights are weaker than expected.

Alternative interface charge transfer principles have been proposed [LEW 02], [WIN 98], but they remain unverifiable. The analysis of macroscopic behavior is clearly not the most relevant way to identify the interface elementary mechanisms because it does not permit the interface effects to be separated from the volume ones, especially when the volume transport processes are not firmly established. Moreover, macroscopic techniques collect and average the information on a certain sample surface (or even volume), like the current injection models do. The point is that we do not have definitive proof that charge generation phenomena are homogenous processes.

We therefore need specific descriptions of the interfaces concerning the property gradient of the insulator near the surface, and the property limits at the metal (or other)/insulation interface. While methods exist to generate carriers in precise regions of the insulator (volume or near the surface) without activating the injection itself (through implantation by electron beam, for example), there are no routine means for looking at what happens when charges cross the interface. Promising results have been obtained, for organic semi-conductors, by near field potentiometry [SIL 96]; the application of such methods to insulators, and more generally of near field imagery techniques, must be envisaged because they could bring determining information about the interfaces. Interfaces are clearly crucial topics: understanding them will lead to strategies to improve the insulators' performances by specific processes (surface treatment, etc.) [HAY 90]. The interfaces can also be internal interfaces, as is the case of multilayer dielectrics [SUH 96], [BOD 04], or in the world of composite dielectrics with nanometric charges [LEW 04].

3.4.3. *Physical models for transport in volume*

3.4.3.1. *Identification of the nature of carriers*

It is still not clear what the nature of carriers is, as pointed out in [WIN 03], and it could remain thus for a long time considering the complexity and diversity of the phenomena coming into play, in generation and as well as in transport: electro-dissociation, electrochemical reactions, generation of electrically active species of low molecular weight, the role of humidity in transport, trapping, influence on a trapped charge, are all supplementary processes to those described in the previous paragraph, and are all to be envisaged. Further, it can be tricky to distinguish between the electronic and ionic carriers: we are used to considering that electronic transport is carried out without mass transfer, which is true, but the transport of protons can also be carried out without significant mass effects. Another aspect is the coupling of carriers and media. In polar media, a charge can move with its own polarization field, thus behaving like a heavy particle having an apparent charge which does not correspond to an electron charge. Even if many questions still remain unanswered, the important point to consider is the vector dominating the transport for a range of experimental conditions. We have available some experimental evidence concerning the electronic processes in insulating polymers under high fields, and some others involving ionic transport under low fields. These questions could be treated by better focusing the experimental conditions around the methods permitting the evaluation of the spatio-temporal distribution of charges [FLE 05], with good spatial and temporal resolutions. This has not necessarily been the case so far, the methods having been applied more to solve practical questions than to study the fundamental transport processes. For example, when the charge measured under DC stress is of the same sign as the polarity of the adjacent electrode, we suppose it corresponds to the injection and we turn towards electronic transport by hopping. However, species of weak molar weight could stabilize the injected charge and give rise to ionic transport in volume. At the other extreme, when charges accumulate at an electrode of opposite polarity, ionic transport is often invoked, but it can also result from the transit of electronic carriers from the opposite electrode.

3.4.3.2. *Trap identification*

The nature of traps are always debated, and we have argued chiefly according to the philosophy stemming from semi-conductor physics and dielectrics theory, in which the polaron is relevant. A set of techniques have been implemented to identify the nature of traps, but they do not necessarily appear consistent [WIN 90], [WIN 03]. There is still no consensus on the interpretation of measurements in terms of localized states formed by the polymer itself, by opposition to impurity states [WIN 03]. Here, a powerful tool was found in molecular simulation. As we saw in the first part of this chapter, it is perfectly possible to foresee the electrical properties of a given physical or chemical defect of a polymer chain, and in particular to define its propensity to stabilize an electron or a hole. More refined calculations can inform on the polarization of the medium generated by the charge, thus giving effective trap depths, which can be larger than estimated without consideration of conformational rearrangements.

3.4.4. *Degradation induced by a charge and/or a field*

This question is marginal with respect to the scope of this chapter, which has focused on the simulation of transport. We have, however, underlined in the introduction that the development of accurate models based on established physical approaches are needed. These models put down their roots in transport models, because factors of influence in the ageing are the local values of charge density or field, whatever the electrical ageing scenario considered. A local field would control the generation probability of hot carriers [ZEL 84] or the electromechanical strain of the medium [LEW 96]. A local charge density would control the irreversible evolution probability of the entities stabilizing these charges and involved in the ageing [DIS 01], as well as the recombination rate of charges in connection with chemical degradations (through electronic excitation of the recombining centers) [LAU 99]. Among relevant and promising simulation activities, we can distinguish two different areas pertaining to macroscopic and microscopic scales. A good example of the macroscopic approach of the problem is proposed in [MAZ 05]. Here, the molecular simulation has a strong added value. The chemical degradations induced by impact of carriers or by recombination of charges involve excited states of electrons or atoms. Simulations on chemical interactions made by Stoneham *et al.* [STO 01] show that in the excited state, the polymer chains evolve in time, through simple vibrations or rotations, or, in a more expected way, can even break. Thus, the excited states in the polymer can give rise to chain scission or radical recombination phenomena (see also [SAN 93]). These processes compete with energy relaxation in the form of light or heat. Molecular simulations of the excited states of the recombining centers deserve to be made. In a similar way, an understanding of the conformational changes associated with charge trapping could go through molecular simulations and be a way to study the steps from free volume formation to that of meso-voids, micro-voids, etc. This field is practically unexplored in the insulating polymers domain, and could be a way of improving the performances of insulators by controlling their molecular architecture.

3.4.5. *Contribution of the physics of non-insulating organic materials*

We have, essentially, been interested so far in insulating polymers. However, in support of the description of different mechanisms at work, we have resorted to research concerning semi-conductor materials, doped polymers and/or ionic conductors. It is very conceivable that the same set of mechanisms could apply to different classes of materials: the problem is to determine the relative importance of these different mechanisms. This is particularly delicate in the case of insulators which, unlike other materials, are not optimized, except for their not very conductive character from a macroscopic point of view and, in certain cases, by the orientation of their dielectric permittivity. An important fact here is the existence of scientific problems and common techniques for different classes: for example, in the case of organic electroluminescent devices, the common worries with insulators are the poor definition of the interface mechanisms, the influence of the space charge on behavior, the understanding and management of degradation under electrical stress. Further, most organic semi-conductors are more similar to insulators than to inorganic semi-conductors: with a relatively large band gap (2 eV and more), and transport controlled by *hopping* rather than according to a band mobility [WAL 02]. In the same way, we should be inspired by what was learned from the transport in ionic conductors to approach the case of insulators. This, therefore, provides a large scope to establishing physical bases common to different classes of organic materials.

3.5. Conclusions

Although the nature of conduction processes in insulating polymers is still a controversial subject, it appears to us that all the conditions – concerning the comprehension of mechanisms, multi-scale simulation approaches, numerical techniques, development of advanced characterization techniques – are available to develop transport and degradation models from different angles of the physics of phenomena. The use of electrical, physical and chemical characterization techniques, with good sensibilities and spatial and temporal resolutions, must be encouraged to tackle fundamental problems and determine the relevant physical parameters to incorporate into the meso and macroscopic models. Finally, molecular description has a high application potential in this field and must be developed in order to better capture the consequences of chemical and/or physical defects on charge stabilization in their environment and on the connection between charge transport, trapping and degradation mechanism – this latter point involving reconfigurations of chains and/or excited states.

3.6. Bibliography

[ABK 95] ABKOWITZ M.A., MIZES H.A., FACCI J.S., "Emission limited injection by thermally assisted tunneling into trap-free transport polymer", *Appl. Phys. Lett.*, vol. 66, no. 10, p. 1288–1290, 1995.

[ABR 92] ABRAMCZYK H., WERNER B., KROH J., "Absorption spectra of the solvated electron in hydrocarbons", *J. Phys. Chem.*, vol. 96, no. 24, p. 9674–9677, 1992.

[ALI 94] ALISON J.M., HILL R.M., "A model for bipolar charge transport, trapping and recombination in degassed crosslinked polyethene", *J. Phys. D: Appl. Phys.*, vol. 27, no. 6, p. 1291–1299, 1994.

[ALI 98] ALISON J.M., "A high field pulsed electro-acoustic apparatus for space charge and external circuit current measurement within solid insulators", *Meas. Sci. Technol.*, vol. 9, no. 10, p. 1737–1749, 1998.

[ANT 02] ANTA J.A., MARCELLI G., MEUNIER M., QUIRKE N., "Models of electron trapping and transport in polyethylene: current-voltage characteristics", *J. Appl. Phys.*, vol. 92, no. 2, p. 1002–1008, 2002.

[ARK 98] ARKHIPOV V.I., EMELIANOVA E.V., TAK Y.H., BÄSSLER H., "Charge injection into light-emitting diodes: theory and experiment", *J. Appl. Phys.*, vol. 84, no. 2, p. 848–856, 1998.

[BAI 89] BAIRD J.K., RUSSELL D.P., "Physical interpretation of the disappearance of dielectric screening in the case of electrons in nonpolar fluids", *Phys. Rev. A*, vol. 39, no. 8, p. 4295–4297, 1989.

[BAL 01] BALDO M.A., FORREST S.R., "Interface-limited injection in amorphous organic semiconductors", *Phys Rev B*, vol. 64, no. 8, p. 85201–1/85201–17, 2001.

[BAM 03] BAMFORD D., REICHE A., DUBLEK G., ALLOIN F., SANCHEZ J.Y., ALAM M.A., "Ionic conductivity, glass transition, and local free volume in poly(ethylene oxide) electrolytes: Single and mixed ion conductors", *J. Chem. Phys.*, vol. 118, no. 20, p. 9420–9432, 2003.

[BAS 93] BASSLER H., "Charge transport in disordered organic photoconductors. A Monte Carlo simulation study", *Phys. Stat. Sol. B.*, vol. 175, no. 1, p. 15–56, 1993.

[BAU 03] BAUER S., BAUER-GOGONEA S., "Current practise in space charge and polarization profile measurements using thermal techniques", *IEEE Trans. Dielectr. Electr. Insul.*, vol. 10, no. 5, p. 883–902, 2003.

[BLA 98] BLAISE G., SARJEANT W.J., "Space charge in dielectrics: energy storage and transfer dynamics from atomistic to macroscopic scale", *IEEE Trans. Dielectr. Electr. Insul.*, vol. 5, no. 5, p. 779–808, 1998.

[BLA 01] BLAISE G., "Charge localization and transport in disordered dielectric materials", *Journal of Electrostatics*, vol. 50, no. 2, p. 69–89, 2001.

[BLO 74] BLOSSEY D.F., "One-dimensional Onsager theory for carrier injection in metal-insulator systems", *Phys. Rev. B*, vol. 9, no. 12, p. 5183–5187, 1974.

[BOD 04] BODEGA R., MORSHUIS P.H.F., SMIT J.J., "Space charge signal interpretation in a multi-layer dielectric tested by means of the PEA method", *IEEE Int. Conf. on Solid Dielectrics*, Toulouse, France, p. 240–243, 2004.

[BOG 05] BOGGS S., "Very high field phenomena in dielectrics", *IEEE Trans. Dielectr. Electr. Insul.*, vol. 12, no. 5, p. 929–938, 2005.

[BOT 78] BÖTTCHER C.J.F., BORDEWIJK P., *Theory of Electric Polarization*, Elsevier, New York, vol. II, 1978.

[BOU 04] BOUFAYED F., LEROY S., TEYSSEDRE G., LAURENT C., SEGUR P., COOPER E., DISSADO L.A., MONTANARI G.C., "Numerical resolution of charge transport in cross-linked polyethylene by means of a bipolar model with a distribution of traps", *IEEE Int. Conf. on Solid Dielectrics*, Toulouse, p. 562–566, 2004.

[BOU 05] BOUFAYED F., LEROY S., TEYSSEDRE G., LAURENT C., SEGUR P., DISSADO L.A., MONTANARI G.C., "Simulation of bipolar charge transport in polyethylene featuring trapping and hopping conduction through an exponential distribution of traps", *IEEE Int. Symp. Electrical Insulating Materials*, Kitakyushu, Japan, p. 340–343, 2005.

[BOU 06] BOUFAYED F., TEYSSEDRE G., LAURENT C., LE ROY S., DISSADO L.A., SÉGUR P., MONTANARI G.C., "Models of bipolar charge transport in polyethylene", *J. Appl. Phys.*, vol. 100, no. 10, p. 104105-1 - 104105-10, 2006.

[CAM 02] CAMPUS A., CARSTENSEN P., FARKAS A.A., MEUNIER M., "Chemical defects and electron trapping relevant to cable dielectrics", *Proc. IEEE Conf. Electrical Insulation and Dielectric Phenomena*, Cancun, Mexico, p. 155–158, 2002.

[COE 93] COELHO R., ALADENIZE B., *Les diélectriques*, Hermès, Paris, 1993.

[CON 97] CONWELL E.M., WU M.W., "Contact injection into polymer light-emitting diodes", *Appl. Phys. Lett.*, vol. 70, no. 14, p. 1867–1869, 1997.

[CUB 02] CUBERO D., MARCELLI G., QUIRKE N., "Electronic states of excess electrons in polyethylene", *IEEE Conf. Electrical Insulation and Dielectric Phenomena*, Cancun, Mexico, p. 430–433, 2002.

[CUB 03a] CUBERO D., QUIRKE N., COCKER D., "Electronic states for excess electrons in polyethylene compared to long-chain alkanes", *Chem. Phys. Lett.*, vol. 370, no. 1–2, p. 21–25, 2003.

[CUB 03b] CUBERO D., QUIRKE N., COKER D.F., "Electronic transport in disordered n-alkanes: from fluid methane to amorphous polyethylene", *J. Chem. Phys.*, vol. 119, no. 5, p. 2669–2679, 2003.

[DIS 92] DISSADO L.A., FOTHERGILL J.C., *Electrical Degradation and Breakdown in Polymers*, Peter Peregrinus, London, 1992.

[DIS 01] DISSADO L.A., MAZZANTI G., MONTANARI G.C., "Elemental strain and trapped space charge in thermoelectrical aging of insulating materials. Part I: elemental strain under thermo-electrical-mechanical stress", *IEEE Trans. Dielectr. Electr. Insul.*, vol. 8, no.6, p. 959–965, 2001.

[DUK 78] DUKE C.B., "Polymers and molecular solids: new frontiers in surface science", *Surface Science*, vol. 70, no. 1, p. 674–691, 1978.

[DUN 95] DUNLAP D.H., "Explanation for the $\sqrt{E}$ -dependent mobilities of charge transport in molecularly doped polymers", *Phys. Rev. B*, vol. 52, no. 2, p. 939–954, 1995.

[DWY 73] O'DWYER J.J, *The Theory of Electrical Conduction and Breakdown in Solid Dielectrics*, Clarendon Press, Oxford, 1973.

[DYR 88] DYRE J.C., "The random free-energy barrier model for AC conduction in disordered solids", *J. Appl. Phys.*, vol. 64, p. 2456–2468, 1988.

[EKS 68] EKSTROM A., WILLARD E., "Effects of matrix polarity on the optical and electron spin resonance spectra of trapped electrons in organic glasses", *J. Phys. Chem.*, vol. 72, no. 13, p. 4599–4603, 1968.

[EMT 66] EMTAGE P.R., O'DWYER J.J., "Richardson–Schottky effects in insulators", *Phys. Rev. Lett.*, vol. 16, no. 9, p. 356–358, 1966.

[FLE 05] FLEMING R.J., "Space charge profile measurement techniques: recent advances and future directions", *IEEE Trans. Dielectr. Electr. Insul.*, vol. 12, no. 5, p. 967–978, 2005.

[FOT 00] FOTHERGILL J.C., "Space charge in dielectrics: Old theories and new measurements", *Proc. 8th Int. Conf. Dielectric Materials, Measurements and Applications,* Edinburgh, UK, p. 1–5, 2000.

[FRÖ 63] FRÖHLICH H., "Introduction to the theory of the polaron", in *Polarons and Excitons,* GOETHE KUPER, C. and WHITFIELD, G.D. (Eds), Oliver & Boyd, London, 1963.

[FUK 94] FUKUMA M., NAGAO M., KOSAKI M., "Computer analysis on transient space charge distribution in polymers", *Proc. 4th Int. Conf. on Properties and Applications of Dielectric Materials*, Brisbane, Australia, p. 24–27, 1994.

[FUK 95] FUKUMA M., NAGAO M., KOSAKI M., "Numerical analysis on transient space charge distribution in XLPE", *5th IEEE Int. Conf. on Conduction and Breakdown in Solid Dielectrics*, Leicester, UK, p. 139–143, 1995.

[GIL 92] GILBERT R., CRINE J.P., NOIRHOMME B., PÉLISSOU S., "Measurement of organic and inorganic ions in cable insulation and shields", *IEEE Proceedings of the Conference on Electrical Insulation and Dielectric Phenomena (CEIDP),* Leesburg, USA, p. 235–240, 1992.

[HAR 75] HARTMANN G.C., LIPARI N.O., "Investigation of transient space-charge-perturbed currents", *J. Appl. Phys.*, vol. 46, no. 7, p. 2821–2827, 1975.

[HAY 90] Hayashi H., Nakano T., Ishii K., Ohki Y., "Effect of plasma surface-modification on the electrical conduction in high-density polyethylene", *IEEE Proceedings of the Conference on Electrical Insulation and Dielectric Phenomena CEIDP),* Pocono Manor, USA, p. 190–195, 1990.

[HEL 62] HELFRICH W., MARK P., "Raumladungsbeschränkte ströme in Anthrazen als Mittel zur Bestimmung der Beweglichkeit von Defektelektronen", *Z. Physik*, vol. 166, no. 4, p. 370–385, 1962.

[HIB 86] HIBMA T., ZELLER H.R., "Direct measurement of space charge injection from a needle electrode into dielectrics", *J. Appl. Phys.*, vol. 59, no. 5, p. 1614–1620, 1986.

[HO 77] HO K.K., KEVAN L., "Temperature dependence of the recombination fluorescence of photoionized indole and N, N, N', N'-Tetramethyl-p-phenylenediamine in organic glasses. Consequences of electron tunnelling and diffusion", *J. Phys. Chem.*, vol. 81, p. 1865–1871, 1977.

[HO 01] HO Y.F.F., CHEN G., DAVIES A.E., HAMPTON R.N., SWINGLER S.G., SUTTON S.J., "Do semicons affect space charge?"*Proc. Int. Conf. Solid Dielectrics*, Eindhoven, p. 105–108, 2001.

[HOF 91] HOFFMANN R., JANIAK C., KOLLMAR C., "A chemical approach to the orbitals of organic polymers", *Macromolecules*, vol. 24, p. 3725–3746, 1991.

[HOZ 94] HOZUMI N., SUZUKI H., OKAMOTO T., WATANABE K., WATANABE A., "Direct observation of time-dependent space charge profiles in XLPE cable under high electric fields", *IEEE Trans. Dielectr. Electr. Insul.*, vol. 1, p. 1068–1076, 1994.

[HOZ 98] HOZUMI N., TAKEDA T., SUZUKI H., OKAMOTO T., "Space charge behavior in XLPE cable insulation under 0.2–1.2 MV/cm DC field", *IEEE Trans. Dielectr. Electr. Insul.*, vol. 5, p. 82–90, 1998.

[HUG 80] HUGHES R.C., "The electronic properties of the metal-insulator contact: Space-charge induced switching", *J. Appl. Phys.*, vol. 51, p. 5933–5944, 1980.

[IWA 96] IWAMOTO M., "Transient current across insulating films with long-range movements of charge carriers", *J. Appl. Phys.*, vol. 79, p. 7936–7943, 1996.

[JON 67] JONSCHER A.K., "Electronic properties of amorphous dielectric films", *Thin Solid Films*, vol. 1, p. 213–234, 1967.

[JON 71] JONSCHER A.K., "Electronic conduction in amorphous semiconductors", *J. Vac. Sci. Technol.*, vol. 8, p. 135–144, 1971.

[JON 96] JONSCHER A.K., *Universal Relaxation Law,* Chelsea Dielectrics Press, London, 1996.

[KAH 67] KAHN D., MAYCOCK J.N., "Difference-equation model for ion conductivity. I. Step voltage response", *J. Chem. Phys.*, vol. 46, p. 4434–4440, 1967.

[KAN 99] KANEKO K., MIZUTANI T., SUZUOKI Y., "Computer simulation on formation of space charge packets in XLPE films", *IEEE Trans. Dielectr. Electr. Insul.*, vol. 6, p. 152–158, 1999.

[KLA 72] KLASSEN N.V., GILLIS H.A., TEATHER G.G., "Spectral shifts of trapped electrons in alkane glasses at 76 K", *J. Phys. Chem.*, vol. 76, p. 3842–3850, 1972.

[LAM 70] LAMPERT M.A., MARK P., *Current Injection in Solids*, Academic Press, New York, 1970.

[LAU 99] LAURENT C., "Optical pre-breakdown warnings in insulating polymers", *IEEE Electr. Insul. Mag.*, vol. 15, no. 2, p. 5–13, 1999.

[LER 04a] LE ROY S., SEGUR P., TEYSSEDRE G., LAURENT C., "Description of bipolar charge transport in polyethylene using a fluid model with a constant mobility: model prediction", *J. Phys. D: Appl. Phys.*, vol. 37, p. 298–305, 2004.

[LER 04b] LE ROY S., TEYSSEDRE G., SEGUR P., LAURENT C., "Modelling of space charge, electroluminescence and current in low density polyethylene under DC and AC field", *IEEE Proceedings of the Conference on Electrical Insulation and Dielectric Phenomena (CEIDP),* Boulder, USA, p. 29–32, 2004.

[LER 05] LE ROY S., TEYSSEDRE G., LAURENT C., "Charge transport and dissipative processes in insulating polymers: experiments and model", *IEEE Trans. Dielectr. Electr. Insul.*, vol. 12, p. 644–654, 2005.

[LES 73] LESS K.J., WILSON E.G., "Intrinsic photoconduction and photoemission in polyethylene", *J. Phys. C: Solid State Phys.*, vol. 6, p. 3110–3120, 1973.

[LEW 90] LEWIS T.J., "Charge transport, charge injection and breakdown in polymeric insulators", *J. Phys. D: Appl. Phys.*, vol. 23, p. 469–1478, 1990.

[LEW 96] LEWIS T.J., LLEWELLYN J.P., VAN DER SLUIJS M.J., FREESTONE J., HAMPTON R.N., "A new model for electrical ageing and breakdown in dielectrics", *Proceedings of the 7th Int. Conf. Dielectric Materials, Measurements and Applications*, Bath, UK, p. 220–224, 1996.

[LEW 98] LEWIS T.J., "The micro-physics of charge in solid dielectrics", in *Space Charge in Solid Dielectrics*, FOTHERGILL, J.C. and DISSADO, L.A. (Eds), The Dielectrics Society, Leicester, 1998.

[LEW 02] LEWIS T.J., "Polyethylene under electric stress", *IEEE Trans. Dielectr. Electr. Insul.*, vol. 9, p. 717–729, 2002.

[LEW 04] LEWIS T.J., "Interfaces are the dominant feature of dielectrics at the nanometric level", *IEEE Trans. Dielectr. Electr. Insul.*, vol. 11, p. 739–753, 2004.

[LEW 05] LEWINER J., HOLÉ S., DITCHI T., "Pressure wave propagation methods: A rich history and a bright future", *IEEE Trans. Dielectr. Electr. Insul.*, vol. 12, p. 114–126, 2005.

[LI 91] LI Y., YASUDA M., TAKADA T., "Influence on spatial charge distribution of cross-linking agent residues in XLPE", *Proceedings of the 3rd IEEE Conf. on Properties and Applications of Dielectric Materials*, Tokyo, Japan, p. 1210–1213, 1991.

[LI 92] LI Y., TAKADA T., "Experimental observation of charge transport and injection in XLPE at polarity reversal", *J. Phys. D: Appl. Phys.*, vol. 25, p. 704–716, 1992.

[MAN 62] MANY A., RAKAVY G., "Theory of transient space-charge-limited currents in solids in the presence of trapping", *Phys. Rev.*, vol. 126, p. 1980–1988, 1962.

[MAR 02] MARCELLI G., MEUNIER M., QUIRKE N., "Electronic traps in polymer insulators: I(V) characteristics", *IEEE Conf. Electr. Insul. Dielectr. Phenomena* (CEIDP), Cancun, Mexico, p. 40–43, 2002.

[MAR 62] MARK P., HELFRICH W., "Space-charge-limited currents in organic crystals", *J. Appl. Phys.*, vol. 33, p. 205–215, 1962.

[MAZ 05] MAZZANTI G., MONTANARI G.C., DISSADO L.A., "Electrical ageing and life models: The role of space charge", *IEEE Trans. Dielectr. Electr. Insul.*, vol. 12, p. 876–890, 2005.

[MEU 97] MEUNIER M., QUIRKE N., BINESTI D., MARTIC G., FOURMIGUE J.M., "Identification of space charge in cable insulation: a new approach using molecular simulation", *IEEE Proceedings of the Conference on Electrical Insulation and Dielectric Phenomena (CEIDP),* Minneapolis, p. 68–71, 1997.

[MEU 98] MEUNIER M., QUIRKE N., BINESTI D., "Electron trapping in polymer insulators: a new approach using molecular modeling", *Proc. Intern. Conf. on Electric Charge in Solid Insulators*, Tours, p. 66–72, 29 June–3 July 1998.

[MEU 00a] MEUNIER M., QUIRKE N., "Molecular modeling of electron trapping in polymer insulators", *J. Chem. Phys.*, vol. 113, p. 369–376, 2000.

[MEU 00b] MEUNIER M., Quirke N., Aslanides A., "Characterization of charge carrier traps in polymeric insulators", *IEEE Proceedings of the Conference on Electrical Insulation and Dielectric Phenomena*, Victoria, Canada, p. 21–24, 2000.

[MEU 01] MEUNIER M., QUIRKE N., ASLANIDES A., "Molecular modeling of electron traps in polymer insulators: chemical defects and impurities", *J. Chem. Phys*, vol. 115, p. 2876–2881, 2001.

[MIA 96] MIAO M.S., VAN CAMP P.E., VAN DOREN V.E., LADIK J.J., MINTMIRE J.W., "Conformation and electronic structure of polyethylene: a density-functional approach", *Phys. Rev. B*, vol. 54, p. 10430–13435, 1996.

[MIZ 04] MIZUTANI T., "High field phenomena in insulating polymers", *Proceedings of the 2004 IEEE International Conference on Solid Dielectrics*, vol. 1, p. 11–16 July 2004.

[MOR 62] MORRISON J.A., EDELSON D., "Solution of the space charge problem for a pulsed Townsend discharge", *J. Appl. Phys.*, vol. 33, p. 1714–1720, 1962.

[MOT 64] MOTT N.F., GURNEY R.W., *Electronic Processes in Ionic Crystals,* Clarendon, Oxford, 1948, Reprinted by Dover, New York, 1964.

[MUL 94] MÜLLER-PLATHE F., "Permeation in polymers – A computational approach", *Acta Polymerica*, vol. 45, p. 259–293, 1994.

[NAT 02] NATALI D., SAMPIETRO M., "Field-dependent mobility from space-charge-limited current-voltage curves", *J. Appl. Phys.*, vol. 92, p. 5310–5318, 2002.

[PAI 75] PAI D.M., "Electric-field-enhanced conductivity in solids", *J. Appl. Phys.*, vol. 46, p. 5122–5126, 1975.

[PAI 75] PAI D.M., ENCK R.C., "Onsager mechanism of photo-generation in amorphous selenium", *Phys. Rev. B*, vol. 11, p. 5163–5174, 1975.

[PIR 76] PIREAUX J.J., SVENSSON S., BASILIER E., MALMQVIST P.A., GELIUS U., CAUDANO R., SIEGBAHN K., "Core-electron relaxation energies and valence-band formation of linear alkanes studied in the gas phase by means of electron spectroscopy", *Phys. Rev. A*, vol. 14, p. 2133–2145, 1976.

[RAC 87] RACZ I., OHKI Y., "High energy electron transport in hydrocarbons", *J. Appl. Phys.*, vol. 26, p. 209–215, 1987.

[RAJ 91] RAJAGOPAL A.K., NGAI K.L., TEITLER S., "Theoretical aspects of coupling model schemes of slow relaxation in complex correlated systems", *J. Non-Crystalline-Solids*, vol. 131, p. 282–288, 1991.

[RAS 02] RASMUSSEN C. N., HOLBØLL J.T., HENRIKSEN M., "Dynamic space charge behaviour in polymeric DC cables", *Proceedings of the 11th IEEE International Symposium on Electrets (ISE),* IEEE, Brisbane, Australia, p. 23–26, 2002.

[REI 90] REI VILAR M., SCHOTT M., "Vibrational and electronic excitation of hexatriacontane thin films by low energy electron impact", *J. Chem. Phys.*, vol. 92, p. 5722–5730, 1990.

[RIG 01] RIGHI M.C., SCANDOLO S., SERRA S., IARLORI S., TOSATTI E., SANTORO G., "Surface states and negative electron affinity in polyethylene", *Phys. Rev. Lett.*, vol. 87, p. 768021–1 – 768021–4, 2001.

[RIT 82] RITSKO J.J., Electronic states and triboelectricity, In *Electronic Properties of Polymers*, MORT, J and PFISTER, G., (Eds), Wiley, New York, p. 13–57, 1982.

[ROS 55] ROSE A., "Space-charge-limited currents in solids", *Phys. Rev.*, vol. 97, p. 1538–1544, 1955.

[SAN 93] SANCHE L., "Electronic aging and related electron interactions in thin-film dielectrics", *IEEE Trans. Electr. Insul.*, vol. 28, p. 789–819, 1993.

[SCH 75] SCHER H., MONTROLL E.W., "Anomalous transit-time dispersion in amorphous solids", *Phys. Rev. B*, vol. 12, p. 2455–2477, 1975.

[SCO 99] SCOTT J.C., MALLIARAS G.G., "Charge injection and recombination at the metal-organic interface", *Chem. Phys. Lett.*, vol. 299, p. 115–119, 1999.

[SER 98] SERRA S., TOSATTI E., IARLORI S., SCANDOLO S., ALBERTINI M., "Interchain states and the negative electron affinity of polyethylene", *IEEE Conf. Electrical Insulation and Dielectric Phenomena*, Atlanta, USA, 25–28, p. 19–22, 1998.

[SER 00] SERRA S., TOSATTI E., IARLORI S., SCANDOLO S., SANTORO G., "Interchain electron states in polyethylene", *Phys. Rev. B*, vol. 62, p. 4389–4393, 2000.

[SES 97] SESSLER G.M., "Charge distribution and transport in polymers", *IEEE Trans. Dielectr. Electr. Insul.*, vol. 4, p. 614–628, 1997.

[SHE 98] SHEN J., YANG J., "Physical mechanisms in double-carrier trap-charge limited transport processes in organic electroluminescent devices: A numerical study", *J. Appl. Phys.*, vol. 83, p. 7706–7714, 1998.

[SHI 72] SHIDA T., IWATA S., WATANABE T., "Electronic absorption spectra of excess electrons in molecular aggregates. Trapped electrons in gamma-irradiated amorphous solids at 77 K", *J. Phys. Chem.*, vol. 76, p. 3683–3694, 1972.

[SHO 71] SHOUSHA A.H.M., YOUNG L., "Space charge contribution to the polarization current in thin amorphous dielectric films", *Thin Solid Films*, vol. 88, p. 383–396, 1971.

[SIL 96] Silveira W.R., Marohn J.A., "Microscopic view of charge injection in an organic semiconductor", *Phys. Rev. Lett.*, vol. 33, no. 11, p.116104.1–116104.4, 2004

[STE 05] STEVENS G., "Nano and meso measurement methods applied to dielectrics study: what to expect?", *IEEE Trans. Dielectr. Electr. Insul.*, vol. 12, No. 5, p. 979–992, 2005.

[STO 01] STONEHAM A.M., HARKER A.H., RAMOS M.M.D., "Charge and the unexpected: solid state processes involving charge species", *4th Int. Conf. on Electric Charges in Solid Insulators*, IEEE, Tours, France, p. 15–22, 2001.

[SUH 96] SUH K.S., KIM J.Y., NOH H.S., LEE C.R., "Interfacial charge in polyethylene/ethylene vinylacetate laminates", *IEEE Trans. Dielectr. Electr. Insul.*, vol. 3, p. 758–764, 1996.

[SUZ 98] SUZUOKI Y., KON H., KANEKO K., MIZUTANI T., ITAYA T., "Packet-like space charge behavior in insulating polymers", in *Space Charge in Solid Dielectrics*, FOTHERGILL, J.C. and DISSADO, L.A. (Eds), The Dielectrics Society, Leicester, UK, 1998.

[TAK 99a] TAKADA T., "Acoustic and optical methods for measuring electric charge distributions in dielectrics", *IEEE Trans. Dielectr. Electr. Insul.*, vol. 6, p. 519–547, 1999.

[TAK 99b] TAKADA T., HOZUMI N., SUZUKI H., OKAMOTO T., "Factors of hetero space charge generation in XLPE under DC field of 20 kV/mm", *Electr. Engg. Jpn.*, vol. 119, p. 13–21, 1999.

[TAN 05] TANAKA T., "Dielectric nanocomposites with insulating properties", *IEEE Trans. Dielectr. Electr. Insul.*, vol. 12, No. 5, p. 914–928, 2005.

[TAY 92] TAYLOR P.L., WANG B.C., OLIVEIRA F.A., DOER T.P., "Dielectric screening of Coulomb interactions in polymers", *Macromolecules*, vol. 25, p. 1694–1698, 1992.

[TEY 01a] TEYSSEDRE G., LAURENT C., ASLANIDES A., QUIRKE N., DISSADO L.A., CAMPUS A., MARTINOTTO L., "Deep trapping centers in cross-linked polyethylene investigated by molecular modeling and luminescence techniques", *IEEE Trans. Dielectr. Electr. Insul.*, vol. 8, p. 744–752, 2001.

[TEY 01b] TEYSSEDRE G., LAURENT C., MONTANARI G.C., PALMIERI F., SEE A., FOTHERGILL J.C., DISSADO L.A., "Charge distribution and electroluminescence in XLPE under DC field", *J. Phys. D: Appl. Phys.*, vol. 34, p. 2830–2844, 2001.

[TEY 02] TEYSSEDRE G., TARDIEU G., LAURENT C., "Characterization of cross-linked polyethylene materials by luminescence techniques", *J. Materials Sci.*, vol. 37, p. 1599–1609, 2002.

[TOR 84] TORPEY P.A., "Double-carrier injection and recombination in insulators, including diffusion effects", *J. Appl. Phys.*, vol. 56, p. 2284–2294, 1984.

[VAN 62] VAN ROGGEN A., "Electrical conduction of polymer single crystals", *Phys. Rev. Lett.*, vol. 9, p. 368–370, 1962.

[WAL 02] WALKER A.B., KAMBILI A., MARTIN S.J., "Electrical transport modeling in organic electroluminescent devices", *J. Phys.: Condens. Matter*, vol. 14, p. 9825–9876, 2002.

[WIL 73] WILLARD J.E., "Trapped free radicals and electrons in organic glasses", *Science*, vol.180, p. 553–561, 1973.

[WIN 03a] WINTLE H.J., "Charge motion in technical insulators: Facts, fancies and simulations", *Proc. IEEE Conf. on Electrical Insulation and Dielectric Phenomena*, Albuquerque, USA, p. 1–15, 2003.

[WIN 03b] WINTLE H.J., "Charge motion in technical insulators: facts, fancies and simulations", *IEEE Trans. Dielectr. Electr. Insul.*, vol. 10, p. 826–841, 2003.

[WIN 88] WINTLE H.J., "The discharge of charged dipolar insulating materials", *J. Appl. Phys.*, vol. 63, p. 1705–1713, 1988.

[WIN 90] WINTLE H.J., "Basic physics of insulators", *IEEE Trans. Electr. Insul.*, vol. 25, p. 27–44, 1990.

[WIN 98] WINTLE H.J., "The motion of electric charges on and near insulator surfaces", *Proc. 3rd Int. Conf. on Electrical Charges in Solid Insulators*, Tours, France, p. 49–59, 1998.

[WIN 99] WINTLE H.J., "Charge motion and trapping in insulators: surface and bulk effects", *IEEE Trans. Dielectr. Electr. Insul.*, vol. 6, p. 1–10, 1999.

[ZAH 74] ZAHN M., TSANG C.F., PAO S.C., "Transient electric field and space charge behavior for unipolar ion conduction", *J. Appl. Phys.*, vol. 45, p. 2432–2440, 1974.

[ZEL 84] ZELLER H.R., PFLUGER P., BERNASCONI J., "High-mobility states and dielectric breakdown in polymeric dielectrics", *IEEE Trans. Electr. Insul.*, vol. 19, p. 200–204, 1984.

Chapter 4

Dielectric Relaxation in Polymeric Materials

4.1. Introduction

Over this last decade, the technological development of dielectric spectroscopy on the one hand [MCC 67], [BLY 79], [KRE 02], and dielectric thermal analysis on the other [VAN 75], [HED 77], [RUN 97], have allowed the development of dielectric relaxation studies in a wide frequency (from 10^{-6} to 10^{+11} Hz) and temperature range, thus leading to a new approach of behavior laws of dielectric relaxation in polymeric materials. In the same way, the development of physico-chemistry and polymer physics [CAR 99], [FON 02], [RAU 02], [HAL 06], [PER 92], [STR 97] has improved the interpretation of the mechanisms observed at a molecular scale.

4.2. Dynamics of polarization mechanisms

The polarization mechanisms are of different origins: (1) from the modification of spatial distributions of electronic and ionic elementary particles and (2) from the reorientation of dipolar entities. These mechanisms are multi-scale and the kinetics to establish the polarization exists at very different time and frequency ranges. We are going to cite them in order of increasing establishment time.

Chapter written by Eric DANTRAS, Jérôme MENEGOTTO, Philippe DEMONT and Colette LACABANNE.

4.2.1. *Electronic and ionic polarization*

Under the influence of an external electric field, the electronic orbits are distorted with very fast kinetics (10^{-15} s) causing electronic polarization (see Figure 4.1). The ionic displacements will be made with slower kinetics (10^{-13} s): they are responsible for the ionic polarization (see Figure 4.1).

Considering the establishment time of all of these mechanisms, they behave as instantaneous phenomena in the time domain, and are observed at much higher frequencies than those of electric fields used in dynamical dielectric spectroscopy. This polarization will be designated by P_∞ and is related to the permittivity at infinite frequency ε_∞ by the equation:

$$P_\infty = (\varepsilon_\infty - 1)\varepsilon_0 E \qquad [4.1]$$

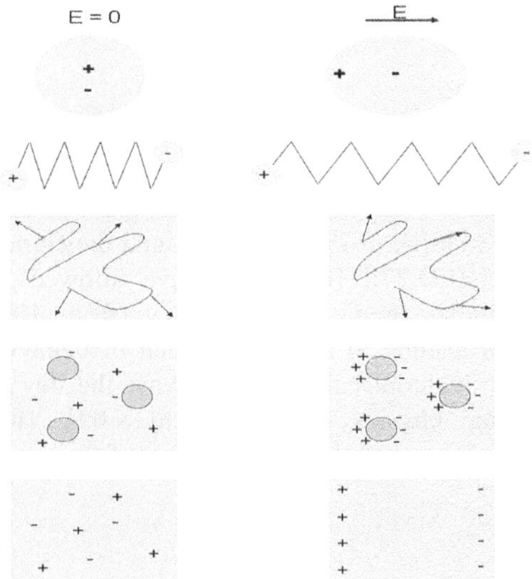

Figure 4.1. *Schematic representation of polarization mechanisms. From top to bottom: electronic, ionic, dipolar, Maxwell–Wagner–Sillars, and inter-facial polarizations*

4.2.2. *Dipolar polarization*

Polymeric materials are made of more or less polar entities, of different (sub- or super-nanometric) scales. Their reorientation, which tends to align them in the direction of the electric field, is the cause of the dipolar orientation polarization P_d.

The time/frequency scale is now of the order of that covered by dielectric spectrometries. This dipolar polarization P_d is associated with the permittivity ε_d by the relation:

$$P_d = \varepsilon_d \varepsilon_0 E \quad [4.2]$$

P_s is the total aquired polarization, in established regime:

$$P_s = (\varepsilon_s - 1)\varepsilon_0 E \quad [4.3]$$

where ε_s is the static permittivity.

The variation amplitude of the dipolar polarization is then:

$$P_s - P_\infty = (\varepsilon_s - \varepsilon_\infty)\varepsilon_0 E \quad [4.4]$$

4.2.3. *Maxwell–Wagner–Sillars polarization*

Any heterogeneity of the dielectric can lead to the accumulation of charges at the interfaces if the structural entities (amorphous and crystalline phases in the semi-crystalline polymers) or morphological (polymer matrix and reinforcement in the composites) have different conductivities (see Figure 4.1). It is then made of relaxing entities likely to turn towards long/low frequency times.

4.2.4. *Interfacial polarization*

The charges trapped on the surface of the sample make a macro dipole whose characteristic dimension corresponds to the thickness of the sample (see Figure 4.1). For small thicknesses, the relaxation of this macro dipole takes place at very long/very low frequency times, with a very strong intensity. Like the Maxwell–Wagner–Sillars polarization, the inter-facial polarization is distinguished from the orientation polarization by a much stronger intensity.

4.3. Orientation polarization in the time domain

To have a simple qualitative description of the establishment of the orientation polarization, the hypothesis of first order kinetics is widely used. After recalling the basic equation of this analogical modeling, the principal analytical equations will allow a quantitative description of the experimental results [BOT 78].

4.3.1. *Single relaxation time model*

τ is the relaxation time characteristic of the polarization kinetics establishment. The dipolar polarization is defined by the first order differential equation:

$$\tau \frac{dP_d(t)}{dt} + P_d(t) = \left(\varepsilon_s - \varepsilon_\infty\right)\varepsilon_0 E \quad [4.5]$$

Under static field E = E_o, considering the boundary conditions on P_d, the solution is written:

$$P_d = (\varepsilon_s - \varepsilon_\infty)\varepsilon_0 E\left(1 - \exp(-\frac{t}{\tau})\right) \quad [4.6]$$

In this model, the dielectric permittivity is given by:

$$\varepsilon(t) = \varepsilon_\infty + (\varepsilon_s - \varepsilon_\infty)(1 - \exp(-\frac{t}{\tau})) \quad [4.7]$$

This variation is generally represented as a function of log t. It is a sigmoid which presents no element of symmetry. The description of the experimental results is correct qualitatively and not quantitatively.

4.3.2. *Discrete distribution of relaxation times*

The dielectric polymers are heterogenous, at different scales, both nanometric and micronic. To refine the description of experimental results, a discrete distribution of relaxation time (τ_j) with a dispersion $(\varepsilon_s - \varepsilon_\infty)_j$ represents an interesting approach. The dielectric permittivity is then written:

$$\varepsilon(t) = \varepsilon_\infty + \sum_j \left(\varepsilon_s - \varepsilon_\infty\right)_j \left(1 - \exp - \frac{t}{\tau_j}\right) \quad [4.8]$$

– with: $\sum_j \left(\varepsilon_S - \varepsilon_\infty\right)_j = \left(\varepsilon_S - \varepsilon_\infty\right)$

4.3.3. *Continuous distribution of relaxation times*

If an homogenization of the medium at a certain scale is realistic, the use of a continuous distribution function φ (lnτ) is relevant; it allows us to work with global responses. The dielectric permittivity is given by:

$$\varepsilon(t) = \varepsilon_{\infty} + (\varepsilon_{s} - \varepsilon_{\infty}) \int_{0}^{\infty} \varphi(\ln\tau)(1 - \exp - \frac{t}{\tau}) d\ln\tau \qquad [4.9]$$

with: $\int_0^{\infty} \phi(\ln\tau)\, d\ln\tau = 1$

4.3.4. *Stretched exponential: Kohlrausch–Williams–Watts equation*

In the same way as the homogenous medium, the use of the stretched exponential is interesting because it is analytically simpler:

$$\varepsilon(t) = \varepsilon_{\infty} + (\varepsilon_{s} - \varepsilon_{\infty}) \left\{ 1 - \exp - \left(\frac{t}{\tau} \right)^{\beta_{KWW}} \right\} \qquad [4.10]$$

with: $0 < \beta_{KWW} \leq 1$

4.4. Orientation polarization in the frequency domain

The description of the dynamic orientation polarization is made with the same methodology as for the transitory polarization since they are related by the Laplace transformation.

4.4.1. *Single relaxation time model: the Debye equation*

In the presence of a dynamic electric field $E = E_o \cos \omega t$, represented by the complex expression $E^* = E_o \exp(i\omega t)$, the solution, in an established regime, of differential equation [4.5] is:

$$\varepsilon_{d}^{*}(\omega) = \frac{(\varepsilon_{s} - \varepsilon_{\infty})}{(1 + i\omega\tau)} \qquad [4.11]$$

The total complex permittivity:

$$\varepsilon^{*}(\omega) \equiv \varepsilon^{'}(\omega) - i\varepsilon^{''}(\omega) \qquad [4.12]$$

is then defined by:

$$\varepsilon'(\omega) = \varepsilon_{\infty} + (\varepsilon_s - \varepsilon_{\infty}) \times \frac{1}{1+\omega^2\tau^2} \qquad [4.13]$$

$$\varepsilon''(\omega) = (\varepsilon_s - \varepsilon_{\infty}) \times \frac{\omega\tau}{1+\omega^2\tau^2} \qquad [4.14]$$

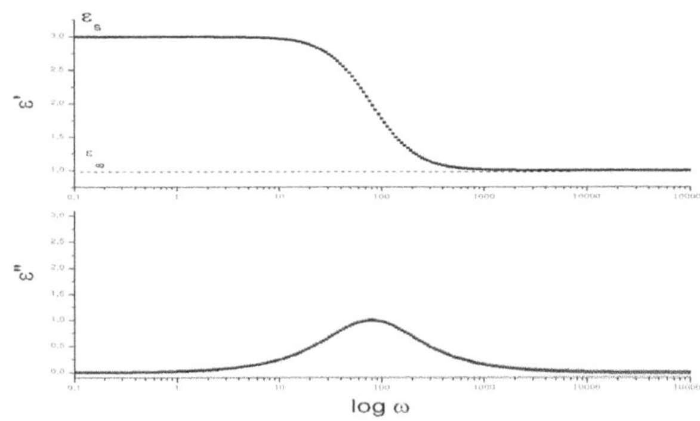

Figure 4.2. *Complex permittivity predicted by the Debye equation: ε' (logω) and ε'' (logω)*

These expressions are designated in the literature by the Debye equations. The variation laws of ε' and ε'' as a function of log ω are represented in Figure 4.2. They present elements of symmetry which are not observed in the experimental variations.

4.4.2. *Discrete distribution of relaxation times*

For heterogenous materials, the complex permittivity is well described by:

$$\varepsilon^{*}(\omega) \equiv \varepsilon_{\infty} + \sum_{j} \frac{(\varepsilon_S - \varepsilon_{\infty})_j}{(1 + i\omega\tau_j)} \qquad [4.15]$$

By using fractional polarizations, the analysis of thermo-stimulated currents allows the parameters $(\varepsilon_s - \varepsilon_\infty)_j$ and τ_j to be accessed.

4.4.3. *Continuous distribution of relaxation times*

For homogenous materials, the approach using a continuous distribution of relaxation times is interesting:

$$\varepsilon^*(\omega) = \varepsilon_\infty + (\varepsilon_s - \varepsilon_\infty)\int_0^\infty \frac{\varphi(\ln\tau)}{1 + i\omega\tau}\, d\ln\tau \qquad [4.16]$$

4.4.4. *Parametric analytical expressions*

To describe the experimental results, numerous parametric expressions have been proposed [JON 83]. To take a didactic approach, we shall only cite those resulting in the Havriliak and Negami equation, which is currently the most widely used.

4.4.4.1. *Cole–Cole equation*

The equation proposed by Cole–Cole allows a widening of the relaxation zone and thus permits a better description of the experimental results.

$$\varepsilon^*(\omega) = \varepsilon_\infty + \frac{(\varepsilon_s - \varepsilon_\infty)}{1 + (i\omega\tau)^{\beta_{cc}}} \qquad [4.17]$$

with: $0 < \beta_{cc} \leq 1$

4.4.4.2. *Havriliak and Negami equation*

The experimental results show an asymmetry of the variation law which is not taken into account by equation [4.17]. The Havriliak and Negami equation allows the description to be optimized by the introduction of an additional parameter [HAV 67], [HAV 97]:

$$\varepsilon^*(\omega) = \varepsilon_\infty + \frac{(\varepsilon_s - \varepsilon_\infty)}{\left(1 + (i\omega\tau)^{\beta_{HN}}\right)^{\gamma_{HN}}} \qquad [4.18]$$

with: $0 < \beta_{HN} \leq 1$ and $0 < \gamma_{HN} \leq 1$

4.4.4.3. *Cole–Cole representation*

By eliminating the frequency between the two Debye equations, [4.13] and [4.14], a relation between ε' and ε'' is obtained. It can be written as [COL 41]:

$$\left\{\varepsilon' - \frac{\varepsilon_s + \varepsilon_\infty}{2}\right\}^2 + \varepsilon''^2 = \left\{\frac{\varepsilon_s - \varepsilon_\infty}{2}\right\}^2 \qquad [4.19]$$

This variation law which corresponds to the model of a single relaxation time is represented in the complex plan in Figure 4.3. We have also reported, for comparison, the variation law ε'' (ε') corresponding to the Havriliak and Negami equation with the parameters β_{HN} = 0.6 and γ_{HN} = 0.4. The advantage of this representation is to give an estimation of the parameters from the geometric construction.

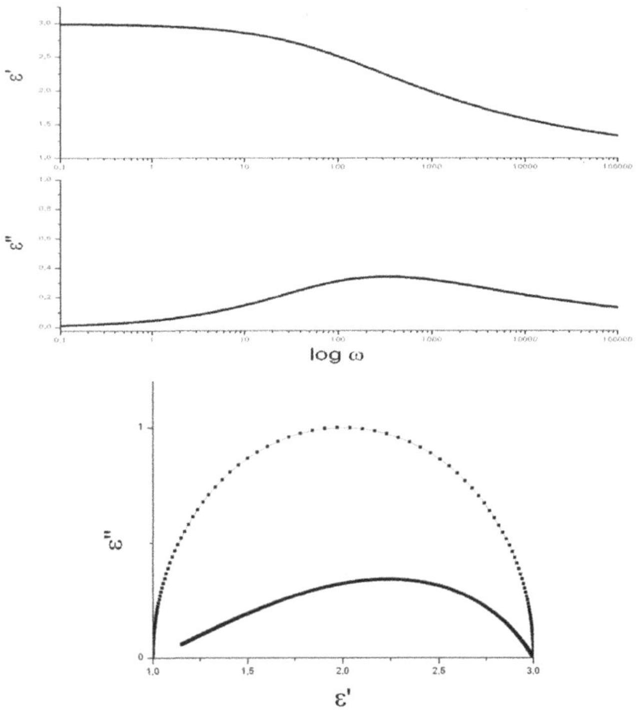

Figure 4.3. *Complex permittivity predicted by the Havriliak and Negami equation with β_{HN} = 0.6 and γ_{HN} = 0.4: ε' (logω); ε'' (logω); ε'' (ε') in the continuous line – the dotted line corresponds to the Debye equation*

4.4.5. *Kramers–Kronig relations*

ε' and ε'' are related by the Kramers–Kronig relationships:

$$\varepsilon'(\omega) - \varepsilon_\infty = \left(\frac{2}{\pi}\right)\int_0^\infty \frac{u\varepsilon''(\omega)}{\left(u^2 - \omega^2\right)} du \qquad [4.20]$$

$$\varepsilon''(\omega) = \left(\frac{2}{\pi}\right)\int_0^\infty \frac{\varepsilon'(\omega) - \varepsilon_\infty}{\left(u^2 - \omega^2\right)} du \qquad [4.21]$$

These equations were first used in the optical frequency domain. For dielectric relaxation, they are very useful to identify the dielectric energy loss of dipolar origin which can be hidden by electrical conduction.

4.5. Temperature dependence

The temperature dependence of the relaxation time is always exponential; the variation and dispersion (ε_s - ε_∞) is a powerful law. For a first approximation, we shall uniquely consider the temperature variation of the relaxation time.

4.5.1. *Shift factor*

In this approximation, temperature causes a simple translation of spectra along the axis logt/logω. The shift factor a_T defines this geometric operation which transforms the isotherm T_o into isotherm T [FER 70].

4.5.1.1. *Time domain*

The shift factor will be designed by a_{Tt} in the time domain. This is:

$$\log a_{Tt} = \frac{\log \tau(T)}{\log \tau(T_0)} \qquad [4.22]$$

4.5.1.2. *Frequency domain*

The shift factor will be designated by $a_{T\omega}$ in the frequency domain. We can easily verify that:

$$\log a_{Tt} = -\log a_{T\omega} \qquad [4.23]$$

4.5.2. *Crystalline or vitreous phases: Arrhenius equation*

For the crystalline or vitreous phases, the temperature dependence of the relaxation time has the same phenomenological behavior. It can be described by the barrier theories already used by Fröhlich [FRO 58] to describe the dielectric relaxation in the crystals. They have been adjusted to molecular crystals by Hoffman [MCC 67].

4.5.2.1. *Thermal activation mechanism*

In the barrier theories, the environment of the relaxing entity is represented by the variation of the Gibbs free enthalpy G. The orientation of the relaxing entities corresponds to the crossing of energetic barriers separating the different minima of G, by thermal activation. A classical description by Boltzmann equations gives, for the transition probability p, a variation as:

$$p \propto \exp\frac{-\Delta G}{RT} \qquad [4.24]$$

where ΔG represents the height of the enthalpic barrier which separates the two sites and R the ideal gas constant.

The relaxation time, which varies in the opposite direction to the probability, is in the form:

$$\tau = \tau_0 \exp\frac{\Delta G}{RT} \qquad [4.25]$$

To obtain an explicit equation as a function of temperature, we must express ΔG as a function of the activation enthalpy ΔH and of the activation entropy ΔS:

$$\tau = \tau_0 \exp - \frac{\Delta S}{R} \exp\frac{\Delta H}{RT} \qquad [4.26]$$

So: $\tau_{0a} = \tau_0 \exp - \frac{\Delta S}{R}$

$$\tau = \tau_{0a} \exp\frac{\Delta H}{RT} \qquad [4.27]$$

This is the Arrhenius equation. The corresponding variation of $\log\tau$ as a function of T^{-1} is represented in Figure 4.4.

4.5.2.2. *Interpretation of the activation parameters*

The activation enthalpy is directly related to the cohesion of the phase in which the relaxation is done: in a crystalline phase it will be higher than in a vitreous phase. It also depends on the size of the relaxing entity. The activation entropy comes from the Boltzmann equation:

$$\Delta S = R \ln \Omega \qquad [4.28]$$

if Ω is the number of sites accessible to the relaxing entity. τ_o can be deduced from Eyring's chemical activation theory [EYR 36]:

$$\tau_0 = \frac{h}{kT} \qquad [4.29]$$

In the crystalline phases, Ω is near unity; τ_{oa} will then be of the order of time constants associated with infrared frequencies. In the vitreous phases, Ω is very high so τ_{oa} very low. This permits the observed relaxation phenomena in semi-crystalline polymers to be localized in one or other of these phases.

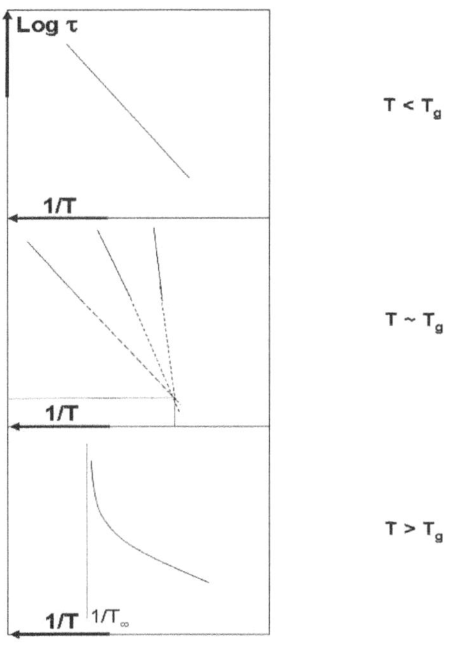

Figure 4.4. *Temperature dependence of relaxation times in the three indicated temperature zones*

4.5.3. *Vitreous phases in the transition zone: the Hoffman–Williams–Passaglia equation*

To describe the dielectric relaxations in a series of paraffins with different lengths, Hoffman–Williams–Passaglia (HWP) [HOF 66] have made the following hypothesis: the activation enthalpies ΔHj and the activation entropies ΔSj of the paraffin composed of j constitutive units are linear functions of the activation enthalpies ΔHo and entropies ΔSo for each constitutive unit. ΔHj and ΔSj are therefore expressed by:

$$\Delta H_j = j\Delta H_0 + \Delta H_1 \qquad [4.30]$$

$$\Delta S_j = j\Delta S_0 + \Delta S_1 \qquad [4.31]$$

ΔH_1 and ΔS_1 represent the activation enthalpies and entropies of the chain extremities. ΔH_j is therefore a linear function of ΔS_j.

with $T_c \equiv \Delta H_o/\Delta S_o$, equation [4.2] is now written:

$$\tau_j = \tau_c \exp\left\{\frac{\Delta H_j}{R}\left(T^{-1} - T_c^{-1}\right)\right\} \qquad [4.32]$$

where the constant τ_c represents the relaxation time at temperature T_c. This "compensation law" which has been established for the relaxation times of paraffins made up of j constitutive units, also describes the different relaxation times isolated in a polymer in the vitreous transition zone. In this case, they reflect the relaxation of entities with different lengths of polymer chain sequences. The behavior of these relaxation times is represented in Figure 4.4.

4.5.4. *Liquid phases: Vogel–Fulcher–Tammann equation (VFT)*

For liquid phases, the dielectric relaxation time has the same temperature dependence as the viscosity. It is described by the free volume theories proposed by Cohen Turnbull [COH 59] for inorganic amorphous materials as well as by the thermodynamic theory of Adam and Gibbs [ADA 65]. We shall content ourselves here with presenting the free volume concept.

4.5.4.1. *Free volume concept*

The specific volume v_s of a polymer is composed of two terms:

– the volume occupied by the constitutive groups of the macromolecule v_o, due to steric hindrance and thermal agitation; and

– the free volume v_f

The basic hypothesis of volume theories is the following: the molecular mobility is liberated when the free volume is greater than a critical value v^*. The relaxation time is then dependent on the fraction of free volume $f \equiv v_f / v^*$. From the free volume theory, we obtain:

$$\tau = \tau_{0V} \exp\left(\frac{1}{f}\right) \qquad [4.33]$$

where τ_{ov} is independent of the temperature.

To explain the dependence on temperature, another hypothesis is necessary; above the critical temperature T_∞, the fraction of free volume is expanded according to the relation:

$$f = \alpha_f (T - T_\infty) \qquad [4.34]$$

where α_f represents the expansion coefficient of the free volume fraction.

The explicit equation of the relaxation time is then:

$$\tau = \tau_{0v} \exp\left\{\frac{1}{\alpha_f (T - T_\infty)}\right\} \qquad [4.35]$$

Figure 4.4 shows this variation law on an Arrhenius diagram. This equation has been widely used in the literature [VOG 21], [FUL 52], [TAM 26] and is known by as the Vogel–Fulcher–Tammann equation. Let us note that T_∞ is then the temperature under which the molecular mobility is fixed. In the thermodynamic theory of Gibbs and Di Marzio, it is the temperature at which the activation entropy is null. Analytically, it is defined as the temperature T_∞ which linearizes the variation of $\ln\tau$ as a function of $1/(T- T_\infty)$.

4.5.4.2. *Williams–Landel–Ferry empirical expression (WLF)*

The shift factor has been widely used in this temperature zone to describe molecular mobility, in particular by Williams–Landel–Ferry [FER 70] who have proposed an empirical equation:

$$\log a_{Tt} = -\frac{c_{10}(T - T_0)}{(c_{20} + T - T_0)} \quad [4.36]$$

c_{10} is a dimensionless constant; c_{20} has the same dimension as temperature. Considering the definition of the displacement factor (see equation [4.23]), this empirical equation is equivalent to the Vogel–Fulcher–Tammann equation. For amorphous polymers having a vitreous transition temperature of less than 100°C, for a reference temperature $T_o = T_g$, universal values are obtained for both constants: c_{1g}= 17.44 and c_{2g} = 51.6°. The corresponding values of the VFT equation are $\alpha_f = 4.84\ 10^{-4}\ (°)^{-1}$and $T_\infty = T_g - 51.6°$.

4.6. Relaxation modes of amorphous polymers

The dielectric relaxations associated with the dipolar reorientation reflect the different organization levels of the polymer. We shall consider them in the order of increasing frequencies on an isochronous spectrum: the primary relaxation mode, thus designated because generally the most intense, is often called relaxation α; the secondary relaxation modes are multiple (β, γ, δ). To simplify, we shall adopt the same nomenclature for amorphous and semi-crystalline polymers, which is not necessarily the case in the literature. The monophasic amorphous polymers present the simplest relaxation spectra.

4.6.1. *Primary relaxation mode*

4.6.1.1. *Complex relaxation in an homogenous liquid medium*

When it is observed after the vitreous transition, this relaxation mode can be treated like a complex relaxational event in an homogenous medium. As shown in the example in Figure 4.5 for an amorphous PET [MEN 99], the Havriliak and Negami equation allows the variation of the dissipative component of the permittivity ε'' (logω) to be described correctly.

The relaxation time τ_{HN} obeys a VFT equation with a critical temperature of Tg – 40°C, not very far from WLF values. This result is coherent with the fact that, considering the frequency range covered in dielectric spectrometry; this relaxation mode is observed in the liquid state. This temperature behavior of the relaxation time is typical of "weak" liquids [ANG 95] i.e. those whose cohesion in the amorphous phase implies weak physical bonding. In the opposite case, the relaxation time obeys an Arrhenius equation: the poly(methyl methacrylate) PMMA has the typical behavior of "strong" liquids.

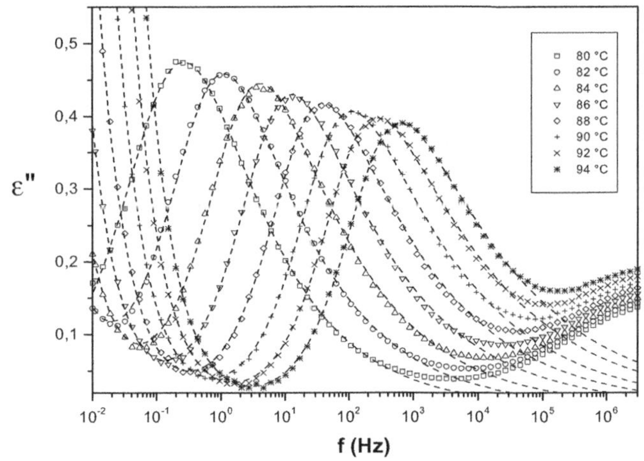

Figure 4.5. *Isothermal dynamic spectra of dielectric loss: the primary relaxation mode of an amorphous polymer (PET)*

4.6.1.2. *Discrete spectrum of simple relaxations in a heterogenous vitreous medium*

This mode has also been studied in thermo-stimulated currents: by the weak equivalent frequency of this technique (10^{-2} Hz), the relaxation is then produced in the vitreous state. This mode can then be resolved by fractional polarizations, in a discrete series of elementary relaxational phenomena represented in Figure 4.6; the activation parameters of each one are reported in Figure 4.7.

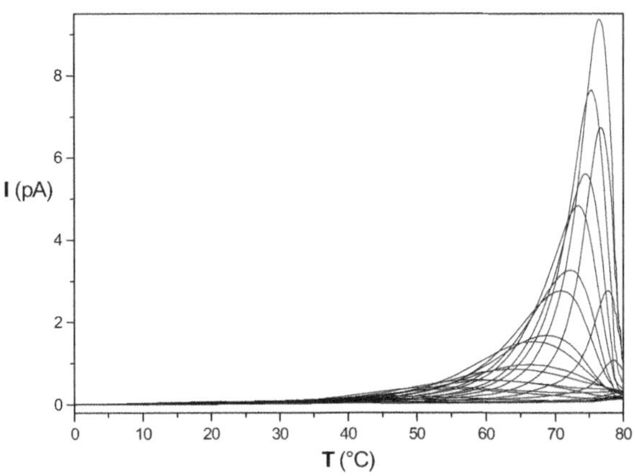

Figure 4.6. *Elementary thermograms of thermo-stimulated current for a primary relaxation mode: example of an amorphous polymer (PET)*

The observation of a compensation law (equation [4.32]) indicates the cooperativity of this molecular mobility. It is important to note that the compensation temperature is of the order of the vitreous transition temperature, and the compensation time τ_c is about 10 seconds, i.e. the time constant values associated with the vitreous transition. By analogy with the Hoffmann–Williams–Passaglia model, the elementary mechanisms have been associated with the mobility of chain sequences (of the order of nanometres) with different lengths which are liberated by the gradual rupture of the physical bonding in the vitreous transition mechanism.

It is important to note the universality of this behavior, independently of the structure and, in particular, the architecture of the polymer chain. Moreover, the activation parameter values reflect the cohesion of the amorphous phase.

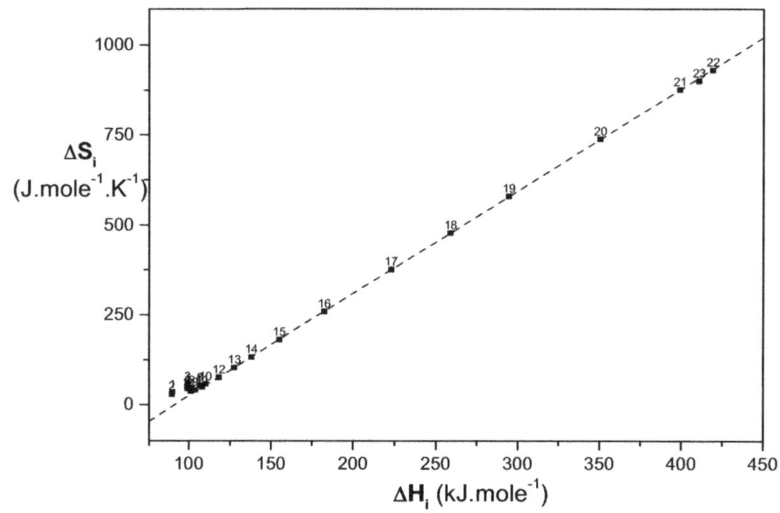

Figure 4.7. *Compensation diagram deduced from analysis of thermo-stimulated current for an amorphous polymer (PET)*

4.6.2. *Secondary relaxation modes*

4.6.2.1. *Specific mobility of the chemical structure*

The polymers which have a polar lateral chain present a relaxation mode specific to this lateral chain. As the dielectric surface of the atactic PMMA reported in Figure 4.8 shows, it is sometimes more intense than the primary relaxation. This is produced in the vitreous phase and is then associated with a localized mobility.

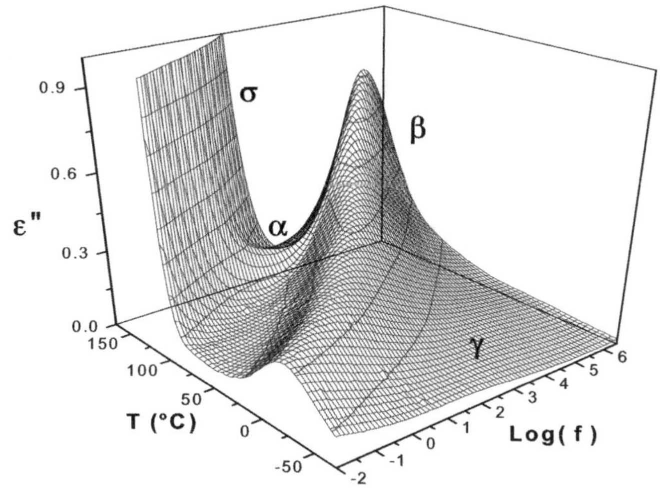

Figure 4.8. *Dielectric surface deduced from dynamic spectroscopy for an amorphous polymer with lateral chain (PMMA)*

4.6.2.2. *Mobility of the main chain*

The vitreous phase presents a localized molecular mobility of the main chain responsible for a generic secondary relaxation, which is illustrated in Figure 4.9 by the isochrone ε'' (T) recorded for an amorphous PET. This is a mobility of sub-micronic sequences which is the cause of this relaxation mode.

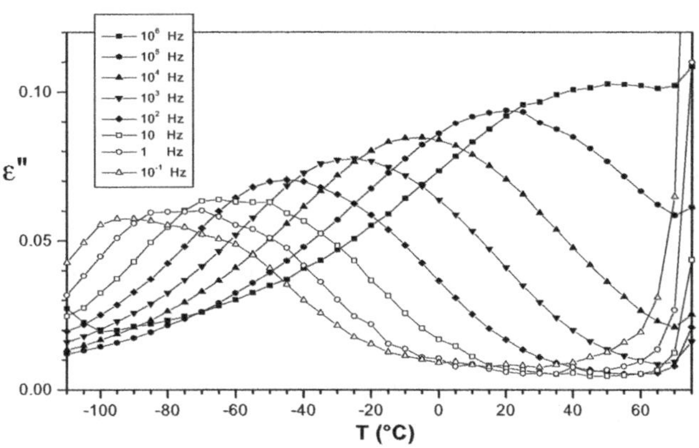

Figure 4.9. *Isochronous dynamic spectroscopy of the secondary relaxation of an amorphous polymer (PET)*

4.7. Relaxation modes of semi-crystalline polymers

In general, semi-crystalline polymers present a primary relaxation which is the most intense, as shown by the dielectric surface of the PVDF represented in Figure 4.10 [MEN 99].

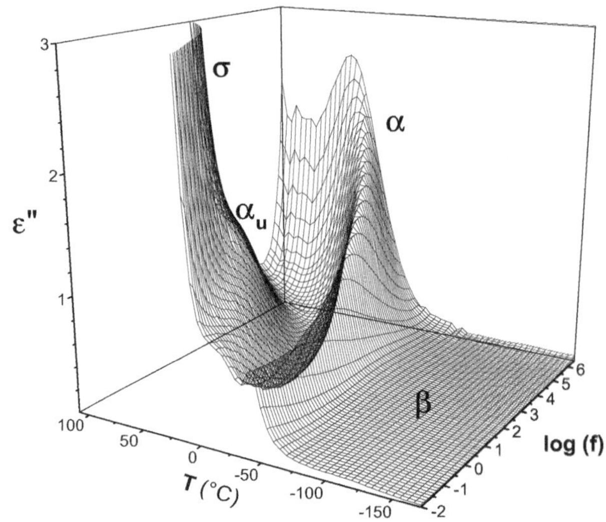

Figure 4.10. *Dielectric surface deduced from dynamic spectroscopy for a semi-crystalline polymer (PVDF)*

These results were obtained from dielectric spectroscopy. The increase of energy losses towards high temperatures is explained by the existence of a significant electrical conduction brought to the fore because the measurements are taken under an electric field. As for amorphous polymers, a secondary relaxation β is observed. In the zone where the amorphous phase is liquid, an additional relaxation α_u is produced.

The shift kinetics of the different relaxation modes is reported in Figure 4.11. To illustrate relaxation kinetics characteristic of a polymer, the figure shows the Arrhenius diagram of the relaxation times for the PVDF. The experimental points corresponding to the relaxation times shorter than 10 seconds come from measurements in dielectric spectrometry; those corresponding to the relaxation times longer than 10 seconds come from measurements in TSC. The elementary processes are spotted by indices, in order of increasing temperatures, as in Figure 4.12.

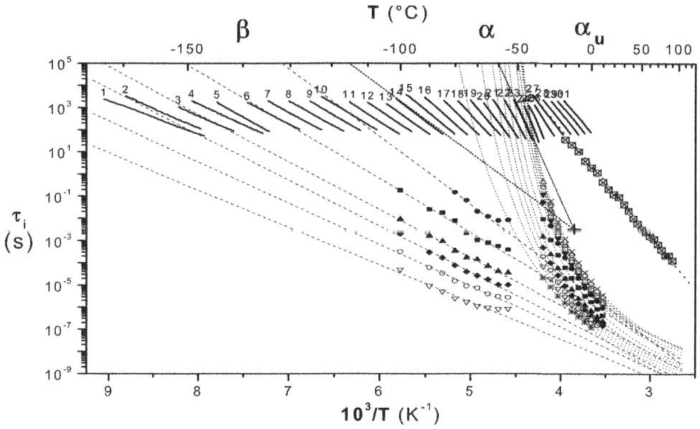

Figure 4.11. *Arrhenius diagram of the relaxation times for a semi-crystalline polymer (PVDF)*

4.7.1. *Complex relaxation in an homogenous medium*

For the relaxation mode β, the constitutive processes always behave in a way which obeys an Arhenius equation: this molecular mobility localized at the sub micronic scale is thermally activated in semi-crystalline polymers, as in amorphous polymers.

4.7.2. *Discrete spectrum of elementary relaxations in a heterogenous medium*

The results of the TSC analysis, in terms of the discrete distribution of relaxation time, for the PVDF are reported on a compensation diagram (see Figure 4.12).

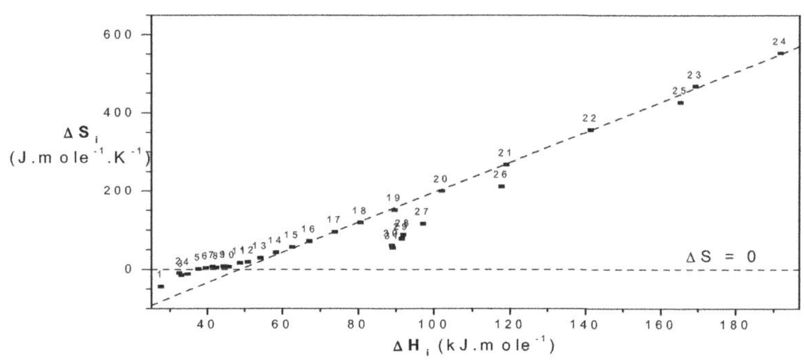

Figure 4.12. *Compensation diagram deduced from analysis of thermo-stimulated current for a semi-crystalline polymer (PVDF)*

Three types of behaviors appear. For the processes of weaker enthalpy, the activation entropy remains pratically null. Following Starkweather's nomenclature [STA 83], these constitutive processes of the β mode are due to a non-cooperative localized molecular mobility. The processes which follow the compensation law with linearly dependent enthalpy/entropy activation parameters correspond to the cooperative molecular mobility liberated at the vitreous transition. They correspond to the primary relaxation. The processes which are characterized by activation parameters of the same order have been localized in the para crystalline phase: the processes are initiated by the conformational defects.

4.7.3. *Universality of the behavior laws in semi-crystalline polymers*

In semi-crystalline polymers, the molecular motions can be made either in the amorphous phase, or in the crystalline phase. From this fact, the relaxation modes are in general more complex than in amorphous polymers. This complexity affects, in particular, the secondary relaxation modes. Thus the β mode of the amorphous PET cleaves into two components in the semi-crystalline PET [MEN 99]. Whatever the localization of secondary relaxations, the constitutive processes are always thermally activated with energetic barriers of less than 50 kJoules/mole.

The primary relaxation has the same phenomenological behavior in semi-crystalline polymers as in amorphous polymers. It is obviously very sensitive to physical ageing [STR 75]. For "weak" liquid polymers, like the PVDF, a discontinuity of the behavior law is observed with cooperative thermally activated processes below Tg and a complex mode governed by the free volume above Tg. For semi-crystalline polymers with "strong" liquid-like polyamids, the behavior above Tg remains thermally activated.

Complementary relaxation modes are observed above Tg in most semi-crystalline polymers. For polymers with a semi-rigid chain whose crystallization kinetic is slow, like the PET and the PEEK, the primary relaxation mode has an additional high temperature component around Tg +20°C. It corresponds to a thermally activated molecular mobility of the "rigid" amorphous phase [BOT 75]. For polymers with a flexible chain like the PVDF or the PEBD, it is the thermally activated molecular mobility of the para-crystalline zones which is observed above vitreous transition.

4.8. Conclusion

In polymeric materials, the localized molecular mobility at the origin of the secondary relaxations is obviously conditioned by the chemical structure (polar/apolar, flexible/rigid, etc.) of the constitutive unit. The delocalized molecular mobility at the origin of the primary relaxation is directly related to the evolution of the physical structure (glass/liquid) and has, therefore, a kinetic character. To control this phenomenon, we have to take into account the thermodynamic history of this material. As for the structural heterogeneities, they obviously have an essential role at low frequency. In the future, studies will allow dielectric relaxations in polymeric materials to be predicted and controlled.

4.9. Bibliography

[ADA 65] ADAM G., GIBBS J.H., *J. Chem. Phys.*, 43, 1, p. 139, 1965.

[ANG 95] ANGELL C.A., *Science*, 267, p. 1924, 1995.

[BLY 79] BLYTHE A.R., *Electrical Properties of Polymers*, Cambridge University Press. 1979.

[BOT 78] BOTTCHER C.J.F., BORDEWIJK P., *Theory of Electric Polarization*, vol. 2, Elsevier, Amsterdam, 1978.

[BOY 75] BOYER R.F., *J. Polym. Sci. Symp*, no. 50, 1, p. 182, 1975.

[CAR 02] CARREGA M., *Les polymères: de la molécule à l'objet*, EDP Sciences, 1999.

[COH 59] COHEN M.H., TURNBULL D., *J. Chem. Phys.*, 31, 5, p. 1164, 1959.

[COL 41] COLE K.S., COLE R.H., *J. Chem. Phys.*, 9, 4, p. 341, 1941.

[EYR 36] EYRING E., *J. Chem. Phys.*, 4, 4, p. 283, 1936.

[FER 70] FERRY J.D., *Viscoelastic Properties of Polymers*, John Wiley, 1970.

[FON 02] FONTANILLE M., GNANOU Y., *Chimie et physico-chimie des polymères,* Dunod, 2002.

[FRO 58] FRÖLICH H., *Theory of Dielectrics*, Oxford University Press, 1958.

[FUL 25] FULCHER G.S., *J. Am. Ceram. Soc.*, 8, 6, p.339, 1925.

[HAL 06] HALARY J.L., LAUPRETRE F., *De la macromolécule au matériau polymère*, Belin, Paris, 2006.

[HAV 97] HAVRILIAK S., HAVRILIAK S.J., *Dielectric and Spectroscopy of Polymeric Materials*, James P. Runt (ed.), Americal Chemical Society, Washington DC, 1997.

[HAV 67] HAVRILIAK S., NAGAMI S., *Polymer*, 8, p. 161, 1967.

[HED 77] HEDVIG P., *Dielectric Spectroscopy of Polymers*, A. Hilger Ltd, 1977.

[HOF 66] HOFFMAN J.D., WILLIAMS G., PASSAGLIA E., *J. Polym. Science: Polym. Symp.*, 14, 1, p. 73, 1966.

[JON 83] JONSCHER A.K., *Dielectric Relaxation in Solids*, Chelsea Dielectric Press, London, 1983.

[KRE 02] KREMER F., SCHONHALS A., *Broad Band Dielectric Spectroscopy*, Springer, Berlin, 2002.

[MCC 67] MC CRUM N.G., READ B.E., WILLIAMS G., *Anelastic and Dielectric Effects in Polymeric Solids*, Wiley, London, 1967.

[MEN 99] MENEGOTTO J., Etude de la Mobilité Moléculaire dans les Polymères Linéaires à l'Etat Solide par Spectroscopies Diélectriques, Doctoral Thesis, University of Toulouse, 1999.

[PER 92] PEREZ J., *Physique et mécanique des polymères amorphes*, Tech. et Doc, Lavoisier, 1992.

[RAU 02] RAULT J., *Les polymères solides*, Cepadues, 2002.

[RUN 97] RUNT J.P., FITZGERALD J.J., *Dielectric Spectroscopy of Polymeric Materials* ACS, 1997.

[STA 81] STARKWEATHER H.W., *Macromolecules*, 14, 5, p. 1277, 1981.

[STR 97] STROBL G., *The Physics of Polymers*, Springer, 1997.

[STR 78] STRUIK L., *Physical aging in amorphous polymers and others materials*, Elsevier, 1978.

[TAM 26] TAMMANN, Hesse G., ANORG. Z., *Allgem. Chem.*, 156, 1, p. 245, 1926.

[VAN 75] VAN TURNHOUT J., *Thermally Stimulated Discharge of Polymer Electrets,* Elsevier, 1975.

[VOG 21] VOGEL H., *Physik Zeitung*. 22, p. 645, 1921.

Chapter 5

Electrification

5.1. Introduction

It is now well known that setting a charged body in motion (in the case of solid particles in a flowing gas) creates currents and electric potentials which can sometimes reach significant values.

All the research already undertaken confirms the important role played by the nature and surface state of the materials involved in the transfer and accumulation of charges. In the next section, we shall attempt to analyze the processes which cause these phenomena.

5.2. Electrification of solid bodies by separation/contact

The generation of charges between two bodies which have been in contact is the fruit of the process known as electrification, i.e. the separation of charges with an opposite sign, combined by displacement or transport. Next, we shall examine the electrification process.

Three types of electrification exist, contact/separation electrification, electrification by influence, electrification by corona effect.

Chapter written by Gérard TOUCHARD.

5.2.1. *The process*

The goal of this section is to recall the process of electrification of two solids during their contact, then their separation and to understand the mechanisms which come into play in order to take all the necessary precautions from a practical point of view to obtain reproducible results.

The contact electrification process always appears at an interface. It is due to the different nature of the two constituents. In all cases, a transfer of ions, due to their physico-chemical difference, or a transfer of peripheral electrons due to the contact potential difference between the surfaces, is produced.

Thus, two large categories can be distinguished according to whether the charge transfer is made by electrons or by ions which flow from one material to the other. The transfer by electrons allows most of the electrification process which appears from contact between solids to be explained. Ionic theory is better suited to contacts between solids and liquids.

Let us take two different materials, and assume they have a perfectly smooth surface (Figure 5.1).

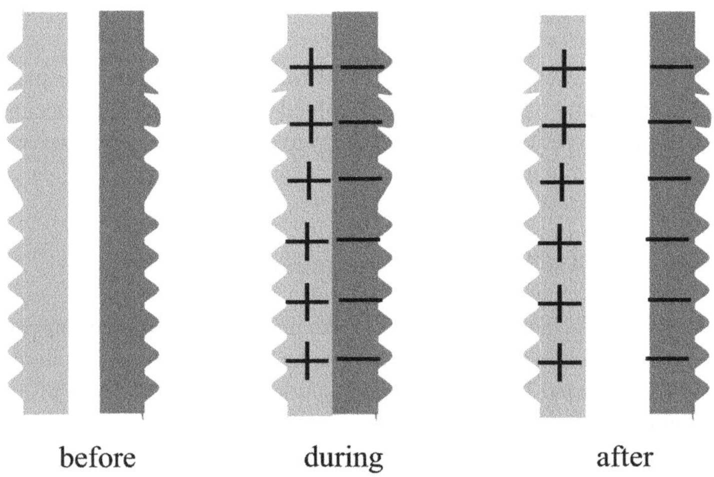

Figure 5.1. *Principle of contact electrification*

Before being in contact along this surface, the two materials are electrically neutral. When they are in contact, a charge transfer occurs, then again during their separation if the materials are sufficiently insulated (so that the total transferred charge does not relax through the last contact point); each of them conserves the same amount of electricity but with an opposite sign (see Figure 5.1). We shall look more closely at this process.

The outbreak of electrostatic charges on a given solid medium is often imputed to the process of electrification by friction. The first electrostatic experiments confirmed this: if we rub a glass stick with cat's fur, for example. In reality, even when we simply place two bodies of different materials in contact and then separate them, we observe a separation of electrical charges (electrification) even though there has not been any friction (tangential motion from one contact surface to the other). In this case, the intensity of the process is always much less than that observed for objects rubbed against each other. Friction, however, does not add anything fundamental causing the electrification; it only amplifies the process which is already present in the simple contact between two neutral bodies of different materials.

Electrification by friction, often called triboelectrification, is in fact a derivative of contact electrification. Friction increases the contact surface and temperature. The deformations generated by the friction in a polymer material create vacant energy levels in which electrons from the contact surface can be captured. In general, friction favors the charge transfer process.

Electrification by impact derives from the two previous types. The impact adds additional parameters to the process, such as pressure.

By a misuse of language, all of these charge generation processes are often called triboelectrification.

We shall now develop a theory which will permit the origins of electrification (of electron transfer) resulting from the simple contact (without friction) of two bodies of a different nature to be understood. We have to look for an explanation for the outbreak of electrification in the electronic characteristics of the materials, and it is these that we shall now study, with the aid of theories stemming from quantum mechanics and the composition of atoms. Several cases have to be distinguished.

5.2.2. *Charge transfer mechanism by the separation contact of two different conductors*

The electrons in conductors (metals) are free; no force acts on them, and they form an "electron gas". There are, however, at the surface of the metal, forces which prevent them from escaping and send them back inside (see Figure 5.2). The electrons in a metal are in a potential well [COE 93]. The kinetic energy of the electrons is lower than the Fermi energy level of the metal which would allow them to leave. An additional energy is required to get over the metal surface. This additional energy, called work function or extraction energy, is of a few electronvolts.

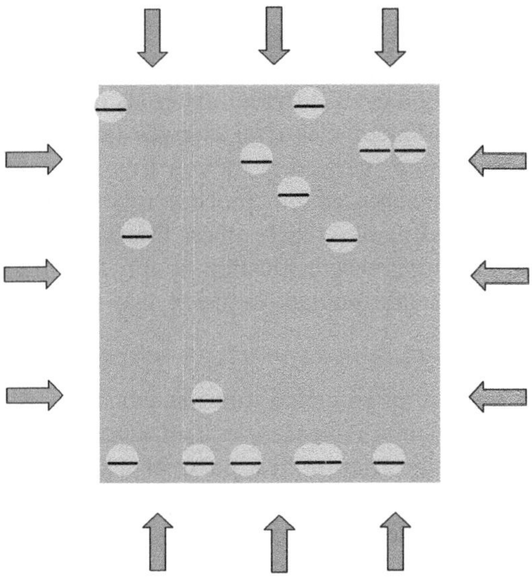

Figure 5.2. *The free electrons within a conductor, unable to freely escape from it*

We can remove the electrons by several processes:

– by photoelectric effect;

– by electron beam (secondary emission);

– by cold emission, creating an electric field.

An electron could escape from the influence of the atom to which it belongs if it acquires a higher energy than the extraction energy. This extraction energy φ is the energy which a peripheral electron must acquire to escape from the influence of the atom (to become free). Two different metals have different Fermi levels (see Figure 5.3).

Extraction energy is an important characteristic of metal, and ranges between 4 and 5 eV for many metals. (1 eV is an energy unit equal to 1.6 10^{-19} joules).

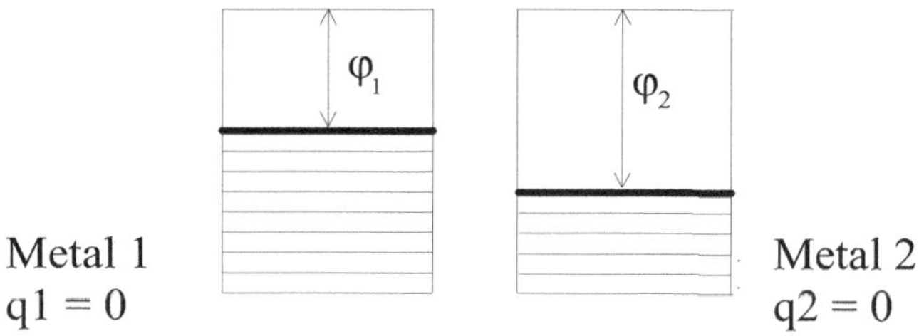

Figure 5.3. *Energetic diagrams for two different bodies*

In Figure 5.3 we have represented the work functions φ1 and φ2 of Metals 1 and 2. When the two metals are separated, no electron transfer can be made from one metal to the other. If, on the other hand, the two metals are in contact, we shall observe on one side of the interface an excess of positive charges and on the other side an excess of the same amount of negative charges (the interface obviously remained neutral as a whole). Indeed, when both metals are in contact, their respective surfaces are separated only by a few Angströms. Their energy levels then become equal through quantum tunneling. Consequently, a potential difference settles; this is the contact potential. In other words, the explanation of this transfer lies in the fact that, in this contact zone, the electrons of Metals 1 and 2 easily acquire a sufficient energy to escape from the influence of the metal to which they belong. The electrons of Metal 1, once freed from the influence of 1, could flow into 2 (there is contact), and vice versa. In this ceaseless motion of electrons between Metals 1 and 2, knowing that it is easier (less energy is required) for an electron 1 to escape 1 than for an electron 2 to get out of 2, the net overall results of the exchanges will show a higher number of electrons in 2 than in 1, hence polarization. Equilibrium will occur when the maximum energetic level occupied by an electron is identical in both materials. For this purpose we need a transfer of electrons from 1 to 2; therefore, there is an outbreak of a negative charge on 2 and a positive charge on 1 (see Figure 5.4).

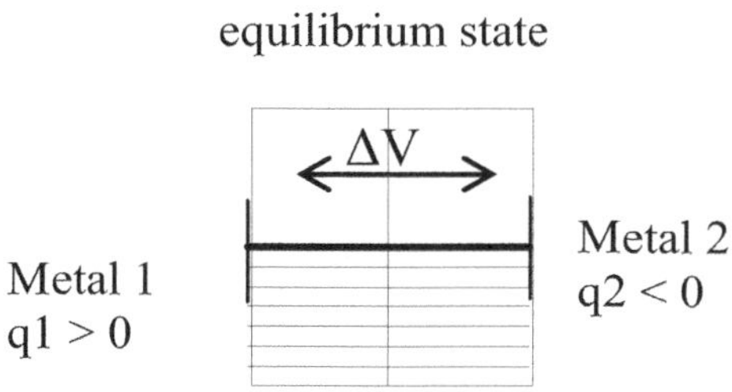

Figure 5.4. *Equilibrium diagram*

In equilibrium, there is a potential difference ΔV between the two materials, equal to:

$$\Delta V = \frac{\varphi_2 - \varphi_1}{e} \qquad [5.1]$$

This is the contact potential difference (e is the charge of the electron).

When the contact breakdown is simultaneous at all points on the surface common to both materials, and if we assume that both metals remain insulated after separation from the ground, Metal 2 conserves the excess of electrons gained from Metal 1 and is negatively charged ($q_2 < 0$), while 1 has lost electrons and is positively charged ($q_1 < 0$). Harper [HAR 51] showed that the transferred charge is proportional to this potential difference ΔV:

$$Q = K\frac{\varphi_2 - \varphi_1}{e} \qquad [5.2]$$

where K is a factor which depends on the contact surface area and the experimental conditions.

Coste [COS 84] showed that this theory is also applicable to the case of metal/semi-conductor and semi-conductor/semi-conductor contacts.

5.2.3. *Polymer–metal contact*

Contact between metals and insulators is still very poorly analyzed, at the present time. Different viewpoints exist on the nature of charge carriers, which is due, on the one hand to the presence of impurities in insulators, and on the other hand, to the difficulty for a direct measurement of the potential difference [COE 93] between a metal and an insulator, when this latter has a resistivity higher than $10^{12}\,\Omega$.m.

For these diverse reasons, current studies are restricted to polymer–metal contact and could give an analysis similar to the previous ones in certain atmospheric conditions (in the case of relatively weak humidities) and for certain polymers whose surface state densities are uniformly spread out.

5.2.4. *Contact between two polymers*

The analysis in this case is not made directly. The technique used consists of first undertaking two studies on the polymer–metal contact for both polymers, and then in deducing the polymer contact between them.

5.2.5. *Triboelectric series*

During contact between two materials, the charge transfer is of an electronic nature when one of the materials is a metal or a semi-conductor. The transferred charge generally increases as a function of the extraction work difference of the materials in contact; it is linear in certain cases of contact (two metals, semi-conductor metal and polymer metal) and seems to have an exponential growth as a function of the apparent extraction works when both materials are polymers. In any case, the transfer is always made from the material with weak extraction energy to the material with strong extraction energy. In most of these cases, the only published results [LÜT 97] consist of triboelectric series (see Figure 5.5), i.e. a listing of the differents bodies with respect to one another concerning the charges they produce in contact. Unfortunately, these series are often subjected to caution and they depend heavily on the experimental conditions in which they have been made.

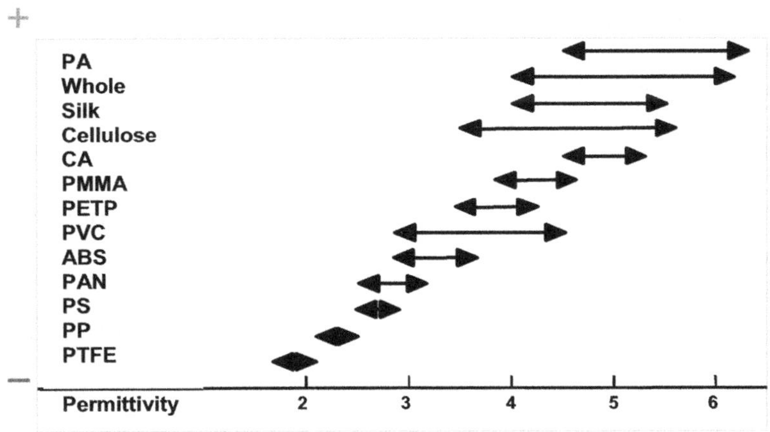

Figure 5.5. *Triboelectric series*

After separation, the charge present on both constituents depends on the ability of the produced charges to flow on the surface towards the last contact point. Indeed, as the bodies get separated, the mobile charges of an opposite sign present at the surface of both bodies tend to recombine. For two metals, the electrons in excess on a surface will tend to move back to the zone where they are lacking (the other surface). The more the charges are mobile at the surface of the materials, the easier this could happen. This is the case for electrons in metals; it is why the total charge after separation of two metals is always weak. On the other hand, when one of the materials (or both) are insulators (polymers, for example), the charges migrate with great difficultly to the last contact point between the surfaces because they are trapped on (or inside) the insulating material. It is this ability of insulators to trap the charge excess transferred to their surfaces for a long period of time which causes certain electrification issues.

Industrial electrification generally come from phenomena involving friction. The collected charges are then more numerous than those obtained by simple contact. This is explained because surface friction favors contact and improves its quality, thus increasing the transfer of electrons. Most investigations made on this subject show that the transferred charge increases with pressure between both materials in contact and with friction speed. On the other hand, it decreases with the roughness of the surfaces. Nevertheless, the process causing the phenomenon remains the same: electrification by contact separation.

5.3. Electrification of solid particles

Let us assume a solid particle in displacement within a solid pipe. At instant t the particle has never been in contact with the solid wall, and is then electrically neutral, just like the solid wall. After the particle hits the solid wall, there is a charge on the particle and an opposite charge on the solid wall; this is the charge transferred during the impact (see Figure 5.6).

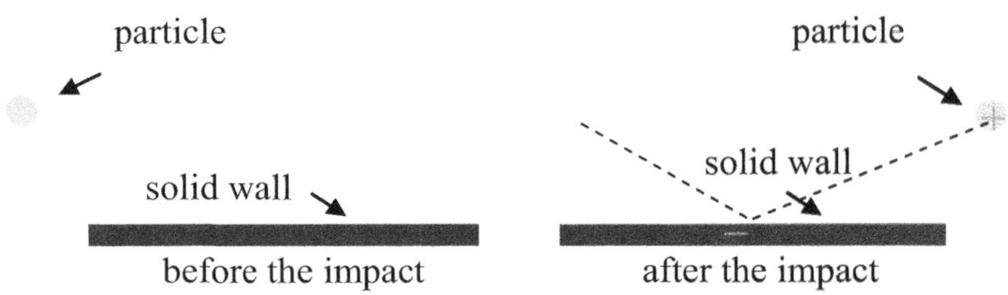

Figure 5.6. *Exchange of charge when a particle hits a solid wall*

A lot of research has been undertaken on the electrification of particles during their impact with walls. We will now examine some of this.

5.3.1. *Theoretical work by Masuda et al.*

Masuda *et al.* [MAS 78] proposed equations to estimate the charge exchanged during an impact. They considered two types of contact, elastic and plastic contact, and proposed an expression for each case.

For an elastic impact:

$$\left|\frac{\Delta q}{m}\right| = -2{,}50\frac{\varepsilon V_c \Delta t}{z\tau \overline{D}p}\left[\frac{\beta^4}{\rho_p^3}\left(\frac{E_1+E_2}{E_1E_2}\right)^2\right]^{1/5}\left[|U|\cos\theta\right]^{4/5} \qquad [5.3]$$

where ε is the permittivity of the medium in which the contact occurs (air, in most cases), z the thickness of the equivalent capacitor during contact, Δt the duration of contact, τ the time constant of the equivalent capacitor (much greater than the contact time), V_c the contact potential, $\overline{D}_p$ the diameter of the sphere having the same volume as the particle, ρ_p the volumic mass of the particles, $\beta = 2r_o/\overline{D}_p$ (where r_o is the radius of curvature of the contact surface of the particle), E_1 Young's modulus of the plate, E_2 that of the particle, $|U|$ the velocity modulus and θ the impact angle with respect to the normal of the plate.

In the case of a plastic impact:

$$\left|\frac{\Delta q}{m}\right| = 2{,}45\frac{\varepsilon V_c \Delta t}{z\tau \overline{D}_p}\sqrt{\frac{\beta}{\rho_p p}}\left[|U|\cos\theta - 0{,}04\sqrt{\frac{p^5\beta^3}{\rho_p}}\left(\frac{E_1+E_2}{E_1E_2}\right)^2\right] \qquad [5.4]$$

Cole and Baum [COL 69] established a relationship for when a particle undergoes several successive collisions with a wall. We can then establish the following relationship:

$$q_t = (q_s - q_i)(1-\exp(-k_c n_c)) \qquad [5.5]$$

where k_c is a coefficient, q_t is the charge transferred at the moment of the impact n_c, q_s is the saturation charge of the particle (i.e. after an infinite number of impacts) and q_i is the initial charge of the particle (that it had before impact n_c).

5.3.2. *Experimental work by Touchard et al. [TOU 91]*

Touchard *et al.* analyzed the impact charge generated on copper, black polyethylene and yellow polyethylene according to:

– the speed U of the particles at the moment of the impact;

– their original charges Q_0;

– the angle θ under which they hit the surface of a material;

– the nature of the material;

– the diameters D_p of the particles.

The physical quantity measured was the electric current generated by impact of a steady flux of particles on a material; this measurement allowed the impact charge of the particles to be obtained, by unit mass. In the case of copper, the material directly constituted the electrode. In the case of polyethylene, the electrodes clung to one side of the material (that opposed to the impact); thus, in this case, the current measured was capacitive.

The device used is represented in Figure 5.7. It is inside a cage (12) surrounded by a tank (13). An injection pipette (1) provided with a vibrator (2) and a tap (14) permits particles to be dropped by gravity. By means of friction on the injection pipette, the particles are ejected with an initial charge Q_0.

These particles then pass through an electric field perpendicular to their trajectory. This field is supplied by a high tension supply (11) connected to two plane electrodes (3). The whole assembly is inside a tank. The field plays the role of charged particle deflector. Thus, coming out of the field, the particles have several trajectories depending on their initial charge. An electromagnetic screen (4) separates the deflector device from the target. A target electrode (6) is situated on a PTFE support (5) which is mounted on a support permitting adjustment of the angle of the target (7) and, at the same time, its position on a tilted ruler (8) vertically.

The current due to the impact charge is measured with the aid of a Keithley 610C electrometer (9) connected to a data acquisition system (10). The particles are then collected in a receptacle (15).

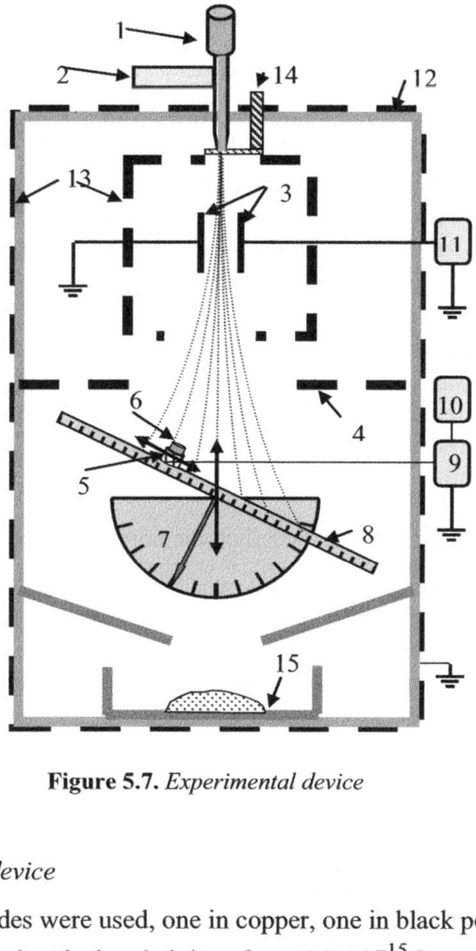

Figure 5.7. *Experimental device*

5.3.2.1. *Experimental device*

Three target electrodes were used, one in copper, one in black polyethylene (with 2% black carbon) with electrical resistivity of $\rho \approx 0{,}8\ 10^{-15}\,\Omega\text{m}$ and one in yellow polyethylene (without black carbon), with resistivity $\rho \approx 2\ 10^{-15}\,\Omega\text{m}$. The particles were sodocalcic glass microbeads with different diameters (with electrical resistivity $\rho \approx 10^{10}\,\Omega\text{m}$ and volumic mass $\rho_{mp} = 2{,}300\ kg/m^3$). Depending on their original charge Q_0 the particles were deviated differently by the electric field existing between the electrodes (3). Particles of 4 different sizes were used: 500 µm, 340 µm, 200 µm, 110 µm.

5.3.2.2. *Results*

5.3.2.2.1. Influence of the impact angle

For these experiments, we determined the positions of the target in a way that the original charge of the particles which hit it take 4 different values: 0C/g, $0{,}66\,10^{-8}\,C/g$, $1{,}63\,10^{-8}\,C/g$ and $2{,}63\,10^{-8}\,C/g$ (a charge with no origin corresponds to the target positioned vertically below the pipette). We also position the target in a way that the modulus $|U|$ of the impact speed is constant; it must also be noted that the positions of the three zones traversed by the particles (2 cm before the field, 4 cm in the field and 11 cm after the field) are practically on the same horizontal axis. The modulus of the impact speed obviously corresponds to normal impact components U_N varying according to the impact angle.

For the three targets, the impact charge strongly decreases as a function of the impact angle. An example is given in Figure 5.8 for the copper target. This decrease is greater than the reduction of the normal component of speed. The reason of this fast decrease is probably due to rebounds more frequent on the target when the impact angle is small.

5.3.2.2.2. Influence of the impact speed on the normal component

The evolution of the charge by unit mass generated as a function of the speed of the particles is similar for the three targets; an example is presented for the black polyethylene target in Figure 5.9. As predicted by Masuda *et al.*, we see that for a given sample and for particles having the same original charge, the charge generated by unit mass increases with the normal component of the speed U_N. This increase is, however, more important than that predicted by Masuda *et al.*

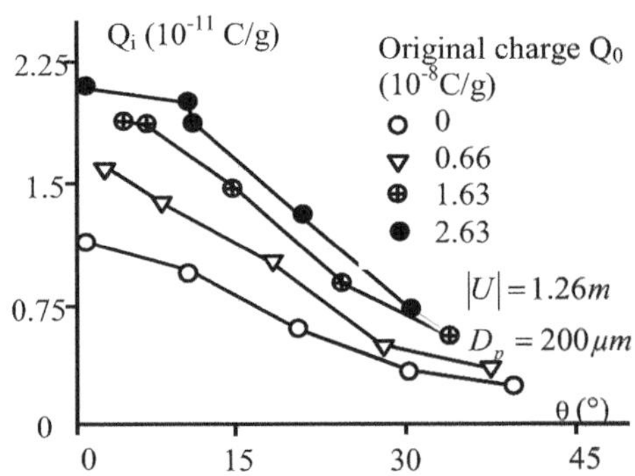

Figure 5.8. *Impact charge as a function of the impact angle*

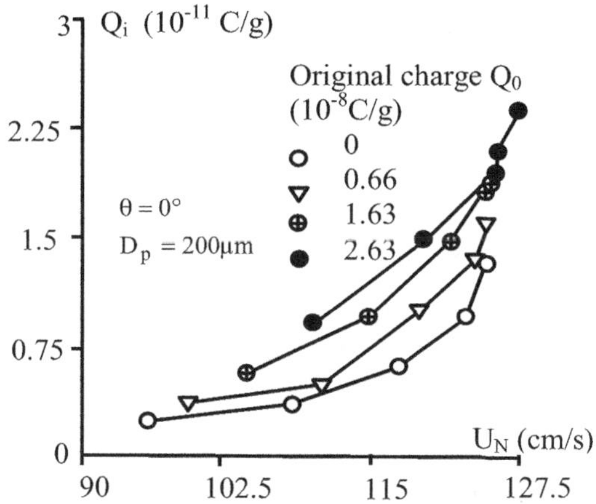

Figure 5.9. *Impact charge as a function of the normal component of the speed*

5.3.2.2.3. Influence of the size of the particles

The samples of particles used for this study have respective average diameters of 110, 200, 340, and 500 μm. The results are presented in Figure 5.10. The normal component of the impact speed is obviously different because air friction is dependent on the size of the particles. Although the impact speed is less for smaller particles, we see that the impact charge decreases with the dimensions of the particles.

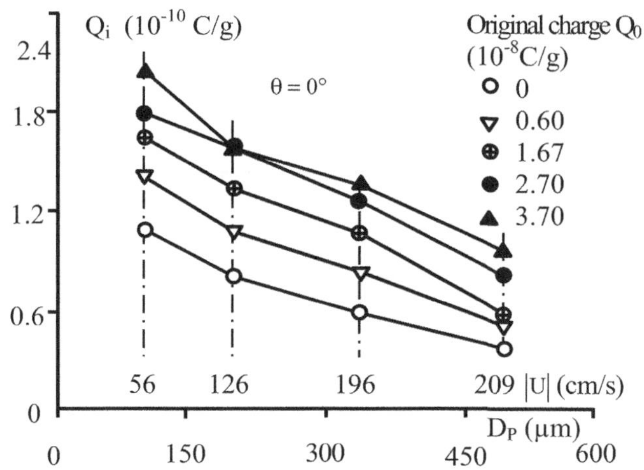

Figure 5.10. *Impact charge as a function of the size of the particles*

5.3.2.2.4. Comparison of results obtained on the three targets

We see in Figure 5.11 a comparison for the three targets. It is clear that the impact charge is weakest for copper and highest for yellow polyethylene.

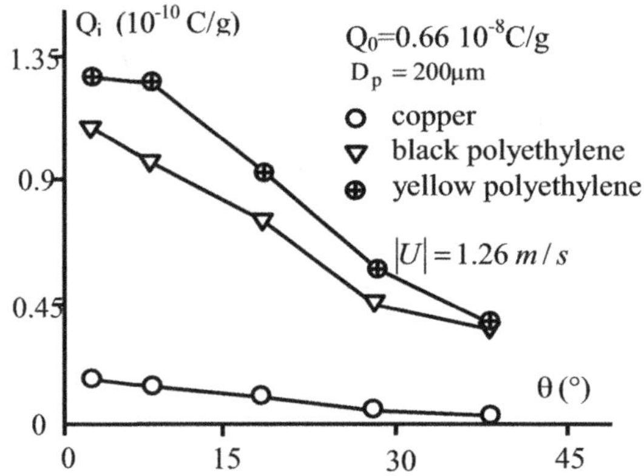

Figure 5.11. *Comparison between the materials of the three different targets*

5.3.2.2.5. Evolution of the total charge of a particle according to the number of impacts

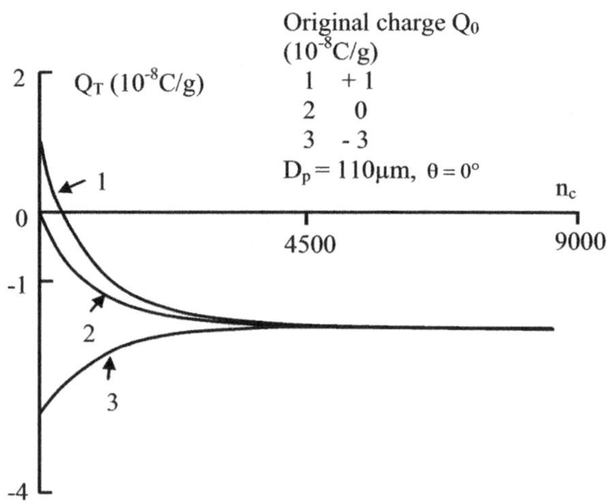

Figure 5.12. *Comparison between the materials of the three different targets*

It is possible to calculate the different parameters which play a part in the expression proposed by Cole *et al.* [COL 69] from these experimental results. We can then deduce the evolution of a particle charge according to the number of impacts on a given target. In the case of a particle of diameter 110µm and for a copper target, we see, in Figure 5.12, this evolution for 3 different original charges. In the three cases, the final charge of the particle tends towards the same value. Its total charge decreases when the particle is initially charged to a value greater than the limit value, whereas it increases when it is initially charged to a weaker value.

5.4. Conclusion

The charge generated by separation contact still remains difficult to predict for most materials. Consequently, the charge generated by the impact of particles on a target and, even more certainly, that generated in the flows of dusty gases or pneumatic transports, can only be obtained experimentally in most cases (see bibliography, section 5.5).

5.5. Bibliography

[ART 97] ARTANA G., TOUCHARD G., MORIN M.F., "Contribution to the analysis of the flow electrification process of powders in pneumatic conveyers", *Journal of Electrostatics*, vol. 40 & 41, p. 277–282, 1997.

[BAI 01] Bailey A., "The charging of insulator surfaces", *Journal of Electrostatics*, vol. 51–52, p. 82–90, 2001.

[BEN 87] BENMADDA M., TOUCHARD G., BORZEIX J., "Etude de la charge engendrée par le choc d'une particle solide sur une plaque métallique", *Rev. Phys. App.*, 22, p. 1071–1074, 1987.

[BOU 91] BOUDALAA M., TOUCHARD G., "Mesure du débit massique de particles dans un transporteur pneumatique travaillant en écoulement en phase dense", *R.G.E.*, vol. 8, p. 10–14, 1991.

[CAS 95] CASTLE G.S.P., SCHEIN L.B., "General model of sphere-sphere insulator contact electrification", *Journal of Electrostatics*, vol. 36, no. 2, p 165–174, 1995.

[COE 93] COELHO R., ALADENIZE B., *Les diélectriques – Propriétés électriques des matériaux isolants*, Hermès, Paris, 1993.

[COL 69] COLE B.N., BAUM M.R., MOBBS F.R. "An investigation of electrostatic charging effect in high-speed gas–solid pipe flows", *Proc. Instn. Mech. Engrs.* 184 (3c), p. 77–83, 1969–1970.

[COS 84] COSTE J., "Idées actuelles sur l'origine des charges électrostatiques – Phénomènes de contact et de frottement", *J.E. SEE Electrostatique*, 1984.

[CRO 87] CROSS J.A., *Electrostatics, Principles, Problems and Applications*, Adam Hilger IOP Publishing, 1987.

[GLO 88] GLOR M., *Electrostatic Hazards in Powder handling*, Research Studies Press, 1988.

[GRO 01] GROSS F.B., GREK S.B., CALLE C.I., LEE R.U., "JSC Mars-1 Martian Regolith stimulant particle charging experiments in a low pressure environment", *Journal of Electrostatics* vol. 53, p. 257–266, 2001.

[HAR 51] HARPER W.R., "The volta effect as a cause of static electrification", *Proc. Roy. Soc. A.*, 205, p. 83–103, 1951.

[JON 91] JONES T..B., KING J.L., *Powder Handling and Electrostatics*, Lewis Publishers, 1991.

[KWE 94] KWETKUS B.A., "Contact electrification of coal and minerals", *Journal of Electrostatics*, vol. 32, no. 3, p. 271–276, 1994.

[LEO 02] LEON-ESCALANTE S.G., TOUCHARD G., DOMINGUEZ G., "Electrification study in dielectric material fluidized beds for different fluidization regimes", *Proc. IEEE-CEIDP*, p. 694–697, 2002.

[LÜT 97] LÜTTGEN G., WILSON N., *Electrostatic Hazards*, Butterworth-Heinemann, 1997.

[MAS 78] MASUDA H., KOICHI I., IINOYA Y., "Electrification of particles by impact on inclined metal plates", *AIChe Journal*, vol. 24, no. 6, p. 950–956, 1978.

[MEH 05] MEHRANI P., BI H.T., GRACE J.R., "Electrostatic charge generation in gas–solid fluidized beds", *Journal of electrostatics,* vol. 63, p. 165–173, 2005.

[SCH 99] SCHEIN L.B., "Recent advances in our understanding of toner charging", *Journal of Electrostatics*, vol. 46, no. 1, p. 29–36, 1999.

[SUG 95] SUGIHARA M., DASCALESCU L., TOUCHARD G., ROMAT H., GRIMAUD P.O., WATANABE S., "Charge generated by impact of balls on a metallic wall", *Journal of Electrostatics*, vol. 35, p. 125–132, 1995.

[TOU 91] TOUCHARD G., ZERGHOUNI A., WATANABE S., "Evolution de la charge électrique d'une particle heurtant une paroi solide", *J. Phys. III France 1,* p. 1233–1241, 1991.

[TOU 90] TOUCHARD G., WU Z.H., WANG W.Q., NASANI, A., WATANABE S., "Electrostatic charge generation in pneumatic conveyers", *Materials Science*, vol. 16, no. 4, p. 5–10, 1990.

[TOU 99] TOUCHARD G., ARTANA G., PUTIER F., "Electrical characteristics and mechanical behaviour of powders of the animal feeding industry", *Journal of Electrostatics* 47 (1–2), p. 3–12, 1999.

Chapter 6

Space Charges: Definition, History, Measurement

6.1. Introduction

Under the action of numerous factors (electric field, temperature, humidity, radiation, etc.), the properties of dielectric materials become degraded over time in an irreversible way. The mechanisms causing this phenomenon (called ageing), bring in all of the stresses which the insulator must obey, and are subjected to approaches described later in this chapter. Ageing generates unfavorable effects on electrical components, which can manifest themselves by the appearance of defects under much weaker stresses than those originally considered.

Due to economic and environmental factors, the past several years in the energy and electronic fields have witnessed a race for miniaturization of components and systems, and for the growth of the power densities they convey. This has led to an increase of electrical and thermal stresses to which the insulators are submitted. In order to guarantee the reliability of components and systems, it is of fundamental importance to know, in a more precise manner, the utilization limits of the insulators used and their long-term evolution.

An ideal insulator should not contain charges other than atomic cores, electrons in the internal layers and valence electrons. In reality, however, there is no such thing as an ideal insulator.

Chapter written by Alain TOUREILLE, Petru NOTINGHER, Jérôme CASTELLON and Serge AGNEL.

Thus, when an insulator is subjected to high stress (high electric field, irradiation, important mechanical efforts and friction, variations of temperature), an excess of electric charges, called *space charges*, appear in certain regions of the insulator. It seems more and more probable today that the issues of performance decrease and insulator breakdown are due to the presence and accumulation of space charges in bulk and on the surface. Indeed, the electric field caused by a space charge (known as a "residual field") can locally increase the electrical stress, by over-soliciting the insulator. Moreover, during any massive detrapping of the accumulated charge, the energy released is considerable and can (either locally or completely) damage the material.

To illustrate this, remember that, when we apply an electric field to an insulator, any macroscopic discontinuity in the material (permittivity variation in a composite, for example) gives rise to classic interfacial (Maxwell–Wagner) polarization. This phenomenon also affects heterogenous materials at a microscopic scale. Moreover, following the internal order of the material, we notice the presence of traps (localized levels) at the molecular scale. These traps are likely to capture charges, producing polarization at a microscopic scale. Thus, one more or less electronic charge on a macromolecule of 100 Å of cross-linked polyethylene gives a density of 10^{24} electrons/m^3 in three dimensions, i.e. a charge density $\rho = 1.6 \cdot 10^5$ C/m^3, which corresponds (integrated on a thickness of 0.1 μm) to a field of 1,000 kV/mm. Even at a lower degree, it is then very obvious that this type of trapping can cause an early dielectric breakdown or/and ageing of the material.

For the reasons described above, these electric charges which tend to accumulate in the insulators have been subjected to thorough studies for 20 years. In the following pages, we are going to tackle some fundamental and historical aspects related to the space charge, summarize the different methods developed to measure these charges and present current trends and perspectives in this area.

6.2. History

In the 1960s, notable advances were realized in the understanding of phenomena related to the application of solid-state physics: amongst others, we could mention the field of diodes and transistors. These important advances mostly touched on orderly materials, such as crystals, used in microelectronics. Theories – subsequently confirmed – showed the strong influence of charge gradients stored near interfaces, with outbreaks of local electric fields and the possible diffusion of carriers. Could these achievements – subsequently developed with the great success we now know – have a beneficial influence on knowledge about mineral insulating materials? And are these theories applicable to organic insulators?

These questions naturally cropped up because, at the same time that microelectronics emerged, there were also notable developments in the world of polymers, first in their mechanical properties (as light and cheap materials), then their electrical properties (as excellent insulators: the first polyethylene-insulated cables were put into service in 1962).

Also, during the 1960s and 1970s, important dielectric studies concerning mineral and organic insulators were undertaken, supported by knowledge of solid-state physics which had been so successful in electronics. Among the main works published on this subject, we can mention: Rose [ROS 63], Coelho [COE 66], Gutmann [GUT 67], Lampert and Mark [LAM 70], Mott [MOT 71], Many [MAN 71], O'Dwyer [ODW 73], Lewis [LEW 77]. However, the insulating nature of these materials on the one hand, and their internal defects due to disorder on the other, led researchers to develop new methods or to improve certain existing techniques.

Among the noteworthy "key" innovations or publications were:

– thermally stimulated currents [CRE 70], which revealed the existence of deep traps in these materials;

– the role of these traps in conduction currents [WIN 73];

– electrode charge injections underscoring Schottky or Fowler–Nordheim laws [SIM 71];

– volume-limited currents (Poole–Frenkel effect [CAS 69] [HAL 71]).

However, certain authors revealed "anomalies" or current instabilities leading to oscillations [FAL 72], [TOU 74], [TOU 76]. These phenomena confirm the essential role of the "electrode-material" couple in injection dynamics and in charge flow.

All this work leans on the existence of both a forbidden gap (8 to 12 eV for an insulator) and trap levels due to structural defects in the material, or to impurities. There is a forbidden gap (a.k.a. a band gap) if we can consider that at least 12 atoms are orderly [AMB 62], [POH 67]. "Localized" levels or traps [LEN 66], [LIL 68] appear in the band gap. Their depths with respect to the conduction band may reach several eV, depending on the local disorder in the molecular arrangement. The most frequent mode of transport then seems to be from trap to trap, with or without passage through the conduction band.

Thus, we see in the early 1980s confirmation of all of these complex theories in thick insulators, where the same type of reasoning as for N or P structures can be used locally, with different N and P structures alternating at a nanometric scale.We therefore have a succession of interfaces; now, microelectronics shows that everything is governed by the interface charges. Facing this evidence, the international community of dielectricians became aware of the necessity to create measurement systems for this local charge called a "space charge".

We then see the appearance of several methods, using either pressure or temperature as a stimulus to reveal charges. The functioning of each of these techniques is different (see Chapters 11, 12, 13 & 14), but their basic principle is the same: to cause variation of influence charges at the electrodes, which results in a transient current or voltage which can be recorded and analyzed.

The LIPP (*Laser Induced Pressure Pulse*, Chapter 13) technique, developed in Paris by Alquié [ALQ 81], then resumed in Darmstadt by Sessler [SES 81], uses a laser to send a pressure pulse to a sample. This technique was used in Paris by Lewiner and his team [ALQ 81], who applied it to polymer samples of the order of a millimeter of thickness. The thermal pulse technique, developed in Washington by Collins [COL 80], and then by De Reggi [DER 82], uses a laser to create a thermal pulse. The LIMM (*Laser Impulse Modulation Method*), developed by Lang, uses a laser to create a thermal modulation [LAN 86]. The *Thermal Step Method* (TSM), developed in Montpellier by Toureille and his team (Chapter 14) [TOU 87], [TOU 91], uses a constant temperature front, thus allowing an analysis of large thicknesses such as those of insulated power cables.

The 1980s then saw the presence of two distinct schools:

– that using thermal stimuli: these are high sensitivity methods, of slow dynamics (the temperature moves at the speed of several hundred microns per second in the insulators studied);

– that using pressure: these are fast methods, requiring sophisticated equipment (the speed of sound is of several kilometres per second in the studied insulators).

In the 1990s, numerous results given by all these methods gave rise to new developments or extensions. With regard to the thermal methods, Franceschi and his team [FRA 97], [FRA 00] developed the FLIMM (*Focused Laser Intensity Modulation Method*, see Chapter 12) by focusing a laser beam in order to obtain space charge distributions in three dimensions on samples of several dozen microns. Reboul [REB 00] uses the ATWM (alternative thermal wave method) on thicknesses of 100 microns. The TSM is industrially applied on cables [CHE 92], [ABO 97], [AGN 00], [NOT 01].

As for methods using pressure, there has been a Japanese innovation; Maeno [MAE 88] developed the PEA (Pulsed Electro Acoustic) method (Chapter 11). This technique uses the pressure pulse created by an electric pulse applied to a sample.

A different method to the previous ones (collectively known as stimulus methods) has also made its appearance: it is a mirror method, which uses a scanning electron microscope (SEM). It has been developed at the CEA by Le Gressus, who uses a reflected beam to measure charges implanted in an insulator [LEG 95], [VAL 95].

Thus, the multiplication of space charge measurement methods has created an increasing number of publications, congresses and international networks on the subject, especially from the 1990s onwards. First, this increase gave rise to more tools and a better scientific knowledge of insulators, and, second, in the industrial sector, it led to the development of smaller and smaller electric and electronic components with improved performances.

If we now consider the scientific results obtained over the last 20 years, we must take for granted (by trying to put the qualitative aspects into focus first) that:

– there are space charges in all insulators, even after manufacture [TOU 98];

– we can inject and accumulate space charges in all insulators, even under alternative current (AC) [AGN 99], [NOT 99];

– these space charges evolve with stress (mechanical, thermal, electrical, chemical, radiative, temporal, etc.) [MAL 97];

– these space charges evolve with ageing [BER 99];

– these space charges are related to lifetime [MAZ 01];

– these space charges reflect the state of the microstructure being considered [RAM 01], and therefore represent one of the essential parameters of the material.

6.3. Space charge measurement methods in solid insulators

After presenting the history of space charges, we will now describe the methods used for their measurement in a more detailed way.

The techniques developed can be classified as "destructive" and "non-destructive" methods. "Destructive" techniques are so called because they "destruct" the charge state of the sample by emptying the charges stored. These methods do not provide information on the localization of carriers, but allow the total charge and the energies required to evacuate the different types of carriers ("activation energies") to be estimated. Since the sample is discharged at the end of the measurement, destructive techniques do not permit the evolution of charges to be studied. The "non destructives" methods do not remove the accumulated charges and allow their distribution throughout the material to be determined. These techniques have the advantage of allowing the charge evolution to be monitored over time.

Finally, the coupling of "destructive" and "non-destructive" methods allows the complete characterization of the processes of charge trapping and/or polarization.

6.3.1. *Destructive methods*

6.3.1.1. *The thermally stimulated current method*

The most widespread method is that of thermally stimulated depolarization currents, or *thermostimulated currents (TSC)*. Developed in the 1970s [BUC 64], [CRE 70], [VAN 75], [VAN 79], it is one the oldest methods and has numerous variants. The principle of TSC consists of warming up a short-circuited sample by applying a temperature ramp to it and, at the same time, measuring the "thermostimulation" current, which appears in the external circuit because of the liberation of charges and/or the disorientation of dipoles (see Figure 6.1).

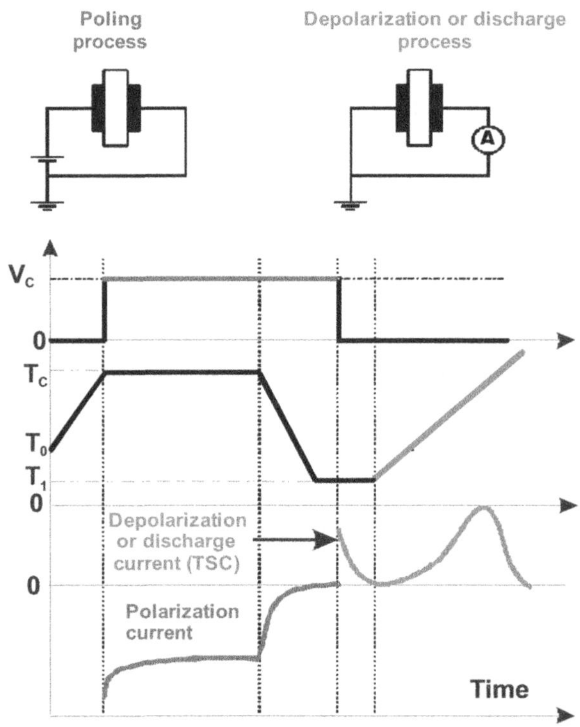

Figure 6.1. *The thermally stimulated currents (TSC) method*

Thus, a sample polarized under a voltage V_C at a temperature T_C is at first cooled down to T_I (often under voltage) in order to "freeze" the charges. The sample is then short circuited and its temperature, considered constant over the whole sample, is linearly increased (by a few degrees/minutes), whilst the current which appears between the electrodes is measured at the same time. This current gives rise to a spectrum composed of one or several peaks, representative of the different discharge or depolarization phenomena.

The spectra obtained in practice are often complex and poorly separated. The analysis of thermostimulated currents allows, especially in the case of dipolar mechanisms, relaxation times, total equivalent charge and activation energies to be measured. On the other hand, this technique is not so suitable for the measurement of other types of charges, notably because of the possibility that the carriers will recombine in the material without giving rise to a current measurable in the external circuit.

6.3.1.2. *The mirror method*

This method [LEG 95], [VAL 95] permits the analysis of the amount of charges likely to be stored in an insulator, as well as their evacuation dynamics ("detrapping") following the application of different stresses (thermal, electrical, mechanical, etc.) Its principle lies in the implantation of charges in the material with the aid of a high energy electron beam. The dynamics of the implanted charge is then studied with the help of a lower energy electron beam.

The mirror method is not really classifiable in one of the two large categories mentioned above. It has been placed in the destructive techniques section since the implantation of charges modifies the material.

6.3.2. *Non-destructive methods*

As previously described, these methods have been developed since the 1980s and are based on the application of a mechanical or thermal stimulus to the material, non-homogeneous in time and in space. The stimulus locally perturbs the electrostatic equilibrium of the sample by giving rise to an electrical response measured in an external circuit (voltage or current). The analysis of this signal permits the electric field distribution and space charge density in the insulator to be calculated.

Most of these techniques lie in the relative displacement of charges with respect to the electrodes, and in the local variation of the permittivity with temperature. Indeed, in the absence of the stimulus, in both electrodes in the sample containing a space charge Q_i, influence charges (called Q_1 and Q_2 in Figure 6.2) appear with respect to the total influence principle. During the application of the stimulus, these charges are modified by *dQ*, resulting (if the sample is in short circuit, as in Figure 6.2) in a transient current in the external circuit, or in a transient modification of the voltage across the sample if the latter is in open circuit.

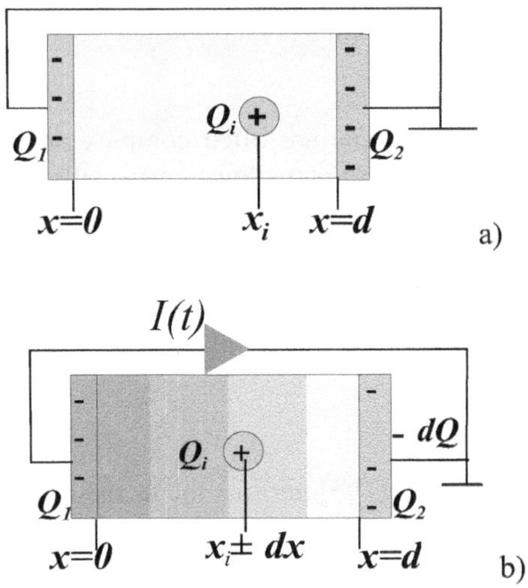

Figure 6.2. *Principle of a stimulus method (example for a sample in short circuit): a) sample in the absence of stimulus; b) sample during the application of the stimulus*

Most of the techniques used today call on thermal or mechanical stimuli. We also need to mention here the *Kerr effect technique* [COE 93], based on the birefringence induced in the insulators by an electric field, but which is therefore exclusively applicable to transparent insulators, and particularly those presenting a high electro-optical birefringence. The sensitivities of the techniques used today range between 0.001 C/m^3 and 1 C/m^3, and their resolutions between 100 nm and 25 mm, according to the characteristics of each technique, and on the nature and thickness of the samples studied.

6.3.2.1. *Thermal methods*

In this case, the sample is subjected to an abrupt temperature rise on one of its sides, which leads to a non-uniform dilation of the material, and to a local variation of its permittivity. Note that the variation factors are very weak (of the order of 10^{-4} to 10^{-6}), which does not prevent these techniques displaying better sensitivities (1 mC/m^3) and resolutions (< 0.1 μm), the reason being a much better thermal transfer efficiency than those of energy transfers of mechanical origins. They are generally "slow" methods, the duration of the measured signals being longer than one second. As a result, the charge monitoring for durations less than a minute is limited; on the other hand, they provide significant signal-to-noise ratios and an excellent reproducibility.

In general, a "thermodilation" signal is measured, whose expression is in the form (for a measurement in short circuit):

$$I(t) = \pm \frac{S}{d} \int_0^d E(x)\,\alpha(x,t) \frac{\partial \Delta T(x,t)}{\partial t} dx \qquad [6.1]$$

with S surface of the sample, $E(x)$ distribution of the electric field throughout the sample, $\alpha(x,t)$ function containing the variation laws and the permittivity thickness of the material with the temperature, and $\partial \Delta T(x,t)/\partial t$ variation law of the temperature over time on the x-axis in the material. Knowledge of the measured current, temperature distribution and thermal parameters of the material allow $E(x)$ to be calculated and, subsequently, the charge distribution. The resolution of the integral equation can be difficult, but mathematical techniques are now much improved and today allow remarkable results to be obtained.

The difference between the thermal methods is most notable at the level of the means of achieving localized heating, which has a direct influence on the amplitude of thermal perturbation (and so on the measurable thickness) and on the equipment to bring into play. According to the technique used to produce a "heat wave", we distinguish:

– the *Thermal* Pulse Method [COL 80] [DER 82]), which leans on the application of a heat pulse (realized by a laser pulse) on a metallic sample whose electrodes are in open circuit. The pulse propagates by moving the charges by local dilation of the material. This method ensures an excellent resolution (< 1 μm) and is notably suitable for the study of thin samples (< 100 μm), because of heat wave damping in thicker materials;

– the Laser Intensity Modulation Method [LAN 86] consists of heating a side of the sample with a laser beam, modulated in intensity, at a variable frequency, and in measuring the resulting pyroelectric current. As for the thermal impulse method, the LIMM suits thin samples (< 100 μm), its resolution being less than 10 μm;

– the FLIMM (Focused Laser Intensity Modulation Method) and the FLAMM (Focused Laser Acoustic Modulation Method [FRA 97], [FRA 00]) use the same principle as the LIMM, but the laser beam is focused and its displacement allows a 3D mapping of the sample. The resolutions of these techniques are of the same order as the LIMM. The insulating thicknesses that can be characterized by FLIMM and FLAMM are also in the same gap as the LIMM;

– the *Thermal Step Method* [TOU 91] [NOT 01] consists of the measurement of the thermodilation current of the sample after application of a *temperature step* on one of its sides. Because of the significant quantity of heat transferred to the material, this latter method is as suitable for thin as for thick samples (between 100 nm and 25 mm), whether the structures are plane or cylindrical. As for other techniques, this method's resolution depends on the thickness of the sample and can get down, in the case of very thin samples such as those used in microelectronics, to 100 nm, or even less. A version of this technique (called the alternative thermal wave method (ATWM [REB 00]) was recently developed, and uses a sinusoidal thermal perturbation to analyze samples whose thickness is of the order of 100 μm.

6.3.2.2. *Pressure methods*

These techniques use an elastic wave to quantify the charges trapped in materials. Because of the high speed of sound in insulators, these are "fast" methods: the duration of signals does not exceed a few μs. This makes them particularly appealing for the monitoring of charge dynamics, notably under sinusoidal electrical stress, and also under thermal stress. On the other hand, the energetic efficiency of the stimulus being more reduced than for thermal methods, the signal-to-noise ratio is generally weak. Consequently, achieving reproducible results fairly often requires the need for an average of several dozen (even hundreds) of successive measurements to be made. The spatial resolution of these techniques can get down to 1 μm, their sensitivity being in the order of 10 mC/m^3.

Today, we mainly use:

– methods calling on a *pressure wave* [ALQ 81], [SES 81], whose principle is the compression of the material by a transient wave with a stiff front. Different means are used to generate this pressure wave (shockwave tube, piezoelectric transducer, laser pulse), and hence the existence of several versions. The most developed method is the *LIPP* (*Laser Induced Pressure Pulse* [SES 81]), which uses a laser impulse of a duration of less than ten nanoseconds. The equations relating the signal measured to the internal electric field in the sample are analogous to those of thermal methods. For example, for a sample in short circuit, the measured current is expressed as:

$$I(t) = \lambda C G(\varepsilon) \int_0^d E(x) \frac{\partial P(x,t)}{\partial t} dx \qquad [6.2]$$

where d is the thickness of the sample, χ the compression factor of the material, C the capacitance of the sample before compression, $G(\varepsilon)$ the coefficient related to the permittivity variation of the material with pressure, $E(x)$ the distribution of the electric field throughout the sample, and $\partial P(x,t)/\partial t$ the variation law of the pressure with time on the x-axis in the material. Knowledge of the measured current, pressure distribution and parameters of the material allows the calculation of $E(x)$, and, subsequently, the charge distribution. If the attenuation of the pressure wave is negligible (in the case of thin samples), we can obtain a reading of the charge distribution without mathematical processing, which constitutes an important advantage. However, consideration of issues related to pressure wave reflection at the interfaces often requires a complex analysis of signals and the need to resort to deconvolution;

– the *Pulsed Electroacoustic Method* [MAE 88] originates in a slightly different principle, and consists of placing a piezoelectric transducer on one side of the sample and exciting the test-tube by a brief high voltage impulse. The Coulomb force produced by the charges under the effect of this stimulus generates a pressure wave, collected by the transducer after a time t_A. This time allows the charges to be located. The PEA method is today very widely adopted around the world and seems well suited for samples of thicknesses greater than 100 µm.

It is quite difficult to compare the different methods. However, thermal methods could be compared, and the mirror method could be positively correlated to the TSM. Comparisons have also been made between the LIPP and the PEA method. On the other hand, genuine comparisons between thermal and pressure techniques cannot be made for now, mostly because the two classes of methods have different sensitivities and because their dynamics are different.

6.4. Trends and perspectives

It seems undeniable that these methods will play an important role in scientific as well as industrial aspects in the future. The "low energy" thermal methods (LIMM, FLIMM, ATWM) can claim thin 3D type measurements on thicknesses from 10 to 100 microns, with a "material" or component orientation. The "high voltage" TSM, with industrial applications, contributes to the choice of material for polymer-insulated cables for direct current power transport, and characterizes the cables aged on site under alternative stress (50 Hz). Further, the TSM recently showed its possibilities for new characterization methods for micro or nanoelectronics. Considering their fast response time, basic pressure methods can directly be used under alternative current.

In conclusion, because the space charge reflects the microstructural state of a material, it has been important, for 30 years now, to develop non-destructive methods for measurement of this fundamental parameter which represents the history of the material or of the component. These methods – in the form of two distinct approaches – have each progressed very far in their respective fields, to arrive today at a protocol [TAK 06] which can act as a base for more and more numerous scientific and industrial studies.

6.5. Bibliography

[ABO 97] ABOU DAKKA M., BAMJI S.S., BULINSKI A.T., "Space-charge distribution in XLPE by TSM using the Inverse Matrix Technique", *IEEE Transactions on Dielectrics and Electrical Insulation*, vol. 4, no. 3, p.314–320, 1997.

[AGN 99] AGNEL S., TOUREILLE A., PLATBROOD G., GEERTS G., "Study of AC charging ability of XLPE insulating for power cables", *Proceedings of JICABLE 99*, Versailles, France, p. 656–661, 1999.

[AGN 00] AGNEL S., NOTINGHER P., TOUREILLE A., "Space charge measurements under applied DC field by the thermal step method", *IEEE CEIDP Annual Report 00*, Victoria, Canada, p. 166–170, 2000.

[ALQ 81] ALQUIÉ C., DREYFUS G., LEWINER J., "Stress wave probing of electric field distribution in dielectrics", *Physical Review Letters*, vol. 47, no. 20, p. 1483–1487, 1981.

[AMB 62] AMBROSI L.E., "Structural dependence of the electrical conductivity of polyethylene terephtalate", *Journal of Polymer Science*, vol. 62, p. 331–346, 1962.

[BER 99] BERTIN C., GUASTAVINO J., MAYOUX C., MARTINEZ-VEGA J.J., "Low density polyethylene microstructure observation under AC or DC high stress", *Proceedings of JICABLE 99*, Versailles, France, p. 785–790, 1999.

[BUC 64] BUCCI C., FIESCHI R., "ionic thermoconductivity method for the investigation of polarization in insulators", *Physical Review Letters,* vol. 12, no. 1, p. 16–19, 1964.

[CAS 69] CASERTA G., RISPOLI B. AND SERRA A., "Current and band structure in amorphous organic films", *Physica Status Solidi*, no. 35, p. 237–248, 1969.

[CHE 92] CHERIFI A., ABBOU-DAKKA M., TOUREILLE A., "The validation of the thermal step method", *IEEE Transactions on Dielectrics and Electrical Insulation*, vol. 27, no. 6, p. 1152–1158, 1992.

[COE 66] COELHO R., "Considérations théoriques sur le contact Métal-Isolant", *Revue générale d'électricité*, no. 6, p. 759–768, 1966.

[COE 93] COELHO R., ALADENIZE B., *Les diélectriques*, Hermès, Paris, 1993.

[COL 80] COLLINS R.E., "Practical application of the thermal pulsing technique to the study of electrets", *Journal of Applied Physics*, vol. 51, p. 2973–2986, 1980.

[CRE 70] CRESWELL R.A., PERLMANN M.M., "Thermal currents from corona charged mylar", *Journal of Applied Physics,* vol. 46, no. 6, p. 2365–2375, 1970.

[DER 82] DEREGGI A., MOPSIK F.I., "Numerical evaluation of the dielectric polarisation distribution for thermal pulse data", *Journal of Applied Physics*, vol. 53, p. 4333–4339, 1982.

[FAL 72] FALLOU B., PERET J., JOCTEUR R., "Résistance négative dans le polyéthylène", *Revue générale d'électricité*, vol. 81, no. 11, p. 757–764, 1972.

[FRA 97] FRANCESCHI J.L., HASS V., "Laser thermo-acoustic modulation method for space charge measurements", *Applied Physics Letters*, p. 2236–2237, 1997.

[FRA 00] FRANCESCHI J.L. *et al.*, "Profils de charges d'espace dans des isolants polymères minces par une méthode d'excitation thermique avec laser focalisé", *Proceedings of SFE 2000,* Montpellier, France, p. 22–32, 2000.

[GUT 67] GUTMANN F., LYONS L.E., *Organic Semiconductors*, Wiley, New York, 1967.

[HAL 71] HALL R. B., "The Poole–Frenkel effect", *Thin Solid Films*, no. 8, p. 263–271, 1971.

[LAM 70] LAMPERT M.A., MARK P., *Current Injection in Solids*, Academic Press, New York, 1970.

[LAN 86] LANG S.B., DAS GUPTA D.K., "Laser intensity modulation method: a technique for determination of spatial distribution of polarisation and space charge in polymer electrets", *Journal of Applied Physics*, vol. 56, p. 2151–2160, 1986.

[LEG 95] LE GRESSUS C., "Space charge characterisation: state of the art and further needs", *Proceedings of CSC'2*, Antibes, France, p. 304–313, 1995.

[LEN 66] LENGYEL G., "Schottky emission and conduction in some organic insulating materials", *Journal of Applied Physics*, vol. 37, p. 807–810, 1966.

[LEW 77] LEWIS T.J., *Dielectric and Related Molecular Processes*, vol. 3, The Chemical Society, London, 1977.

[LIL 68] LILLY A.C., MC DOWELL J., "High field conduction in films of mylar and teflon", *Journal of Applied Physics*, vol. 39, no. 1, p. 141–147, 1968.

[MAE 88] MAENO T., FUTAMI T., KUSIBE H., TAKADA T., COOKE C.M., "Measurements of spatial charge distribution in thick dielectrics using the pulsed electroacoustic method", *IEEE Transactions on Electrical Insulation,* vol. 23, p. 433–439, 1988.

[MAL 97] MALRIEU S., CASTELLON J., "Space charge measurements by the thermal step method: results in some polymers", *Journal of Electrostatics*, vol. 40–41, p. 283–288, 1997.

[MAN 71] MANY A., GOLDSTEIN Y., GROVER N.B., *Semiconductor Surfaces*, Holland Publishing Company, Amsterdam, 1971.

[MAZ 01] MAZZANTI G., MONTANARI G.C., PALMIERI F., "Mobility estimation from space charge measurements as an indicator for insulation ageing", *Proceedings of CSC'4,* Tours, p. 388–391, 2001.

[MOT 71] MOTT N.F., DAVIS E.A., *Electronic Processes in Non-Crystalline Materials*, Clarendon Press, Oxford, 1971.

[NOT 99] NOTINGHER P., TOUREILLE A., SANTANA J., ALBERTINI M., "Space charge in AC poled XLPE for HV cables", *Proceedings of JICABLE 99,* Versailles, p. 701–706, 1999.

[NOT 01] NOTINGHER P., AGNEL S., TOUREILLE A., "The thermal step method for space charge measurement under applied DC field", *IEEE Transactions on Dielectrics and Electrical Insulation*, vol. 8, no. 6, p. 985–994, 2001.

[ODW 73] O'DWYER J.J., *The Theory of Electrical Breakdown in Solid Dielectrics*, Clarendon Press, Oxford, 1973.

[POH 67] POHL H.A., "Theories of electronic behavior in macromolecular solids", *Journal of Polymer Science Part C*, no. 17, p. 13–40, 1967.

[RAM 01] RAMOS M.D., ALMEIDA A.M., CARNEIRO J.P.M., STONEHAM A.M., "Atomistic and mesoscopic modelling of electric charge evolution in polymers", *Proceedings of CSC'4, Tours, France*, p. 108–113, 2001.

[REB 00] REBOUL J.M., CHERIFI A., CARIN R., "Mesure de charges d'espace dans les films diélectriques par des méthodes thermiques. Etude des performances de nouveaux dispositifs expérimentaux", *Proceedings of SFE 2000*, Montpellier, France, p. 51–60, 2000.

[ROS 63] ROSE A., *Concepts in Photoconductivity and Allied Problems*, Interscience Publishers, New York, 1963.

[SES 81] SESSLER G., WEST J.E., GERHARD G., "Measurement of charge distribution in polymer electrets by a new pressure–pulse method", *Polymer Bulletin*, vol. 6, p. 109–111, 1981.

[SIM 71] SIMMONS J.G., *DC Conduction in Thin Films*, Mills & Boon Ltd, London, 1971.

[TAK 06] TAKADA T., HOLBOELL J., TOUREILLE A., DENSLEY J., HAMPTON N., CASTELLON J., HEGERBERG R., HERIKSEN M., MONTANARI G.C, NAGAO M., MORSHUIS P., "Guide pour la mesure des charges d'espace dans les diélectriques et les matériaux isolants", *Electra*, no. 224, p. 53–63, 2006.

[TOU 74] TOUREILLE A., Sur les phénomènes de conduction à champ électrique élevé dans les hauts polymères, Thesis, Montpellier University, January 1974.

[TOU 76] TOUREILLE A., "High field conduction and oscillations in polymers", *Journal of Applied Physics*, vol. 47, no. 7, p. 2961–2965, 1976.

[TOU 87] TOUREILLE A., "Sur une méthode de détermination de densité spatiale de charge d'espace dans le polyéthylène", *Proceedings of JICABLE 87,* Versailles, p. 98–103, 1987.

[TOU 91] TOUREILLE A., REBOUL J.P., MERLE P., "Détermination des densités de charges d'espace dans les isolants solides par la méthode de l'onde thermique", *Journal de Physique III*, no. 1, p. 111–123, 1991.

[TOU 98] TOUREILLE A., VELLA N., "On the origin of space charge in polyethylene: a study by the thermal step method", in *Space Charge in Solid Dielectrics*, FOTHERGILL J.C., DISSADO, L.A., Dielectrics Society, 1998.

[VAL 95] VALLAYER B., TREHEUX D., "Influence de l'usinage sur les propriétés de charge de monocristaux de MgO et d'Al_2O_3", *Proceedings of CSC'2,* Antibes, France, p. 578–583, 1995.

[VAN 75] VAN TURNHOUT J., *Thermally Stimulated Discharge in Polymer Electrets*, Elsevier, Amsterdam, 1975.

[VAN 79] VANDERSCHUEREN J., GASIOT J., "Field induced thermally stimulated currents", *Topics in Applied Physics*, vol. 37, Springer Verlag, Berlin, 1979.

[WIN 73] WINTLE H.J., "Absorption current, dielectric constant and dielectric loss by the tunnelling mechanism", *Journal of Applied Physics*, vol. 44, no. 6, p. 2514–2518, 1973.

PART 2

Phenomena Associated with Environmental Stress – Ageing

Chapter 7

Dielectric Materials under Electron Irradiation in a Scanning Electron Microscope

7.1. Introduction

The study of electron irradiation of dielectric materials is of considerable interest from a fundamental point of view as well as from the point of view of applications. Research has covered a wide domain of applied physics, from the behavior of satellites subject to cosmic radiation in space research, to electrets used in radiobiology; we try to avoid the charging up of dielectric materials on board satellites in space and subject to solar winds, while we use charging up of dielectric materials to produce electrets. Dielectric material irradiation by energetic electrons is accompanied by numerous complex phenomena, notably secondary electron emission, photon emission, trapping of charges, luminescence, etc. In the case of conductors, secondary electron emission is a source of issues for particle accelerators, while it is the base of the contrast observed in electron microscopy and exploited in the manufacture of electron multipliers. In the case of irradiated insulators, the situation is different: their secondary electron emission contributes to strange effects, commonly called charging effects, which are irregular and hard to predict.

Although these complex effects are qualitatively well understood and have been described in a proliferation of articles (the complete list of which is impossible to detail here but which is previewed in this special issue [JES 92]), the physical

Chapter written by Omar JBARA, Slim FAKHFAKH, Sébastien RONDOT and Dominique MOUZE.

mechanisms of the elementary physical phenomena brought into play are far from being explained.

This chapter deals with the different aspects of charge effects of insulators under a low energy electron beam (between 300 eV and 30 keV) in a scanning electron microscope (SEM) with, in particular, the analysis of parameters implied, notably the sample, its geometry and its environment (i.e. a vacuum) as well as the irradiation conditions (energy, intensity, beam size, etc.). Before tackling the charge phenomena resulting from the electron bombardment of an insulating material, we shall first inspect the basic elements associated with the electron-matter interaction in general and electron-insulator in particular. The different physical processes which accompany the injection of a charge into an insulator and the behavior of this latter following injection will be described. The general properties of insulators, such as the role of disorder in charge localization, the influence of defects, either intrinsic or induced by electron irradiation, and the processes of charge transport will also be presented. The influence of charging phenomena on emitted signals, notably electron emission andemission of X-rays, will then be tackled. The amplitude of these phenomena is often difficult to predict, because they are not only a function of the chemical composition of the sample but also of unknown structural parameters, such as the density of traps as well as the often complex conduction surface mechanisms. They will then be studied by developing some recent experimental techniques and illustrated by numerous examples.

7.2. Fundamental aspects of electron irradiation of solids

7.2.1. *Volume of interaction and penetration depth*

During the progress of an electron of energy E_0 in a solid sample, we usually distinguish two types of concomitant interactions: elastic and inelastic interactions [SEI 84]. The interaction is elastic when an electron penetrating the influence zone of the potential which reigns around an ionic core is deviated without any appreciable energy loss. If the energy of the electron varies in a significant way, for example by transmitting part of its energy to an atomic layer electron, the interaction is said to be inelastic. According to these two types of interactions, incident electrons take random routes in a material until total loss of their kinetic energy. An example of these trajectories obtained by numerical simulations, according to the Monte Carlo method, is given in Figure 7.1(a); the place where all of these interactions occurs is called the "volume of interaction". The lateral dimensions z_m (average penetration depth, from a few nm to a few μm) of this volume depend mainly on the incident energy and the physico-chemical properties of the target [FIT 74].

7.2.2. *The different emissions*

In the volume of interaction, multiple emissions with different physical natures are generated. Mainly, these are the electron emission, photon emission, cathodoluminescence, plasmons (due to collective oscillations of valence electrons), phonons (associated with the thermal agitation of a periodic structure), as well as the creation of electron-hole pairs. Most of these signals, illustrated in Figure 7.1(b) in the case of a thin target, are produced simultaneously and make possible the observation and analysis of a chosen object in a scanning electron microscope.

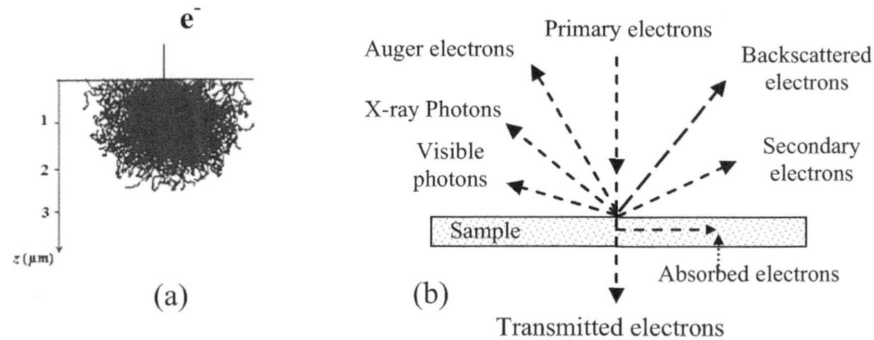

Figure 7.1. *(a) Electron trajectories simulated by the Monte Carlo method in a grounded gold coated Al_2O_3 at 20 keV [JBA 96b]. (b) Electron bombardment of a thin target and resulting signals*

For simplification, and for the needs of this chapter, electron emission (i.e. secondary and backscattered electrons) and photon emission (X-ray photons) are described very basically below.

7.2.2.1. *Electron emission*

7.2.2.1.1. Backscattered electrons

These electrons from the primary beam come out of the irradiated material after having given up (or not) a more or less large part of their energy. They come from an average depth estimated at R/2 (R maximal penetration depth) by Niedrig *et al.* [NIE 82], and their spectral distribution coming out of the target ranges from 50 eV to the energy of primary electrons. We often characterize backscattered electrons by what we call the backscattering coefficient η defined by the ratio of the current intensity due to the backscattered electrons, I_r, to the primary current intensity I_0. For literature concerning the importance of backscattering, see [NIE 82] (and included references); here we content ourselves with a brief reminder of some characteristic notions of this emission [JOY 95]:

– for a primary energy ranging between 10–30 kV, the backscattering coefficient is practically independent of the primary energy;

– for low energies, the coefficient decreases as a function of energy for targets composed of light elements and increases for targets composed of heavy elements.

7.2.2.1.2. Secondary electron emission

The phenomenon of secondary electron emission was subjected to numerous theoretical as well as experimental research concerning the mechanisms and elementary processes which play a part in this emission. At the present time, the most complete and satisfying interpretation on the emissive properties of metal targets can be found in the articles of Dionne [DIO 75], Schwarz [SCH 90] and Shih [SHI 97].

Mechanism

The secondary emission phenomenon occurs according to three different processes: generation, transport towards the surface of the sample and, finally, the crossing of this latter. The generation of secondary electrons takes place along the course of primary electrons in the volume of interaction and happens isotropically. By losing part of its energy, the incident electron excites the electrons in the external layers of the atoms in the target which are weakly bonded. The depth distribution of the secondary electron production rate is proportional to that of the energy losses. The emission being isotropic, only a fraction (to a first approximation, half) of the electrons generated migrate in the direction of the target/vacuum interface. However, because of the interactions with the electrons of the conduction band and with the phonons, only the electrons created between the surface and what we call the escape depth reach this interface. According to the works of Seiler [SEI 67] the maximum escape depth of secondary electrons is about 5λ (λ being the mean free path of the secondary electrons). Let us note that the transport of secondary electrons towards the target/vacuum interface is weaker in an insulator than in a metal because $\lambda \approx 0.5$–1.5 nm in metals, while in insulators $\lambda \approx 10$–20 nm [SEI 67]: secondary electrons interact exclusively with phonons. Further, to cross the target/vacuum interface, the secondary electrons must have at least an energy equal to the work function in the case of a metal or to the electron affinity (minimum energy required to be supplied to an electron situated at the bottom of the conduction band to eject it in a vacuum) in the case of an insulator or a semi-conductor. As an example, the electron affinities χ of alumina (Al_2O_3) and silicium (SiO_2), both monocrystalline, are close to $\chi = 1$ eV [ALI 78]. The emitted secondary electrons are not very energetic (between 0 and 50 eV, according to conventional limit) and come from regions near the surface (of the order of a few nm). The secondary electron emission rate or secondary electron yield, δ, is defined as being the ratio of the intensity of the secondary electrons current to that of the primary current.

Total emission yield

The total electron emission yield σ is defined by the following relation:

$$\sigma = \delta + \eta \tag{7.1}$$

where δ is the secondary emission yield and η the backscattering coefficient.

Figure 7.2 schematically represents the characteristic shape of this yield and its two components δ and η as a function of the energy E_0 of the primary beam.

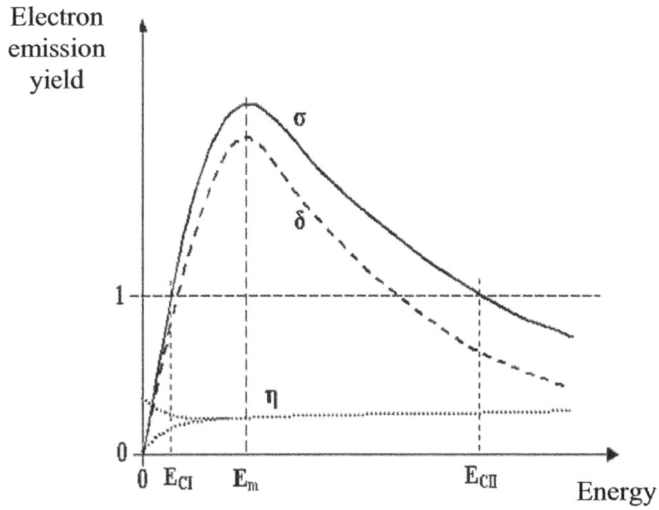

Figure 7.2. *Schematic layout of total electron emission yield σ as a function of primary electron energy*

The shape of this curve is the same for all materials. With the backscattering coefficient η being practically constant and its variation very weak (compared with δ), the shape of σ is consequently dictated by the yield variation δ, except at very low energies. To explain the total electron yield variation σ as a function of the primary energy E_0 and the influence of the incidence angle on the secondary emission yield, see [SCH 90] and [JOY 95].

7.2.2.2. Emission of X-ray photons

We have seen that when electrons make their way in a material medium, they interact with it, undergoing elastic and inelastic collisions. These latter inelastic collisions can give rise to vacancies in the internal layers of the atoms which are thus found in an excited state. The return to an equilibrium state (de-excitation) occurs either by radiative emission (X-ray emission) or by non-radiative emission (Auger emission). In the first case, the created vacancy is filled by an electron from an upper layer (the transition obeys the selection rules of quantum mechanics) and the energy difference between the two levels brought into play is transformed into an X-ray photon (Figure 7.3).

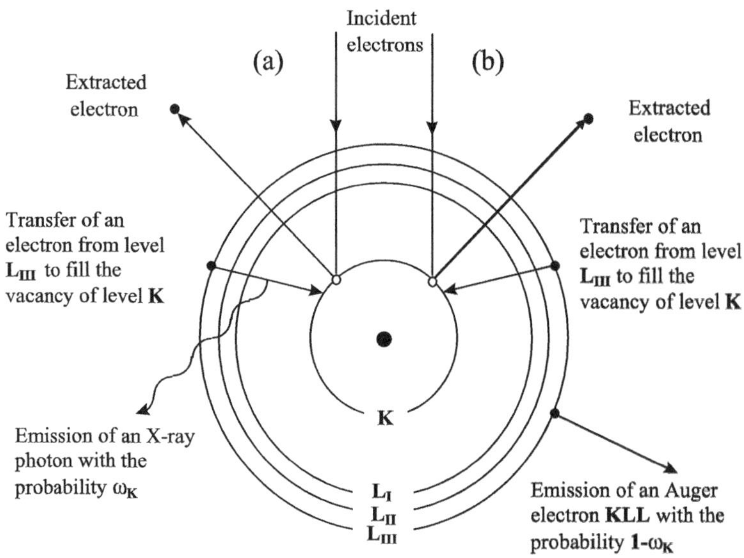

Figure 7.3. *Schematic illustration (Bohr model) of the emission process of (a) characteristic X-ray photons or of (b) Auger electrons emissions*

The electron irradiation of a material is then accompanied by the emission of X-ray photons with the well-defined energies and characteristics of the atoms composing the target. This mechanism constitutes the basic principle of X-ray microanalysis. Besides this characteristic emission, the electron bombardment of a solid target generates a continuous radiation related to the slowing down of incident electrons in the strong electromagnetic fields of the atom cores in the target. In X-ray microanalysis, this radiation (*Bremsstrahlung*) constitutes the continuous background on which the characteristic lines will appear. Please note that the maximum energy of this continuous X-ray radiation, which could not exceed the energy of the primary electrons, represents a limit which we call the "Duane–Hunt limit" and which we often use to control the acceleration tension of the scanning electron microscope on metallic targets, or to measure the surface potential when we deal with insulating targets (see section 7.4).

7.3. Physics of insulators

7.3.1. *General points*

Materials known as insulators, which we come across everywhere in our environment (glasses, ceramics, metal oxides, micas, rubber, paraffin, paper, polymers, composites, etc.) have resistivities much greater than those of metals and semi-conductors, even intrinsic resistivitities with a large band gap (see Figure 7.4). As was shown in Chapter 2, (section 2.1 and Table 2.1), this is mainly due to a very weak density of mobile charge carriers, although the mobility of these carriers can often approach that of metals, or even exceed it. Insulators have a very weak conductivity because of the absence of free charges and not because they are unable to conduct these charges. Another property is that, unlike metals, insulators trap charges because of the defects and the impurities they contain. This leads to the "charge phenomena" described in Chapter 2. Insulators can also get polarized when an electric field is applied to them.

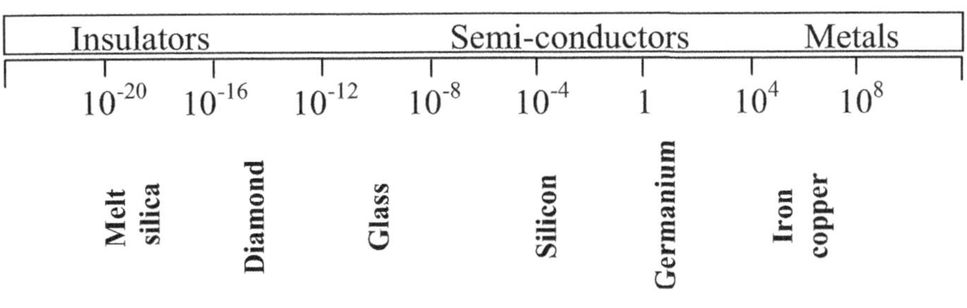

Figure 7.4. *Range of electric conductivities of insulators compared with those of metals and semi-conductors (in $\Omega^{-1}.cm^{-1}$)*

7.3.2. *Insulators under electron irradiation*

7.3.2.1. *Microscopic phenomena*

7.3.2.1.1. Secondary electron emission of insulators

During the electron irradiation of an insulator, electron-hole (e.h.) pairs are created along the path of the incident electron, until its maximal penetration depth R.

This creation essentially results in two processes, one is direct (Valence Band (VB)-Conduction Band (CB) transition) and the other is indirect (relaxation of plasmons). The number of e.h. pairs created by the incident electron is n(e.h.)=E_0/E(e.h.) (with E_0 energy of incident electrons, and E(e.h.) creation energy of a pair which ranges between 10 and 25 eV in the materials being considered) [CAZ 99 and some references included]. The very slow electrons (a few eV from the bottom of the CB) resulting in the creation of these pairs and which do not spontaneously recombine with the holes, propagate in the insulator and can escape in vacuum if their final energy is greater than the affinity of the insulator. Withing the framework of this study, it is interesting to compare the secondary emission of metals with those of insulators under electron bombardment, knowing that this comparison is difficult because, as the insulators get charged, the yield varies. This comparison is illustrated in Figure 7.5a.

The emission peak of metals has a width larger than that of insulators. It is of the order of 10 eV for metals and 5 eV for insulating materials [WHE 64]. The weakness of electron density in the conduction band leads to a maximal secondary emission δ_m which, for uncharged insulators, can be of an order greater than that of metals. The maximal secondary emission of metals generally ranges between 1 and 2; that of insulators is a bit more than one order of magnitude greater; that of polymers is often limited to the interval $3 < \delta_m < 7$ [WIL 73]; for monocrystalline MgO, $\delta_m \approx 22$ [WHE 64] and this yield can reach extremely high values such as recently obtained with a diamond sample ($\delta_m \approx 84$) [SHI 97] at an energy of 3 keV.

The differences observed in the secondary emission of the two classes of materials can be discussed in a qualitative manner from the study of secondary electron generation and their transport until exit at the surface of the samples. Several theoretical discussions [SHI 97, KHA 82, AKK 94] attribute the observed experimental results to the lack of electron–electron interaction during the transport of secondary electrons generated in insulators, and to their very low or equally negative affinity. This lack of electrons allows an increase in the mean free path (in uncharged insulators) and a weak affinity often allows the generated secondary electrons to escape in a vacuum [SHI 97]. The escape depth (the mean depth in the insulator from which the electrons are emitted, which governs the value of δ) of the secondary electrons ($\lambda \cong 50$–100 nm) is greater than that of metals ($\lambda \approx 5$–10 nm). The interactions of secondary electrons with phonons or with crystalline defects explain the reduction of this depth when the temperature increases (see Figure 7.5b) or when a polycrystalline sample (rather than a monocrystalline sample) is irradiated [CAZ 93]. For irradiated insulators, the secondary emission in a vacuum in principle leaves a positively charged zone at the surface of the sample, and the resulting charging effects lead to self-regulation processes which limit the secondary emission (see section 7.3.2.2.2).

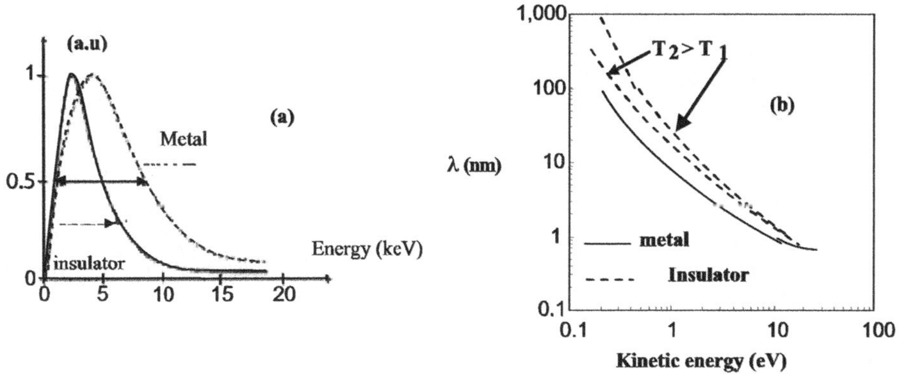

Figure 7.5. *(a) Comparison of secondary emission yields of insulators and metals. (b) Escape depth λ of secondary electrons as a function of their kinetic energy [CAZ 95a]*

7.3.2.1.2. Induced defects and desorption of species

The interaction of incident particles with target electrons and atomic cores leads to the creation of defects which are generally classified in two categories: those which come from the bond breakdown due to ionization of the atomic levels and those which come from lattice distortion due to an atomic displacement. The latter is unlikely in SEM irradiation conditions because it occurs when the high energy incident particles interact with the lattice.

The bond breakdown is due to the atomic level ionization of the target components by the incident electrons which leads, by de-excitation, to the emission of X-ray photons with probability ω_{ij} and the emission of Auger electrons with complementary probability a_{ijk}. Following the electron vacancy created (in just a few 10^{-15} seconds), cascade transitions, implying more and more external electron levels, are generated. The final electron levels thus become depleted of electrons. In the case of metals, the electronic charge of each atom is restored by conduction electrons, whereas in the case of insulators restoration is not possible since these electrons are virtually nonexistent.

Thus, in irradiated oxides the ionization of oxygen at the 1s level will (temporarily) transform the 0^{2-} ion into 0° (by the most probable emission, $a_{ijk} > 99\%$, of the Auger electron O_{KLL}) while in irradiated fluoride, the ionization of the 1s level of fluorine will transform the F^- ion into F^+ (probably due to the emission of the Auger electron F_{KLL}). There can be electron rearrangements with neighboring atoms but inversion of the polarity of certain "halogeneous" ions will lead to the inversion of their Madelung energy and their (coulombian) expulsion from their initial site [SAT 91], [CAZ 95b], [JBA 96b]. The most spectacular result then concerns the possibility of atom displacements (and their desorption in vacuum) of atom species (initially negative) by simple irradiation of insulators with the aid of incident particles (electrons or X-ray photons) whose initial kinetic energy cannot exceed 1 keV [BEN 97].

Silicium, alumina [JON 00] and quartz [VIG 85] are three examples of materials whose defects induced by irradiation are fairly well described in the literature. These defects are often identified with the aid of techniques such as electron paramagnetic resonance (EPR) and the mechanisms proposed for their interpretation are different to those based on the Auger mechanism discussed above.

In the case of silicium, the defects are essentially of three types: the first is oxygen vacancy (neutral, denoted as $\equiv Si^- \ldots {}^-Si \equiv$, or positive, called the E' center, and denoted as $\equiv Si^- \ldots {}^+Si \equiv$) accompanied by a relaxation of neighboring atoms. This type of process is favored if the hole is generated in the part of the valence band corresponding to the Si-O bonds [WEE 56], [GRI 80], [FEI 74], [EDW 82]. The second type is radical peroxide ($\equiv Si - O - O^-$), which can result from either an E' center or a peroxide bridge ($\equiv Si - O - O - Si \equiv$) [STA 78], [GRIS 81]. The third type is a non-bonding oxygen or silicium (NBO) [STA 79]. These defects are associated with bond breakdowns in the sequence between silicium and oxygen atoms, characteristic of a local disorder. These breakdowns can be created by irradiation or during the elaboration of silicium. In the case of quartz, the induced defects are essentially crystalline lattice defects induced by irradiation which favor the formation of excitons responsible for the creation of oxygen-bond peroxide vacancy pairs [VIG 85].

In the case of alumina, in the fundamental state, the electron configuration of oxygen is near O^-. At an oxygen vacancy position, we would expect an equivalent electron distribution of about one electron. Owing to the absence of the atomic core, this distribution is strongly coupled with the vibration modes of the lattice. Such a center is called an F^+ center. An electron state localized in the band gap, partially occupied by electrons from the lattice, is associated to an F^+ center. The trapping of an electron by an F^+ center gives an F center (oxygen vacancy occupied by two electrons). The occupied electron state associated to the F center is less strongly bonded than that associated to the F^+ center. With each F^+ and F center are associated excited levels. The trapping of an electron is exclusively made through weakly bonded levels, and therefore situated very near the bottom of the conduction band. For an F^+ center, it is made in strongly excited states.

Based on an interpretation of the Auger mechanism, as well as the trapping phenomenon (see section below), one of the consequences of the desorption initiated by this mechanism is the fact that the O^{2-} ion of the oxides becomes neutral, leading to a reduction of associated metals (or semi-conductors). The desorption can get amplified when the excited atoms belong to the third line of the periodic table, from Na to Cl, because the Auger mechanism can lead to the loss of 4 electrons (KLL transition followed by two LVV transitions). Other mechanisms can be implied in the creation of defects, the production of charged ions in partially ionic solids. These effects, which allow explanation of the electrical migration of negative species and their ejection in vacuum for electron irradiated insulators by electrons, have been subjected to numerous works in the literature [JBA 95], [GED 99].

7.3.2.1.3. Trapping and charge transport

The different processes

Several approaches have been proposed to interpret the phenomena of conduction and trapping of electrons in an insulator. The first approach originates in the high resistivity of the insulator to electric current, which amounts to saying that the charges are fixed by an ohmic process. The injected charges could slowly flow, and, after thermalization, appear virtually motionless. The constant $\tau=\varepsilon.\rho$ then represents the characteristic time of this type of propagation, obeying an exponential law $e^{-t/\tau}$. The time constant values τ of most insulators are of the order of 10^{-7}s to several days. This approach, however, comes up against two problems: the first is due to the fact that there are insulators, notably monocrystalline ceramics such as MgO, Al_2O_3 (sapphire), which may not get charged at room temperature after an appropriate thermal process. The second concerns the fact that the resistivity cannot show the ability of the insulator to trap electrons through an ohmic process. Indeed, the notion of resistivity is applied to intrinsic charge carriers of the material, while the charge phenomena are due to external charge carriers injected into the solid.

Another approach is based on the band structure by assuming the existence of localized energy levels in the band gap due to defects (impurities, dopants, etc.) and electron states localized at the bottom of the bands (CB or VB) (Anderson states) (see Chapter 2, section 2.3.2). This latter seems incomplete because it neither takes into account the polarization effects of the material around a trapped charge, nor the thermodynamic property changes brought to the fore by Moya [MOY 93]. Finally, the approach which gives a more global explanation of the trapping process is that made from the polaron theory described in Chapter 2.

After this description of different trapping approaches, we can say that, once it penetrates a bombarded sample, an electron loses its kinetic energy, either by giving it up to the lattice, or by exciting electrons from the medium. Electron-hole pairs are created in the insulator but the majority of these pairs quickly recombine and only a few electrons and holes move in the conduction and valence bands before getting trapped (again) at the localized levels of the band gap, representing the impurities and the structure defects produced under electron irradiation and/or pre-existing in the material. The ability of electrons to get trapped is an increasing function of the density of the trapping centers and their trapping energy [CAZ 96 and references included]. This ability then increases with crystalline disorder, i.e. by passing from the monocrystalline state to the amorphous one. The distribution of different trapping centers characterized by their trapping energy, well known for silicium, can partly be revealed by cathodoluminescence experiments, underscoring the relation between trapped charges and color centers [JAR 95]. In the ideal case of a perfectly monocrystalline insulator without structural defects (neither pre-existing, nor induced by incident radiation) and without impurities, the trapping of electrons can only be conceived on the surface of the sample [CAZ 90], [CAZ 91], [REM 98]. If the ideal insulator is earthed, the injected excess electron will quickly be evacuated and no charge localization (or trapping) is possible.

Mobility of charge carriers in disordered insulators

The effects of disorder on the localization and transport of electrons injected in an insulator was recently reviewed by Blaise [BLA 01] and described in Chapter 2. Although this analysis does not include effects due to the electric field induced by a trapped charge itself on electron transport, that is to say the detrapping of trapped charges and the slowing down or acceleration of charge carriers [CAZ 99], [WIN 97], [PEP 98], we can say that, according to the degree of disorder, there is a competition between the conductivity due to mobility in extended states and the conductivity due to mobility in localized states (at the bottom of the conduction band). It should be noted that, for a certain number of insulators and from a certain degree of disorder, the conduction in localized states is activated at room temperature.

7.3.2.2. *Macroscopic phenomena: charging effects*

The only common characteristic of the diverse varieties of insulators under electron irradiation is that their weak electron density in the conduction band forbids them from spontaneously compensating the electric charge deficiency which results from this irradiation. As a result of this, two types of phenomena inherent to the electric field created by the excess electric charge (+ or -) are addressed below: on the one hand, the effects of direct charges which lead in particular to the deflection of the incident beam (responsible for the distortion of the electron image in an SEM) and the modification of the physical parameters of electron-matter interactions which leads notably to the modification of the secondary emission and the X-ray emission, and on the other hand the possible chemical modification of the sample which results from the migration of mobile ions or the stimulated desorption of species. The literature abounds with descriptions of charge phenomena and it is impossible here to draw up an exhaustive list; readers are directed to a few notable articles which deal with these phenomena: [VID 95], [CAZ 99b] and [VID 01].

7.3.2.2.1. General law of conservation of current and induced charge

The total accumulated charge in an insulator results from the competition between electron emission, trapping and charge transport.

If I_0 and I_L are, respectively, the primary current and the leakage current resulting from the transport of a few electrons towards the ground, the law of conservation of current is then written as:

$$I_o = \sigma I_o \pm I_L + \frac{dQ}{dt} \qquad [7.2]$$

where σ is the total electron emission coefficient and dQ/dt the trapping rate during the charging process.

Conceptually, the total charge, Q, can be positive or negative depending on whether the secondary emission prevails or not over the incident electrons implanted and evacuated. If Q is > 0, the current I_L will be ohmic (due to conductivity) whereas when Q is < 0, this current is non-ohmic (related to I_0).

The positive charge which results from the secondary electron emission, δ, extends on a layer of thickness corresponding to the emission depth of the secondary electrons. The negative charge resulting from the injection in the sample of incident electrons (less the backscattered electrons) extends on a layer of thickness R approximately corresponding to the maximal penetration depth of the primary electrons. The global charge of the sample will then depend on the initial primary energy value, E_0.

7.3.2.2.2. Self-regulation and conventional approach: sign of the induced charge

During electron irradiation we see a competition between the positive charges resulting from the secondary emission and the negative charges resulting from the trapping of incident charges in the insulator. In the absence of leakage current, the compensation phenomena take place and lead to a final state characterized by a total unitary yield which corresponds to the identity between the current entering and the current exiting the sample. If the initial state of the sample is characterized by $\delta+\eta>1$, the positive excess charges lead to a positive surface potential V_S which reattracts the slowest secondary electrons and freezes their emission (once Vs reaches a few volts). Conversely, if the initial state is characterized by $\delta+\eta<1$, the excess negative charges lead to a surface potential which slows down the incident electrons outside the sample; their effective energy at the surface (landing energy) is then E_0+qVs (with $Vs < 0$ and $q = 1.6 \cdot 10^{-19}$ C) and the secondary emission increases towards a stationary state for which its yield is equal to one. Self-regulation phenomena (freeze or increase of the secondary emission) lead to the conventional approach $\delta = f(E_0)$ based on arguments taken from the total electron emission yield curve, generally obtained by irradiating the target with short electron pulses, the "pulse mode technique" [BRU 54], which allows nearly all charging phenomena to be avoided. This approach predicts the sign of most charges implanted in insulators irradiated with electrons. Indeed, the sample gets positively charged when $E_{CI}<E_0<E_{CII}$ and negatively charged when $E_0>E_{CII}$ (or $<E_{CI}$). E_{CI} and E_{CII} are the critical energies for which $\delta+\eta=1$. For most insulators, the use of the impulsion method leads to a total electron emission yield without trapping, often several orders of magnitude greater than that of metals and, consequently, to a critical energy E_{C2} greater than or equal to 10 keV [WHE 64, SHI 97, CAZ 01].

In the case of permanent irradiation, a negative charge is often observed for primary energies as weak as $E_{CII} \sim$ 2–3 keV, corresponding to electron emission yield obtained by the pulse method, almost at their maximum [CAZ 01 and included references]; the conventional approach predicts a positive charge in this case. Thus, the predictions of the conventional approach are in contradiction with numerous experimental results [CAZ 01 and included references] notably those about monocrystalline alkaline halogens studied with the impulsion method and for which the maximum of the secondary electron emission yield δ_m is systematically greater than 5 (most often 10) for primary energies in the interval 1–2 keV. In the case of permanent irradiation of LiF and NaCl, under normal circumstances, the obtained critical energies E_{CII} are, respectively, 2 keV and 2.5 keV [REI 92], which correspond to a negative charge.

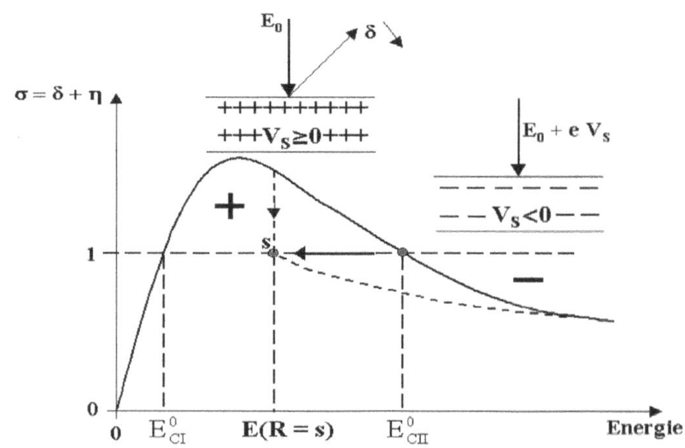

Figure 7.6. *Displacement from the critical energy E^0_{CII} to the energy E(R = s) under permanent irradiation. The point S being the new equilibrium state (from [CAZ 01]). In the dotted line: schematic curve $\delta = f(E)$ of the charged insulator*

To discard this ambiguity, an approach recently proposed by our laboratory [CAZ 99], [CAZ 01] permits the nominal yield δ^0 (corresponding to the critical energy E^0_{CII}) obtained with the aid of the pulse method of the practical yield δ (corresponding to the critical energy E_{CII}) obtained using a permanent irradiation to be distinguished. In the case of permanent irradiation, laboratory analysis suggests that a surface potential equal to zero can only be reached when the penetration depth of the primary electrons is close to the escape depth of the secondary electrons. This state corresponds to the new critical energy E_{CII}=E(R=3s) (Figure 7.6), slightly higher than the energy corresponding to the maximum of the function δ, and therefore less than E^0_{CII}.

In conclusion, the conventional approach of the total electron emission yield can be applied in conditions of use of a permanent irradiation only by replacing the critical energy E^0_{CII} with E(R ≈ 3s). The consideration of current leakage leads to a partial evacuation of the excess negative charge, and consequently a reduction of the negative surface potential. In this case, the effective energy of the primary electrons can be greater than E(R ≈ 3s) [CAZ 01].

7.3.2.2.3. Deflection of the primary beam

The electric field created within an irradiated insulator is a direct consequence of the trapped charge in that insulator. The deflection of the primary beam (schematically represented in Figure 7.10a) which results from it can be either partial or total. In the first case, it is responsible for the distortion of the electron image and the pseudo mirror contrast [BEL 00b] and, in the second case, it leads to the mirror effect. Although these two effects present disadvantages at both observation and analysis levels (section 7.4.2.2), they carry quantitative and qualitative information on the charging state of the insulator.

Partial deflection

The formation of images in scanning electron microscopy results from synchronous scanning of the sample and the visualization screen whose Wehnelt is modulated by the signal emitted by the different points of the sample under electron irradiation. Each point on the screen ($G.x_0$, $G.y_0$) corresponds to a point on the target (x_0, y_0), with G being the proportionality factor which defines the magnification of the image (see Figure 7.7a).

However, because of the presence of a "perturbing" electric field $\vec{E}$ in the vacuum of the specimen chamber, the primary beam initially aiming at coordinates (x_0, y_0) will be deviated. The primary electrons will then hit the surface sample at a distinct point (x, y). As the scanning of the visualization screen remains synchronous with that of the deflective coils, the intensity affected to the pixel ($G.x_0$, $G.y_0$) on the visualization screen will then be associated with point (x, y), not point (x_0, y_0). The effect of this desynchronization is an apparent distortion of the scanned zone of the sample (see Figure 7.7(b)). From this distortion, it is possible to determine the quantity of trapped charges and its evolution over time, as well as the corresponding surface potential V_S (section 7.4.2.2.1).

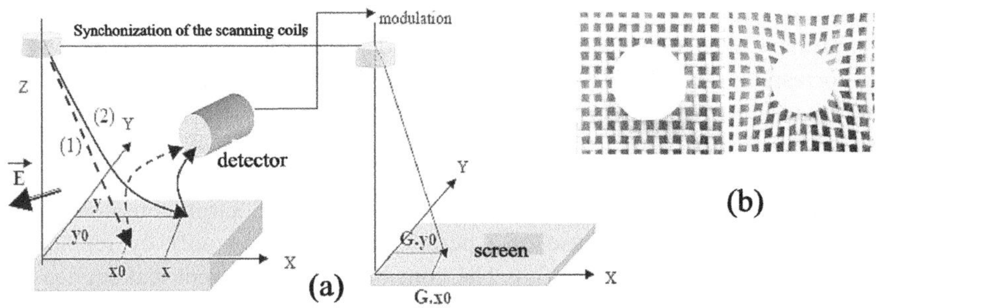

Figure 7.7. *(a) Principle of formation of the image in a scanning electron microscope. (1) Trajectory of primary electrons in the absence of perturbing electric field. (2) Trajectory of primary electrons in the presence of perturbing electric field E. (b) Dynamic distortion of the electron image of a sphere-grid system (left, at t = 0s; right, at t = 70s) (from [BEL 01b])*

Total deflection: mirror effect

In the extreme case of total deflection of the primary beam, we observe a spectacular effect called the mirror effect. This deflection is produced when the primary energy of the incident electrons is less than the surface potential resulting from the trapped charge in the irradiated insulator. This latter then behaves like an electrostatic mirror and, instead of observing the surface sample, we see the specimen chamber of the microscope (see Figure 7.8 and section 7.4.1).

7.3.2.2.4. Chemical modifications and other irradiation effects

One of the possible consequences of charging effects is the chemical modification of the sample. Besides the external consequences (deflection of the beam), these effects are translated into the establishment of an internal electric field which can be sufficiently large to result in a dielectric breakdown, but they can also induce the chemical modifications of the surface *via* the migration of mobile ions under the influence of the field. This is particularly the case in the analysis of glass where the concentration of sodium evolves during irradiation [JBA 95], [GED 99]. The electron stimulated desorptions (ESD) of species naturally lead to a chemical modification of the irradiated surface. Amorphous silicium for example can transform into silicium under irradiation [JAR 88] because the ion O^{2-} becomes neutral, leading to a reduction of the silicium. This mechanism could explain the cold recrystallization of amorphous silicium [SAT 91], [CAZ 95b]. This reduction remains valid for metals when dealing with the irradiation of, notably, LiF alkali halides [JBA 96]. Superficial morphological transformations can also be produced when the carbon of organic materials has some electronic bonds with its neighbors which are suddenly cut off.

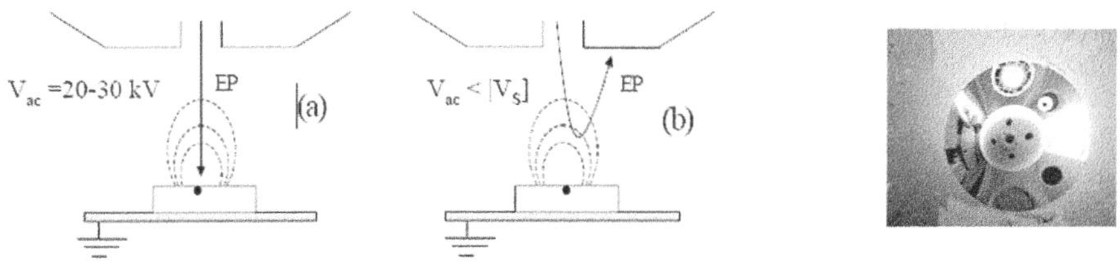

Figure 7.8. *Principle of the mirror method: (a) injection stage of the charge; (b) measurement stage of the charge (obtaining of the mirror image) Vac < Vs (from [BEL 01b]); (c) mirror image from [JBA 04]*

It is interesting to mention the fact that charging effects and the phenomenon of species desorption often coexist, as illustrated by the results obtained for SiO_2 by Jardin [JAR 88]. Both phenomena can equally interact with one another. Indeed, the electron irradiation (at 1 keV) of KCl leads to the formation of an insulated metallic layer which results from the ejection of chlorine and which thus favors the formation of a negative surface charge [SZY 92].

7.3.2.3. *Parameters governing the charge phenomena*

The charging effects in close relation to the ability of an insulator to charge are the result of several dynamic physics processes, complex and closely related, depending on numerous "parameters" not only related to the structure and the nature of the insulator, but also to its environment and the irradiation conditions. The organogram below [BEL 01b] shows a list of some of the most relevant parameters when an insulator is in an environment such as that of a scanning electron microscope.

Among the experimental and environmental parameters studied and likely to affect the ability of the insulator to charge, some of them, like the primary current density and scanning frequency, can easily be controllable and fixed. The effects of the parasite source (electrons coming from polar pieces which result from the acceleration and focus of the secondary electrons emitted by the insulator [BEL 00a]) are hardly controlable but can be reduced by covering the polar pieces with light materials. It should also be noted that a sample which has been irradiated can conserve its negative trapped charge for a long time (from several days to several months) even after an input of air into the chamber of the SEM. The neutralization of this charge before a new irradiation can be ensured either by heating the sample (if this sample permits it), or by irradiation with the aid of an auxiliary electron gun of energy E_0 such that $E_{CI} < E_0 < E(R = 3s)$ (see section 7.3.2.2.2) bringing positive charges, or with the aid of the primary beam whose energy would have decreased [MOR 76], (in this latter case, the sample is not discharged in volume).

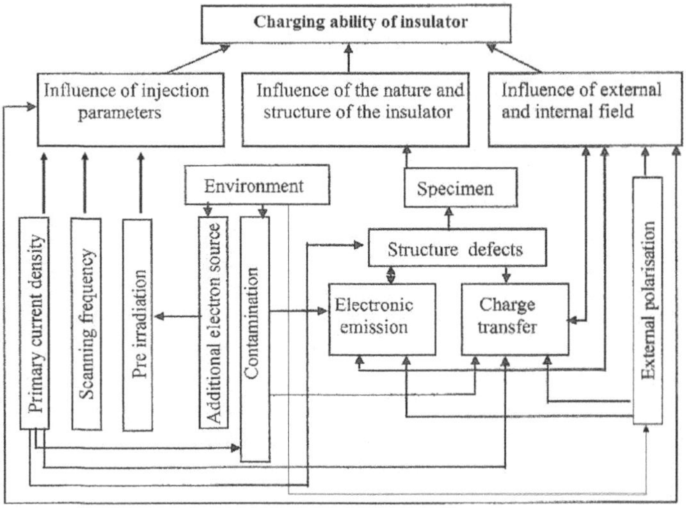

Figure 7.9. *Summary of parameters for charging of an insulator in an SEM*

On the other hand, contamination remains a major issue whose consequences are important and unpredictable. To completely eliminate this latter parameter, the use of equipment under ultra-high vacuum (i.e. Auger equipment) presents an advantage compared with the use of a SEM; however, decontamination plasma devices especially designed for SEM are being developed. Indeed, apart from the quality of the vacuum, it is possible to check the absence or not of the contamination in real time by means of the Auger spectrum of the irradiated zone. When a contamination is observable, it is possible to eliminate it by steaming and/or ion erosion. In these conditions, it will then be possible to more easily study the relation between the nature of the insulator and its electrical behavior. This relation is conditioned by the regulation mechanisms which are constituted by the secondary electron emission and the charge transport phenomena. These two mechanisms, which are interdependent because of the electric field generated by the trapped charge, must be dissociated. The simultaneous measurement of the influence current and the leakage current allows not only the evolution of the trapped charge during irradiation to be obtained, but also that of the electron emission yield (see next section). The study of the physical parameter effects of the insulator (defects, structure, dimensions, etc.) on electron emission and charge transport can then be undertaken independently.

7.4. Applications: measurement of the trapped charge or the surface potential

7.4.1. *Introduction*

The electron irradiation within an SEM constitutes a privileged tool for injecting an electric charge into an insulator because it permits on the one hand the circumvention of the disadvantages related to the use of an electrode on the contact of the surface sample (the influence of the nature of the electrode and the additional defects induced at the insulator/metal interface) and, on the other hand, the good spatial resolution of an SEM allows the electron beam to be focused on a small zone and consequently the phenomena which develop in this zone to be explored. Further, on-board experiments to understand the origin of the problems of charging and discharging suffered by the dielectric materials which cover satellites (as a result of their irradiation in space) remain expensive and rare in practice, and electron irradiation in an SEM remains an interesting alternative to best reproduce the space environment in a laboratory.

The reproduction in an SEM chamber of the environment, dimensions and conditions to which an insulator will be constrained in its application medium, also permits us to see which intrinsic parameters of the insulator (composition, structure, dimensions) we must act on during its development so that it can adjust to a desired application.

Diverse methods allowing the evaluation of the ability of an insulator to charge have been developed and implemented within an SEM. They are based on the measurement of the trapped charge in the insulator and its evolution during irradiation. A considerable amount of work has been undertaken to extract information from these measurements concerning the structural, chemical and electrical characteristics of insulators [BRU 83], [VAL 99], [ONG 98], [SON 96], [BEL 00a], [BIG 97], as well as the mechanisms of associated trapping and detrapping.

In this study, we shall not only describe methods using an electron beam or the signals they generate. These methods allow the measurement of the trapped charge (or the resulting surface potential) and its evolution over time. One of these techniques is based on surface potential measurement at the equilibrium state, either by X-ray spectroscopy, or by electron spectroscopy [BEL 01a], [JBA 01]. The second uses the mirror effect. The third is based on the measurement of the dynamic distortion of electron images induced bytrapped charges under electron irradiation [BEL 00a]. The last, which allows the measurement of trapped charge evolution, not only during irradiation but also after irradiation stops, is based on the measurement of the influence current.

Techniques allowing access to the spatial distribution of internal charges in an insulator, notably the thermal and accoustic methods (described by [LAU 99] and references included) are not studied here. Although it is not impossible to adjust them to the chamber of an SEM, these methods, however, present the disadvantage of metallization of the sample surface (electrode measurement). For a more rigorous approach notably to the charging conditions of the dielectric covering of spatial instruments in orbit, it would be useful to develop a measurement system without contact (an idea which worries some laboratories of SEEDS (Système d'Energie Electrique dans leur Dimension Sociétale) dielectrics).

7.4.2. *Static methods*

These methods are described as static because the measurement time is relatively long compared with the time constants of the observed phenomena. Nevertheless, in certain cases they can be considered as dynamic.

7.4.2.1. *Mirror method*

The method which uses the mirror effect, the "mirror method" [LEG 91], [VAL 95], [VAL 99] operates in two stages, injection followed by measurement. The first stage (injection) consists of injecting primary electrons through a focused or defocused electron beam, under accelerating voltage generally of the order of 10 to 30 kV and with variable injection times (from a few ms). During this injection phase, it is possible to dynamically measure the influence and conduction currents (see section 4.2.2). During the second stage (measurement), the previously irradiated zone is scanned by an electron beam whose accelerating voltage is reduced to values generally ranging from 100 V to a few kV. If the charges resulting from the injection are sufficiently trapped locally to the injection point, the primary electrons are submitted to a total deflection in vacuum before reaching the surface of the sample. They then strike the walls of the microscope chamber. The electron image of this latter (mirror image) is substituted for that of the sample surface. The dimension measurement of some remarkable elements of the chamber on the mirror image (particularly the image diameter, d, of the output diaphragm of the column) allows, using the linear part of the curve 1/d=f(V), the quantity of the locally trapped charges Q_p to be determined. If this charge can be considered punctual (i.e. at a point in space) and the material isotropic, the curve is rigorously linear and obeys the law:

$$\frac{1}{d} = K\,\frac{2\pi\varepsilon_0\,(\varepsilon_r+1)}{Q_p}V \qquad [7.3]$$

where K is a constant depending on the geometry of the SEM chamber, ε_r is the dielectric constant of the material and ε_0 is the vacuum dielectric constant.

For more complex charge distributions (lateral spreading, in depth homoidal distribution, etc.) or for an anisotropic material, there are more or less complex modeling of the curve 1/d=f(V) [BIG 01], [TEM 06].

It is important to note that the mirror method must necessarily be associated with the measurement of the influence and conduction currents (see section 7.4.2.2) so as to know the totality of the charges Q_t trapped in the whole sample during the injection, and possibly those which are relaxed after the injection.

7.4.2.2. *Spectroscopic methods*

7.4.2.2.1. X-ray spectroscopy

This method is based on the maximal energy measurement of the photons in the continuous background of the emitted X-ray spectrum (Bremsstrahlung: see section 7.2.2.2) by the insulator. This energy, called the Duane–Hunt limit (DH), is situated at E_0+qVs with $Vs < 0$, not at E_0. The surface potential, V_S, is deduced from the displacement, towards low energies, of the limit at high DH energies of the continuous background, due to the slowing down of primary electrons caused by the electric field induced in vacuum by the trapped charge [BEL 01a]. Knowing the accelerating voltage, V_{ac}, and the upper limit of the background, E^X_{sup}, (effective energy of incident electrons), the surface potential is given by:

$$|V_S| = V_{ac} - \frac{E^X_{sup}}{q} \qquad [7.4]$$

where q is the elementary charge.

One of the disadvantages of this method lies in the fact that a spectrum of generally acquired X-ray photons is distorted (especially when the surface potential is high) because it results from the super-imposition of two spectra: that created by a primary beam and that created by electrons from the polar pieces (a parasite source) irradiating the immediate neighborhood of the sample. Consequently, the position of the DH limit is greater than q (V_{ac} -V_S) and the surface potential is underestimated [BEL 01a].

7.4.2.2.2. Electron spectroscopy

Within an SEM, it is also possible to measure the surface potential with the aid of the energetic distribution of electrons emitted by an insulator (electron spectroscopy). Indeed, the use of a new spectrometer developed by Rau [RAU 96] allows this measurement; it is a toroidal electron spectrometer (see Figure 7.10a) especially fitted for an SEM [JBA 01]. The value of the surface potential is deduced from the displacement measurement in energy of the distribution maximum of the secondary electrons [JBA 01] as shown in Figure 7.10b, giving the electron spectrum of an irradiated alumina for a 16 kV acceleration voltage. The surface potential is estimated at 11.5 kV in this case.

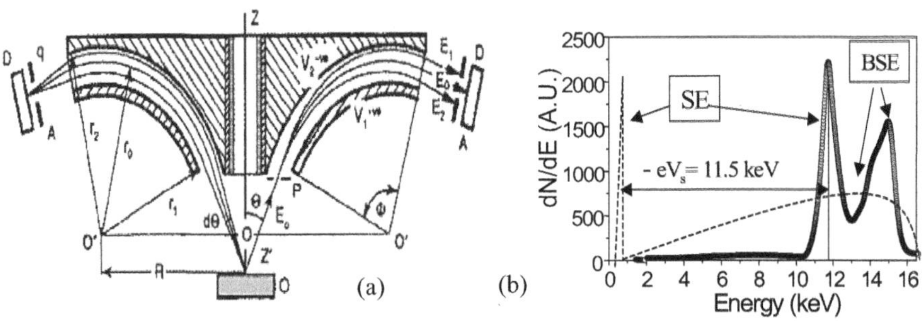

Figure 7.10. (a) *Section of the toroidal electron spectrometer [RAU 96] and (b) distribution of secondary and backscattered electrons [BEL 01b]*

7.4.3. *Dynamical methods*

7.4.3.1. *Method based on the deflection of the primary beam*

This method, which requires a sample with regular geometry (a disk, sphere, etc.) is based on the calculation of electron trajectories by using a uniform charge distribution. The determination of the trapped charge quantity uses a comparison procedure of the experimental electron image of the irradiated zone, with simulated electron images (or, more simply, a comparison of the experimental deflection with the simulated deflection) by adopting as an adjustable parameter the value of the trapped charge [BEL 01b].

The comparison procedure is based on the search for a maximum resemblance between images compared by using the least squares method [BEL 00a]. Unlike other methods developed (the mirror method, for example), the advantage of the primary beam deflection method is its ability to study the kinetics of charge trapping. An example of time evolution of a trapped charge (related to the surface density of trapped electrons) in alumina under electron irradiation for different acceleration voltages is given in Figure 7.11.

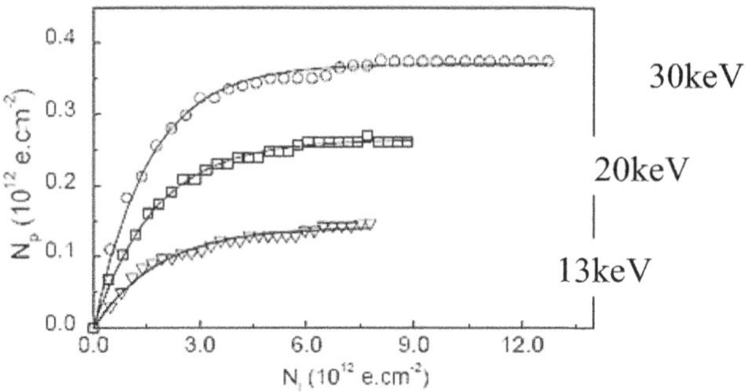

Figure 7.11. *Variation of the surface density of trapped electrons as a function of the surface density of injected electrons. The used primary currents are (o): I_0 = 9.4 nA ;(□): I_0 =3.5 nA and (∇):I_0 = 3 nA [BEL 01b]*

7.4.3.2. *Method based on electrostatic influence*

7.4.3.2.1. Principle

If we place an electric charge Q near a bonded conductor, a charge Q' proportional to Q (Q'= KQ) with an opposite sign develops at the surface of this conductor. During the establishment of the image charge Q', the electrons travel from the conductor to ground if this charge is negative and in the opposite direction if the charge is positive (Figure 7.12). The variation of the charge over time then induces a current I_{inf}, proportional to this variation dQ/dt.

$$I_{inf} = dQ'/dt = K\, dQ/dt \qquad [7.5]$$

This measurable current between the conductor and ground is commonly called an electrostatic influence current and the proportionality coefficient K is the electrostatic influence factor. In the context of a scanning electron microscope, the charge Q represents the charge trapped by the insulator under electron irradiation.

7.4.3.2.2. Experimental device

The sample is placed in a metal cage (A), earthed, and provided with a hole on its upper side. The diameter of this hole can be adjusted according to the size of the samples. A metallic disk (C), earthed through a picoammeter and electrically insulated from the cage, is placed under the sample but without electrical contact with it. This avoids a current due to the electrons emitted by the chamber from being collected in the disk (C) and consequently being superposed on the influence current. The conduction current is measured by means of the electrode (B). This electrode, provided with a square opening (photo frame-shaped), is closely connected to the edge of the lower side of the sample. The primary current is measured with a Faraday cup, placed near the sample on the metal cage, linked through an electrometer. This method was applied in the case of polycrystalline alumina whose typical shape of the influence current measured during and after the irradiation is represented in Figure 7.12. The variation over time of the image charge obtained by integration of the curves $I_d(t)$ is represented in the insert of this figure.

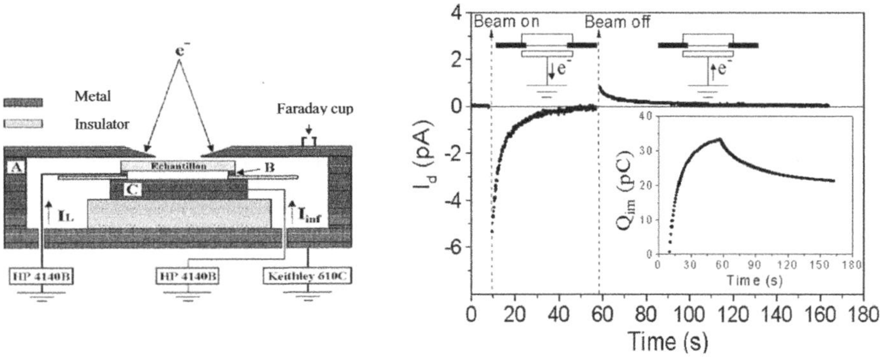

Figure 7.12. *Left: section of the experimental device used to measure the influence and conduction currents. Right: influence current during the charge and the discharge of an insulator, the image charge is represented in the insert [BEL 01b]*

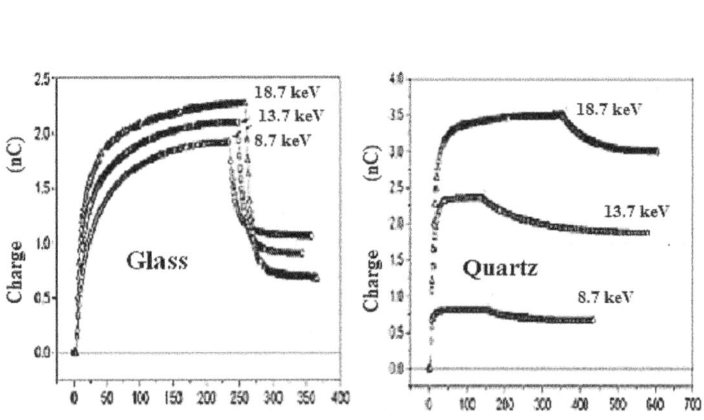

Figure 7.13. *Variation of the trapped charge as a function of time during and after the irradiation under an accelerating voltage ranging from 8.7 to 18.7 kV and a primary current fixed at 2 nA [FAK 03], [FAK 04]*

To determine the time evolution of the trapped charge, it is fundamental to calculate or to measure the influence coefficient K. In this latter case, K is estimated, either by using the method based on deflection of the primary beam, or by using the toroidal spectrometer, or even by using the influence and conduction currents measured during the discharge period (from the primary beam) [FAK 03], [FAK 04]. Figure 7.13 represents the time evolution of the trapped charge in the case of two materials (glass and quartz) having very different behaviors under irradiation.

This method can also be applied to study the charge kinetics of the insulators covered with a metallic layer under electron irradiation [JBA 02].

7.5. Conclusion

The electron irradiation of insulating materials can essentially lead to two types of effects: a charging effect and the chemical modification of the sample. These two phenomena often coexist and can interact with one another equally. The charging effect is a consequence of the electric field created by the space charge resulting from the trapping of incident electrons by the preexisting defects, and/or induced in the irradiated insulating material. The chemical modification results from the migration of mobile ions or the stimulated desorption of species, and probably implies the Auger mechanism (among other mechanisms) on the one hand and certainly the induced electric field on the other.

Both effects often depend on the crystalline structure of the sample, its geometry, its environment and irradiation conditions. Consequently, it is difficult to predict and to quantify their amplitude. Thus, the aim of this chapter was to contribute in some way to the physical understanding of these phenomena and to underline their importance on a fundamental and from a practical point of view. This chapter also sheds light on the issues associated with the irradiation of an insulator to, on the one hand remedy the latter at least partially by limiting the frequent errors of interpretation during the implementation of the analysis methods and, on the other hand benefit from it by developing techniques of electrical characterization (measurement of the charge, the time constants of charge or discharge, the surface potential and the electric field, etc.) within an SEM. These techniques are based on the mirror effect, the measurement of the Duane–Hunt limit and the measurement of both the influence and the conduction currents. A few succinct but fundamental notions concerning the secondary emission of charged insulators were also given and the importance of this phenomenon was underlined.

7.6. Bibliography

[AKK 94] AKKERMAN A., BOUTBOUL T., BRESKIN A., CHECK R., GIBREEKHTMAN A., "Low energy electron transport in alkali halides", *J. Appl. Phys.* 76, p. 4656–4662, 1994.

[ALI 78] ALIG R.C., BLOOM S., "Secondary electron escape probabilities", *J. Appl. Phys.* 49, p. 3476–3480, 1978.

[BEL 00a] BELHAJ M., ODOF S., MSELLAK K., JBARA O., "Time dependent measurement of trapped charge in electron irradiated insulators: Application of Al2O3-sapphire", *J. Appl. Phys.,* 88, p. 2289–2294, 2000.

[BEL 00b] BELHAJ M., JBARA O., ODOF S., MSELLAK K., RAU E.I., ANDRIANOV M.V., "An anomalous contrast in scanning electron microscopy of insulators: The pseudo mirror effect" *Scanning* 22, p. 352–356, 2000.

[BEL 01a] BELHAJ M., JBARA O., FILIPPOV M.N., RAU E.I., ANDRIANOV M.V., "Analysis of two methods of measurement of surface potential of insulators in SEM: Electron spectroscopy and X-ray spectroscopy methods", *Appl. Surf. Sci.*, 177, p. 58–65, 2001.

[BEL 01b] BELHAJ M., Contribution à l'étude des effets de charge dans les isolants soumis à une irradiation électronique, Thesis, University of Reims Champagne-Ardennes, 2001.

[BEN 97] BENNEWITZ R., REICHLING M., MATTHIAS E, "Force microscopy of cleaved and electron-irradiated CaF2 (111) surfaces in ultra high vacuum", *Surf. Interf*, p. 387, p 69–77, 1997.

[BIG 97] BIGARRÉ J., FAYEULE S., PAULHE O., TRÉHEUX D., "Characterisation of the trapping of charges in polystyrene", *IEEE Tran. Dielectr Electr Insul Annual Report-Conference on Electrical Insulation and Dielectric Phenomena*, p. 101–104, 1997.

[BIG 01] BIGARRÉ J., ATTARD C., HOURQUEBIE P., MATTALANA J., "SEM-mirror methods and application to insulating characterization", *IEEE Trans Dielectric and Electrical Insulation*, 8, p. 942–952, 2001.

[BLA 91] BLAISE G., LE GRESSUS C., "Charging and Flashover induced by surface polarization relaxation process", *J. Appl. Phys.* 69, p. 6334–6339, 1991.

[BLA 01] BLAISE G., "Charge localization and transport in disordered dielectric materials" *Journal of Electrostatics*, 50, p. 69–89, 2001.

[BRU 54] BRUINING H., in *Physics and applications of secondary Electron Emission*, McGraw-Hill, New York, 1954.

[BRU 83] BRUNNER M., MENZEL E,. "Surface potential measurements on floating targets with a parallel beam technique", *J Vac Sci Technol B1*, p. 1344–1347, 1983.

[CAZ 90] CAZAUX J, KIM K.H., JBARA O., SALACE G., "Charging effects of MgO under electron bombardment and nonohmic behavior of the induced specimen current", *J.Appl.Phys*, 70, p. 960–965, 1990.

[CAZ 91] CAZAUX J., LE GRESSUS C, "Phenomena relating to charge in insulators macroscopic effects and microscopic causes", *Scanning Microscopy*, 5, p. 17–26, 1991.

[CAZ 93] CAZAUX J, in "Ionisation of solids by heavy Particles", In *NATO ASI Series B 306 Physics*, BARAGIOLA, R.A. (ed.), Plenum, New York, 1993.

[CAZ 95a] CAZAUX J., "Correlations between ionisation radiation damage and charging effects in transmission electron microscope", *Ultramicroscopy*, 60, p. 411–425, 1995.

[CAZ 95b] CAZAUX J., "The role of the Auger mechanism in the radiation damage of insulators", *Microsc. Microanal. Microstruct.* 6, p 345–352, 1995.

[CAZ 96] CAZAUX J., "Electron probe microanalysis of insulating materials", *X-Ray Spectrometry*, 25, p. 265–280, 1996.

[CAZ 99a] CAZAUX J., "Mechanisms of charging in electron spectroscopy", *J. Elect. Spect. Rel. Phen*, 105, p.155–185, 1999.

[CAZ 99b] CAZAUX, J., "Some considerations on electron emission from e-irradiated insulators", *J. Appl. Phys.*, 85, p.1137–1147, 1999.

[CAZ 01] CAZAUX J., "About the secondary electron yield and the sign of charging of electron irradiated insulators", *European Physics Journal, AP*, 15, p. 167–172, 2001.

[DIO 75] DIONNE G.-F., "Origin of secondary electron emission yield curve parameters", *Journal of Applied Physics*, 46, p. 3347–3351, 1975.

[EDW 82] EDWARDS A. H., FOWLEr W. B., "Theory of the peroxy-radical defect in a-SiO_2" *Phys. Rev. Bulletin.* 26, p. 6649–6660, 1982.

[FAK 03] FAKHFAKH S., JbARA O., BELHAJ M., FAKHFAKH Z., KALLEL A., RAU E., "An experimental approach for dynamic investigation of trapping properties of glass-ceramic under electron beam irradiation from a scanning electron microscope", *European Journal Applied Physics*, 21, p. 137–146, 2003.

[FAK 04] FAKHFAKH S., Etude des phénomènes de charge des matériaux diélectriques soumis à une irradiation électronique permanente, Doctoral Thesis, University of Reims Champagne-Ardennes, 2004.

[FEI 74]. FEIGL F. J,. FOWLER W. B., YIP K.-L., "Oxygen vacancy model for the E_1' center in SiO_2", *Solid State Commun*, 14, p. 225–229, 1974.

[FIT 74] FITTING H.J., "Transmission, energy distribution, and SE excitation of fast electrons in thin solid films", *Phy. Stat. Sol.* (a) 26, p. 525–535, 1974.

[GED 99] GEDEON O, JUREK K, HULINSKY V, "Fast migration of alkali ions in glass irradiated by electons", *Journal Non-Crystalline Solids*, 246, p. 1–8, 1999.

[GRI 78] GRISCOM D. L., "Defects in amorphous insulators", *J.Non-Cryst. Solids*. 31, p. 241–266, 1978.

[GRI 80]. GRISCOM D. L, E' "Center in glassy SiO_2: ^{17}O,^{1}H, and "very weak" ^{29}Si superhyperfine structure", *Phys. Rev. Bulletin* 22, p. 4192–4202, 1980.

[GRI 81] GRISCOM D. L., FRIEBELE E. J., "Fundamental defect centers in glass: ^{29}Si hyperfine structure of the nonbridging oxygen hole center and the peroxy radical in a-SiO_2", *Phys. Rev. Bulletin*, 24. p. 4896–4898, 1981.

[HAC 59] HACHENBERG O., BRAUER W., "Secondary electron emission from solids", in *Advances in Electronics and Electron Physics,* MARTON, L. (ed.), vol. 11, p. 413, Academic Press, New York, 1959.

[JAR 88] JARDIN C., "Manifestation de la polarisation superficielle d'un isolant en spectroscopie d'électrons Auger et de pertes d'énergie", *Le vide, les couches minces*, Société Française du Vide, No. spécial, p. 95–99, 1988.

[JAR 95] JARDIN C., DURUPT P., RIVIERE C., ROBIN M., LE GRESSUS C., "Simultaneous measurements of the surface potential and the cathodo-luminescence of Al_2O_3 samples" *Le vide, science, technique et applications*, supplément n. 275, p. 557–561, 1995.

[JBA 95] JBARA O., CAZAUX J., TREBBIA P., "Sodium diffusion in glasses during electron irradiation", *J. Appl. Phys.*, 78, p. 868–875, 1995.

[JBA 96a] JBARA O., CAZAUX J., REMOND G., GILLES C., "Halogen ion electric field assisted diffusion in fluorite and polyvinyl chloride during electron irradiation", *J Appl Phys* 79, p. 2309–2313, 1996.

[JBA 96b] JBARA O., PORTRON B., MOUZE D., CAZAUX J., "Electron probe microanalysis of insulating materials: Monte Carlo simulations", *X-ray Spectrometry*, 26, p. 291–302, 1997.

[JBA 01] JBARA O., BELHAJ M., ODOF S., MSELLAK K., RAU E.I., ANDRIANOV M.V., "Surface potential measurements of electron irradiated insulators using backscattered and secondary electron spectra from an electrostatic toroidal spectrometer adapted for scanning electron microscope applications", *Rev. Sci. Instrum.*, 72, p.1788–1795, 2001.

[JBA 02] JBARA O., FAKHFAKH S., BELHAJ M., CAZAUX J., RAU E.I., FILIPPOV M.V., ANDRIANOV M., "A new experimental approach for characterizing the internal trapped charge in ground coated insulators during their e irradiation" *Nucl. Instrum. Methods. Phys. Res., B* 194, p.302–10, 2002.

[JBA 04] JBARA O., FAKHFAKH S., BELHAJ M., RONDOT S., "Charge Implantation Measurement on Electron – Irradiated Insulating Materials by Means of a SEM Technique", *Nucl Microsc. Microanal* 10, p. 697–710, 2004.

[JES 92] *Numéro spécial de J. Electr. Spectrs.* Rel. Phen. 59, 1992.

[JON 00] JONNARD P., BONNELLE C., BLAISE G., REMOND G., ROQUES-CARMES C., "F^+ and F centers in a-Al_2O_3 by electron-induced X-ray emission spectroscopy and cathodoluminescence", *J. Appl. Phys*, 88, p. 6413–6417, 2000.

[JOY 95] JOY D.C., "A Database on Electron-Solid Interactions", *Scanning* vol.17, p. 270–275, 1995.

[KHA 82] KHAIRI I., GRAIS L., BASTAWROS A.M., "A study of secondary electron emission in insulators and semiconductors", *J. Appl. Phys.*, p. 5239–5242, 1982.

[LAU 99] LAURENT C., "Diélectriques solides et charge d'espace," *Tech. Ing. Traité analyse et caractérisation*, D3, D 2305-1-D 2305-13, 1999.

[LEG 91] LE GRESSUS C., VALIN F., HENRIOT H., GAUTIER M., DURAND J.P., SUDARSHAN R.G., BOMMAKANTI R.G., BLAISE G., "Flashover in wide-band-gap high-purity insulators: Methodology and mechanisms", *J. Appl. Phys*, 69, p.6325–6332, 1991.

[MOR 76] MORIN P., PITAVAL M., VICARIO E., "Direct observation of insulators with a SEM", *J. Physics E*, 9, p. 1017, 1976.

[MOY 93] MOYA-SCIESSE D., SAMI A., MOYA G., "Characterization of charge trapping/detrapping phenomena by calorimetric measurement in charged dielectrics" in *Proceeding CEIDP; IEEE Annual Report*, p.104–109 & p.146–149, 1993.

[NIE 82] NIEDRIG H., "Electron backscattering from thin film", *J. Appl. Phys.*, 53, p .4, 1982.

[ONG 98] ONG C.K., SONG Z. G., OH K. H., GONG H., LE GRESSUS C., "Variation of the relative permittivity of charged dielectrics", *Appl. Phys. Let.* 72, p.317–319, 1998.

[PEP 98] PÉPIN M.P., WINTLE J.H., "Charge injection and conduction on the surface of insulators", *J. Appl. Phys.*, 83, p. 5870–5879, 1998.

[RAU 96] RAU E.I., ROBINSON V.N.E., "An annular toroidal electron energy analyser for use in scanning electron microscopy", *Scanning,* 18, p. 556–561, 1996.

[REI 92] REIMER L., GOLLA V., BONGELER R., KASSENS M., SCHINDLER M., SENKEL R., "Charging of bulk specimens, insulating layers and free-supporting films in scanning electron microscopy", *Optik*, 92, p.14–22, 1992.

[REM 98] REMOND G., "Importance of the specimen preparation on the reliability of quantitative x-ray analysis", *Proceedings of EMAS 98*, Barcelona, LIOVET, X.C, MERLET, C. SALVAT, F. (eds), p. 249–278, 1998.

[SAT 91] SATO F., GOTO K., CHIKAWA J.I., "Solid-Phase Epitaxy with X-ray irradiation to grow dislocation-free silicon films at low temperatures", *Jap. J. Appl. Phys*, 30, p. 205–208, 1991.

[SCH 90] SCHWARZ S.A., "Application of semi-empirical sputtering model to secondary electron emission", *J. Appl. Phys.* 68, p. 2382–2391, 1990.

[SEI 67] SEILER H., "Some problems of secondary electron emission", *Angew Z Phys.* 22, p. 249–263, 1967.

[SEI 83] SEILER H., "Secondary electron emission in scanning electron microscope", *J. Appl. Phys.* 54, p. R11–R18, 1983.

[SEI 84] SEILER H., "Electron Beam Interaction with Solids" in *SEM,* KYSER, D.F., NIEDRIG, H., NEWBURY, D.E. and SHIMIZU, R. (eds), p. 33, Chicago, 1984.

[SHI 97] SHIH A. J., YATER J., HOR C., ABRAMS R., "Secondary electron emission", *Appl. Surf. Sci.*, 111, p. 251–258, 1997.

[SON 96] SONG Z.G., ONG C.K., GONG H., "A time-resolved current method for the investigation of the charging ability of insulators under electron beam irradiation", *J. Appl. Phys.,* 79, p.7123–7128, 1996.

[STA 79] STAPELBROEK M., GRISCOM D.L., FRIEBELE E.J. AND SIGEL, G.H. "Oxygen-associated trapped-hole centers in high-purity fused silicas" J.Non-Cryst. Solids. 32, p. 313–326, 1979.

[SZY 92] SZYIMONSKI M., PORADZISZ A., CZUBA P., KOLODZIEJ J., Piatkowski P., Fine J., Tanovic L., Tanovic N., "Electron stimulated desorption of neutral species from (100) KCl surfaces", *Surface Science,* 260, p. 295–330, 1992.

[TEM 06] TEMGA T., JUVÉ D., TRÉHEUX D., GUERRET-PIÉCOURT C., JARDIN C., "Conduction and trapping of electric charges in an anisotropic material after irradiation with an electron beam: application to TiO_2 single-crystal" *Nucl. Instr. and Meth. in Phys. Res B,* 245, p. 519–527, 2006.

[VAL 94] VALLAYER B., OH K.H., TRÉHEUX D., "Charging of MgO", *Proceedings CEIDP; IEEE annual report,* p. 161–166, 1994.

[VAL 95] VALLAYER B., "Développement d'une méthode de caractérisation des matériaux isolants: la méthode du miroir. Application à l'étude des propriétés de charge d'oxydes", Thesis from the Ecole Centrale, Lyon, 1995.

[VAL 99] VALAYER B., BLAISE G., TRÉHEUX D., "Space charge measurement in dielectric material after irradiation with a 30 kV electron beam: Application to single-crystals oxide trapping properties", *Rev. Sci. Instrum* 70, p. 3102–3112, 1999.

[VID 95] *Le vide, science, technique et applications,* Supplément to No. 275, 2th International Conference on space charges in solid dielectrics, Antibes, 1995.

[VID 01] *Le vide, science, technique et applications,* numéro special, 4th International conference on space charges in non conductive materials, Tours, 2001.

[VIG 85] VIGOUROUX J. P., DURAD J. P., LE MOEL A., LE GRESSUS C., GRISCOM D. L., "Electron trapping in amorphous SiO_2 studied by charge build up under electron bombardment", *J. Appl. Phys.* 57, p. 5139–5144, 1985.

[WEE 56] WEEKS R. A., "Paramagnetic resonance of latice defects in irradiated quartz", *J. Appl. Phys.* 27, p. 1376–1382, 1956.

[WHE 64] WHETTEN N.R, "Cleavage in high vacuums of alkali halide single crystals-secondary electron emission", *J. Appl. Phys.* 35, p. 3279–3282, 1964.

[WIL 73] WILLIS R. F., SKINNER D.K, "Secondary electron emission yield behaviour of polymers" *Solid State Commun.* 13, p. 685–688, 1973.

[WIN 97] WINTLE H.J., "Surface conduction on insulators: Analysis and interpretation of Faraday cage experiment", *J. Appl. Phys.* 81, p. 2682–2685, 1997.

Chapter 8

Precursory Phenomena and Dielectric Breakdown of Solids

8.1. Introduction

The physical stresses to which dielectric materials are subjected do not, *a priori,* contribute to the early ageing of these materials or to their failure, and it is the fundamental reason for the undertaking of a test which guides the choice of test made. Experiments show that the joint action of physical or physico-chemical parameters leads to an unexpected evolution and breakdown of dielectrics used as electrical insulating materials (as we shall subsequently call them).

The evolution of the materials is related to the growth of stresses, such as temperature (which can have a positive as well as negative effect), high voltage, electrical field, environment (whether on earth or in space, where humidity and pollution in a broad sense can play a part), and radiation across the whole range of frequencies. New terms such as very high voltage and reliability appear and, because of these requirements, the use of natural materials such as wood, natural rubber, paper, etc, is significantly reduced.

Please note, however, that the use of paper-oil (PO) remains, particularly in the insulation of direct current high voltage cables. Due to its physical and chemical properties and thermal stability up to 700°C, muscovite mica is always used for high voltage systems and rotating machines.

Chapter written by Christian MAYOUX, Nadine LAHOUD, Laurent BOUDOU and Juan MARTINEZ-VEGA.

Obviously, chemistry has always played a large part in this evolution and as an example, cables (very well described by Arrighi [ARR 86] in the introduction to the Jicable conference in 1984). Indeed, this chapter notes the exponential growth over the years of the extent of high voltage cables insulated with chemically cross-linked polyethylene (XLPE); this polymer has the advantage of keeping up sufficient mechanical resistance at operating temperatures of 90°C to 100°C in exceptional circumstances, and 250°C in short-circuit conditions.

However, several causes of failure can appear: the manufacturing of the basic material as well as interfaces, accidental working conditions such as overvoltage, reverse voltage and the presence of inorganic impurities or water molecules [DEN 94]. Tests and standards can help avoid problems, and research can aid understanding and help in overcoming them. It should be stressed, however, that damage prevention in a closed system represents a major difficulty for the user. Readers could usefully consider Densley's [DEN 01] work on cables, an approach which we believe is applicable to other systems.

In the following sections, particular attention will be paid to organic solid materials, considering the important role they play in the wide world of electrical insulation: for the production of nuclear energy, energy transport, rotating machines, electronics and in space (where we include aeronautics and satellites).

8.2. Electrical breakdown

The fundamental aspects of electrical breakdown have been dealt with elsewhere so here we shall here only consider the final aspect: the destruction of the insulating material. However, first we will mention important syntheses published on the subject: very early on by Von Hippel [HIP 37], whose originality was to compare the breakdown of the three states of matter, then those by Whitehead [WHI 53], Straton [STR 61] and O'Dwyer [O'DW 79], whilst a particular interest was shown in alkaline halogens by Cooper [COO 63], [COO 66]; organic materials and their industrial applications have been studied by Mason [MAS 59], Fava [FAV 77] and Ieda [IED 80]; the different aspects of the dielectric breakdown of solids are detailed in more recent books, such as those of Coelho [COE 79], Nelson [NEL 83], and Dissado and Fothergill [DIS 92].

The dielectric breakdown of a system is an established fact which reveals either the simple puncture of the material or a perforation in the carbonization of all (or part) of the material, and sometimes damage to the conductive part which it was meant to be insulating. Carbonization is related to the current supplied by the source, as long as the incident did not send it off-circuit; obviously the flammable character of the decomposition products of some organic materials can contribute to the final degradation. At the microscopic scale, a perforation without carbonization can affect a weak part of the material if, locally, the melt temperature is reached. The phenomenon is particularly seen in both low density polyethylene (LDPE) and high density (HDPE) polyethylene films, chemically reticulated polyethylene (XLPE) or polypropylene (PP) (which is itself meant for the manufacture of capacitors). A simulation can be realized in a laboratory for bulky materials as well as thin films. The degradation of the material, however much hidden, locally removes or at least diminishes the dielectric quality in volume, and consequently could diminish the lifetime of the insulation.

As far as experiments are concerned, the notion of *intrinsic breakdown or strength* remains academic, as Blythe very rightly underlines [BLY 79] in his introduction on polymer breakdown, despite certain conceptual approaches which cannot ignore the insulating material, strictly speaking, and its particular utilization [CUD 87].

The maximum electrical voltage an insulating material can withstand before getting perforated defines its *dielectric strength*, which is expressed (under the International System of Units (SI)) in MV/m, kV/mm, or V/µm, depending on the type of insulating material or dielectric concerned, i.e. depending on whether we are dealing with bulky materials (cables) or thin layer (films, semi-conductors etc.). Indeed, the breakdown voltage V_b, of a dielectric depends on its thickness e, and the resulting breakdown field F_b is the ratio V_b / e. Consequently, the growth of V_b should in principle follow that of e, (remembering that, for a cylindrical structure such as a cable, the relationship becomes $F_b = V_b / \ln(r_0 / r_1)$, ($r_0$ and r_1 being respectively the external and internal radii). But what contradicts this simple relationship? Dielectric breakdown tests have a long history, as pointed out by Bartnikas in Chapter 3 of his book [BAR 87], such that they take the form of a rite if we want to obtain a correct value of the rigidity of a dielectric. Depending on the application, the tests can be undertaken under direct or alternating voltage for different frequencies, since the growth rates of the test voltage can be variable [MAS 73]. The specimen thickness is to be specified when it is related to the electric field value. An example is given in Table 8.1 for vinyl acetate (EVA), an elastomer used in the insulation of photovoltaic cells [CUD 87].

e (mm)	F_b (kV / mm)
0.12	98.3
0.15	85.5
0.40	44.2

Table 8.1. *Values of the electrical field as a function of the thickness*

Cables (kV)	Field (kV / mm)
$60 < V < 110$	$4 < E < 6$
$120 < V < 245$	$8 < E < 10$
$V > 300$	$12 < E < 15$

Table 8.2. *Stress values for cables*

In agreement with one of our introductory remarks, tests have shown that this characteristic was: $200 > F_b > 260$ kV/mm for muscovite mica and $140 > F_b > 200$ kV/mm for phlogopite, the former being recognized as the better of the two materials [HEP 00]. For comparison only, Table 8.2 gives the stress values for high voltage (HV) and very high voltage (VHT) cables.

8.3. Precursory phenomena

Because the breakdown of a dielectric is produced in a very short time (in less than a second), the events which precede it particularly deserve our interest.

8.3.1. *Definition*

Following the Latin adjective præcursor (forerunner), the term precursory phenomena must be applied to all phenomena which we can detect before an irreparable breakdown. When dealing with a real system of solid insulation, two situations are presented: either the phenomenon appears at the surface and is therefore optically or electrically observable, or it is created inside the volume and only detectable by electronic detection. These are, for example, the respective cases of an insulating material, a cable or a motor.

Another approach has been proposed for cables used in the instrumentation and control of nuclear power stations. The insulating materials concerned are vinyl polychlorure (PVC), XLPE and ethylene copolymer-vinyl acetate (EVA), for example. Samples, in the form of a test cable, were introduced inside a reactor in sufficient quantities to allow periodic removal of samples over about 40 years. The samples are tested chemically, mechanically and electrically [BAM 99a].

The problems which arose have led to pragmatic studies, which have influenced manufacturers and equipment, and have had more fundamental influence on materials sometimes far from practical or commercial interests, such as ion crystals. This has obviously led very quickly to extending the range of materials studied to those used in construction, such as polyepoxy or polyethylenes for thick materials, polyethylene terephthalate (PTFE), polypropylene (PP) or polyethylene naphthalene (PEN) for films. Consequently, the term precursor, previously defined, will fortunately be applied to observable phenomena in the laboratory and allow advances in the understanding of the degradation and breakdown processes.

We thus move from investigating the black box constituting a transformer, or a motor or cable in service, a cable insulator under pressure, or a capacitor, towards consideration of the component itself: the insulating material or dielectric. The observations are made at the microscopic level and, owing to modern analysis techniques as well as mathematical models and techniques, we can imagine the processes of ageing and breakdown of the material. The term *precursor* then takes a wider meaning, because it reveals the origin of the problem which could be a gaseous cavity, or an impurity, which can be defined, or simply a semi-crystalline structure whose distribution is random, especially given that interfaces are created between materials of a different nature (metal/insulating material being an example).

8.3.2. *Potential precursors*

8.3.2.1. *The material*

The physical, mechanical or electrical properties which we expect from an organic insulating material depend particularly on its microstructure: thus, the empty volume hypothesis for polyethylene which was proposed by Matsuoka [MAT 61]. Depending on the degree of branching (CH3/100C), their concentration was estimated between 18 x 10^{-3} cm^3/g and 23 x 10^{-3} cm^3/g. At a scale of several tens of nanometers, a lamellar structure appears due to the withdrawal of polymer chains. The lamellae are organized in microstructures and constitute spherulites. The water tree problems in buried and XLPE insulated cables thus give dimensions to voids with diameters ranging between 5 and 10 µm for the biggest [MUC 76]. Observations situate the voids at the junction of two or three spherulites. The latter, of average diameter equal to 0.1 µm, can appear bigger in LDPE than in XLPE, but depend on the presence of antioxidants which play the role of nucleant agents (see Figure 8.1).

We shall note, on the one hand, that the analysis reveals voids of weaker sizes and, on the other hand, that in the case of polymers obtained by the chemical process of crosslinking, some residue such as dicumyl or acetophenone peroxide, for example, could be found in these spaces, despite a thermal process intended for their extraction at the end of manufacture. The concentration of these volatile products is estimated at between 500 and 2,000 ppm [DEN 94].

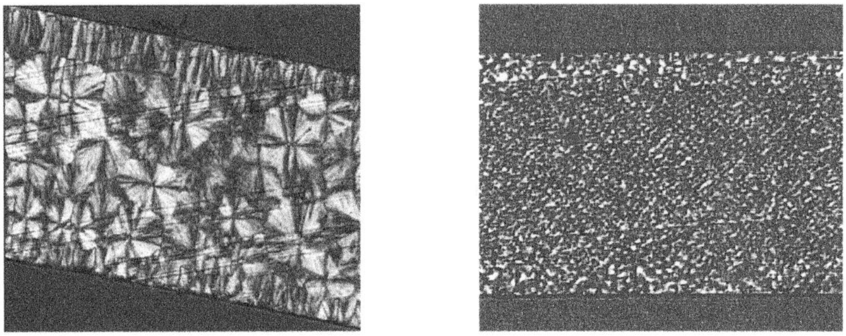

(a) (b)

Figure 8.1. *Spherulites in an LDPE (a) and an XLPE containing an antioxidant (b) [BER 98]*

Spherulites have an orderly structure and a high density compared with the space between them. If the space is dense, the field of electrical or mechanical stress is higher. The reduction of the spherulites' size therefore leads to the growth of the insulating material's breakdown field, as shown by Kolesov for PP and HDPE [KOL 80]. The presence of both amorphous and crystalline phases favors the diffusion of gas towards the amorphous phase and *a fortiori* towards empty spaces. With the exception of helium and hydrogen, gases contained in the atmosphere have neighboring diffusion constants [LAU 83]; this is another characteristic which can reveal the harmfulness.

Material is sensitive to temperatures and their variation, either during manufacture or whilst in use. Just to take XLPE as an example, this influence has been shown in laboratory models in miniature cables placed between plaque samples and the sample under investigation. Thus investigations in this area have shown annealing effects due to temperature rise, which leads to a densification of the polymer, and the formation of micro-cavities whose dimension are of the order of 100 nm. This non-homogeneous evolution obviously depends on the thermal gradient. In fact, considering the evolution of local stresses, competition takes place between quenching and annealing of the material, with an influence on the number and size of voids [KAR 89].

During manufacture, an important effort has been made to reduce the number of cavities and their size [YOD 73]. Thus from 10^6 cavities/mm^3 with diameters ranging between 1 and 5 µm in 1975 [KAG 75] we reached, about ten years later, a concentration of 10^3–10^7 cavities/mm^3 of the same size in cables insulated by XLPE [JAC 78], [BAL 84]. In his interpretation of dielectric breakdown, Artbauer drew a clear distinction between what are conventionally called cavities and holes or voids, these latter (in the amorphous phase of the polymer) having dimensions of a molecular order, therefore very much less than those of cavities which favor a specific degradation mode, as we shall analyze later [ART 96]. It is this theory of *free volume* that the author relies on to explain the electrical breakdown of a polymer.

8.3.2.2. *Impurities*

Despite the precautions taken to achieve good insulation, a post-mortem after an incident can reveal the presence of many sorts of impurities [MIN 84]. Tests in the laboratory have shown that metallic inclusions were less dangerous than mineral impurities [MOR 88]. However, the implementation of cables with more and more significant voltages (400–500 kV) has required a reduction of the use of organic materials (XLPE for this type of application), because of contamination with metallic impurities whose size can range between 10 and 100 µm, for example [BOS 03]. The site of manufacture also needs to be under close supervision; one example shows that the failure of a 275 kV cable was as a result of a joint contaminated by fibers. Laboratory tests comparing cotton, polyester, nylon and silk, have revealed the double danger of a hollow natural fiber such as cotton [KAM 92]. Indeed, this foreign body can have a hydrophilic character and most of all behave like a cavity since it is hollow. Tests carried out on polyethylene films within which silver microspheres had been moulded, have shown an important decrease of the breakdown field of the order of 37% when this type of contaminant reached a diameter of 20 µm, and 58% for 80 µm [STA 84].

When making connections and at the ends of high voltage cables, modern technology has led to the making of interfaces which can be in the presence of XLPE, a silicon rubber or polyepoxy. Generally speaking, an interface can be a weak point when next to an applied field, on the one hand because of the discontinuity of the physical properties of the insulation created (for example, a heterogenous response to thermal stress) or, on the other hand, because of the possibility of gas trapping. It is at this point in the insulation that the tangential component of the field is the most noxious, giving rise to the creation and propagation of an electrical tree. The example of a silicone–XLPE interface failure on a 150 kV cable in operation is given by Ross [ROS 99]. The diffusion of ionic impurities originated from semi-conductor screens and the ground in which certain high voltage cables are buried could be observed. Analyzes on samples extracted from non-aged cables or placed under voltage for several years, have revealed the presence of different types of ions (Al, Si, Fe, P, Mn, S, K, Fe, etc.) due to the diffusion of water coming from the ground in the case of cables not protected by a sheath [BOG 94], [CRI 87], [MAS 87], [MAS 92].

In the same way, conductive particles have been detected, no longer in the volume but on the surface of the charged resin playing the role of an insulating spacer. In general, bi-phenol A or cycloalyphatic resins are used, in which mineral charges representing about 70% of their weight are mixed. These charges are chosen from quartz, melted silicium, tri-hydrated alumina and sometimes combined mixture. These mechanical supports play a part in transport lines and circuit-breakers whose insulation is maintained by SF_6 under pressure. This type of contamination reduces the by-passing voltage by 50%, underlined by Cookson in a large, heavily referenced, review of gaseous insulation cables [COO 85]. The influence of free particles in transmission lines, sub-stations and circuit-breakers operating at 60 Hz, and insulated by sulfur hexafluoride at pressures of 300 and 500 kPa, shows the decreased conditions of the breakdown voltage related to the displacement of the impurity [CHA 89]. More recently, another well-documented analysis underlined the difficulty of removing the contamination of metallic particles in transmission lines and circuit-breakers insulated by SF_6 which can operate up to 750 kV. A continuous control is required for such systems; ecological considerations mean that, in certain countries, the use of nitrogen instead of sulfur hexafluoride should be envisaged [MOR 00]. Laboratory studies, with direct voltage, have been undertaken, during which metallic particles and insulators with different profiles were introduced in identical systems filled with N_2 or an N_2/O_2 mix. The generation of discharges and the breakdowns observed were determined by the displacement of particles [HAR 87]. A first review was proposed on the dielectric standard of gases related to the contamination by particles in 1981 [LAG 81]; a second, which considered the by-passing mechanisms along dielectrics in gases under pressure, appeared five years later [SUD 86].

As far as insulators maintaining the support of external high voltage conductors are concerned, their environment favors the same process. This process is observable from the outside, but if the erosion (by short-circuit) is at an advanced stage, we can only see the destruction. The discharge mode is dependent on the type of voltage (whether direct or alternating). In dry periods, dust constitutes sediment which, under the effect of rain, is transformed into a conductive film. The conductivity of this film is clearly modified by the formation of saline solutions in coastal regions; it is also variable as a function of the contents brought by surrounding pollution from industrial or agricultural spaces. Recent studies have shown the efficiency of cycloaliphatic resins (already mentioned), but comprising two dispersed phases. The improvement of mechanical properties is obtained by introduction of trihydrated alumina or silicium; we have noted the efficiency of a preliminary treatment of these charges with silane towards by-passing the mechanism on one hand and the preservation of the dielectric constant (~ 4) in a humid medium on the other. Damping measurements show there is no spreading of the conductive film previously seen [BEI 01]. The influence of non-soluble contaminants on external porcelain insulators gave rise to a modeling applicable to direct and alternating voltages, and in good correlation with the experiment it permits to limit the tests and maintain *in situ* control [SUN 96]. Finally, we should mention the problem encountered by certain external insulators, made of porcelain or a composite (EPDM) which are subjected to low temperatures. A more or less excessive crystallization covers the elements with a density varying from 0.3 to 0.9 g/cm^3, and the leak current doubles and by-passing probability increases. The parameters playing on this type of contamination are the wind and a saline atmosphere [FAR 95]. Despite its publication date, we could usefully refer to the synthesis of a workshop on transmission lines held in 1981 [EPR 81]. The recent use of optical fibers for the communication industry, and installed along the towers supporting high voltage lines, could be added to the previous research; indeed, on the surface of the fibers, by-passing and erosion phenomena appear (the example provided has LDPE coverings, charged with black carbon, insulating the fibers which are mounted along line supports of 161 kV subjected to sub-tropical climatic conditions [KAI 00]).

8.3.3. *Induced precursors*

So far we have not taken into account the influence of stresses undertaken in the system by the insulating material itself, which is important. Although these stresses could be considered when looking at the elements themselves, we are going to consider the evolution of the part they play during the operation of the system.

8.3.3.1. *Outgassing of the insulating material*

A number of experiments have been carried out on inorganic and organic materials, showing that the presence of a strong field could lead to their outgassing. We compared the behavior of these materials under vacuum (~ 10^{-4} Pa) and submitted them to a direct voltage of up to 150 kV. It should be remembered that the behavior of different organic materials varies; only methyl polymethacrylate (PMMA) is close to ceramic and reveals a gas emission from 120 kV. At the same time we noted a current increase of 10^{-7}–10^{-8} A preceding breakdown [AVD 67]; this behavior is in accordance with PMMA's listing, where it appears within the highest values for the breakdown field and for cohesion energy density (CED). The parameter δ^2 (vaporization energy/molecular volume) characterizes the energy required to separate a compact structure in a large number of separated molecules [SAB 76].

A gas emission was also detected by mass spectrometry when films of polyethylene fluoride, polyethylene terephtalate and polyvylidene fluoride (PVDF) were submitted to fields ranging from 40 to 100 V/µm. This gas emission (H_2, CH_3,

HF, etc.) is produced rapidly, so cannot be imputed to an increase in temperature [BIH 87]. A polarization of PVDF under fields of the order of 100 V/µm has revealed emissions (HF) during the suppression of the field at both polarities, a phenomenon which was not observed during the PET study [EBE 93]. In experiments whose goal was to study the conductivity of Nylon 66, it is interesting to note the synchronism of gas emission which stopped with the polarization application [SEA 68]. Despite the particular choice that these materials represent, experiments emphasize the influence of the electrical field and permit, for example, at least partly, the conductivity evolution of certain insulating surfaces to be justified, whatever the type of pressure surrounding them, as we have already seen. The behavior of the materials is particularly noxious to the vacuum-packed systems strength under high voltage. The avalanche of secondary electrons at the surface of the material under stress and preceding the breakdown has often been evoked to justify the adsorption of gas molecules preceding breakdown. These observations have led to the proposition of models in close agreement to the experiment [AND 80]. When the gas molecules remain in the material, we have also underlined the existence of free volumes likely to receive them; indeed, the local field, the defect size and the nature of the gas can contribute to the apparition and the preservation of the ionization process, which we shall see in the following section.

8.3.3.2. *Mechanical deformations*

The action of a unidirectional mechanical stress, an elongation for example, can generate ellipsoidal-shaped defects comparable to gaseous cavities, considering the little matter they contain. Wendorff chose polyoxymethylene to demonstrate this phenomenon; while the average dimension of defects, from 8 to 120 nm according to the direction of the stress, is independent of that stress, their concentration increasing exponentially with it [WEN 79]. In the domain which concerns us, it is the stress generated by the electrical field which causes the initiation of defects and possible fractures [BLO 69]. The joint action of charges and operating voltages leads to the deformation of the material and the bond break; and charges generate local fields getting combined to the macroscopic field in an additional or opposite manner. We consider the easiest of the deformations presented by the amorphous phase of the semi-crystalline material when dealing with a polymer. Amongst the three relaxation processes (generally named α, β, γ, in decreasing order of time), α plays a dominating role on the free volume and the creation of microcavities. A component perpendicular to the direction of the field contributes to the local dilation of the structure and therefore to an increase of the free volume [LEW 02], [JON 05]. When taking into account only the electromechanical force Maxwell applied at the interface of the conductor and the insulating material, the stress undergone by this latter is proportional to the square of the applied field ($P = \frac{1}{2} \varepsilon_0 \varepsilon E^2$). The presence of induced stresses has been shown by optical measurements on obviously transparent materials, i.e. free from mineral charges. We can mention for example polyepoxy [SAB 76], LDPE [LEW 93], XLPE [STA 55] [DAV 92] [DAV 94], and PET [MAM 04b]. We note, however that the purely mechanical effect subordinated to the action of the field and its frequency [ARB 86], depends on the nature of the insulating material, its chemical and physical structure, its crystallinity and its working temperature.

8.3.3.3. *The frequency*

The influence of the working voltage frequency of an electrical system on the ageing of insulating materials has been revealed in several commercial polymers. In dry materials, we have observed the breakdown voltage drop as well as the decrease of the lifetime, making the frequency vary between 60 Hz and 1 kHz. On the other hand, experiments show that the development of water trees present in cable insulating materials such as LDPE, XLPE and EPR increases in a significant way with the frequency and can lead to an early breakdown [CRI 98]. The limit frequency for tests on insulated cables by XLPE was estimated at 1,700 Hz [BAH 82].

8.3.3.4. *Irradiations*

In this section we consider two important areas: space and nuclear applications. In the latter, irradiation conditions are not very severe but long lasting or, conversely, contain important doses for a short time, pressurized water reactors being characteristic of both situations. The insulating materials used in cables appear on top of the polymer listing being subjected to irradiation tests [SCH 79], since the ethylene-propylene-hexadiene terpolymer 1-4 (EPR) insulating conductors, and chlorosulfonated polyethylene (HYPALON) play the role of internal and external sheath. These materials are usable until doses of 1 MGy (in the SI system, 1gray = 1 J/kg = 100 rads). Certain studies have shown that for outputs of doses less than 500 $Gy.h^{-1}$, degradation is controlled by oxygen diffusion and accelerated by temperature. We observe, for the EPR, an oxidated superficial layer and a cross-linking process in volume; this latter is unique in the case of HYPALON. We believe that the structure modification generated by irradiation and amplified by temperature leads to the loss of electrical properties for this type of material [GUE 92].

In space, in a vacuum environment, the parameters participating in the degradation of electrical insulating materials are even more numerous (electrons, protons, ions, photons, etc.) [FRE 82], [ROS 87], [LAG 90]; but the electrons and protons are of great toxicity considering their deepness of penetration, the most important one being electrons, and therefore of the creation of a space charge as a function of the stopping ability of the used materials [BAL 83], [BEE 81]. The breakdown processes concern the volume as much as the surface of the material and, in this latter case, we find again the conditions of surface discharges or by-passing which we have previously evoked [RID 82].

The term *precursor* for dielectric solids, used in embedded systems, can be applied when experiments are carried out on earth, in order to simulate at least one part of life conditions; however, it can only be used to attempt to explain the failure phenomena which have appeared in orbit, like in the case of experimental satellites such as SCATHA (Spacecraft Charging at High Altitudes) [REA 87]. More recent experiments called SPEAR (Space Power Experiments Aboard Rockets) have been carried out in order to verify the efficiency of devices capable of inhibiting the breakdown processes generated, and kept by outgassing of the materials and the effluent of the vehicle [RUS 92].

8.3.4. *Observed precursors*

Chapter 6 states the injection and the storage of charges in insulating materials under stresses, as well as the consequences. We shall examine here some behaviors of materials subjected to the electrical field stress and subordinated to induced precursors previously mentioned.

8.3.4.1. *Electroluminescence*

Although the light emission phenomenon of polymer films subjected to an alternating electrical field was observed over fifty years ago, a greater interest has been shown in this phenomenon much more recently to explain the degradation process of organic electrical insulating materials.

To avoid any discharge phenomenon on the surface of films placed under uniform stress, the tests were first carried out, under vacuum, on samples of LDPE, Nylon, PET, etc., one of the electrodes being composed of a conductor glass. This technique allowed the simultaneous recording of the applied voltage and the signal delivered by a photomultiplier detecting the light through the transparent electrode. The presence of light was imputed to the presence of carbonyle groups [HAR 67]. It was difficult to distinguish between the groups created because of the oxygen diffusion in the material and the groups inherent to the structure. In the following years, the same approach was undertaken with polyepoxy to study the dielectric breakdown phenomenon [COÏ 78]. More recently, in a synthesis article, Laurent [LAU 99] presented the radiative processes (fluorescence or phosphorescence) generated in the materials, particularly underlining the importance for ageing, of the presence of oxygen molecules to which an excitation can be transferred, giving rise to other reactions. This electroluminescence phenomenon is well associated with degradation reactions; it is an *indicator* of pre-breakdown and, even if it cannot be discovered on a real system, its study contributes to the understanding of breakdown mechanisms of solid materials. Detection thresholds differ according to whether the applied stress is direct or alternate: for the XLPE 80 kV/mm in direct voltage to 10 kV/mm in alternating voltage was achieved, and from 400 kV/mm to 70 kV/mm under the same conditions for the PET. Bamji estimates the highest detection sensibility for electroluminescence if we compare it to that of partial discharges, which is shown by the conversion of water arborescence into electrical arborescence [BAM 99b].

8.3.4.2. *Pre-breakdown currents*

The polarization of a certain number of organic insulating materials (PE, PET, etc.) in the form of films identical to those destined for the manufacture of capacitors, has often revealed current oscillations as the electrical field increases; the threshold field of the phenomenon decreasing when the testing temperature increases. This deviation from Ohm's Law could manifest itself for fields as weak as 25 V/µm applied to PET or LDPE films for example [PRE 71], [TOU 74], [GOF 82]. The warning signals of an impending breakdown can be interpreted without resorting to local defect hypotheses [JON 80]. In certain experiments, it is a rapidly increasing pre-breakdown current, of the order of 10^{-3} s before perforation, which has been recorded on several polymer films; LDPE and EVA make an exception to this behavior [MIZ 87], [HIK 90]. More recently, current measurements on LDPE plates of 1 mm thickness, submitted to direct voltages ranging from 10 to 60 kV, have revealed instabilities of transient current when the voltage was greater than 50 kV. These instabilities can be attributed to an injection of electrons at the cathode of this experimental capacitor. Simultaneous measurements of the space charge subordinated to the increase of the applied field show the coincidence of these instabilities with the period preceding the breakdown [VEL 96].

8.3.4.3. *Arborescence*

Two types of these defects can appear in insulating materials (and amongst energy transport cables in particular), with electrical arborescence perfecting the degradation achieved by water arborescence when this latter exists; statistical studies on cables found faulty after several years have demonstrated it [CHA 90]. We shall evoke here the observation which was first made on weak voltage cables (8 kV), insulated from XLPE, having aged 8 years. Physico-chemical analyzes have revealed oxidation marked in the regions where water arborescences were present; this same oxidation could be activated on samples placed in contact with high-temperature water. The outbreak of microcavities was observed, which become the seat of discharges, and lead to the breakdown by electrical arborescence [GAR 87]. This latter observation gets even more important when dealing with higher voltages and therefore operating temperatures of the order of 90°C. A second observation relative to this transition phenomenon calls into question impulsion due to lightning, as happened when 80 kV struck a 15 kV cable insulated with XLPE. The modeling of such a circumstance suggests a consequent warming-up of the water contained in water trees, the creation of an electromechanical stress due to the effect of the local field, and the creation of cavities becoming the seat of electrical discharges [BOG 98].

8.3.4.4. *Presence of electrical discharges*

In this section we will present a few representative examples of *observed precursors* taken from the numerous articles published on the subject. We also recommend reading reference books on measurements and their interpretation, in which it is noted (particularly by Timpe) that the first experiments relative to the toxic nature of gas bubbles in solids took place at the end of the 19th Century.

Timpe also records cable failures from as early as 1902 [TIM 79]. In 1936, Robinson, whose interest was focused on oil-impregnated high-voltage cables, already called into question the presence of cavities and ion bombardment of their walls [ROB 36]. In fact, ions, electrons and UV radiation appear when under the influence of alocal field, and the gas trapped in these cavities gets ionized. More recently, Bartnikas [BAR 02] records the history of what were first called *crown discharges*, and then *partial discharges*, and chiefly stresses the efforts realized over the years to detect this *precursor* of degradation and breakdown in all types of insulation. Bartnikas recalls that progress in electronics has enabled the use of detectors in variable ranges of frequency, from 30 to 400 kHz for routine measurements, and from 800 kHz to 1 GHz according to whether we are dealing with cables, rotating machines or gas insulated cables (SF_6) under pressure. The very fruitful association of the computer with analysis on current impulsions related to discharges must also be mentioned. These tracks are then correlated in the form of cavities, their number and their localization [GUL 92], [VOL 92]. Let us note finally that techniques have been developed to get a numerical analysis of the transient electromagnetic field [POM 02].

Laboratory studies have strongly influenced this progress by the approach of physical phenomena in ionized gases, as well as in degradation processes of the insulating material under the effect of discharges.

We would remark as an example that the evolution underlined at the beginning, i.e. the replacement of paper by a plastic film to constitute the dielectrics of a power capacitor, leads to an increase of the operating field until 60 V/µm, but limits the acceptable level of partial discharges to 20%. This same level is at 200% when the dielectric is constituted by a paper-oil association [HAN 92], [HAN 93]. The diagnostic test of a 15 kV cable insulated by EPR shows that the cavity of about 1 mm will generate a signal of the order of 10%, the minimum detection threshold. When the cable is the seat of arborescence, the form of impulsions, their width and their growth time are different from spherical cavities. However, the time separating the detection from the breakdown can only be defined from the amplitude of discharges, a large amplitude not being synonymous with harmfulness, nor that their outbreak is subordinated to the presence of a floating metallic particle. The experiment shows that the detection is specific to the equipment concerned and the insulating material [BOG 00]. We note that practical detections are much greater than those realizable in a laboratory, a level of 0.01% having been reached and the detection of the first arborescence branch (< 10 μm) made from 0.05% [LAU 93]. Let us note that amongst the disadvantages constituted by the discharges, certain types are difficult to detect; these are the *luminescent* or *pseudo luminescent* discharges which can appear in cavities trapped in the insulation of turbogenerators. These cavities containing hydrogen subjected to a neighboring field of the ionization threshold field could be the seat of this type of non-detected discharges [BAR 92]. Recent laboratory research has focused on characteristics of discharges appearing in electrical arborescence created in XLPE and at the EPOXY-EPR interface. The characteristics found are identical in both types of insulation; however, the threshold and the voltage of the discharge outbreaks appears to depend on the homogeneity of the bonding of both materials, and therefore of the presence or not of trapped gas at the moment of manufacture. This conclusion from laboratory tests is applicable to the manufacture of a cable. These are also the characteristics of discharge (growth time, width, etc.) which distinguish the origin of the defect (interface or core of the material) [DEN 01]. An investigation of about 25 years concerning the detection of discharges and its interest in the domain of rotating machines and particularly stator coils has recently been published by Stone which shows the possibilities of following the behavior of a system during its lifetime and of intervening before a breakdown [STO 04].

8.3.4.5. *The case of transformers*

The analysis of gases dissolved in power transformers in service, amongst them being paper and oil, can lead to their origin. A list of detectable defects has recently been presented by Duval [DUV 02]. It includes in particular the defects observed in service from the Defects of thermal origin are classified according to temperature, with 300°C < T < 700°C or > 700°C. A comparison is made with laboratory models in order to confirm the nature of analyzed hydrocarbons and their correlation with the amplitude of detected discharges. The decomposition products of cellulose must obviously appear as late as possible and, in order to avoid the oxidation of oil, the extraction of oxygen appears essential in the same way as particles suspended in the liquid phase [FER 02]; the prevention of this oxidation is generally secured by the addition of an antioxidant in the transformer oil.

8.4. Conclusion

We have seen that by taking the term precursor in the broadest sense, we have been led to define the causes of the breakdown of electrical insulating materials.

They have thus shown the difficulties of preventive detection, system failures and their possible life end. The revelation of the role of precursors before the material undergoes the stresses of operation has underlined the precautions required in the development of materials and equipment. During the use of this latter, parameters such as environment, temperature, direct or alternative voltages with their frequencies of use, are separately or in association with the actors of the lifetime. Of the three types of insulation (solid, liquid and gaseous), only the first presents the disadvantage of having an irreversible breakdown process. As we have seen, the control methods are not numerous but most of them are efficient owing to their analysis techniques. The contribution of laboratories is fundamental in this domain even though caution is required; different schools of thought are sometimes opposed to or otherwise distinguish themselves from ageing or breakdown in their relative outlines. This point was well made by Cygan and Laghari about ageing models in 1990 [CYG 90].

8.5. Bibliography

[ADV 67] ADVIENKO A.A., KISELEV A.V., "Outgassing from insulator surfaces in a strong electric field in vacuum", *Soviet Physics – Technical Physics,* vol. 12, no. 3, p. 381–384, 1967.

[AND 80] ANDERSON R.A., BRAINARD J.P., "Mechanism of pulsed surface flashover involving electron-stimulated desorption", *J. Appl. Phys.*, vol. 51, no. 3, p. 1414–1421, 1980.

[ARB 86] ARBAB M.N., AUCKLAND D.W., "The influence of vibration on the initiation of trees in dielectrics", *IEEE Proc. – A,* vol. 133, no. 9, p. 618–622, 1986.

[ARR 86] ARRIGHI R., "From impregnated paper to polymeric insulating materials in power cables", *IEEE Trans. Electr. Insul.,* vol. 21, no. 1, p. 7–18, 1986.

[ART 96] ARTBAUER J., "Electric strength of polymer", *J. Phys. D Appl. Phys.* vol. 29, no. 2, p. 446–456, 1996.

[BAH 82] BAHDER G., GARRITY T., SOSNOSKI M., EATON R., KATZ C., "Physical model of electric ageing and breakdown of extruded polymeric insulated power cables", *IEEE. Trans. PAS,* vol. 101, no. 6, p. 1379–1390, 1982.

[BAL 84] BALL E.H., HOLDUP H.W., SKIPPER D. J., VECILLO B., "Development of cross-linked polyethylene insulation for high voltage cables", *Inter. Conf. Large High Volt. Elec. Sys., Proceedings of the 30th session, CIGRE (International Council on Large Electric Systems)*, 1984, p. 21, 1984.

[BAL 83] BALMAIN K.G., HIRT W., "Dielectric surface discharges: effects of combined low energy and high energy incident electron", *IEEE Trans. Electr. Insul.,* vol. 18, no. 5, p.498–503, 1983.

[BAM 99a] BAMFORD H.M., FOURACRE R.A., "Nuclear technology and ageing", *IEEE Electrical Insulation Magazine,* vol. 15, no. 5, p. 19–27, 1999.

[BAM 99b] BAMJI S.S., "Electroluminescence – A technique to detect the initiation of degradation in polymeric insulation", *IEEE Electrical Insulation Magazine,* vol. 15, no. 3, p. 9–14, 1999.

[BAR 87] BARTNIKAS R., "High voltage measurements", *Engineering Dielectrics*, vol. II-B, ch. 3, BARTNIKAS R., (Ed.), ASTM, STP 926, 1987.

[BAR 92] BARTNIKAS R., NOVAK J.P., "On the spark to pseudoglow and glow transition mechanism discharge detectability", *IEEE Trans. Electr. Inssul.*, vol. 27, no. 1, p. 3–14, 1992.

[BAR 02] BARTNIKAS R., "Partial discharges. Their mechanism, detection and measurements", *IEEE Trans. Electr. Insul.*, vol. 9, no. 5, p. 763–808, 2002.

[BEE 81] BEERS B.L., PINE V.W., IVES S.T., "Internal breakdown of charged spacecraft charging", *IEEE Trans. Electr. Nucl. Sci.*, vol. NS-28, no. 6, p. 4529–4534, 1981.

[BEI 01] BEISELE C., KULTZOW B., "Experiences with new hydrophobic cycloaliphatic epoxy outdoor insulation systems", *IEEE Electrical Insulation Magazine*, vol. 17, no. 4, p. 33–39, 2001.

[BER 98] BERTIN C., Phénomène de prérupture des polyoléfines utilisées dans l'isolation électrique – Relation avec la structure, Doctoral thesis, Paul Sabatier University, Toulouse, 1998.

[BIH 87] BIHLER E., HOLDIK K., EISENMENGER W., "Electric field-induced gas emission from PVDF films", *IEEE Trans. Electr. Insul.*, vol. 22, no. 2, p. 207–210, 1987.

[BLO 69] BLOK J., LE GRAND D.G., "Dielectric breakdown of polymer films", *J. Appl. Phys.*, vol. 40, no. 1, p. 288–293, 1969.

[BLY 79] BLYTHE A.R., *Electrical Properties of Polymers*, Cambridge University Press, Cambridge, 1979.

[BOG 94] BOGGS S.A., MASHIKIAN M.S., "Role of semiconducting compounds in water treeing of XLPE cable insulation", *IEEE Electrical Insulation Magazine*, vol. 10, no. 1, p. 23–27, 1994.

[BOG 98] BOGGS S.A., DENSLEY J., KUANG J., "Mechanism for impulse conversion of water trees to electrical trees in XLPE", *IEEE Trans. Pow. Deliv.*, vol. 13, no. 2, p. 310–315, 1998.

[BOG 00] BOGGS S.A., DENSLEY J., "Fundamentals of partial discharge in context of field cable testing", *IEEE Electrical Insulation Magazine*, vol. 16, no. 5, p. 13–18, 2000.

[BOS 03] BOSTROM J.O., MARSDEN E., HAMPTON R.N., NILSON U., LENNARTSSON H., "Electrical stress enhancement of contaminants in XLPE insulation used for power cables", *IEEE Electrical Insulation Magazine*, vol. 19, no. 4, p. 6–12, 2003.

[CHA 89] CHAKRABARTI A.K., VAN HEESWIJK R.G., SRIVASTAVA K.D., "Free particle-initiated 60 Hz breakdown at a spacer surface in a gas-insulated Bus", *IEEE Trans. Electr. Insul.*, vol. 24, p. 549–560, 1989.

[CHA 90] CHAN J.C., COMETA E.T., HARTLEY M.D., HIJVALA L.J., "Field failure analysis of medium voltage XLPE-insulated power cables", *Conf. Record, IEEE Intern. Symp.*, CH2727, Toronto, p. 277–280, 1990.

[COE 79] COELHO R., *Physics of Dielectrics for Engineers*, Elsevier, Amsterdam, 1979.

[COÏ 78] COÏSSON R., PARACCHINI C., SCHIANCHJ G., "Electroluminescence in an epoxy resin", *J. Electrochem. Soc.: Solid-State Sc. Tech.*, vol. 125, no. 4, p. 581–583, 1978.

[COO 85] COOKSON A.H., "Gas-insulated cables", *IEEE Trans. Electr. Insul.*, vol. 20, p. 859–890, 1985.

[COO 63] COOPER R., in *Progress in Dielectrics*, vol. 5, J.B. Birks & J. Hart (Eds). Academic Press, New York, 1963.

[COO 66] COOPER R., "The electric strength of solid dielectrics", *Brit. J. Appl. Phys.*, vol. 17, no. 2, p. 149–166, 1966.

[CRI 87] CRINE J.P., PELISSOU S., SAINT-ONGE H., SAINT PIERRE S., KENNEDY G., HOUDAYER A., HINRICHSEN P.F., "Elemental and ionic impurities in cable insulation and shields" *Proc. JICABLE Conf.*, Versailles, 1987.

[CRI 98] CRINE J.P., JOWE J., "A review of the influence of frequency on accelerated ageing of PE and XLPE cables", *RGE*, ISSN 12655–6534, vol. 3, p. 9–14, 1998.

[CUD 87] CUDDIHY E.F., "A concept for the intrinsic strength of electrical insulation materia", *IEEE Trans. Electr. Insul.*, vol. 22, no. 5, p. 573–589, 1987.

[CYG 90] CYGAN P., LAGHARI J.R., "Models for insulation ageing under electrical and thermal multi-stress", *IEEE Trans. Electr. Insul.*, vol. 25, no. 5, p. 923–934, 1990.

[DAV 92] DAVID E., PARPAL J.L., CRINE J.P., "Photoelastic measurements of mechanical stresses induced in XLPE during electrical ageing", *IEEE Intern. Symp. Electr. Insul. Dielec. Phenom.,* Baltimore, USA, 1992.

[DAV 94] DAVID E., PARPAL J.L., CRINE J.P., "Influence of internal stress and strain on electrical treeing in XLPE", *IEEE Conf. Electr. Insul. Dielec. Phenom. (CEIDP)*, p. 575–581, 1994.

[DEN 94] DENSLEY R.J., BARTNIKAS R., BERNSTEIN B., "Multiple stress ageing of solid-dielectric extruded dry-cured insulation systems for power transmission cables", *IEEE Trans. Pow. Deliv.*, vol. 9, no. 1, p. 559–571, 1994.

[DEN 01] DENSLEY J., "Ageing mechanisms and diagnostics for power cables – An overview", *IEEE Electrical Insulation Magazine*, vol. 17, no. 1, p. 14–22, 2001.

[DEN 01] DENSLEY J., KALICKI T., NODOLNY Z., "Characteristics of PD pulses in electrical trees and interfaces in extruded cables", *IEEE Trans. Dielec. Electr. Insul.*, vol. 8, p. 48–57, 2001.

[DIS 92] DISSADO L.A., FOTHERGILL J.C., *Electrical Degradation and Breakdown in Polymers,* Peter Peregrinus Ltd, London, 1992.

[DUV 02] DUVAL M., "A review of faults detectables by gas-in-oil analysis in transformers", IEEE Electrical Insulation Magazine, vol. 18, no. 3, p. 8–17, 2002.

[EBE 93] EBERLE G., SCHMIDT H., EISENMENGER W., "Influence of poling conditions on the gas emission of PVDF", *IEEE Conf. Electr. Insul. Dielec. Phenom. (CEIDP)*, p. 263–268, 1993.

[EPR 81] EPRI, "Transmission line insulators", *IEEE Trans. Electr. Insul.*, Epri Workshop vol. 16, no. 3, p. 153–276, 1981.

[FAR 95] FARZANEH M., KIERNICKI J., "Flashover problems caused by ice build-up on insulators", *IEEE Electrical Insulation Magazine*, vol. 11, no. 2, p. 5–17, 1995.

[FAV 77] FAVA R.A., "Treatease on materials science and technology", in *Properties of Solid Polymeric Materials, Part B*, vol. 10, SCHULTZ, J.M. (ed), Academic Press, New York, 1977.

[FER 02] FERGUSON R., LOBEIRAS A., SABAU J., "Suspended particles in the liquid insulation of ageing power transformers", *IEEE Electrical Insulation Magazine*, vol. 18, no. 4, p. 17–23, 2002.

[FRE 82] FREDERICKSON A.R., WOOLF S., "Electric fields in Kev electron irradiated polymers", *IEEE Trans. Electr. Nucl. Sci.*, vol. NS-29, no. 6, p. 2004–2011, 1982.

[GAR 87] GARTON A., BAMJI S., BULINSKI A., DENSLEY J., "Oxidation and water tree formation in service-aged XLPE cable insulation", *IEEE Trans. Electr. Insul.*, vol. 22, no. 4, p. 405–412, 1987.

[GOF 82] GOFFAUX R., COELHO R., "Sur la rupture thermique filamentaire différée dans des isolants électriques", *Rev. Phys. App.*, vol. 17, no. 2, p. 55–64, 1982.

[GUE 92] GUEGUEN V., Vieillissement d'élastomères utilisés comme isolants électriques en ambiance nucléaires, Thesis, Ecole Normale Supérieure d'Arts et Métiers Centre de Paris, 1992.

[GUL 92] GULSKI E., KREUGER F.H., "Computer-aided recognition of discharge sources", *IEEE Trans. Electr. Insul.*, vol. 27, no. 1, p. 82–92, 1992.

[HAN 92] HANTOUCHE C., "PD in power capacitors", *IEEE Intern. Symp. Eletr. Insul.*, Pub. 92CH3150-0, Baltimore, USA, 1992.

[HAN 93] HANTOUCHE C., FORTUNE D., "Digital measurement of PD in full sized power capacitor", *IEEE Trans. Electr. Insul.*, vol. 28, no. 6, p. 1025–1032, 1993.

[HAR 87] HARA M., ADACHI K., TOBATA H., AKAZAKI M., MAO F.L., "Particle-initiated breakdown of conical insulator in N_2 and N_2/O_2 mixture under DC voltage", *IEEE Trans. Electr. Insul.*, vol. 22, p. 87–96, 1987.

[HAR 67] HARTMAN W.A., ARMSTRONG H.L., "Electroluminescence in organic polymers", *J. Appl. Phys.*, vol. 38, no. 5, p. 2393–2395, 1967.

[HEP 00] HEPBURN D.M., KEMP I.J., SHIELDS A.J., "Mica", *IEEE Electrical Insulation Magazine*, vol. 16, no. 5, p. 19–27, 2000.

[HIK 90] HIKITA M., HIROSE T., ITO Y., MIZUTANI T., IEDA M., “Investigation of electrical breakdown of polymeric insulating materials using a technique of pre-breakdown current measurements”, *J. Phys. D: Appl. Phys.*,vol. 23, no. 12, p. 1515–1527, 1990.

[HIP 37] V-HIPPEL A., “Electric breakdown of solid and liquid insulator”, J. Appl. Phys., vol. 8, no. 12, p. 815–832, 1937.

[IED 80] IEDA M., “Dielectric breakdown process of polymers”, *IEEE Trans. Electr. Insul.*, vol. 15, no. 3, p. 206–224, 1980.

[JAC 78] Jacobsen C.T., Attermo R., Dellby B., “*Experience of dry XLPE-insulated high voltage Cables*”, International Council on Large Electric Systems (CIGRE), Paper 21–06, 1978.

[JON 05] JONES J.P., LEWELLYN J.P., LEWIS T.J., “The contribution of field-induced morphological change to the electrical ageing and breakdown of polyethylene”, *IEEE Trans. Dielec. Electr. Insul.*, vol. 12, no. 5, p. 951–966, 2005.

[JON 80] JONSCHER A.K., “Physical basis of dielectric breakdown”, *J. Phys. D: Appl. Phys.,*.vol. 13, L143–148, 1980.

[KAG 75] KAGEYAMA S., ONO M., CHABATA S., “Microvoids in cross-linked polyethylene insulated cable”, *IEEE. Trans. PAS.,* vol. 94, p. 1258–1263, 1975.

[KAI 00] KAIDANOV F., MUNTEANU R., “Damages and destruction of optical cables on 161 kV overhead transmission lines”, *IEEE Electrical Insulation Magazine,* vol. 16, no. 4, p. 16–23, 2000.

[KAM 92] KAMINAGA K., IDA S., FUKUNAGA S., YATSUKA K., INOUE Y, “Influence of fibrous contaminants on dielectric property of XPLE insulation”, *IEEE Intern. Symp. Eletr. Insul.*, Baltimore, Pub. 92CH3150-0, p. 401–406, 1992.

[KAR 89] KARAKELLE M., PHILLIPS P.J., “The influence of structure on water treeing in cross-linked polyethylene”, *IEEE Trans. Electr. Insul.*, vol. 24, no. 6, p. 1083–1108, 1989.

[KOL 80] KOLESOV S.N., “The influence of morphology on the electric strength of polymer insulation”, *IEEE Trans. Electr. Insul.*, vol. 15, no. 5, p. 382–388, 1980.

[LAG 81] Laghari J.R., Qureshi A.H., “Surface flashover along spacers in compressed gas insulation systems””, *IEEE Trans. Electr. Insul.*, vol. 16, no. 5, p. 373–398, 1981.

[LAG 90] Laghari J.R., Hammoud A.N., "A brief survey of radiation effects on polymer dielectrics", IEEE Trans. Electr. Nucl. Sci., vol. NS-37, p. 1076–1083, 1990.

[LAU 83] LAURENT C., MAYOUX C., NOËL S., “Dielectric breakdown of polyethylene in divergent field : Role of dissolved gases and electroluminescence”, *J. Appl. Phys.*, vol. 54, no. 3, p. 1532–1539, 1983.

[LAU 93] LAURENT C., MAYOUX C., “From initiation to propagation in electric treeing”, *Intern. Conf. Partial Discharges, Proc. IEEE*, no. 378, p. 7–8, 1993.

[LAU 99] LAURENT C., “Optical pre-breakdown warnings in insulating polymers”, *IEEE Electrical Insulation Magazine*, vol. 15, no. 2, p. 5–13, 1999.

[LEW 93] LEWIS T.J., LEWELLYN J.P., VAN DER SLUIJS, "Electrokinetic properties of metal-dielectric interfaces", *IEEE Proc. – A*, vol. 40, p. 385–392, 1993.

[LEW 02] LEWIS T.J., "Polyethylene under electrical stress", *IEEE Trans. Dielec. Electr. Insul.*, vol. 9, no. 5, p. 717–729, 2002.

[MAM 04a] MAMY P. R., MARTINEZ-VEGA J.J., DUPRÉ J.C., BRETAGNE N., "Quantification of mechanical deformation induced by an electric field in a semi crystalline organic insulator", *J. Appl. Polym. Sci.*, vol 93, no. 5, p. 2313–2321, 2004.

[MAM 04b] MAMMY P. R., "Etude de la réponse mécanique des isolants organiques solides au cours de l'application d'un champ électrique : cas du PET semi-cristallin", doctoral thesis, Paul Sabatier University, 2004.

[MAS 87] MASHIKIAN M.S., GROEGER J.H., "Ionic impurities in extruded cable insulation: analytical detection techniques, sources, nature, and effects", *Proc. JICABLE Conf.*, Versailles, 1987.

[MAS 92] MASHIKIAN M.S., GROEGER J.H., "Ageing phenomena at cable-shield interfaces affecting the propensity of the insulation to develop water trees", *RGE,* ISSN 0035-3116, no. 1, p. 47–53, 1992.

[MAS 59] MASON J.H., "Dielectric breakdown in solid insulation", *Progress in Dielectrics*, vol. 1, p. 1, BIRKS, J.B. (ed.), Heywood & Co., London, 1959.

[MAS 73] MASON J.H., "Dielectric breakdown of solids", *Insulation Engineer*, p. 5–12, May–June 1973.

[MAT 61] MATSUOKA S., "Hypothesis of voids in semi-crystalline polymers", *J. Appl. Phys.*, vol. 32, no. 11, p. 2334–2336, 1961.

[MIN 84] MINTZ J.D., "Failure analysis of polymeric-insulated power cable", *IEEE Trans. PAS*, vol. 103, no. 12, p. 3448–3453, 1986.

[MIZ 87] MIZUTANI T., KANNO I., HIKITA M., IEDA M., "Pre-breakdown currents due to filamentary thermal breakdown in polyimide film", *IEEE Trans. Electr. Insul.*, vol. 22, no. 4, p. 473–477, 1987.

[MOR 00] MORCOS M.M., WARD S.A., ANIS H., SRIVASTAVA K.D., GUBANSKI S.M., "Insulation intergrity of GIS/GITL systems and management of particle contamination", *IEEE Electrical Insulation Magazine*, vol. 16, no. 5, p. 25–37, 2000.

[MOR 88] MORSHUIS P.H.F., KREUGER F.H, LEUFKENS P.P, "The effect of different types of inclusions on PE cable life", *IEEE Trans. Electr. Insul.*, vol. 23, no. 6, p. 1051–1055, 1988.

[MUC 76] MUCCIGROSSO J., PHILLIPS P.J., "The morphology of cross-linked polyethylene insulation", *IEEE Trans. Electr. Insul.*, vol. 13, no. 3, p. 172–178, 1976.

[NEL 83] NELSON J.K., "Breakdown strength of solids", in *Engineering Dielectrics,* ASTM, STP 783, vol. II-A, ch. 5, 1983.

[O'DW 79] O'DWYER J.J., *The Theory of Electrical Conduction and Breakdown in Solid Dielectrics*, Oxford University Press, 1979.

[POM 02] POMMERENKE D., SAGAGUCHI S., "Application of Maxwell solvers to PD propagation-Part I: Concepts and codes", *IEEE Electrical Insulation Magazine*, vol. 18, no. 5, p. 15–21, 2002.

[PRE 71] PREDECKI P., SWAROOP N., "Anomalous conduction in polymer films at high DC fields", *J. Polym. Sci., Part B: Polymer Letters,* vol. 9, no. 1, p. 43–47, 1971.

[REA 87] REAGAN J.B., MEYEROTT R.E., GAINES E.E, NIGHTINGALE R.W., FIBERT P.C., IMHOF W.L., "Space charging currents and their effects on spacecraft systems", *IEEE Trans. Electr. Insul.*, vol. 18, no. 3, p. 354–365, 1987.

[RID 82] RIDDEL J.D., CHAVENAK J.G., VAN LINT V.A.J., "Ionization induced breakdown and conductivity of satellite dielectrics", *IEEE Trans. Electr. Nucl. Sci.*, vol. NS-29, p. 1754–1759, 1982.

[ROB 36] ROBINSON D.M., *Dielectric Phenomena in HV Cables*, Chapman & Hall, London, 1936.

[ROS 87] ROSE M.F., "Electrical insulation and dielectrics in space environment cable", *IEEE Electrical Insulation*, vol. 22, no. 5, p. 555–571, 1987.

[ROS 99] ROSS R., "Dealing with interface problems in polymer cable terminations", *IEEE Electrical Insulation Magazine*, vol. 15, no. 4, p. 5–9, 1999.

[RUS 92] RUSTAN J.R. P.L., "SIDO space insulation technologie", *IEEE Intern. Sympos. Electr. Insul.*, Baltimore, 92CH3159-0, p. 2–12, 1992.

[SAB 76] SABUNI H., NELSON J.K., "Factors determining the electric strength of polymeric dielectrics", *J. Material Science Letters*, no. 11, no. 8, p. 1574–1576, 1976.

[SCH 79] SCHÔN BACHER H., STOLARZ-IZICKA A., Compilation of radiation damage test data. Part I: cable insulating materials. European Organization for Nuclear Research, Report CERN 79-04, Geneva, 1979.

[SEA 68] SEANOR D.A., "Electronic and ionic conductivity in Nylon 66", *J. Polymer Sc.*, vol. 6, no. 3, p. 463–477, 1968.

[STA 84] STANNET A.W., JACKSON R.J., "Some comments on the impulse electric strength of polythene", *Dielec. Mater. Meas. and Appl.*, Conf. No. 239, Lancaster, 1984.

[STA 55] STARK K.H., GARTON C.G., "Electric strength of irradiated polythene", *Nature*, vol. 176, p. 1225–1226, 1955.

[STO 04] STONE G.C., WARREN V., "Effect of manufacturer, winding age and insulation type on stator winding partial discharge levels", *IEEE Electrical Insulation Magazine*, vol. 20, no. 5, p. 13–17, 2004.

[STR 61] STRATON R., "The theory of dielectric breakdown in solids", *Progress in Dielectrics*, J.B. BIRKS (Ed.), Wiley, New York, vol. 3, 1961.

[SUD 86] SUDARSHAN T.S., DOUGAL R.A., "Mechanisms of surface flashover along solid dielectrics in compressed gases: a review", *IEEE Trans. Electr. Insul.*, vol. 21, no. 5, p. 727–746, 1986.

[SUN 96] SUNDARARAJAN R., GORUR R.S., "Role of non-soluble contaminants on the flashover voltage of porcelain insulators", *IEEE Trans. Electr. Insul.*, vol. 3, no. 1, p. 113–118, 1996.

[TIM 79] TIMPE N.P, "Corona measurements and interpretation", *Engineering Dielectrics*, Bartnikas/Mc Mahon, vol. I, ch. 5, ASTM, STP 669, 1979.

[TOU 74] TOUREILLE A., Sur les phénomènes de conduction à champ électrique élevé dans les hauts polymères, (PhD thesis), University of Montpellier, 1974.

[VEL 96] VELA N., TOUREILLE A., ZEBOUCHI N., HOANG T.G., "Precursor phenomena of electric breakdown in polyethylene", RGE, ISSN 1265–6534, vol. 1, p. 102–106, 1996.

[VOL 92] "The Volta Colloquium on Partial Discharge Measurements", *IEEE Trans. Electr. Insul.*, vol. 27, no. 1, 1992.

[WEN 79] WENDORFF J.H., "Submicroscopic defects in strained polyoxymethylene", *Progr. Colloid & Polymer Sc.*, vol. 66, p. 135–142, 1979.

[WHI 53] WHITEHEAD S., *Dielectric breakdown of solids*, Clarendon Press, Oxford, 1953.

[YOD 73] YODA B., MURAKI K., "Development of EHV cross-linked polyethylene insulated power cable", *IEEE Trans. PAS*, vol. 92, p. 501–513, 1973.

Chapter 9

Models for Ageing of Electrical Insulation: Trends and Perspectives

9.1. Introduction

In the world of electrical engineering, device reliability is widely determined by the durability of the insulating materials' properties. Electrical equipment failure is often associated with the *dielectric breakdown* phenomena of the insulation. These phenomena can occur for electrical fields clearly weaker than the estimated breakdown field. This is widely due to the insulating materials' *electrical ageing*, a generic term which groups together all of the mechanisms by which the general characteristics of materials, and particularly their electrical properties, evolve over time under the action of working stresses.

Solid insulating materials have been introduced in a wide range of electrical equipment, from transport and distribution to the use of electrical energy. In particular, polymers are widely used for cables, capacitors, alternators, transformers and motors. We also find them in electronic devices, i.e. as coatings and in power components [LAU 99].

Although the main function of the polymer is to maintain electrical insulation, it also needs to have a good resistance to mechanical and thermal stresses. The understanding of ageing processes leading to either the loss, reversible or not, of functional properties, or to the dielectric breakdown of organic solid insulating

---

Chapter written by Nadine LAHOUD, Laurent BOUDOU, Christian MAYOUX and Juan MARTINEZ-VEGA.

materials must allow the behavior of this type of material to be predicted, according to the system's life (30 to 40 years, for example, for energy transport cables or alternators [LAU 99]). This behavior obviously defines the lifetime of systems in which these materials are integrated.

In this sense, the characterization of the physico-chemical properties of insulating materials and the understanding of complex relationships between their structure, whatever the scale (micro, nano, etc.), their properties and ageing, are key steps in the process of the development of new materials, allowing their improvement and the reliability of systems.

It is therefore fundamental to be able to develop a modeling approach leading to the prediction of materials' long term behavior, and to develop short classical laboratory tests, if possible.

This approach must take into account the fundamental recovering mechanisms typical of amorphous materials, elastic and plastic deformation, and the nonlinear character of these mechanisms when the conditions exceed a certain level, e.g. rate of rise, frequency, temperature. Charge transport phenomena, mechanical or electrical breakdown, bond energy ranges, the probable reversibility of breakdown and reactions also all need to be considered.

We now propose to briefly present existing models which could be good starting points for researchers interested in this area. Readers are obviously invited to refer to the bibliography for more detailed information.

Current trends and perspectives will be revealed by this analysis of the different ageing scenarios which have inspired the models.

9.2. Kinetic approach according to Zhurkov

9.2.1. *Presentation*

This model originated from experiments led by Zhurkov [ZHU 65], [ZHU 66], destined for measuring the breakdown time, under uniaxial mechanical stress on a series of different materials such as metals, crystals or polymers. This study is closely related to that of the application of a purely electrical stress, knowing that the electrical field could generate electromechanical stresses [DIS 97], [PAR 97], [BLA 95].

Taking polymers as an example, the experimental results led the author to write the following relationship:

$$\tau = \tau_0 \exp\left[\frac{U_0 - \gamma\sigma}{kT}\right] \qquad [9.1]$$

with τ the lifetime, τ_0 a pre-exponential factor, k the Boltzmann constant, T the absolute temperature, σ the mechanical tensile stress, U_0 and γ being coefficients subsequently defined for the understanding of the physical sense of the experimentally established relationship [9.1].

The author could thus define, in the same experimental frame, the coefficient τ_0 whose value coincides with the value of the thermal oscillation period of atoms in a solid. τ_0 is of the order of 10^{-13} s independently of the structure and the chemical nature of the material. By writing equation [9.1] in the form:

$$kT \ln(\tau/\tau_0) = U_0 - \gamma\sigma \qquad [9.2]$$

where the left part of this equation is identified with an energy U which represents the barrier to be crossed in order to initiate a breakdown. τ_0 being determined, we can draw the curves from previous results and $U(\sigma)$ shows a linear relation between the effective barrier U and the applied stress. The extrapolation for a null stress gives values of U_0 for materials placed under testing. The comparison of these values with those of interatomic bond dissociation energies deduced from the thermal destruction of polymers, has led Zhurkov to identify U_0 to an atomic bond energy.

The follow-up of the fracture kinetics of the polymer subjected to a mechanical stress has been made by electron spin resonance (ESR) experiments through which the authors have observed the rate of free radical formation increasing in an exponential manner as a function of the applied stress [ZHU 66]. Zhurkov thus estimated that the breakdown of polymers is essentially due to the breaking of covalent bonds [ZHU 65], [ZHU 66], [ZHU 83]. This mechanism is debatable.

In Figure 9.1, Zhurkov evoked the possibility of the creation of a cavity or a free space smaller than a micron, due to chemical bonds breaking which lead to free radicals. Indeed, Figure 9.1(a) illustrates the thermo-mechanical breaking of a covalent bond leading to the creation of two radicals. These two highly reactive radicals, in Figure 9.1(b), will then interact with the adjacent molecule and thus create two new radicals. A chain reaction is produced (see Figure 9.1(c)). The final result of this process is represented in Figure 9.1(d) by a submicronic cavity into which partial discharges may be initiated, after coalescence for example, under size and gas pressure conditions (the Pashen Law), leading to the breakdown of the material.

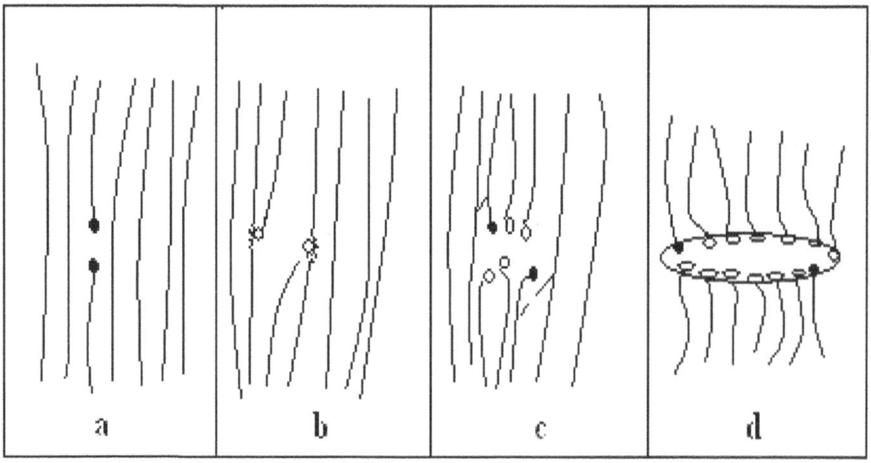

Figure 9.1. *Mechanism of a submicronic cavity generation, from [ZHU 72]*

The coefficient γ represents the volume which undergoes the stress, i.e. an activation volume. In a paper published in 1983, Zhurkov [ZHU 83] expressed the coefficient γ as a function of the parameters suited to the material structure, i.e. the calorific capacity, the linear thermal expansion coefficient, Young's macroscopic modulus and a surcharge coefficient. Unfortunately, the paper did not demonstrate the means by which this term can be expressed as a function of the different variables which are at the same time macroscopic and microscopic.

9.2.2. *Interpretation of the process and introduction to the notion of a dilaton*

The base of this kinetic theory of breakdown is the overloading of some chemical bonds with respect to the constrained polymer volume. The probability of their breakdown is quite high, in such a way that they become initiator sites for the insulating material breakdown [ZHU 69].

The vibration frequency of the molecules of a polymer depends on the amplitude of the strength applied, the atomic bond energy and the valence angles. The deformations of these bonds and angles modify the vibration frequency of the principal chain. From this, Zhurkov and his team [ZHU 69] determined the amplitude of real mechanical stresses acting on the polymer chains by following the vibration frequency changes, through infrared spectrophotometry.

The coefficient γ related to the structure allows, for a given strength and temperature, the activation volume as well as the emergency time of a submicronic breaking to be known. This coefficient is defined by the relationship:

$$\gamma = \frac{C.\chi}{\alpha.Y} \qquad [9.3]$$

where C is the specific heat, χ an overloading coefficient proportional to the mean free path of a phonon Λ, α the linear thermal dilation coefficient and Y the Young's macroscopic modulus.

The kinetic approach presented by Zhurkov makes the local breakdown appear like a thermal fluctuation process, which could bring into play the breaking or the rearrangement of atomic chains, or both simultaneously. The preexisting defects, or those created under stress because of thermal fluctuations, are localized in an element of the material volume. The notion of dilatons was thus introduced and has been tackled by different authors (see below).

9.2.2.1. *Definition according to Zhurkov (1983)*

Zhurkov refers to an ancient idea, that of Smoluchowski (1908), who was the first to define the concept of negative fluctuation of density [ZHU 83]. This phenomenon appears spontaneously in a solid as the consequence of a random atom displacement, or the superimposing of elastic waves describing network vibrations. The interatomic bonds are then stretched and weakened. The density of the region concerned decreases and this zone thus becomes a trapping site for phonons (these latter being defined as energy *quanta* of an elastic wave, their mean free path Λ representing the dimensions of the density fluctuation region, which is called a *dilaton*). The amplitude of Λ and that of the dilaton both depend on the structure of the existing defects, the impurities, etc.

9.2.2.2. *Definition according to Kusov (1979)*

Kusov uses the previous idea but excludes the necessity of the presence of a defect. The defects play the role of catalysers in the breakdown process, by favoring the conditions required for the outbreak of dilatons [KUS 79]. In return, he mentions the importance of anharmonicity in the process of thermal fluctuation formation. This anharmonicity allows trapped energy to favor the thermal dilation of the zone concerned. To better explain this process, Kusov considers a linear chain of carbon atoms, as represented in Figure 9.2 where AB is the fluctuation zone, r is the distance between two neighboring atoms which do not belong to the fluctuation zone, and r' is the distance between two neighboring atoms situated in the fluctuation zone. The chain is at temperature T, and the atoms can only make longitudinal vibrations.

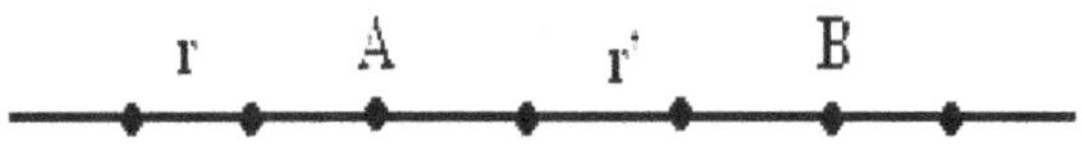

Figure 9.2. *Linear chain of carbon atoms*

AB is the zone where local fluctuations and density are negative. In other words, it is the zone where the atoms are more spread out from one another which we call the dilaton. d is the fluctuation dimension and – Δ_{loc} is the fluctuation intensity, such that:

$$\Delta_{loc} = (r - r')/r \qquad [9.4]$$

Kusov showed that a phonon exchange between the dilaton region and the rest of the chain causes an energy accumulation in the fluctuation region AB. Because of the relaxation phenomena, a competition appears between the increase and the decrease of the fluctuation intensity (Δ_{loc}). If the initial fluctuation density is sufficiently important, thermal expansion prevails ($\Delta_{loc} \geq \Delta_{critical}$) and wins over the relaxation. In the dilaton region, the energy accumulation can lead to the chain breaking.

9.2.2.3. *Definition according to Petrov (1983)*

Oscillations in a solid imply collision processes of phonons. The collisions could take place between phonons, defects or surfaces. Petrov therefore took into account the mean free path Λ of the phonon to define the spatial dimension of the fluctuation. An accumulation of phonons is similar to a local increase of thermal energy which leads to a deformation of the medium. This deformation has a limit when a stress is applied. Petrov views dilatons as sub-systems having spatial characteristics Λ, an interatomic distance a, and a volume $V = \Lambda.a^2$. The dilaton is an elastic deformation region which absorbs the phonons which traverse it [PET 83].

9.2.2.4. *Universal breakdown kinetics*

In conclusion, the approach of a breakdown mechanism of interatomic bonds developed by Zhurkov's team to establish universal breakdown kinetics on metals and polymers deserves attention. However the γ parameter which depends essentially on the structure of the material cannot be used to predict the breakdown time of the insulating material. Within this latter, a vacuole resulting from the submicronic cavities, is created and its insulating properties are damaged in an irreversible manner. It is important to point out in this model that the parameter γ represents an activation volume; this volume is presented in the following model which relies on thermodynamic processes.

9.3. Thermodynamic approach according to Crine

The model proposed by Crine is founded on a simple thermodynamic concept, taking into account thermodynamic parameters such as intermolecular cohesion energy, the amorphous phase thickness of a material [CRI 90], breakdown theory related to the free volume proposed by Artbauer [ART 96], the concept of submicronic cavity initiation, and the fact that the first stage of electrical ageing is essentially a molecular process. This model, applicable to ageing under alternative stress, supposes that the deformation of macromolecular chains is a fatigue process accelerated by frequency. Of course, purely electronic concepts described by numerous models are brought into consideration during the final breakdown mechanism.

The fact that processes such as electrical conductivity, dielectrical or mechanical relaxation depend on temperature, implies that they are most often expressed by the Arrhenius relationship:

$$X = X_0 \exp\left(-E / KT\right) \qquad [9.5]$$

where X_0 is an empirical factor and E the apparent activation energy.

Crine presented a simple thermodynamic interpretation of the compensation law applicable to polymers. An equation is then proposed by Crine relating Gibbs' free energy of activation theory to the fundamental thermodynamic parameters [CRI 84].

From Eyring's theory, the reaction which allows the passing from an active state to an upper state follows the relationship:

$$K_f = (kT / h) \exp(-\Delta G^{\#} / kT) \qquad [9.6]$$

where $\Delta G^{\#}$ is the activation free energy. This free energy relates the activation enthalpy, the activation entropy and the temperature by the relationship:

$$\Delta G^{\#} = \Delta H^{\#} - T.\Delta S^{\#} \qquad [9.7]$$

We can then rewrite relationship [9.5] in the form:

$$X = X_0 \exp(\Delta S^{\#} / k) \exp(-\Delta H^{\#} / kT) \qquad [9.8]$$

A few years ago, a theoretical interpretation was proposed by Lawson [LAW 57] and Keyes [KEY 58] to demonstrate that, for a process having a weak activation volume $\Delta V^{\#}$, the following expression takes place:

$$\Delta H^{\#} = \Delta V^{\#} / K.\beta \quad [9.9]$$

where β represents the isothermal compressibility coefficient and K a dimensionless constant. Subsequently, $\Delta S^{\#}$ can be expressed as:

$$\Delta S^{\#} = \Delta V^{\#}.(\alpha / \beta) \quad [9.10]$$

where α represents the thermal dilation coefficient. We can then deduce the Gibbs free energy as a function of thermodynamic parameters (α and β), the activation free volume $\Delta V^{\#}$, the K parameter and the temperature, which gives:

$$\Delta G^{\#} = \left[(\alpha/\beta).\Delta V^{\#}\right].\left[(K.\alpha)^{-1} - T\right] \quad [9.11]$$

In order to verify this equation, it is important to know the signification of the K parameter. From Keyes [KEY 58] and Barker [BAR 67], this latter can be better understood if associated to the Grüneisen constant (γ) of polymers. This constant measures the anharmonicity of intermolecular vibrations in solids and is related to thermal and mechanical properties by the relationship:

$$\gamma = \alpha.V / \beta.C_V \quad [9.12]$$

where V is the molar volume and C_V, the molar calorific capacity at constant volume. Knowing K, $\Delta S^{\#}$ or $\Delta H^{\#}$, α and β, it is possible to evaluate the activation free volume $\Delta V^{\#}$ for a given process.

An important fact is to note here: the value of the activation volume $\Delta V^{\#}$ is close to the volume obtained by Zhurkov [ZHU 65], [ZHU 66], from the coefficient γ for purely mechanical tests. This reinforces the logic of the approach which was elaborated by using thermodynamic parameters and supposing that the first stage of ageing is a molecular fatigue process.

Parpal and Crine [PAR 97] proposed a model which takes into account a critical electrical field beyond which submicronic cavities can be created and then become the seat of discharges, leading to final polymer breakdown. The value of this field can be determined by knowing the cohesion energy of the polymer; this leads to a simple lifetime relation, which only takes into account two physical parameters: ΔG_0 representing the activation energy of the process distorting the chains being considered, and λ_{max} representing the maximum size of submicronic cavities created.

The major characteristic of this model is the presentation of electrical ageing as a phenomenon which is not necessarily associated with an electron displacement under the action of an electrical field but, rather, with a fatigue phenomenon under an alternative field. The proposed ageing process happens in two very distinct steps:

– the first step is a purely molecular process. It consists of a rearrangement of the polymer's free volume, induced by an electromechanical deformation of macromolecular chains under the action of a field beyond a certain value;

– the second process is an electronic avalanche. When the microcavities are created, the charges injected and accelerated under the action of the electrical field acquire enough kinetic energy (from the electrical field) to break the bonds and consequently increase the size of the microcavities. Bigger cavities then become available, allowing the electrons to break the intramolecular bonds, leading to the final breakdown.

Above a certain critical energy, the weakest bonds (Van der Waals) can be broken under the effect of electromechanical deformations due to an applied alternative electrical field [PAR 97]. This hypothesis is reinforced by work done on cables extruded with PE where a cyclic deformation was observed during the application of the alternating field [DAV 94]. After the breaking of the bonds under consideration, the macromolecular chains can move and create a sort of rearrangement of the free volume.

The breakdown process is controlled by an energy barrier of a height equal to the activation energy $\Delta G^{\#}$. Parpal and Crine [PAR 97] established the fact that this energy barrier does not represent a simple trapping or detrapping phenomenon of entities, but rather the phenomenon of bond breaking. This implies that the displacement of carriers under a high field preferably appears in the void present in the submicronic cavities created, rather than in the hopping phenomenon from one electronic trap to another.

Following the analysis of over 200 publications, it appears that the inverse power law [$t=A.F^{-n}$] does not satisfy ageing data corresponding to long periods [DAN 96]. The results from experiments on LDPE, XLPE, and EPR follow an equation of the type:

$$t = (h/2kT)\exp(\Delta G_0/kT)\operatorname{csc}h(e\lambda F/kT) \qquad [9.13]$$

where t is the time required to reach the state corresponding to broken bonds, or, in other words, the time required to get over the free energy barrier of height ΔG_0; h and k are, respectively, the Planck and Boltzmann constants, λ is the deformation distance of the molecular chains and e is the charge of the electron.

The interpretation of the phenomenon is as follows: when an electrical field F is applied, the macromolecular chains are distorted over a distance λ, and ΔG_0 is then reduced to a quantity equal to the deformation energy applied to the barrier, which is $W = e\lambda F$. We shall see later that, in order to correlate his model with a thermodynamic parameter of the polymer, i.e. its cohesion energy (E_{coh}), Crine calls on the energy brought into play during the application of an electromechanical stress σ on an activation volume ΔV. W can then be written in the form $W = \Delta V.\sigma$.

For fields $F > F_C$ (critical field), equation [9.13] is reduced to:

$$t = (h/2kT)\exp[(\Delta G_0 - e\lambda F)/kT] \qquad [9.14]$$

which means that for high fields, there is a regime where t varies exponentially with F. Knowing ΔG and λ from the exponential regime, it is then possible to calculate t for weak fields ($F < F_C$), i.e. by using equation [9.14]. However the experimental results of Equation [9.14] indicate that λ is constant only in the exponential regime and it varies with F for weak fields. It is therefore fundamental to determine the value of this critical field F_C from experimental data, which are the thermodynamic parameters suitable for polymers.

For a large range of temperatures (from 77 K to fusion temperature), Crine *et al.* showed that the dielectric behavior of numerous polymers depends on the cohesion energy density of the polymer defined as the cohesion energy (E_{coh}) divided by the molar volume (V).

If the Van der Waals bond breaking is the first stage of the mechanical breakdown, it implies that the deformation energy W is greater than the cohesion energy E_{coh} of the polymer:

$$W \succ E_{coh} = \Delta H_{vaporization} - kT \qquad [9.15]$$

where ΔH is the vaporization heat of the polymer [CRI 97].

By hypothesis, W then corresponds to a strain exerted on the polymer, which is:

$$W = \Delta V * \sigma \qquad [9.16]$$

We easily deduce from equations [9.15] and [9.16] the critical stress σ_C, above which the submicronic cavities can be initiated:

$$\sigma_c = E_{coh} / \Delta V \qquad [9.17]$$

The critical electrical field required to significantly distort the Van der Waals bonds is that which brings an energy (or effort) greater than the cohesion energy, so:

$$E_c = E_{coh} / e\lambda_{\max} \qquad [9.18]$$

The most interesting aspect of the Crine model is that it permits the parameter λ to be adapted to the morphology of the polymer. With a weak field, the value of λ deduced from equation [9.13] is weaker than that deduced from equation [9.14]. This allows us to say that λ varies with the field F until a maximum value λ_{max} for $F_C > 20$–40 kV/mm.

According to Crine, we can then:

– deduce the fundamental thermodynamic parameters. The value of the free volume $\Delta V^{\#}$ of an activated process can be calculated by using activation enthalpy and entropy and not the activation energy which is often difficult to identify;

– determine the required time for the breaking of intermolecular bonds from a critical field (F_C), i.e. the required time for the creation of submicronic cavities and therefore breakdown initiator sites. This threshold is also identifiable from thermodynamic parameters such as cohesion energy (E_{coh}) and parameters specific to the polymer;

– affirm that the first electrical ageing process of polymers is a fatigue phenomenon induced by the application of an alternative electrical field. This phenomenon is accelerated by frequency. It essentially constrains the Van der Waals type bonds, i.e. intermolecular bonds; hence, this is the principal difference between this model and that of Zhurkov, which does not show interest in electron displacement because covalent bonds are involved.

Amongst the parameters that the Crine model does not take into account, we particularly note the space charge (see Chapter 6), which other authors have been very interested in [DIS 95], [DIS 92], and the distribution of submicronic cavities, etc.

9.4. Microscopic approach according to Dissado–Mazzanti–Montanari

According to Dissado, Mazzanti and Montanari, any damage is associated with the presence of micro rather than macro defects. Microscopic or mesoscopic defects are associated with the presence of space charge regions. These latter are chiefly trapped at the extremities of cavities present in the insulating material or are localized at the level of inclusions and contaminant, whether conductors or not. They significantly modify the local field, increasing the ageing reaction rate and can introduce new degradation modes. Therefore, the distribution of these space charges in the material creates local fields. This is the basis for the development of the Dissado, Mazzanti and Montanari model (DMM) for ageing related to space charges [DIS 97].

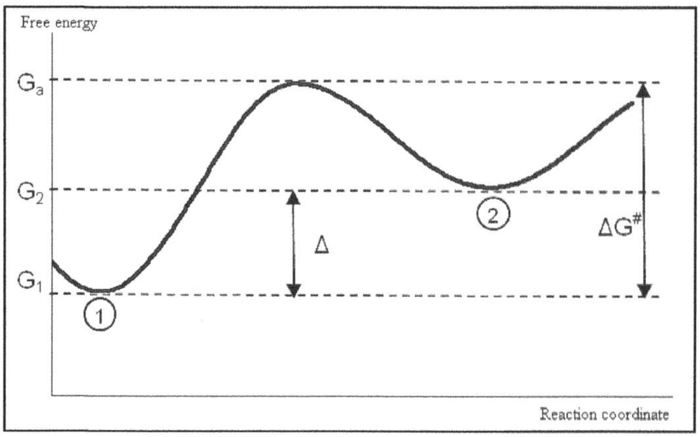

Figure 9.3. *Diagram of the free energy in the absence of electrical current [DIS 95]*

This model is therefore based on a thermodynamic approach describing the degradation rate of insulating materials submitted to thermal and electrical stresses. A polymer insulation is composed of a set of units each subjected to local reactions. The degradation of the material is the consequence of the net evolution of those local reactions which changes the morphology of the polymer in such a way to set off a rapid dielectric breakdown process. The ageing process is described according to a reaction which allows the polymer to pass from a initial non-aged state (state 1) to an aged final state (state 2), the passage from (1) to (2) possibly being reversible. To each of these two states is associated a free energy for each entity (Gibbs energy), G_1 and G_2 respectively (Figure 9.3). These two alternative states are connected by an energy barrier ΔG^* and the two reversible reactions take place with rates depending on species concentrations in both local states and the energy barrier height which links these two states [DIS 95].

9.4.1. *Thermal ageing*

Beyond a critical temperature which allows a thermal degradation phenomenon to be set off, the fraction of concentrations in a dielectric, defined by $A = c_2 / (c_1 + c_2)$ (where c_1 and c_2 are the species concentrations present in states 1 and 2, respectively) undergoes a passage towards a value at the equilibrium Aeq. This value essentially depends on the absolute temperature T and the difference of the free energy for each entity between states 1 and 2:

$$A_{eq} = \frac{1}{1+\exp(\Delta / kT)} \qquad [9.19]$$

with Δ = G2-G1 and k the Boltzmann constant. In the absence of applied external stresses, the conversion to a new equilibrium value is supposed to be weak in insulating materials. The application of external stresses leads to a change in local energies, thus causing the reduction of Δ and favoring a conversion towards the equilibrium state.

The value of A at an ageing time t, $A(t)$ is given by the expression:

$$\ln\left[\frac{1-A(t)/A_{eq}}{1-A(0)/A_{eq}}\right] = -(K_1 + K_2)t \qquad [9.20]$$

With $A(0)$ the initial value of A which will subsequently be supposed to be zero, K_1 and K_2 are the rate constants of the degradation reversible reaction in both directions, for 1 and 2 respectively.

The breakdown takes place when a critical quantity of units passes from the initial to the final state. The fraction of concentrations then reaches a limit value A^*.

Thus equation [9.20] is written in the following way by replacing A by A^* and t by L (the lifetime of the insulating material):

$$L = (K_1 + K_2)^{-1} \ln\left[\frac{1}{1 - A^* / A_{eq}}\right] \qquad [9.21]$$

By introducing the expressions of K_1 and K_2 in equation [9.21], we get:

$$L(T) = \frac{h}{kT}\,\frac{\exp(\Delta_G / kT)\ln[A_{eq} /(A_{eq} - A^*)]}{\exp(-\Delta / 2kT) + \exp(\Delta / 2kT)} \qquad [9.22]$$

where h is the Planck constant, and ΔG is a parameter depending on the activation free energies of states 1 and 2, as follows [DIS 01]:

$$\Delta_G = G_a - \frac{G_1 + G_2}{2} \qquad [9.23]$$

The necessity to reach a critical value of conversion A^* justifies the existence of a threshold temperature for the damaging of insulating materials [DIS 95]:

$$T_{th} = \frac{\Delta}{kT\left[\ln\left(\frac{1-A^*}{A^*}\right)\right]} \qquad [9.24]$$

9.4.2. *Ageing under electrical field: space charges effect*

According to Dissado, Mazzanti and Montanari, when an electrical field is applied, space charges injected from electrodes, semi-conductors or discharge phenomena move towards sites where they can be trapped. In organic materials, the distribution of the charge trapping sites is not uniform within the material. This is the case for PE in which the sites are situated at the interface of the amorphous and crystalline zones. Consequently, local charge concentrations are formed. This local aspect takes place especially around morphological defects, inclusions and impurities. The shape, size and density of microcavities can vary significantly according to the materials and the insulating systems [DIS 95].

The Dissado, Mazzanti and Montanari (DMM) model describes the formation time of defects (microcavities, holes) in the insulating materials as a function of the electrical field and the applied thermal stresses. The model was initially developed in direct regime. The model parameters significantly change when dealing with data in an alternating current. Consequently, an addition, for an alternating regime, was then proposed.

In order to simplify the analytical approach for obtaining the model, the following hypotheses have been taken into account:

– charges accumulated at the charge centers are injected from electrodes such that the injection current is insufficient to generate a direct breakdown. The trapped charges get to equilibrium in a much shorter time than the insulation life;

– close to the charge centers, the external field (Laplacian) E is negligible with respect to the field created by the space charges;

– charge centers are spherical, of radius r_0, with a uniform charge density and all possess the same amount of charge q_c which depends on the applied field, $q_c = CE^b$;

– a number N_m of species or "moieties" is affected on average in each microcavity. These species reside within a thin shell of thickness λ at the cavity wall;

– the material is homogenous at the macroscopic scale [MAZ 05].

In the presence of an electrical field, the system reaches a global disequilibrium state and goes through modifications over time, in order to reach a new thermodynamic equilibrium.

The effect of trapped space charges is to increase the local electrical field which reaches its maximum value at the extremities of the space charge regions. Consequently, elementary electrostatic and electro-mechanical stresses and strains are created in these regions with a local storage of electrostatic and electro-mechanical energy. This effect concerns the species in their initial state whose free energy G_1 increases by a value ΔG_m (see Figure 9.4), according to the equation:

$$G_1(E) = G_1 + \Delta G_m \qquad [9.25]$$

with:

$$\Delta G_m = AW_{es} + BW_{em} \qquad [9.26]$$

where W_{es} and W_{em} are, respectively, the electrostatic and electro-mechanical energies, A and B are proportionality constants with A and B ≤ 1. The detailed expressions of the energy variation are explained in [DIS 95].

The energy barrier of the degradation reaction (direction 1) decreases (whilst that of the reaction in the opposite direction increases) proportionally to the stored electro-mechanical energy and so a net transfer to the reaction products (state 2) takes place. The breakdown is related to the fact that the fraction of the degraded species A exceeds, locally, a critical value A^* which corresponds to the creation of cavities sufficiently large to set off partial discharges, and consequently electrical arborescence phenomena.

Thus, the ageing process is essentially a sort of apparition of local progress, and rather than the time to break down, the DMM model determines the time required for the formation of voids in critical amount to trigger a breakdown.

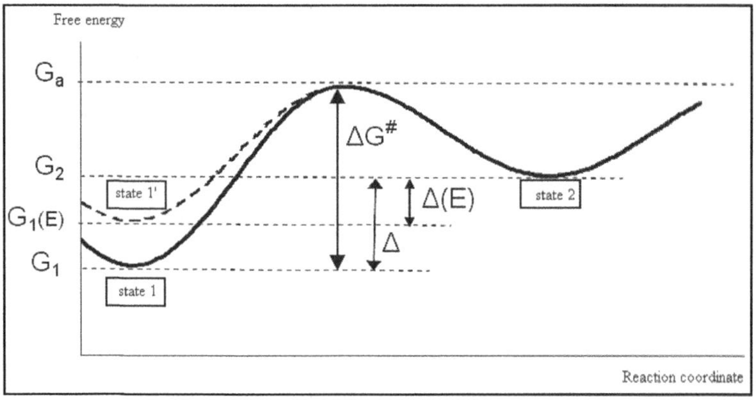

Figure 9.4. *Diagram of free energy in presence of electrical current [DIS 97]*

After the analytical development of these hypotheses, the final expression of the DMM model under a direct field is the following:

$$L(E,T)=\frac{h}{2kT}\exp\left[\frac{\frac{\Delta H}{k}-\frac{C'E^{2b}}{2}}{T}-\frac{\Delta S}{k}\right]\ln\left[\frac{A_{eq}(E)}{A_{eq}(E)-A^*}\right]\left[\cosh\left(\frac{\frac{\Delta}{k}-C'E^{2b}}{2T}\right)\right]^{-1} \quad [9.27]$$

$$\text{with: } A_{eq}(E)=\left[1+\exp\left(\frac{\frac{\Delta}{k}-C'E^{2b}}{T}\right)\right]^{-1} \quad [9.28]$$

with $Aeq(E)$ being the value of A at equilibrium between both the reactions (in opposite directions), T is the absolute temperature, ΔH and ΔS are, respectively, the activation enthalpy and entropy for each species, and Δ is the free energy difference between the reactants and the products. C' is a constant which essentially depends on the properties of the materials:

$$C'=\frac{B\alpha_E\delta\lambda C^2}{16\pi\varepsilon^2 r_0^2 N_m k} \quad [9.29]$$

where B is a proportionality constant between the stored electro-mechanical energy and its contribution to the free energy barrier of the degradation reaction, δ is the elementary deformation and α_E is the electrostriction coefficient. C and B relate the previous expression to the relationship between the stored charges in the charge centers and the applied field [MAZ 05].

To conclude, the proposed thermo-electrical model essentially relies on the idea that, following the application of a thermo-electrical stress, the space charges trapped within an insulating polymer material contribute considerably to the ageing phenomenon of the material. The effect of the space charges is to create local fields concentrated around charge centers. This local aspect generates electro-mechanical stresses which modify the free energies associated with two states brought into play for the definition of the ageing reaction. Since the ageing reaction rate is led by these free energies, it will in turn be modified by them.

9.5. Conclusions and perspectives

All these approaches allow us to consider that the long-term breakdown of insulating materials is not, in general, a direct and simple consequence of the application of a voltage, but a much more complex process related to a synergy of environmental stresses, such as the accumulation of charges, a local temperature increase, direct or induced mechanical stresses, UV radiation, humidity rate, etc.

These many factors contribute to a deterioration of physical properties of the material and, consequently, to its reliability as an electrical insulation.

A high number of mechanisms are involved in the description of the ageing process of organic insulating materials. Two principal approaches have been tackled. First, the model presented by the Russian school, which could be considered as a macroscopic approach. This approach keeps a global description of the degradation process based on the study of solid insulating samples or on systems. Second, the Crine and Dissado, Mazzanti and Montanari models offer a more microscopic approach where, on the one hand, we place ourselves upstream from previous phenomena and, on the other hand, the behavior of the solid insulating material is taken into account at the molecular scale.

The first two theories which we have briefly analyzed have the mechanical behavior of the polymer in common, but with an essential difference concerning the nature of this behavior. The tensile stress studied by Zhurkov leads to the breaking of interatomic chemical bonds, and is revealed in particular by the formation of free radicals. Covalent bonds are then involved and the breakdown is irreversible. This theory does not tackle the study of displacement phenomenon within the material of entities such as electrons, space charges, etc. Tensile stress would cause, according to Crine, deformation and chain rupture in the free volume where the Van der Waals bonds are especially affected. In this case the damage remains reversible until a critical value of the deformation is reached. In both theories, breakdown leads to the initiation of submicronic cavities which could become the site of discharges.

In the theoretical approach tackled in the DMM model, the ageing phenomenon essentially takes place because of the formation of space charge centers during the application of a field. A threshold voltage is then defined as a value from which the local field created by the charge centers is sufficient to accelerate the ageing phenomenon and trigger a rapid breakdown within the material.

The theorical approaches of breakdown mechanisms of electrical insulating materials have, of course, the simple understanding of phenomena as a first objective, then their application to real systems for which security and lifetime problems are posed. The prediction of a long or short-term breakdown is difficult to make if we want to take into account all of the intervening parameters.

Thus, it can be seen that there are numerous attempts at ageing models, each with their own difficulties. A two-site model has been systematically proposed by different authors for the representation of the free energy variation between two states to define the ageing phenomenon. However, the energies in polymers are associated with distribution and relaxation functions. In this case, the two-site model is not representative in defining the nature of entity displacements within a polymer material.

Among the simplifications of the existing models, one consists of considering the structure of the organic material as homogenous. However, most industrial electrical organic insulating materials are semi-crystalline, i.e. constituted of an amorphous and a crystalline phase; hence, the necessity of introducing in these models the notions of anisotropy and heterogeneity. Further, several parameters which could have an influence on the ageing phenomenon are not considered in the already existing models. We could mention, for example, the role of the temperature to which the material is exposed with respect to the characteristic temperatures of that material, such as the glass transition temperature T_g, the melting temperature T_m and the crystallization temperature T_c.

Thus, the issues of ageing and the breakdown of solid organic insulating materials are, and will no doubt remain for many years to come, the subject of long investigations.

9.6. Bibliography

[ART 96] ARTBAUER J., "Electric strength of polymers", *Journal of Physics D, Appl. Phys.*, vol. 29, no. 2, p. 446–456, 1996.

[BAR 67] BARKER R.E., "Grüneisen numbers for polymeric solids", *J. Appl. Phys.*, vol. 38, no. 11, p. 4234–4242, 1967.

[BLA 95] BLAISE G., "Charge Trapping/Detrapping induced lattice polarization/relaxation processes", *IEEE 1995 Annual Report, IEEE Conf. Elect. Insul. Dielect. Phen*, no. 95CH35842, p. 37–39, 1995.

[CRI 84] CRINE J.P., "A thermodynamic model for the compensation law and its physical significance for polymers", *J. Macromol. Sci.-Phys.*, vol. B23, no. 2, p. 201–219, 1984.

[CRI 90] CRINE J.P., "A model of solid dielectric ageing", *Conference Records of the 1990 IEEE International Symposium on EI*, Toronto, Canada, p.25–26, 1990.

[CRI 97] CRINE J.P., "A molecular model to evaluate the impact of ageing on space charges in polymers dielectrics", *IEEE Trans. on Dielec. and Elec. Insul.*, vol. 3, no. 2, 1996.

[DAN 96] DANG C., PARPAL J.L, CRINE J.P., "Electrical aging of extruded dielectric cables – Review of existing theories data", *IEEE Trans. on Dielec. and Elec. Insul.*, vol. 3, no. 2, 1996.

[DAV 94] DAVID E., PARPAL J.L., CRINE J.P, "Influence of internal stress and strain on electrical treeing in XLPE", *IEEE 1994 Annual Report, IEEE Conf. Elect. Insul. Dielect. Phen*, p. 575-581, 1994.

[DIS 01] DISSADO L.A., MAZZANTI G., MONTANARI G.C., "Elemental strain and trapped space charge in thermo electrical ageing of insulating materials. Part 1: Elemental strain under thermo electrical – mechanical stress", *IEEE Trans. on Dielec. and Elec. Insul*, vol. 8, no. 6, p. 959–965, 2001.

[DIS 92] DISSADO L., FOTHERGILL J., "Electrical degradation and breakdown in polymers", In *IEE Materials and Devices, Series 9*, STEVENS, G.C. (eds), 1992.

[DIS 95] DISSADO L., MAZZANTI G., MONTANARI G.C., "The incorporation of space charge degradation in the life model for electrical insulating materials", *IEEE Trans. on Dielec. and Elec. Insul*, vol. 2, no. 6, p. 1147–1158, 1995.

[DIS 97] DISSADO L.A., MAZZANTI G., MONTANARI G.C., "The role of trapped space charges in the electrical aging of insulating materials", *IEEE Trans. on Dielec. and Elec. Insul*, vol. 4, no. 5, p. 496–506, 1997.

[KEY 58] KEYES R.W., "Volume of activation for diffusion in solids", *J. Chem. Phys.*, vol. 29, no. 3, p. 467–475, 1958.

[KUS 79] KUSOV A.A., "Phonon model of breaking of a loaded atomic chain", *Sov. Phys. Solid State*, vol. 21, issue 10, p. 1781–1783, 1979.

[LAU 99] LAURENT C., "Diélectriques solides et charge d'espace", *Techniques de l'Ingénieur, traité Génie électrique*, D 2 305, p. 1–13, 1999.

[LAW 57] LAWSON A.W., "Diffusion of impurities in amorphous polymers", *J. of Chemical-Physics*, vol. 32, no. 1, p. 131–132, 1960.

[MAZ 05] MAZZANTI G., MONTANARI G.C., DISSADO L.A., "Electrical ageing and life models: The role of space charge", *IEEE Trans. on Dielec. and Elec. Insul*, vol. 12, no. 5, p. 876–890, 2005.

[PAR 97] PARPAL J.L., CRINE J.P., DANG C., "Electrical ageing of extruded dielectric cables. Part II - A physical model", *IEEE Trans. on Dielec. and Elec. Insul.*, vol. 4, no. 2, p. 197–209, 1997.

[PET 83] PETROV V.A., "Dilaton model of thermal fluctuation crack nucleation", *Sov. Phys. Solid State*, vol. 25, issue 10, p. 1800–1802, 1983.

[ZHU 65] ZHURKOV S.N., "Kinetic concept of the strength of solids", *Fracture Mechanics*, vol. 1, no. 4, p. 311–323, 1965.

[ZHU 66] ZHURKOV S.N., TOMASHEVSKI E.E., "An investigation of fracture process of polymers by the electron spin resonance method", In *Proceedings of the Conference: Physical basis of yield and fracture*, Oxford, p. 200–208, 1966.

[ZHU 69] ZHURKOV S.N., VETTEGREN V.I., KORSUKOV V.E., NOVAK I.I., "Infrared spectroscopic study of the chemical bonds in stressed polymers", *FTT*, paper 47, p. 545–550, 1969.

[ZHU 72] ZHURKOV S.N., ZAKREVSKYI V.A., KORSUKOV V.E., KUKSENKO V.S., "Mechanism of Submicrocrack Generation in Stressed Polymers", *Journal of Polymer Science*, Part A-2, vol. 10, p. 1509–1520, 1972.

[ZHU 83] ZHURKOV S.N., "Dilaton mechanism of the strength of solids", *Sov. Phys. Solid State*, vol. 25, issue 10, p. 1797–1800, 1983.

PART 3

Characterization Methods and Measurement

Chapter 10

Response of an Insulating Material to an Electric Charge: Measurement and Modeling

10.1. Introduction

In this chapter, we will take an interest in the observable response of an insulator subjected by its environment to different configurations of charge and electric field. The application of a DC stress was favored here, but the results presented could be extended to the case of a low frequency AC stress.

The response generally depends on a superposition of the internal mechanisms presented in the previous chapters. The material displays a specific order at the atomic scale (more or less deducible from its chemical formula), a structure at the nanometric scale due to the interactions between neighboring molecules (in the case of polymers), defects and impurities, a structure at the micrometric scale (spherulites for polyethylene, grains for polycrystals), together with surface phenomena, free surfaces, voids, and possibly interfaces when dealing with composite materials. Phenomena at each of these scales will have an influence on the general behavior of the insulator, at least in given conditions of field, pressure or temperature.

Research scientists know that theory in this domain is generally not able to predict accurately, for a given material, the results of even a simple an experiment. Therefore, our only ambition here is to show, with examples, what the phenomenological descriptions of the most frequent insulation behaviors are and on which physical hypotheses they rely.

Chapter written by Philippe MOLINIÉ.

10.2. Standard experiments

We shall consider the following two classical types of measurements:

– case (a): the environment sets the potential, and the insulator response determines the charge at its boundaries. This is the case of capacitors and of most circuits where an insulator is used to separate conductors. We measure in this case the current flowing through the insulator.

– case (b): the environment deposits a charge (or sets a current) on the surface, and the potential is determined by the properties of the insulator. This situation is the most widespread case when the insulator is not included in an electrical circuit. We measure in this case a surface voltage (a null field being imposed at the surface of the insulator).

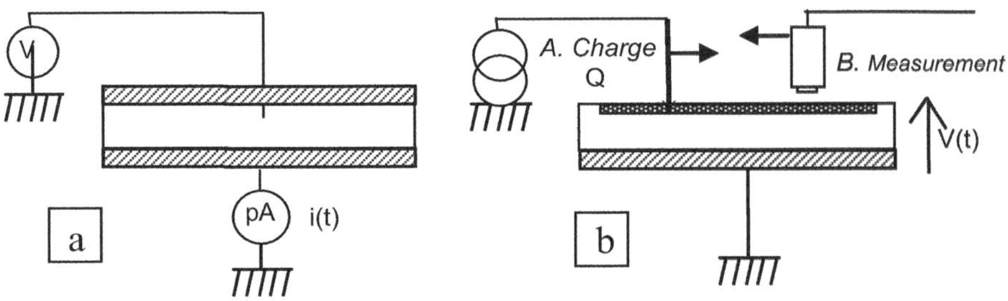

Figure 10.1. *Electrostatic measurements: (a) measurement of i(t) at fixed potential; (b) measurement of V(t) at fixed charge*

Case (a) is generally called a *closed* circuit, and case (b) an *open* circuit, though this definition is vague, as shown by the example of the "voltage decay" experiment, where the charge (by corona, or electron beam) of the open circuit (represented in 10.1(b)) generally begins at a constant current (case b), then continues at a constant potential (case a) after the surface has reached an equilibrium potential and, at the end of charging, the potential decays without any charge supply (case b again).

Similarly, it is generally assumed that case (a) is allowing a permanent regime, while case (b) corresponds to a transient regime. It is often true, but it is possible to measure a steady potential during the application of a current (an example being the mirror method), and current flowing from a short-circuited insulator has indeed to be transient.

10.3. Basic electrostatic equations

10.3.1. *General equations*

In electrostatics, the Maxwell equations relate the electrical displacement D ($D = \varepsilon_0 E + P$, where E is the electric field and P the polarization in the insulator), the charge density ρ and the current density j in each point of the insulator, by:

$$div\vec{D} = \rho \text{ (Poisson equation)} \tag{10.1}$$

$$\frac{\partial \rho}{\partial t} + div\vec{j} = 0 \text{ (continuity equation)} \tag{10.2}$$

We can deduce from this that $div\left(\frac{\partial \vec{D}}{\partial t} + \vec{j}\right) = 0$: the flux of $\frac{\partial \vec{D}}{\partial t} + \vec{j}$ is conservative, and therefore equal to the current measured in the external circuit. The quantity $\frac{\partial D}{\partial t}$ is called the *displacement current density*. It expresses the local polarization variations in the insulator.

The current density includes several components, in the general case, such as diffusion for instance. We will only consider here conduction current, which is equal to the sum, for the different charge carriers, of the products of the electric field by their mobility μi and their density ρi:

$$\vec{j} = \left(\sum_i \mu_i \rho_i\right)\vec{E} \tag{10.3}$$

If the material presents a given intrinsic conductivity σ, we shall separate this term to write $\vec{j} = \left(\sigma + \sum_i \mu_i \rho_i\right)\vec{E}$. The second part of the expression is then relative to a non-homogenous charge – often injected from the outside (a *space charge*). We therefore get:

$$\vec{j} = \frac{\partial \vec{D}}{\partial t} + \left(\sigma + \sum_i \mu_i \rho_i\right)\vec{E} \tag{10.4}$$

In the following sections, we shall examine the influence of these three physical parameters – dipolar polarization, intrinsic conduction, and space charge – on the observables (ground current (a), or surface potential (b)).

10.3.2. *Current measurement at a fixed potential: case (a)*

In this case, the average field is imposed by the voltage difference between the boundaries of the insulator and, from the null field condition inside the electrodes, the surface charge on them may be deduced. The application of a voltage difference on a neutral insulator will impose at each of its points a field initially proportional to the applied potential, and the polarization phenomena of the dielectric will give rise to a displacement current, first as an "instantaneous" peak of charge (whose amplitude only depends on the external circuit) equal to the flux of $\varepsilon_0 \dfrac{\partial \vec{E}}{\partial t}$ and corresponding to the vacuum polarization, and then a slow charging current, equal to the flux of $\dfrac{\partial \vec{P}}{\partial t}$, with response times corresponding to the characteristic polarization time constants of the different material components. The practical result of this gradual polarization, called absorption current, is a current which decreases in time.

The absorption current of dipolar origin will be superposed onto currents related to conduction, but, at low DC polarization fields, it could remain predominant for weeks within the material (Figure 10.3.), while for 50 Hz AC, it will represent the major part of the measured current.

Following insulator polarization, it is also possible to proceed to short-circuit discharge current measurements. The average field in an insulator being brought to zero, a depolarization current flows in the circuit, corresponding to the relaxation of what has been polarized during the charging. Equal charge and discharge currents (in absolute value) indicates a predominance of dipolar phenomena. Above a certain value of field and temperature, we observe a higher value of the charge current, which is a sign of the outbreak of a conduction favored by the field during charging.

10.3.3. *Voltage measurement at a fixed charge: case (b)*

The boundary conditions, in this case, are a zero potential at the lower side of the insulator, and a null field outside the insulator (this can be guaranteed by the use of an electrostatic voltmeter using a voltage feedback loop, but otherwise it remains true as long as the field in the air remains negligible compared with the field inside the insulator).

Integration of equation [10.1] (Gauss's theorem) then implies that the electrical displacement in every point of an insulator will be equal to the surface charge density existing between that point and the outside of the insulator. The charge distribution, being fixed initially from the outside, will then be the input quantity of the system. For a one-dimensional problem (wide plane capacitor of thickness L), the continuity equation may be written for every point as follows:

$$\frac{\partial D}{\partial t}+\left(\sigma+\sum_i \mu_i \rho_i\right)E = j_{ext} \qquad [10.5]$$

In the absence of conduction phenomena, the insulator charging could be described by $\varepsilon_0\frac{\partial E}{\partial t}+\frac{\partial P}{\partial t}=j_{ext}$, i.e. the potential rise will be: $\frac{\partial V}{\partial t}=\frac{1}{\varepsilon_0}\left(Lj_{ext}-\int_0^L\frac{\partial P}{\partial t}dx\right)$. Part of the applied charge will thus be compensated by the polarization phenomena, and the potential rise will then be weaker in the presence of the insulator. When a charging current is interrupted, the circuit is then open, and slow polarization phenomena will lead to a voltage decay $\frac{\partial V}{\partial t}=-\frac{1}{\varepsilon_0}\int_0^L\frac{\partial P}{\partial t}dx$.

In the case where polarization is stabilized, it can be described by a dielectric constant ε, and the measured voltage changes will then be related to conduction effects. We thus get:

$$\frac{\partial V}{V\partial t}=-\frac{1}{\varepsilon_0}\left(\sigma+\sum_i \mu_i \rho_i\right) \qquad [10.6]$$

In the same way as for current measurements, a neutralization of the insulator can be performed. However in this case, it is done by a transient transfer of charges on the insulator surface. This can be compared to a temporary short-circuit. We then measure a voltage return, which will essentially be due to the depolarization phenomena, but could also be the consequence of the return of a dissymmetric space charge towards the electrodes.

10.4. Dipolar polarization

The dielectric relaxation of most insulators, notably polymers, has a very low frequency component, related at the same time to internal molecular reorganizations, and to complex interfacial polarization phenomena. In a linear regime, we can model this response using the dielectric functions $\phi_D(t)$ and $\phi_E(t)$ involved in convolution relations between the electrical displacement and field:

$$\text{(a)}.\ D(t)=\varepsilon_0\int_{-\infty}^{t}E(\tau)\varphi_D(t-\tau)d\tau \qquad \text{(b)}.\ E(t)=\frac{1}{\varepsilon_0}\int_{-\infty}^{t}D(\tau)\varphi_E(t-\tau)d\tau \qquad [10.7]$$

ϕ_D and ϕ_E are not independent, the product of their Laplace transform being 1.

For a homogenous dielectric of thickness L, only surface-charged, the electric field will be constant in the insulator, and related to the potential by $E=V/L$. The displacement will then be equal to the free charge density q on the surface.

In the case of the application at $t=0$ of a voltage step ($V(t)=\Gamma_0(t)V_0$), we may deduce from [10.7a] the insulator absorption current, proportional to ϕD (t) according to:

$$i(t)=S\frac{dD}{dt}=\frac{\varepsilon_0 SV_0}{L}\frac{d}{dt}\int_0^t \varphi_D(\theta)d\theta = C_0V_0\varphi_D(t) \qquad [10.8]$$

(S being the surface of the insulator, and C_0 its geometric capacitance).

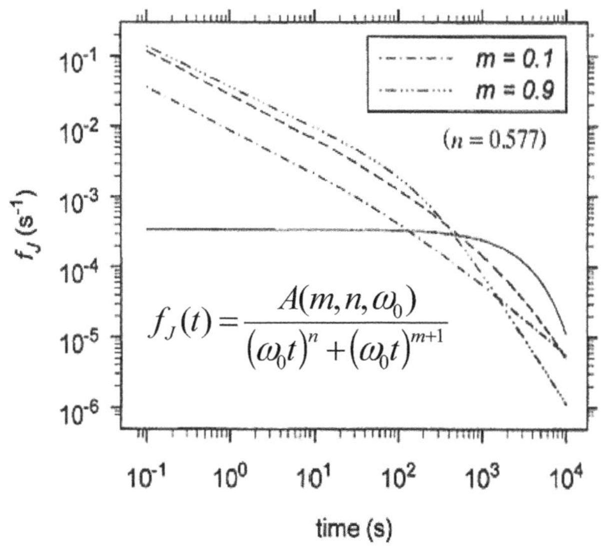

Figure 10.2. *Jonscher's response functions [BER 03]*

In the case of the deposit at $t=0$ of a charge amount q_0 on the surface ($q(t)=\Gamma_0(t)q_0$), we may deduce from [10.7b] the decay rate of the surface potential:

$$\frac{dV(t)}{dt}=\frac{1}{L}\frac{dE}{dt}=\frac{Lq_0}{\varepsilon_0}\frac{d}{dt}\int_0^t \varphi_E(\theta)d\theta=\frac{Lq_0}{\varepsilon_0}\varphi_E(t) \qquad [10.9]$$

For reasons studied by Jonscher [JON 96] in particular, dielectric functions in condensed matter follow time power laws, according to Figure 10.2 (where the function f_J is equal to ϕ_D for t>0).

The $log\phi_E(t)=f(logt)$ curve, and therefore the voltage decay, in the $log(dV/dt)=f(logt)$ plot, also tends to be composed of one or two straight line segments, but with slopes different to $log\phi_D(t)=f(logt)$ or to the absorption current.

The dipolar phenomena described by a dielectric function present a linear dependence on the value and the sign of the voltage or the applied charge. The effects produced during the depolarization experiment, with current or return voltage, are, then, easy to predict from the results of the polarization experiment.

10.4.1. *Examples*

Figure 10.3 shows an example of absorption current measurements on 15 μm thick polypropylene films. It is clear that, for an identical 67 kV/mm field, the charge and discharge currents are superimposed until a certain temperature threshold, between 300 and 340 K, corresponding to the outbreak of a genuine conductivity (probably ionic).

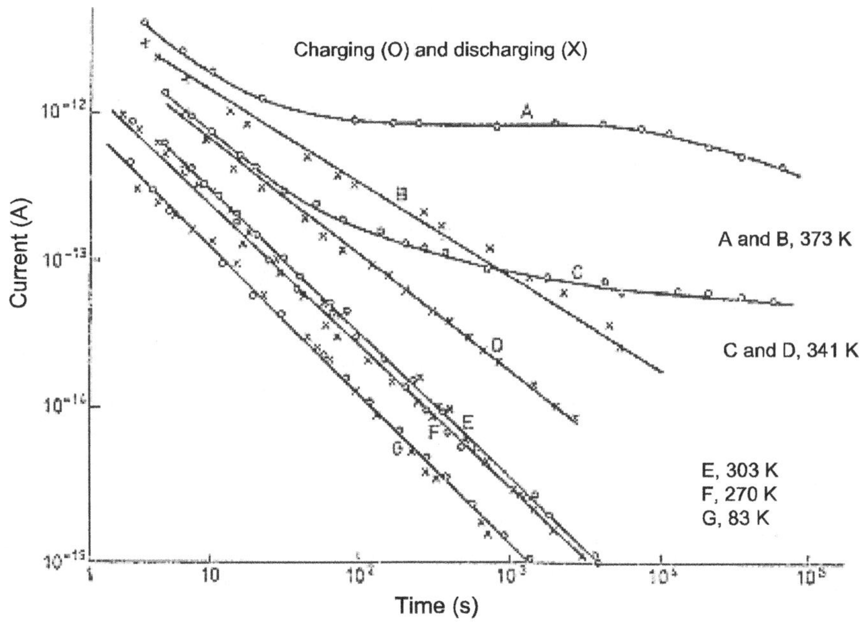

Figure 10.3. *Absorption currents on polypropylene films [DAS 76]*

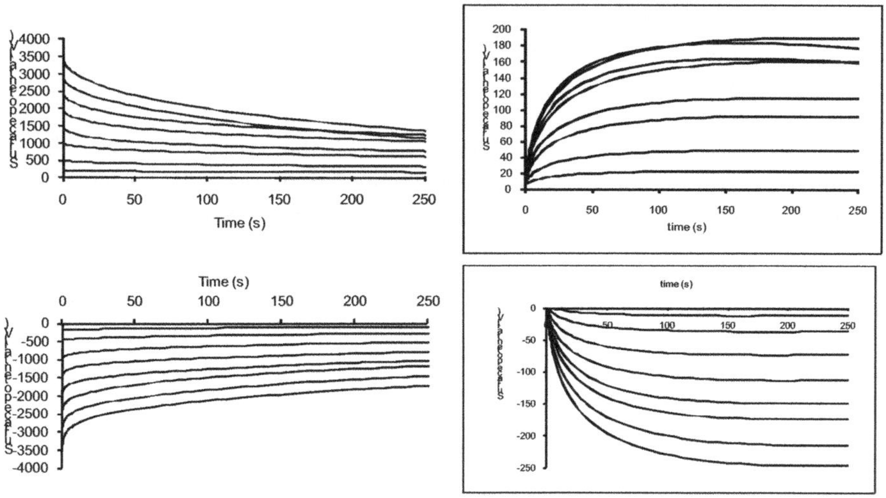

Figure 10.4. *Voltage decay and return for different initial voltages [LLO 02]*

Figure 10.4 shows the result of voltage decay and return experiments at 25°C on epoxy films (190µm thick). The dipolar phenomenon is predominant, and the non-linear effects, attributed to the charge injection, only appear in positive polarity for fields above 15 to 20 kV/mm; this result has been confirmed on the same samples by PEA measurements, detecting a charge injection above 18 kV/mm.

10.5. Intrinsic conduction

The existence of an intrinsic conduction σ induces a shielding of all net charge present in a material because equation [10.2] will in this case be $\frac{\partial\rho}{\partial t}+\sigma divE=0$, which leads to:

$$\frac{d\rho}{\rho dt}=-\frac{\sigma}{\varepsilon} \quad [10.10]$$

A constant intrinsic conductivity must therefore give rise to an exponential decay of any charge density present in the material, with a time constant equal to the product of its resistivity and its permittivity.

Intrinsic conduction is, however, in practice nearly zero at room temperature and a moderated field for most materials used in electrical insulation, whose energy bands present a large gap and, above all, an extremely weak effective carrier mobility. The observables measured in this case will first of all be of dipolar origin, or related to an injected space charge.

The outbreak of a intrinsic conduction within the volume requires the generation of a certain amount of mobile carriers, with a renewal rate which compensates for the recombinations. Only irradiation will be energetic enough to allow generation of electron-hole pairs by hopping over the forbidden band. However, for insulators presenting mainly shallow traps, or for materials at high temperature or subjected to a high field, the thermal detrapping of part of the inner trapped charge can lead to a non-zero intrinsic conductivity. This process, described by the Poole–Frenkel Law, manifests itself by a conductivity proportional to the exponential of the square root of the electric field, divided by kT.

10.5.1. *Example: charged insulator irradiated by a high-energy electron beam*

During a voltage decay measurement, when a genuine conductivity exists in an insulator, the decay must be exponential, following the relationship:

$$\sigma = -\frac{1}{\varepsilon}\frac{dV}{Vdt} \qquad [10.11]$$

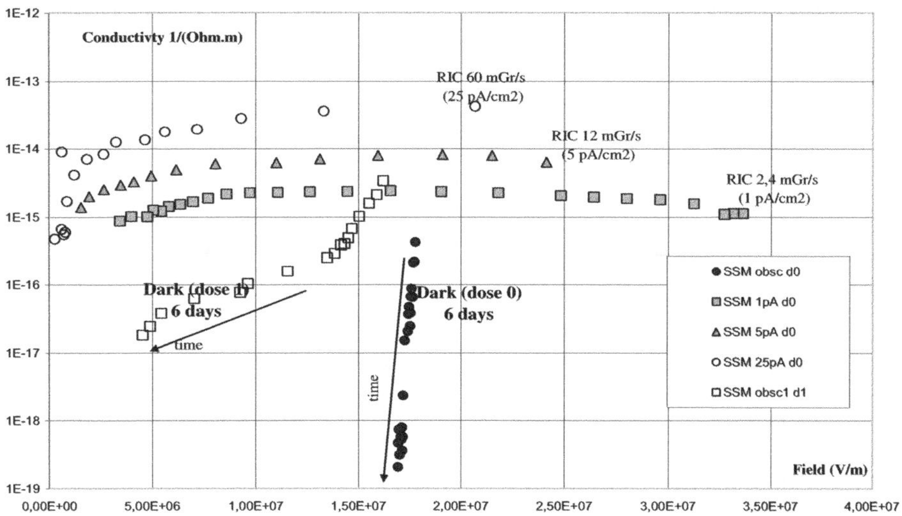

Figure 10.5. *Equivalent conductivity on irradiated SSM Teflon® plates [LEV06]*

This parameter is plotted in Figure 10.5, as a function of the average field in the insulator during a voltage decay measurement on Teflon samples charged by a 25 keV electron beam, and submitted to increasing irradiation doses by a 400 keV beam, flowing through the insulator. The effect of the irradiation is to create a genuine conduction, lacking in the case of a non-irradiated insulator, which is a function of the dose rate. We also see that this effect remains perceptible several days after stopping the ionizing beam.

10.6. Space charge, injection and charge transport

Numerous models concerning the injection and transport of space charges have been developed. We can roughly classify them in four groups, depending on the way the properties of the material and interfaces are taken into account concerning injection and charge transport. The first three groups are models in which the dynamic is assumed to be set by the volume of the material, whereas for the fourth group, it is determined by the interfaces.

10.6.1. *Electrostatic models*

Charge trapping and detrapping are not taken into account here, and the charge transport is only determined by a mobility value μ: the charge motion only depends here on the local electric field $\left(\vec{v}=\mu\vec{E}\right)$.

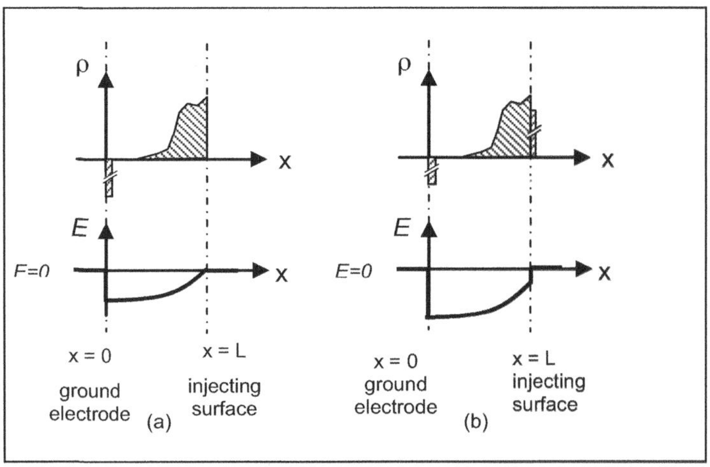

Figure 10.6. *Charge density and field: (a) open circuit, and (b) closed circuit*

Considering case (a), the average field in the insulator is fixed from the outside, but not the local field, which depends on the space charge inside the material. The field is therefore reduced, or even cancelled, in the insulator in the vicinity of the injecting electrode. The corollary of this lower field on one side of the material is an increase of the field – and therefore of the image charge – on the ground electrode: a current measurement on this electrode will hence detect the motion of the charge in the bulk of the material. The transient regime when a DC step voltage is applied is characterized by a rapid increase of the current during the injection phase, followed by its decay, due to the reduction of this injection caused by the decrease of the field at the injection electrode. In steady-state, we observe a characteristic space charge limited current (SCLC) regime, proportional to the square of the applied voltage.

Considering case (b), with a null field outside the insulator, the charge distribution can be considered as charge sheets successively injected into the material. Each of them is subjected to a constant field during its drift, proportional to the charge amount separating it from the surface (Gauss's theorem), so that it will move at constant speed, proportional to this amount; the distribution will therefore gradually broaden in a homothetical way. This type of model predicts a constant initial voltage decay rate dV/dt, and we find (as in the previous case) a quadratic dependence of this parameter as a function of the deposited charge. Further, the image charge on the ground electrode remains constant until complete transit of the first charge sheet (total electrostatic influence of the charge distribution on the ground plane). A current measurement on the ground electrode will therefore not detect the motion of the charges, unlike in the previous case, where the electrostatic influence of each charge is shared between both electrodes, and switches during the drift from the upper to the ground electrode.

10.6.1.1. *Example: transient current measurements on polyethylene films*

Figure 10.7 clearly shows the transition between a decreasing transient current regime, polarization-dominated, and a charge injection regime, above a field of about 15 to 20 kV/mm. The authors hesitate, however, to attribute the current peak to a transit time, as predicted by the theory evoked above. Indeed, we shall see later that trapping phenomena play a large part in the shape of such a curve.

10.6.2. *Models combining electrostatics and thermodynamics: the influence of trapping and dispersive transport*

A correct treatment of charge transport in a disordered material requires taking trapping into account. Trapping is a consequence of material disorder, but this disorder involves many different aspects, leading to a wide range of energies for the traps. The energetically shallow levels are usually consequences of a weak disorder, often more or less periodical: the depth of this kind of trap is typically less than 0.1 eV.

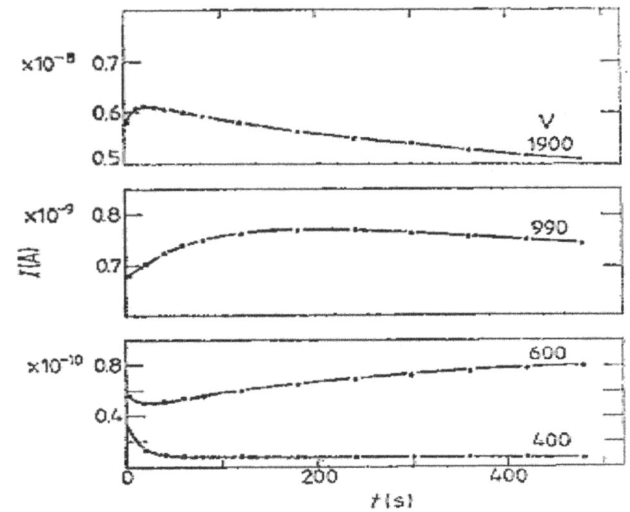

Figure 10.7. *Transient current on 30µm polyethylene films (at 57°C) [HAN 81]*

On the other hand, traps related to chemical defects, like oxygen vacancies in the oxides, generally produce much lower energy levels, around 3eV below the conduction band. This is considered deep trapping. The density of these deep levels is particularly high in polymers.

Modeling can include shallow traps by introducing an effective carrier mobility, representing the hopping conduction processes between the traps. In a closed circuit, the carriers are being renewed by an interface and absorbed by another, so that an equilibrium may rapidly be reached, when the occupation rate of these traps reaches a steady-state value everywhere in the material. This model is the same as the electrostatic model, with a change of the mobility value only. However, in an open circuit, the studied regime is necessarily transient, and we must take into account the dispersive character of the charge transport. This type of modeling has also been undertaken, assuming for instance a constant energy distribution of trap levels, or an exponential distribution.

Trapping in deep levels can be modeled as irreversible trapping, or by including a detrapping process – possibly assisted by the field and temperature. In the first case, the only possible permanent regime in a closed circuit (unless we reach trap saturation) is the complete freeze of the current by the trapped space charge; if we include detrapping, the current flowing into the insulator will be equal to the detrapped charge by a unit of time. This case is detailed in the following section.

10.6.3. *Purely thermodynamic models: current controlled by detrapping*

Time will play a key role in the evolution of charge distribution. At the beginning of the experiment, the charge distribution divides into shallow and superficial trap levels, proportional to the capture probabilities of these levels: the shallow traps being more numerous, the charge will therefore still be quite mobile. Then, it gradually gets trapped in deeper levels where it will become more stable. The charge's average mobility will therefore decrease with time. Over a long time, when the mean transit time of the charge carriers is weak compared with their mean characteristic detrapping time, we can completely eliminate the geometric factor from the models, the detrapping kinetics being then the only factor determining the equilibrium current or voltage.

This progressive charge trapping phenomenon may be described by a demarcation energy:

$$E_d(t) = -kT \ln\left[(N_c/N_t)(t/\tau_0)\right] \qquad [10.12]$$

N_c being the conduction states density, N_t the trap density, and τ_0 the carriers lifetime in the conduction states. This demarcation energy at a given instant t can be considered as the energy below which at this moment the trap emission can be neglected, and above which the trapping levels may be assumed to be in equilibrium with the transport states. This energy is thus the boundary between filled deep traps and empty shallower traps. Within this modeling frame, the evolution of the measured current or potential with time will be directly determined by the shape of the energetic distribution of the insulator traps.

10.6.3.1. *Example 1: short-circuit current during the discharge of a polyethylene film*

During conditioning under stress, we assumed that the various trap levels of the insulator were charged proportionally to their density. We may deduce from the previous reasoning that, during discharge, the traps' emission current at t is related to their energetic density at the corresponding demarcation energy $N(E_d)$:

$$I(t) \propto N(E_d(t))\frac{dE_d}{dt} = N(E_d(t))\frac{kT}{t}(N_t/N_c)\tau_0 \qquad [10.13]$$

For current measurement experiments in closed circuit, we can deduce that plotting $tI(t)=f(\log t)$ provides an image of $N(E)$ as a function of the trapping energy. This plot of the current during the discharge of a short-circuited polyethylene film after a 2h 800 V charge is given Figure 10.8.

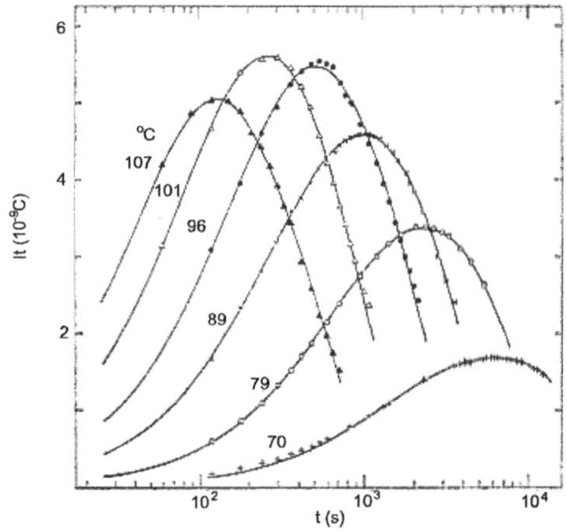

Figure 10.8. *It=f(logt) plot of the discharge current for a polyethylene film [HAN 81]*

10.6.3.2. *Example 2: voltage decay on a polystyrene film charged by an electron beam*

Displaying a voltage decay measurement using a *tdV/dt=f(logt)* plot will also give a representation of the trap density. Figure 10.9 presents measurements for three different values of temperature, which are superimposed by shifting them in a way to represent them directly as a function of energy. We must point out, however, that this calculation mode under-estimates shallow trap density, since it neglects the influence of retrapping during the charge drift.

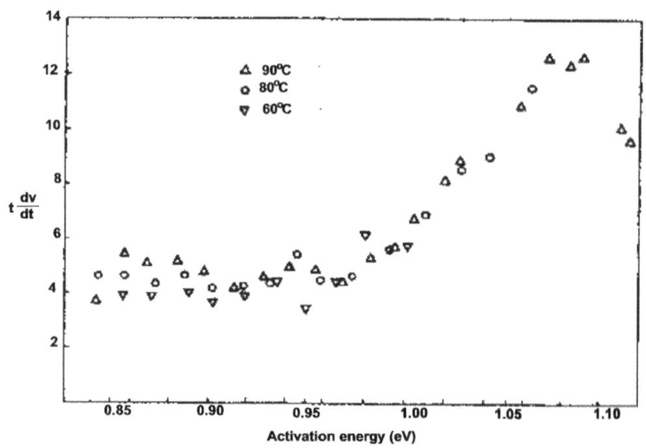

Figure 10.9. *Voltage decay tdV/dt=f(logt) plot for a polystyrene film [WAT 95]*

10.6.4. *Interface-limited charge injection*

The measured signal will be determined by the bulk phenomena evoked in the previous sections only if they are the main obstacle to charge displacement. In numerous cases, however, the measured signal will be determined by interface phenomena. The main reason for this is the energetic barrier existing at the metal–insulator contact, which also exists for surface charging by corona discharge ions.

In the first case (metal–insulator contact), the parameters for the barrier to cross essentially depend on the applied field. For moderated fields and temperatures above the ambient, the barrier will be crossed by means of thermoelectronic emission assisted by the field, and the Schottky law of current through this barrier will apply. For very strong fields, the barrier could be crossed without thermal assistance by field emission, and the law of current will then be the Fowler–Nordheim Law.

In the second case (with a non-metallic surface), the situation is more complex because the energies of deposited charges may be varied and they will evolve in time. We could then reuse models discussed in the case of gradual charge detrapping; the detrapping kinetics of the surface charge could determine the global behavior.

10.6.4.1. *Example 1: stationary current in polypropylene films*

Figure 10.10 shows current measurements on polypropylene films for different levels of field and temperature, using different data treatments. The graph on the left relies on a Schottky law hypothesis, i.e. a current limited by the thermoelectronic injection of carriers into the film. The determination of the permittivity value from the measured slopes makes this mechanism plausible, at least below 70°C and for moderated fields.

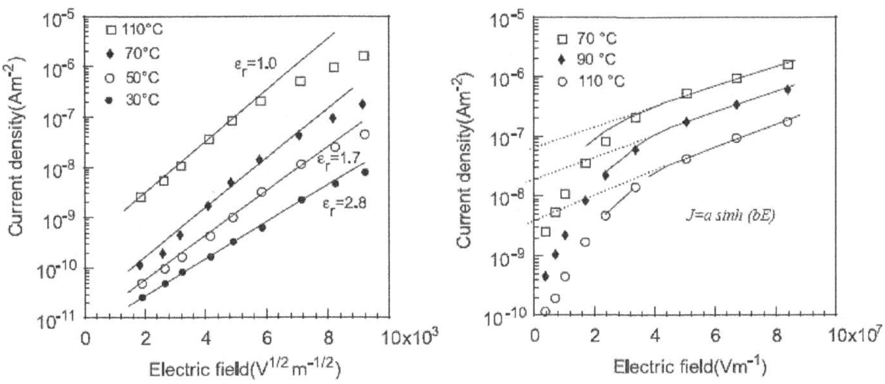

Figure 10.10. *Two representations of a stationary current [KIM 00]*

However, at 110°C as well as for fields exceeding 50kV/mm, the data treatment shown on the right, based on the hypothesis of a current limited by a hopping conduction in the volume of the insulator, is more suitable. Hence we come to the classical hypothesis of a transition, above a critical value of temperature and field, from a regime limited by electrodes to a regime limited by volume.

10.6.4.2. *Example 2: voltage decay on corona-charged films*

To account for voltage decay measurements on 6μm thick polypropylene films charged by corona at fields of about 200kV/mm, a decay limited by the detrapping of the surface charge had to be assumed. In this case, the charge density integrated within the film bulk was assumed to be weak compared with that present at the surface, and the transit time in the film assumed to be negligible compared with the characteristic time of surface charge detrapping [LLO 04].

10.7. Which model for which material?

The key issue to operate the variety of models presented here is to assess their validity for a practical situation with a given material and experimental conditions. The literature presents much experimental data, as current or voltage measurements, associated with the development of a model, followed by a discussion showing a good agreement between computed and experimental data. However, most often this evidence is not convincing because models based on different physical hypotheses, such as polarization of a disordered material, dispersive transport of an injected charge or gradual charge detrapping, may lead to the same time dependence of the measured value, composed of power laws, as shown in Figure 10.2. Nonetheless, the physical significance of the bend in the plot depends on the model: in the first case, it is a characteristic dipolar *relaxation* time; in the second, it is a spatial *transit* time; and in the third, it is a characteristic *detrapping* time. It is important to be aware of this difficulty and to carry out different tests to discriminate the phenomena as precisely as possible. Using sophisticated volume space charge measurement techniques may also provide important additional information, though it does not always allow injected charges from polarization heterogeneities to be distinguished.

In any case, the predominant mechanisms will depend on the stress level applied to the insulator. Thus, a measured current will generally appear, at low fields, with an ohmic appearance and be of dipolar origin. By increasing the field, the following scenario often occurs: a charge injection mechanism leads first, for intermediate fields, to a strong dependence on electrodes and field level described by a Schottky law, until a given field threshold is reached above which the current becomes volume-limited, implying a SCLC or hopping conduction law. At very high fields, electronic avalanches or hot electron phenomena will occur and lead to pre-disruptive pulsed phenomena.

Last but not least, the nature of carriers and microscopic conduction mechanisms is an important and difficult question which has not been considered here, where we favored a more macroscopic approach. The various mechanisms listed here do not rely on a particular hypothesis about the nature of charge carriers (whether electrons, holes, ions, etc.).

Insulator modeling cannot be reduced to considering one aspect or two (such as permittivity or conductivity), but reveals a wide diversity of phenomena. It cannot be done without testing the model with a variety of measurements.

10.8. Bibliography

[BER 03] BERLEZE S. M., ROBERT R., "Response functions and after effect in dielectrics", *IEEE Tr. Diel.&El. Ins.*, vol. 10, no. 4, p. 665–669, 2003.

[COE 93] COELHO R., ALADENIZE B., *Les Diélectriques*, Hermès, Paris, 1993.

[DAS 76] DAS GUPTA D. K., JOYNER K., "A study of absorption currents in polypropylene", *J.Phys. D: Appl. Phys.*, vol. 9, no. 14, p. 2041–2048, 1976.

[FOU 00] FOURNIÉ R., COELHO R., "Diélectriques. Bases théoriques", *Techniques de l'Ingénieur*, D2300, p. 1–18, février 2000.

[FOU 90] FOURNIÉ R., *Les isolants en électrotechnique: essais, mécanismes de dégradation, applications industrielles*, Eyrolles, Paris, 1990.

[HAN 81] HANSCOMB J.R., GEORGE E.P., "Space charge and short-circuit discharge in high-density Polythene", *J. Phys. D: Appl. Phys.*, vol.14, no. 12, p. 2285–94, 1981.

[JON 96] JONSCHER A. K., *Universal Relaxation Law*, Chelsea Dielectrics Press, London, 1996.

[KIM 00] KIM D.W., YOSHINO K., "Morphological characteristics and electrical conduction in syndiotactic polypropylene", *J. Phys. D: Appl. Phys.*, vol. 33, no. 4, p. 464–471, 2000.

[LEV 06] LÉVY L., DIRASSEN B., REULET R., VAN EESBEEK M., MOLINIÉ P., "Dark and radiation induced conductivity on space used external coatings", *10th Int. Symp. on Materials in a Space Environment,* Collioure, France, 2006.

[LLO 02] LLOVERA P., Etude des mécanismes d'injection de charge dans les matériaux isolants au moyen de mesures électrostatiques de déclin et retour de potentiel. Nouveaux outils d'analyse, Doctoral Thesis, Paris XI University, 2002.

[LLO 04] LLOVERA P., MOLINIÉ P., "New methodology for surface potential decay measurements – application to study charge injection dynamics on polypropylene films", *IEEE Tr. Diel.& El. Ins.*, vol. 11, no. 6, p. 1049–1056, 2004.

[MOL 05] MOLINIÉ P., "Measuring and Modeling Transient Insulator Response to Charging: the Contribution of Surface Potential Studies", *IEEE Tr. Diel.& El.Ins.,* vol.12, no. 5, p. 939–950, 2005.

[RAJ 03] RAJU G.G. *Dielectrics in Electric Fields*, Marcel Dekker, New York, 2003

[SEG 00] SÉGUI I. "Diélectriques. Courants de conduction", *Techniques de l'Ingénieur*, D2301, no. 5, p. 1–12, May 2000.

[WAT 95] WATSON P. K., "The transport and trapping of electrons in polymers", *IEEE Tr. Diel.& El. Ins.*, vol. 2, no. 5, p. 915–924, 1995.

Chapter 11

Pulsed Electroacoustic Method: Evolution and Development Perspectives for Space Charge Measurement

11.1. Introduction

The presence of space charges (SC) in materials is a phenomenon which has been studied for many years [IED 84], [SES 97]. However, understanding of the generation, transport, and trapping modes of these charges has not yet been achieved. Amongst the numerous techniques developed to detect the presence of SC in materials [TAK 99], this chapter focuses on the pulsed electroacoustic (PEA) method.

During the past 20 years, much progress has been made concerning charge distribution detection in dielectrics through this technique. How this method is performed is in constant evolution, but the operating principle, which will be described in this chapter, remains unchanged [MAE 88].

In the first part of the chapter, we will endeavor to describe the general operating mode of the PEA system, and then, in part two, we shall list the diverse assemblies developed for specific applications. We shall outline the current systems' limits and also the development perspectives that they open in diverse application areas.

Chapter written by Virginie GRISERI.

11.2. Principle of the method

11.2.1. *General context*

The charges which accumulate during the application of an external excitation (electric field, irradiation, etc.) modify the material's internal electric field and can cause an electrical breakdown in certain cases. The charges accumulation has been studied for a long time by techniques capable of giving information on the global quantity of charges, like the thermally stimulated depolarization current (TSDC) method [ENR 95] or surface potential measurement [MOL 00].

During the 1980s, diverse techniques were developed to study the charge distribution in the thickness of various samples. In most cases, a sample is placed between two electrodes; a charge displacement is imposed by an external stimulation, whose effect is to modify the influence charge on the electrodes. This modification is observed in the external circuit. The detected signal is transformed into a voltage variation through the sample in the case of a measurement made in open circuit, and in a current variation if we work in short-circuits. This type of measurement can only work if the external perturbation, which is applied to probe the material, undergoes a non-uniform modification in the volume direction and if its propagation is known over time.

The charges motion with respect to the electrodes can be controlled by a non-uniform expansion of the medium, as is the case when we heat one side of a sample [CHE 92] or by pressure wave generation [ALQ 81]. These methods, which are respectively known as thermal and acoustic methods, became popular very rapidly in Europe and North America. As its name indicates, the PEA method, whose principle we shall outline in detail later, belongs to the second category. Indeed, under the influence of a pulse voltage, the charges present in the studied material generate an acoustic wave while moving. This wave is transformed into an electrical signal by a piezoelectric captor for the analysis. This measurement technique was initially developed in Japan. Since the 1990s, this method, used in a routine manner in laboratories and on industrial sites, has also been well established in Europe [MON 00], [ALI 98], [BOD 04], [GRI 04].

11.2.2. *PEA device*

In a classic device (see Figure 11.1), the tested material is placed between two electrodes. From the upper electrode, it is possible to polarize the sample and apply voltage pulses which play the role of the probe to make the measurements.

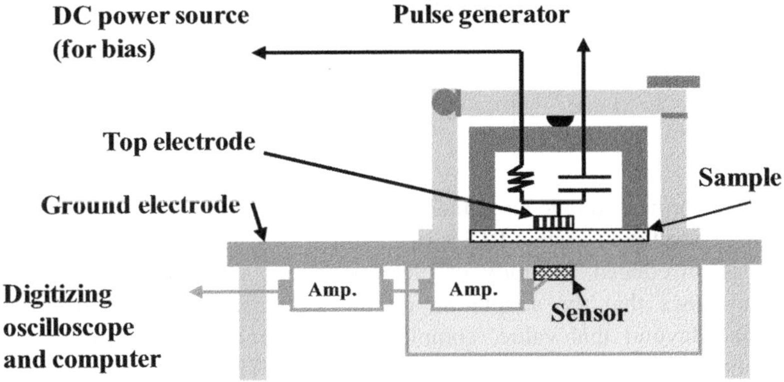

Figure 11.1. *Classic PEA device*

One of the first conditions to obtain a charge profile is to use a pulse width much narrower than the transit time of the acoustic waves through the sample. Under the pulses influence, the volume or surface charges are moved in a sudden way by the Coulomb force. The moment exchange between the electrical charges connected to the atoms of the dielectric creates an acoustic wave, which then moves in the sample at the propagation speed of sound in the considered medium. A piezoelectric captor made of Polyvinylidene (PVDF) or Lithium Niobate (LiNbO3) converts this acoustic wave into an electrical signal, which is then amplified by preamplifiers before being visualized in an oscilloscope. The output signal width is closely related to that of the applied pulses.

Further, the resolution depends on the voltage pulse transmission quality at the sample. For this purpose, an electrical circuit composed of discrete resistances permits the transport line of the pulse to the electrode to be adapted to 50 Ω. A decoupling capacitance (typically C_d=250 pF) is placed between the adaptation circuit and the electrode to allow the simultaneous application of a polarization voltage. This capacitance being much bigger than that of the sample which is typically of the order of C_s=20 pF, the resulting capacitance is:

$$C_r = \frac{C_d C_s}{C_d + C_s} \approx C_s \qquad [11.1]$$

If the voltage applied to the sample is in the form:

$$V(t) = V_o\left[1 - e^{\frac{-t}{RC_r}}\right] \qquad [11.2]$$

with V_0 the applied voltage amplitude, R the resistance of the adaptation circuit with value 50 Ω, the time constant RC_r is of the order of 1 ns, such that a weak distortion of V(t) is introduced. A study of electromagnetic compatibility with a network analyzer shows that the adaptation circuit was effective up to a frequency of 300 MHz. Beyond this value, complex resonance effects disturb the resistance bridge and the circuit impedance is no longer 50 Ω. The voltage pulses are typically supplied by a pulse generator with frequencies of the order of 400 Hz and amplitudes varying from 50 V to a few hundred volts. In general, pulses creating a field of about 1 kV/mm are selected to allow the measurement by perturbing the sample as little as possible.

A semi-conductor film is inserted between the excitation electrode and the sample in order to ensure a better acoustic impedance adaptation.

A high voltage is applied through a resistance of a few MΩ whose role is to limit the current in the case of sample breakdown.

The lower electrode is composed of an aluminum plate with thickness of about one centimeter. It plays the role of delay line between the analyzed acoustic signal and the electrostatic noise coming from the pulse generator. The piezoelectric film with surface of about 1 cm^2 and a few micrometers thick (typically 4 µm) is placed under the electrode. An absorber is positioned under the piezoelectric captor to avoid the signal reflections.

By using a reference signal, recorded beforehand, it is possible to associate the raw signal amplitude with a charge density; the delay giving access to the charges distance with respect to the sample surface. Owing to this non-destructive method, the spatial charge distribution can be determined in the volume perpendicularly to the material's surface.

11.2.3. *Measurement description*

We consider a sample of thickness '*d*' in which there is a negative charge excess 'q_2', situated at the distance '*x*' from an electrode. Neutrality is obtained because of the presence of influence charges q_1 and q_3 at each electrode at x=0 and x=d respectively.

$$q_1 = \frac{x-d}{d} q_2 \qquad [11.3]$$

$$q_3 = \frac{-x}{d} q_2 \qquad [11.4]$$

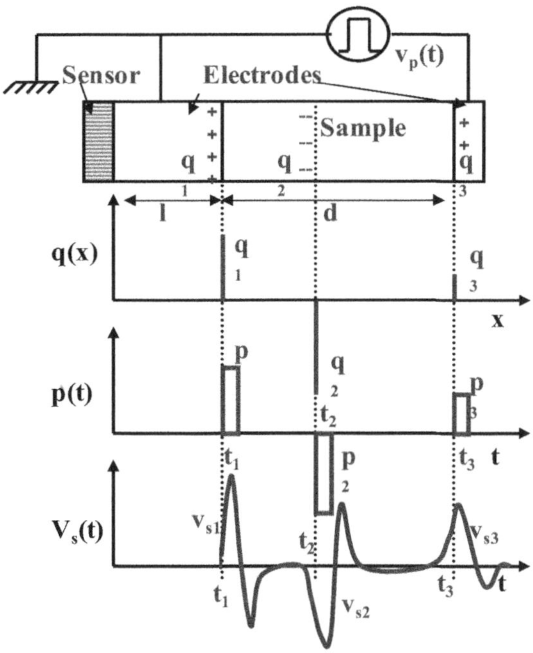

Figure 11.2. *Principle of the PEA method to obtain the charge distribution profile of a sample*

Under the action of short-duration pulses (between 5 and 9 ns), and with a high repetition frequency (400 Hz), charges are solicitated by the Coulomb force around their equilibrium position. Basic pressure waves stemming from each charged zone, with amplitude proportional to the local charge density in the sample, move at the speed of sound. We consider that the form of the pressure wave *p(t)* is identical to that of the pulsed electric field. The piezoelectric captor transforms the acoustic signal into voltage $V_s(t)$, which is characteristic of the pressure considered. The piezoelectric system frequency response is not perfectly constant, so the signal $V_s(t)$ complex form is distorted compared to the former pressure wave *p(t)*. In order to obtain a precise charge distribution, a deconvolution technique was adopted. By processing the acoustic waves received at each instant, it is possible to determine the charge distribution when a voltage is applied but also during the depolarization phase.

11.2.4. *Signal processing*

To obtain a charge profile, the output signal must be deconvoluted. We consider here that the volume charge density of $\rho(x)$ is distributed in the insulator (see Figure 11.2). For the study, we devised a sample in layers of thickness Δx in which we assumed that the charge density was homogenous. When a pulsed electric field $e(t)$ is applied, a pressure $f_\Delta(x,t)$ acts on each charge layer Δx and is given by the expression:

$$f_\Delta(x,t) = \rho(x) \cdot \Delta x \cdot S \cdot e(t) \quad [11.5]$$

The pressure wave, obtained by dividing f by the surface S of the electrode, reaches the piezoelectric sensor without changing form. For this purpose, we hypothesize that the medium in which the acoustic wave propagates is homogenous and perfectly elastic.

The delay time between the moment when the wave is produced and the moment when it is detected depends on the speed of sound of each traversed material. We shall designate the speed of sound in the material as V_m, and the speed of sound in the electrode which is adjacent to the piezoelectric sensor as V_e. This thick electrode is, in most cases, aluminum.

The pressure thus seen by the detector is in the form:

$$p_\Delta(x,t) = e\left(t - \frac{l}{v_e} - \frac{x}{v_m}\right).\rho(x).\Delta x \quad [11.6]$$

The pressure which is exerted on the captor by the set of basic layers is obtained with a summation:

$$p(t) = \int_{-\infty}^{+\infty} p_\Delta(x,t) = \int_{-\infty}^{+\infty} e\left(t - \frac{l}{v_e} - \frac{x}{v_m}\right).\rho(x).\Delta x \quad [11.7]$$

We note that the electric field is null outside the sample, and we can therefore limit the calculation for the interval x [0, d].

If we define $\tau = x/v_m$, and $\rho(x) = \rho(\tau.v_m) = r(\tau)$, we obtain:

$$p(t) = v_m \int_0^d e\left(t - \frac{l}{v_e} - \tau\right).r(\tau).\Delta\tau \quad [11.8]$$

As this latter equation is a convolution, it can be simplified by the application of the Fourier transform:

$$P(\nu) = F\left[p(t)\right] = v_m.R(\nu).E(\nu).\exp\left[-2i\pi\nu.\frac{l}{v_e}\right] \qquad [11.9]$$

Similarly, the output signals $v_s(t)$ and entry signals $p(t)$ of the piezoelectric sensor can be expressed by a convolution law in the Fourier space:

$$V_s(\nu) = H(\nu).P(\nu) \qquad [11.10]$$

where $V_s(\nu)$ and $P(\nu)$ are, respectively, the Fourier transforms of $v_s(t)$ and $p(t)$, and $H(\nu)$ is the piezoelectric sensor transfer function characteristic, comprising the whole of the amplifiers and wave guides.

We recall that the goal is to measure the charge distribution profile in the materials. We just need to determine the transfer function $H(\nu)$, which characterizes the sensor and the electronic components response to obtain $R(\nu)$. This is possible using a reference signal.

The reference signal is recorded by polarizing an initially non-charged sample under a voltage U in the volume. The charge density (we designate here the capacitive charges) which is present at the electrodes is noted as:

$$\sigma_1 = \frac{\varepsilon_1.U}{d_1} \qquad [11.11]$$

We use index 1 to designate the data obtained during the calibration phase.

Since the pulse voltage amplitude u_p is much less than the polarization voltage U used to record the reference signal, we can consider that the detected capacitive charges are only caused by this latter. Thus, the pressure created by the charges at the electrode closest to the detector can be expressed in the following way:

$$p_1(t) = \sigma_1.e\left(t - \frac{l}{v_e}\right) \qquad [11.12]$$

and we obtain:

$$F[p_1(t)] = P_1(v) = \sigma_1.E(v).\exp\left[-2i\pi v.\frac{l}{v_e}\right] \quad [11.13]$$

The properties of the piezoelectric sensor conversion properties always being the same, we have:

$$V_{S_1}(v) = H(v).P_1(v) \quad [11.14]$$

We can thus extract *H(v)* and, by combining the previous equations, we can obtain the expression F[ρ(x)]:

$$R(v) = \frac{\sigma_1}{v_p}.\frac{V_s(v)}{V_{s_1}(v)} \quad [11.15]$$

To obtain the value of $\rho(x)$, we need to work out the following inverse Fourier transform:

$$\rho(x) = F^{-1}\left[\frac{\varepsilon_1.U}{v_p.d_1}.\frac{V_s(v)}{V_{s_1}(v)}\right] \quad [11.16]$$

To summarize, a charge profile determination requires two measurements. The first one allows $v_s(t)$ to be obtained, which will be converted with the Fourier transform into $V_s(v)$. The second measurement (which corresponds to the reference signal) gives $vs_1(t)$, which will be converted into $Vs_1(v)$ by the same process.

11.2.5. *Example of measurement*

Any measurement starts with a reference signal recording. This measurement can be made on the material itself if it is not initially charged or on a material whose characteristics are known (propagation speed of sound, relative permittivity, etc.). This raw signal is visualized in an oscilloscope (see Figure 11.3). For signal processing, only the peak detected on the electrode situated near the piezoelectric sensor and the signal coming from the material volume are used.

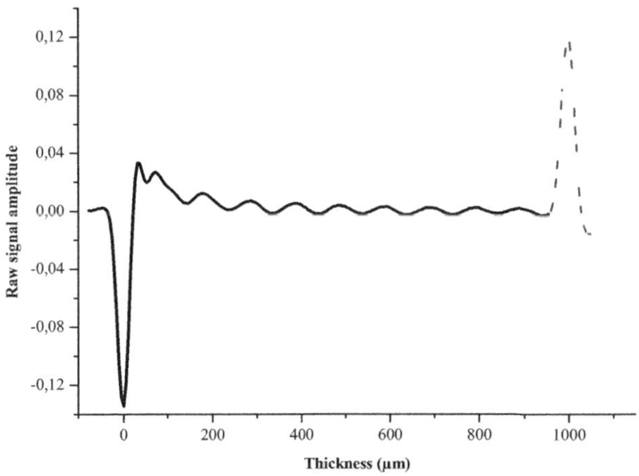

Figure 11.3. *Example of a reference signal recorded on a 1 mm polyimide sample, polarized below 5 kV and probed with 600 V pulses. Only the continuous line of the signal is used for deconvolution*

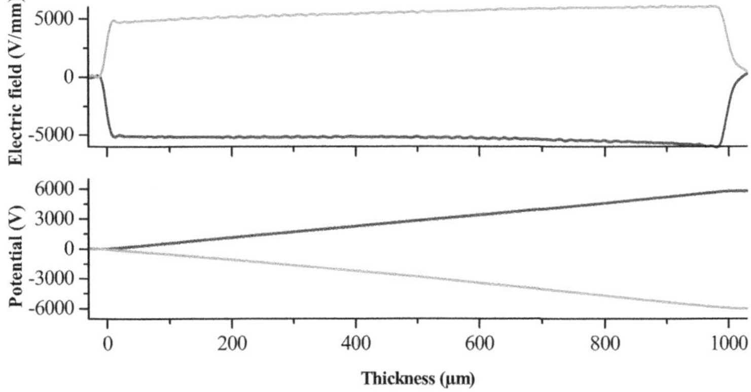

Figure 11.4. *Example of PEA measurements obtained after deconvolution of signals recorded on a 1 mm polyimide sample polarized between -5 kV and 5 kV and probed with 600 V pulses. The reference had been recorded on the sample itself when it was polarized below 5 kV*

In the example of Figure 11.4, the charge profile corresponds to that of a polarized sample, which does not contain any charges in the volume after processing, because of the reference signal recorded beforehand.

Peak charges appear at the sample–electrode interfaces. The peak amplitude depends on the polarization amplitude and the polarity by a double integration. We can verify for each measurement that the electric field and potential profile correspond to the applied polarization.

11.3. Performance of the method

Formerly, this method was widely used to test electrical energy transport cable insulators [LIU 93]. Numerous results obtained on PolyEthylene (PE) samples or its by-products were published [LI 92], [FUK 06]. Since then, it has been commonly used for a wide sample group of materials ranging from solid materials [ALIS 98] to composite materials [TAN 06], to gels for biomedical applications.

The performance criteria lie in the spatial resolution and sensibility. Another important point concerns the speed at which the measurements can be made and repeated. As the resolution is determined by the entire measurement system's signal-to-noise ratio, it is essential to reduce the noise coming from the outside (as, for example, that created by the pulse generator) but also that coming from inside (the thermal noise coming from preamplifiers, etc.).

11.3.1. *Resolution in the thickness*

The sample spatial resolution in the volume lies at first on the way the acoustic wave propagates. This resolution can be improved by reducing the acoustic signal width. As we have already seen, the acoustic wave is generated by application of an electric field; the duration during which the electric field is applied must be less than a few ns to generate an acoustic wave as narrow as possible. However, other factors must be taken into account, such as for example the oscilloscope bandwidth, the sensor geometry, the sample thickness. Indeed, the acoustic wave is detected by a piezoelectric sensor. Its thickness determines the resolution in volume of the signal. Over time, the spatial resolution, initially at 100 μm gets down to 5 μm [MAE 96], then 2 μm [MAE 98] by respectively using a 100 μm piezoelectric captor and an evaporated captor. Using these two factors, it is possible to improve the resolution which we shall call volumic. Classical systems generally offer a resolution of the order of 10 μm.

11.3.2. *Lateral resolution*

The classic PEA system allows a 2D charge distribution analysis in the sample's volume. To obtain an observation in a third direction, we need to move the probe in diverse points on the material surface. The first developed 3D system [IMA 95] permitted the sample surface to be probed by the mechanical motion of the upper electrode. The resolution depended on the electrode area. During the first tests, knowing that the electrode had an area of 1 mm^2, this resolution was estimated at 1 mm. Later, a PEA system equipped with a focus lens [MAE 01] permitted a 100 µm resolution to be achieved, the resolution depth being 5 µm for this system. This system's operating principle and that of the latest developed will be described later.

11.3.3. *Acquisition frequency*

The PEA measurements must be repeated with a high repetition speed to be able to carry the signal and observe the transient phenomena. When a conventional mercury commutator is used, it is possible to observe a signal every 400 ms. The interval between 2 signals is 20 µs if we use a rapid commutator as a pulse generator, which permits the study of charge profiles evolution whose duration is of the same order.

11.3.4. *Signal/noise ratio*

The minimum amount of charge which can be detected by this method is about 0.1 C/m^3. This value is extremely weak since it corresponds to a basic charge for 10^{11} atoms. However, such a charge can disturb an electric field of about 5 kV/mm on a 1 mm thickness.

11.4. Diverse measurement systems

Numerous system modifications were brought over time to obtain 3D measurements, for example, or measurements under constraints (high voltage, high temperature, vacuum-packed and during irradiation, etc.).

11.4.1. *Measurements under high voltage*

After charge distribution measurements validation for coaxial cable insulators [FUK 90], the PEA system was adapted to carry out tests under high voltage (up to 550 kV during breakdown tests) [HOZ 94]. In this device, 2 kV pulses with a 60 ns width are used. The interest in this system is that it allows testing directly on the cable at the manufacturing site (see Figure 11.5). In order to widen the measurements to all cable ranges, new modifications [FU 00], [FU 03] at the electrode level connected to the earth have been made over the past few years.

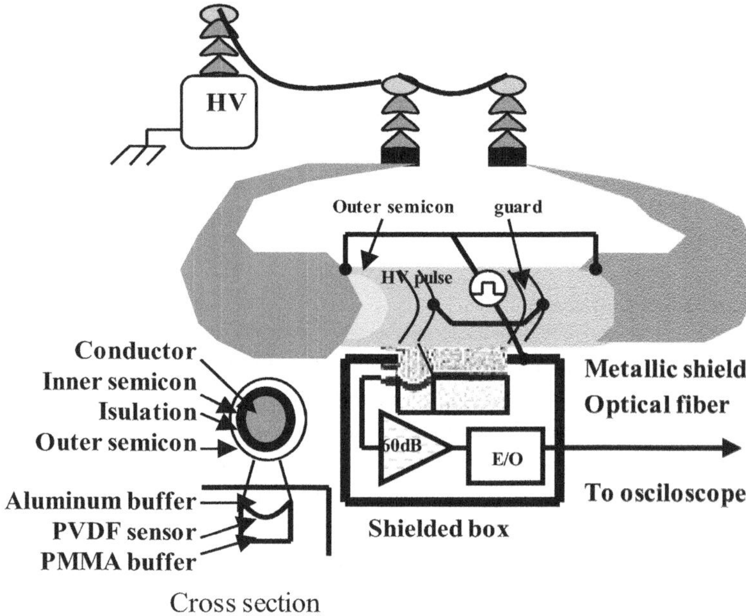

Figure 11.5. *PEA system for high-voltage cable measurements*

In the same way, a more classical system to measure thick planning samples submitted to high voltages (up to 150 kV) was developed for laboratory tests [HOZ 98].

11.4.2. *High and low temperature measurements*

When the temperature increases, the carriers' mobility and the trapping sites nature can vary strongly. This is why a system permitting measurements to be made in these conditions was developed. First, a cell comprising a thermo bath with silicone oil was used for a few studies [SUH 94]. In this case, the high voltage and the pulses were transmitted by a spherical electrode to limit the congestion around the sample and thus allow good temperature diffusion at the sample. The silicone oil bath can disturb certain materials' properties and make their manipulation unsuitable for other tests, so measurements are now regularly carried out by placing the classic device in a thermally controlled enclosure. Tests up to 343 K and 85% humidity are frequently realized [KIT 96], [GAL 05].

The demand for measurements under much higher temperatures, for the study of ceramics for example, prompted the development of a PEA system with electronic parts cooling. Measurements up to 523 K were carried out [MIY 03]. However, as the temperature increases, a signal displacement in the time range is observed. This is caused by a decrease of the speed of sound in the aluminum electrode, which increases the signal transit time towards the sensor. The signals are stretched because the speed of sound is also reduced in the sample. We must take into account the modifications of the propagation speed of the signals for the analysis.

Conversely, dielectric studies made at cryogenic temperature for super conductor cable development allow measurements at 180 K and 290 K [MUR 01] without system modifications.

11.4.3. *Measurements under lighting*

Amongst functional polymers, photoconductor materials are widely used in photocopiers and electroluminescent devices. To complete the surface potential decline of these materials submitted to electric fields and to precise exposure to light, the manufacturers who make these materials rapidly demanded techniques to measure the internal charge distribution during their exposure. The first studies were undertaken by the *Pressure Wave Propagation* (PWP) method [TAN 92], but a PEA cell was rapidly developed to reply to the demand [SAT 96].

For these measurements, a semi-transparent aluminum electrode is sputtered on the irradiated surface, which permits voltage pulses to be applied during the lighting and thus allows the *in situ* evolution of the charge distribution in the volume.

11.4.4. *3D detection system*

The first 3D charge profiles (see Figure 11.6) were obtained by moving the electrode by which the high voltage is applied [IMA 95]. Since this process is very slow, a new device was developed. An acoustic lens was thus introduced [MAE 01]. Only the acoustic waves generated at the focus point are detected by the piezoelectric sensor. To scan the sample surface, we need, in this case, to move the acoustic lens. Reasonably, 4 mm^2 areas can be studied. The acquisition time depends on the number of points chosen for the analysis. The lateral resolution is about 0.5 mm and the volume resolution is 12 μm.

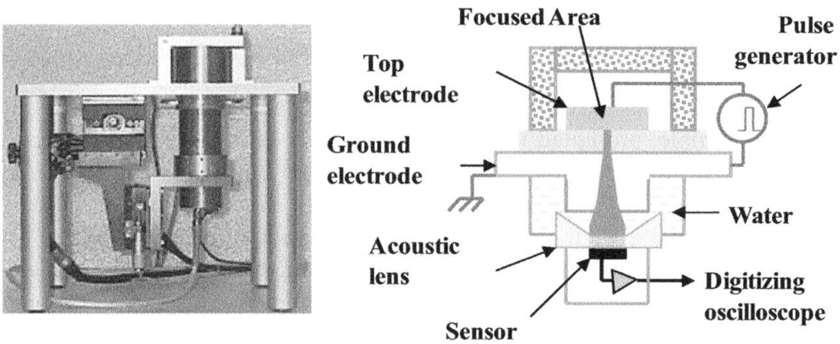

Figure 11.6. *PEA-3D device with focus lens*

More recently, another system composed of multiple sensors (see Figure 11.7) was developed [TAN 03]. These sensors are positioned under the detection electrode and simultaneously receive a signal when probe voltage pulses are applied. The distance (typically 3 mm) between the PVDF sensors (whose dimensions are 2mm*3mm per side and 9 µm thick) determine the lateral resolution. To record the signals coming from the sensors independently, a coaxial switch is used.

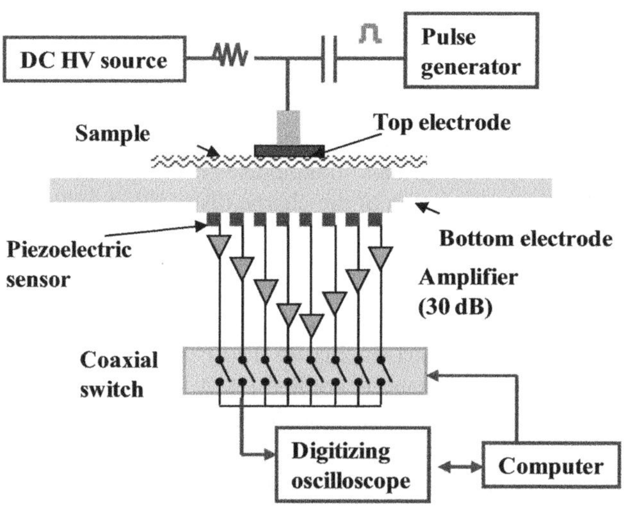

Figure 11.7. *PEA-3D device with multiple sensors*

11.4.5. *PEA system with high repetition speed*

To study transient phenomena such as the charge dynamics during a breakdown in a dielectric, it is fundamental to measure signals with a high repetition speed. Electric field pulses can be obtained by a generator with a rapid semi-conductor switch (available in shops).

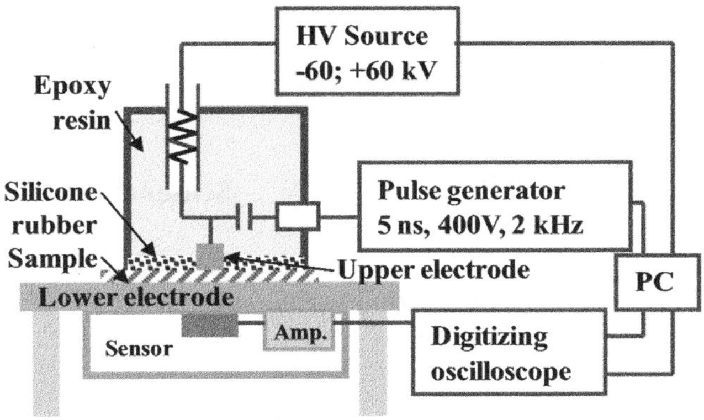

Figure 11.8. *PEA-3H device*

The data recording can be realized by the new generations of oscilloscopes. The problem encountered is that of data storage and therefore their processing.

So far, two systems have been elaborated. The first is the PEA-3H (H for *High*, see Figure 11.8) [MAT 02]. It is a system which allows working under high voltage (in a domain which covers [-60 kV ;+60kV]), with high repetition speed (2 kHz), and a large spatial resolution (10 µm). Faster systems (signal recorded every 20 µs) are restricted by the size of the oscilloscope's memory. For long-lasting ageing tests, it is normal to record about 1,000 signals in 500 ms and to transfer them to a PC before the next set of data. This operation requires about 130 ms. The operation repetition for the whole duration of the test allows the breakdown phenomena to be studied in a fairly precise manner.

The second system is based on a high repetition speed and high temperature [FUK 04] (see Figure 11.9). This system can be used for breakdown studies. Silicone oil is used to guard the surface discharges and the partial discharges from the high voltage electrode. The temperature can be controlled. The generator delivers signals with a frequency of 10 kHz, which allow space charge profiles to be obtained every 10 µs and generally the observation time is limited to 1 ms before the next set of data.

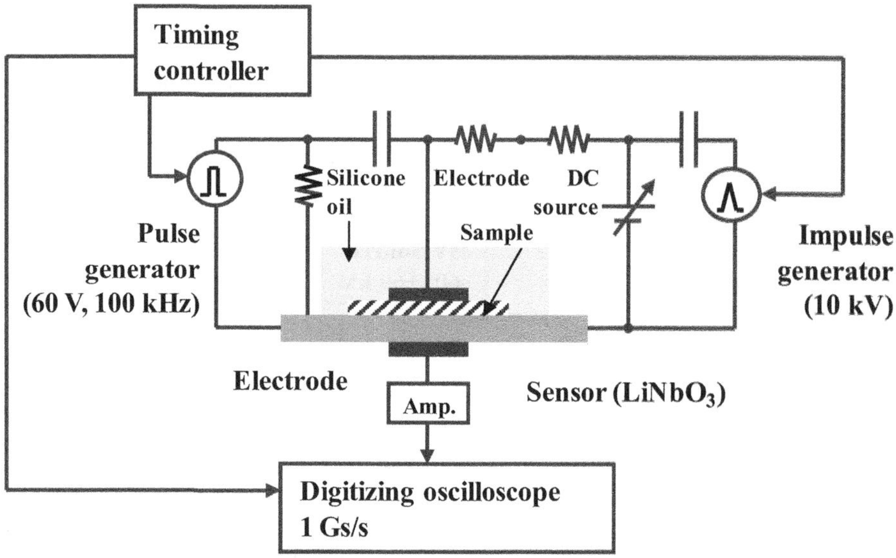

Figure 11.9. *High resolution, high temperature PEA device*

11.4.6. *Portable system*

Since the demand for PEA measurements on industrial sites was increasing, a miniature system was developed. Indeed, the classic system configuration (measurement cell, oscilloscope and computer for data processing) did not predispose this technique to taking measurements at diverse points along a production chain, for example. Besides a size reduction, the form of the voltage pulses was improved to allow observation of direct charge profiles on the oscilloscope [MAE 03] (see Figure 11.10).

11.4.7. *Measurements under irradiation*

The charge distribution study in materials subjected to irradiation has become essential, notably in the space field, to understand electrostatic discharge phenomena which come up after a period of accumulation. The classic system was modified to allow measurements in a vacuum and also measurements during irradiation. The voltage pulse is applied either through a cone-shaped electrode [TAN 04] or an evaporated electrode [GRI 02] (see Figure 11.11).

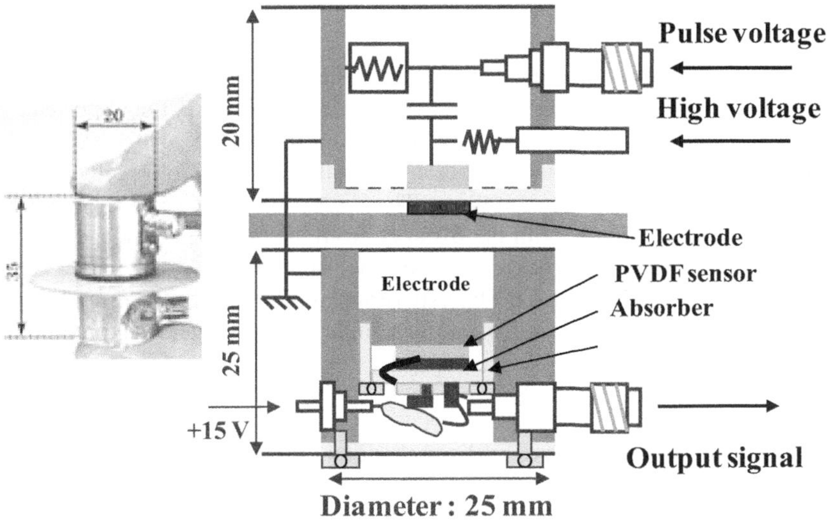

Figure 11.10. *Portable PEA device*

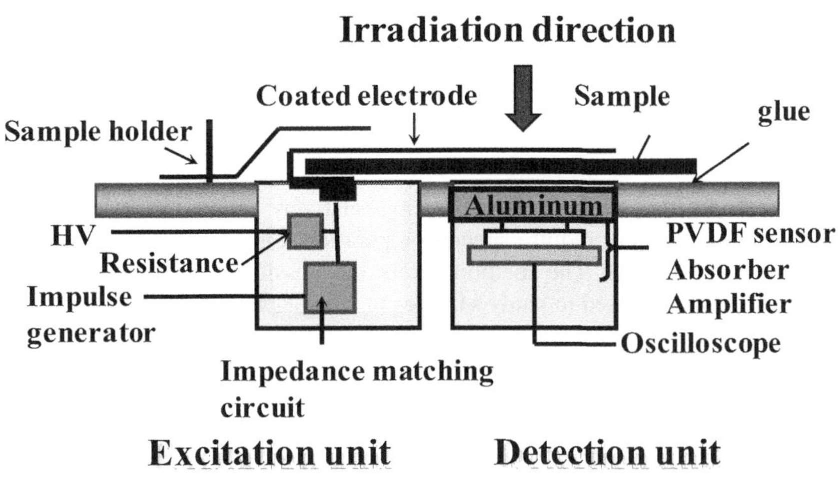

Figure 11.11. *PEA device adapted for measurements under irradiation*

11.4.8. *Contactless system*

In response to ever harder demands, in particular for measurements under irradiation, a contactless system was created. Under this system, the upper electrode is placed at a distance (approximately 1 mm) from the surface of the sample to be studied [IMA 04]. This surface is therefore floating and it is possible to detect charges left without disturbing anything. The disadvantage of such a system is the decrease in signal amplitude with the distance from the upper electrode, which turns out fundamental when the charge state of the studied system increases and the electrostatic discharge risk grows. The experimental protocol for the reference signal recording also has to be modified since the sample can no longer be polarized in the normal way.

11.5. Development perspectives and conclusions

Since its creation, the PEA method has never stopped being modified to allow measurements to be made in numerous configurations, and there is a good chance that efforts will continue, considering the increasing number of users of the method across the world. It is a tool which gives supplementary information to traditional measurements (current measurements for polarized materials, surface potential measurements for irradiated materials, etc.). New modeling axes are being created, owing to the data made accessible by PEA measurements. Indeed, the same material can be studied by the same technique with different measurement configurations and thus supply physical parameters concerning transport, such as mobility, trapping and recombination rates, etc.

This method offers the advantage of being relatively simple to implement in the sense that samples do not require special preparation. Signal processing is made by basic mathematical tools. The method offers the possibility of studying dynamic phenomena and can be used to study samples in which space charges were generated in a complex manner.

11.6. Bibliography

[ALI 98] ALISON J.M., "Pulsed electro-acoustic system for space charge and current density measurements at high fields over a temperature range 25°C–80°C", *Proceedings CSC'3*, p. 711–714, 1998.

[ALIS 98] ALISON J.M., "A high field pulsed electroacoustic apparatus for space charge and external circuit current measurement within solid insulators", *Meas. Sci. Technol.*, vol. 9, p. 1737–1750, 1998.

[ALQ 81] ALQUIÉ C., DREYFUS, G., LEWINER J., "Stress wave probing of electric field distribution in dielectrics", *Phys. Rev. Lett.*, vol. 47, p. 1483–1487, 1981.

[BOD 04] BODEGA R., MORSHUIS P.H.F., SMIT J.J., "Space charge signal interpretation in a multi-layer dielectric tested by means of the PEA method", *Proceedings ICSD'04*, p. 240–243, Toulouse, France, 5–9 July, 2004.

[CHE 92] CHERIFI A., ABOU DAKKA M., A. TOUREILLE, "The validation of the thermal step method", *IEEE Trans. Electric. Insulat.*, vol. 27, no. 6, p. 1152–1158, December 1992.

[ENR 95] ENRICI P., ROPA P., JOUMHA A., REBOUL J.P., "Space charge measurements by the thermal-step-method and TSCD in polymeric materials", *Proceedings IEEE ICSD'95* Virginia Beach, USA, p. 504–507, October 1995.

[FU 00] FU M., CHEN G., DAVIES A.E., TANAKA Y., TAKADA T., "A modified PEA space charge measuring system for power cables", *Proceedings 6th ICPADM'00*, p. 104–107, juin 2000.

[FU 03] FU M., CHEN G., "Space charge measurement in polymers insulated power cables using flat ground electrode PEA system", *IEE Proc.-Sci Meas. Technol.,* vol. 150, no. 2, March 2003.

[FUK 90] FUKUNAGA K., MIYATA H., SUGIMORI M., TAKADA T., "Measurement of charge distribution in the insulation of cables using the pulse electro-acoustic method", *Trans. IEE Japan*, vol. 110-A, no. 9, p. 647–648, 1990.

[FUK 04] FUKUMA. M., MAENO T., FUKUNAGA K., NAGAO M., "High repetition rate PEA system for *in-situ* space charge measurements during breakdown tests", *IEEE Trans. DEI*, vol. 11, no. 1, p. 155–159, 2004.

[FUK 06] FUKUMA M., FUKUNAGA K., LAURENT. C, "Two-dimensional structure of charge packets in polyethylene under high DC fields", *Appl. Phys lett.*, vol. 88, no. 25, June 2006.

[GAL 05] GALLOT-LAVALLÉE O., TEYSSÈDRE G., LAURENT C. ROWE S., "Space charge behaviour in an epoxy resin: the influence of fillers, temperature and electrode nature", *J. Phys. D: Appl. Phys.*, vol. 38, no. 12, p. 2017–2025, 2005.

[GRI 02] GRISERI V., LÉVY L., PAYAN D., MAENO T., FUKUNAGA K., LAURENT C., "Space charge behaviour in electron irradiated materials", *Proceedings IEEE CEIDP'02*, p. 922–925, 2002.

[GRI 04] GRISERI V., FUKUNAGA K., MAENO T., LAURENT C., LÉVY L., PAYAN D, "Pulsed electro-acoustic technique applied to in-situ measurement of charge distribution in electron-irradiated polymers", *IEEE Trans. Dielectr. EI.*, vol. 4, no. 5, p. 891–898, October 2004.

[HOZ 94] HOZUMI N., SUZUKI H., OKAMOTO T., WATANABE K., WATANABE A., "Direct observation of time-dependent space charge profiles in XLPE cable under high electric fields", *IEEE Trans. Diel. EI*, vol. 1, no. 6, p. 1068–1076, 1994.

[HOZ 98] HOZUMI N., TAKEDA T. SUZUKI H., OKAMOTO T., "Space charge behaviour in XLPE cable insulation under 0.2-1.2mV/cm DC fields", *IEEE Trans. Diel EI*, vol. 5, no. 1, p. 82–90, 1998.

[IED 84] IEDA M., "Electrical conduction and carrier traps in polymeric materials", *IEEE Trans. Electr. Insulat.*, vol. EI-19, p. 162–178, June 1984.

[IMA 95] IMAISUMI Y., SUZUKI K., TANAKA Y., TAKADA T., "Three-dimensional space charge distribution measurement in solid dielectrics using pulsed Electroacoustic Method", *Proceedings IEEE ISEIM'95*, p. 315–318, 1995.

[IMA 04] IMAI S., TANAKA Y., FUKAO T. TAKADA T., MAENO T., "Development of new PEA system using open upper electrode", *Proceedings CEIDP'04*, IEEE, p. 61–64, July 2004.

[KIT 96] KITEJIMA M., TANAKA Y., TAKADa, "Measurement of space charge distribution at high temperature using the pulsed electro acoustic method", *Proceedings 7th Int. Conf. DMMA'96*, p. 8–11, Bath, UK, September 1996.

[LI 92] LI Y., TAKADA T., "Experimental observation of charge transport and injection in XLPE at polarity reversal", *J. Phys. D: Appl. Phys.*, vol. 25, p. 704–716, 1992.

[LIU 93] LIU R., TAKADA T., TAKASU N., "Pulsed electroacoustic method for measurement of charge distribution in power cables under both DC and AC electric fields", *J. Phys. D: Appl. Phys.*, vol. 26, p. 986–993, 1993.

[MAE 88] MAENO T., FUTAMI H., KUSHIBE T. COOKE C.M., "Measurement of space charge distribution on thick dielectrics using the pulsed electroacoustic method", *IEEE Trans. Electr. Insul.*, vol. 23, p. 433–439, 1988.

[MAE 96] MAENO T., FUKUNAGA K., TANAKA Y., TAKADA T., "High resolution PEA charge measurement system", *IEEE Trans. Diel. EI*, vol.3, no. 6, p. 755–757, 1996.

[MAE 98] MAENO T., "Extremely high resolution PEA charge measurement system", *Report of Study Meeting IEE Japan, DEI-98-78*, p. 81–86, 1998.

[MAE 01] MAENO T., "Three-dimensional PEA charge measurement system", *IEEE Trans. DEI*, vol. 8, no. 5, p. 845–848, 2001.

[MAE 03] MAENO T., "Portable space charge measurement system using the pulsed electroacoustic method", *IEEE trans. Diel. EI*, vol. 10, no. 2, 2003

[MAT 02] MATSUI K, TANAKA Y., FUKAO T., TAKADA T., MAENO T., "Short-duration space charge observation in LDPE at the electrical breakdown", *Proceedings IEEE CEIDP'02*, p. 598–601, 2002.

[MIY 03] MIYAUCHI H., FUKUMA M., MURAMOTO Y., NAGAO M., MAENO T., "Space charge measurement system under high temperature on PEA method", *Proceedings IEE Jpn. CFM'03*, no. 7–14, 2003.

[MOL 00] MOLINIÉ P., LLOVERA P., "Surface potential measurements: implementation and interpretation", *Proceedings. IEEth Int. Conf. DMMA'00*, p. 253–258, 17–21 September 2000.

[MON 00], MONTANARI G.C., MAZZANTI G., BONI E., DE ROBERTIS G., "Investigating AC space charge accumulation in polymers by PEA measurements", *Proceedings IEEE ICSD'00*, Victoria, Canada, p. 113–116, October 2000.

[MUR 01] MURAKAMI Y., FUKUMA M., HOZUMI N., NAGAO M., "Space charge measurement in EVA film at cryogenic-room temperature region", *Trans. IEE Jpn.*, vol. 121-A, no. 8, p. 758–763, 2001 (in Japanese).

[SAT 96] SATOH Y., TANAKA Y., TAKADA T., "Observation of charge behavior in organic photoconductor using the pulsed electroacoustic method and Pressure wave propagation Method", *Proceedings IEE 7th Int. Conf. DMMA'96*, p. 112–115, 23–26 September 1996.

[SES 97] SESSLER G.M., "Charge distribution and transport in polymers", *IEEE Trans. Dielectr. Electric. Insulat.* vol. 4, no. 5, p. 614–628, octobre 1997.

[SUH 94] SUH K.S., HWANG S.J., NOH J.S. TAKADA T., "Effects of constituents of XLPE on the formation of space charge", *IEEE Trans. Diel. EI*, vol. 1, no. 6, p. 1077–1083, 1994.

[TAK 99] TAKADA T., "Acoustic and optical methods for measuring electric charge distributions in dielectrics", *Proceedings IEEE ICSD'99*, vol. 1, Austin, USA, p. 1–14, octobre 1999.

[TAN 92] TANAKA A., MAEDA M., TAKADA T., "Observation of charge behavior in organic photoconductor using pressure-wave propagation method", *IEEE Trans. on Electric. Insulat.*, vol. 27, no. 3, p. 440–444, June 1992.

[TAN 03] TANABE Y., FUKUMA M., MINODA A., NAGAO M, "Multi-sensor space charge measurement system on PEA method", *Proceedings IEE Jpn. CFM'03*, no. 7–14, 2003.

[TAN 04] TANAKA Y., FUKUYOSHI F., OSAWA N., MIYAKE H., TAKADA T., WATANABE R., TOMITA N., "Acoustic measurement for bulk charge distribution in dielectric irradiated by electron beam", *Proceedings ICSD'04*, p.940–943, 2004.

[TAN 06] TANAKA H., FUKUNAGA K., MAENO T., OHKI Y. "Three-dimensionnal space charge distribution in glass/epoxy composites", *Proceedings 8th IEEE ICPADM'06*, Bali, Indonesia, June 2006.

Chapter 12

FLIMM and FLAMM Methods: Localization of 3-D Space Charges at the Micrometer Scale

12.1. Introduction

Polymer materials are widely used as basic insulator structures in many devices in electrical engineering, such as high voltage cables, transformers or capacitors. They have to resist strong voltage gradients, sometimes up to a few tens of kV/mm. Furthermore, engineers have to take into account important internal and local electric fields due to the presence of space charges which are trapped within the materials. These mainly result from the dissociation of electrically neutral species, charge injections at the surface of the insulator, or due to the orientation of the electric dipole.

In these zones affected by a strong local electric field, dissipative energy phenomena can lead to a premature ageing of the material, and potentially to a faster than expected breakdown of the dielectric. For these reasons, many efforts have been implemented over the past ten years in order to make more reliable the detection and characterization of these charges.

Several space charge measurement techniques have been developed [TAK 99], mostly non-destructive thermal ones. However, these methods present a very poor spatial resolution.

Chapter written by Anca PETRE, Didier MARTY-DESSUS, Laurent BERQUEZ and Jean-Luc FRANCESCHI.

With an aim to increase this resolution, we proposed to develop a technique based on the use of focused thermal excitation as a source, allowing the characterization of space charge profiles at the scale of a few tens of micrometers.

For this purpose, two investigation methods were developed: FLAMM (*Focused Laser Acoustic Modulation Method*) which detects thermo-acoustic waves generated by a laser beam using a piezoelectric sensor stuck on the rear side of the object under study, and FLIMM (*Focused Laser-Intensity Modulation Method*), an evolution of the well-known LIMM technique. In this last case, the interactions between temperature variations induced by the laser beam and the electric charges trapped in the material produce, under specific conditions, an informative signal to be processed. The main specificity of both methods lies in their ability to detect quasi-localized charges. This, combined with the scan of a given area of the sample, results in multidimensional charge cartographies that may be carried out at the micrometer scale.

12.2. The FLIMM method

12.2.1. *Principle*

The FLIMM consists of irradiating the surface of a sample using an intensity-modulated laser diode (45 mW, 658 nm) at a given frequency. This beam is focused at the surface of the sample to create a non-uniform thermal gradient inside the structure (Figure 12.1). The interaction between this thermal wave and the space charge and/or the spatial polarization produces a pyroelectric current.

The pyroelectric current is converted into voltage by a low-noise transconductance preamplifier. Then, the signal is extracted from noise by a lock-in amplifier which gives both the real and the imaginary parts of the pyroelectric current for a given modulation frequency of the beam. A mathematical treatment is then used to calculate the space charge profile. All the system is managed by a computer: laser beam frequency modulation, position of the spot beam at the sample surface, acquisition and treatment of the information signal.

From each position of the beam a current is recorded, and after the mathematical deconvolution, a 2-D representation of the in-depth space charge profile can be made. By scanning the surface with the laser, 3-D representations and cartographies of the trapped space charges are possible.

Samples are metallized on both sides with a thin gold or aluminum layer (20 to 50 nm) and placed in a measurement cell, used at the same time as a sample support and as protection against electromagnetic perturbations.

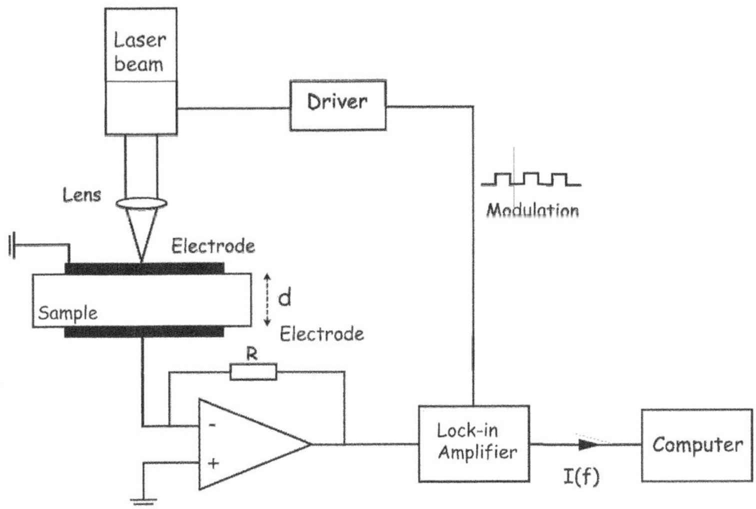

Figure 12.1. *Outline of the FLIMM set-up*

12.2.2. *Characteristic FLIMM equation*

The pyroelectric current I depends on the modulation frequency f of the laser beam, the local value of the temperature $T(z,f)$ in the sample as well as the charge function $r(z)$ (space charge $\rho(z)$ and polarization $p(z)$), z being the direction of the sample thickness at a given position of the laser beam impact.

The expression of the pyroelectric current is:

$$I(f) = j.2\pi.f.\frac{A}{L}\int_0^L r(z)T(z,f)dz$$

$$\text{with } r(z) = p(z) - (\alpha_x - \alpha_e).\varepsilon.\varepsilon_0.E(z) \qquad [12.1]$$

where $A\,(m^2)$ represents the thermal spot size, $r(z)\,(Cm^{-3})$ the total space charge, $p(z)\,(Cm^{-3}K^{-1})$ the pyroelectric coefficient, $E(z)\,(Vm^{-1})$ the internal electric field, α_x, α_ε (K^{-1}) respectively the relative dependencies of the local expansions and the electric permittivity with temperature in the material.

12.3. The FLAMM method

This method is similar to the FLIMM with regard to thermal excitation by a focused laser beam. Instead of directly recording the FLIMM current from an electrode on the sample, the electric charge Q which appears on the electrode through a piezo-electric polymer sensor (PVDF) is measured [FRA 97], [FRA 00], [HAS 97]. This plays the role of insulating capacity when the sample is subjected to a potential difference (Figure 12.2).

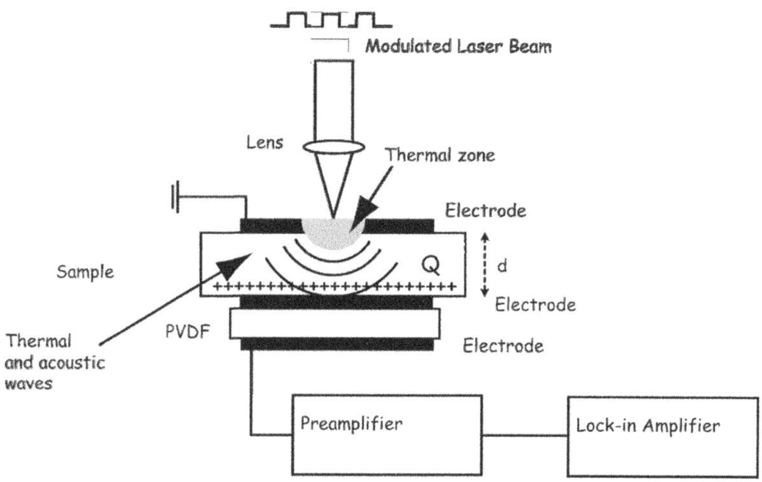

Figure 12.2. *Outline of the FLAMM set-up*

For energies and frequencies used, the contribution of thermal phenomena to the total signal is widely preponderant, compared to acoustic pressure waves phenomena induced effects. The PVDF sensor thus allows the induced charge $Q(f)$ to be measured by capacitive effect, but also records an acoustic component. This is why this technique was named FLAMM (*Focused Laser Acoustic Modulation Method*). Using uni-dimensional modeling of the temperature, the expression of the charge $Q(f)$ was established by Mopsik [MOP 82]. From this, the expression of the potential $V(f)$ measured between the electrodes of the PVDF sensor can then be deduced:

$$V(f)=\frac{\alpha_x-\alpha_\varepsilon}{\varepsilon}\int_0^L \rho(x).g(x,f)dx \text{ avec } g(x,f)=\int_0^x T(u,f)du-\frac{x}{L}\int_0^L T(u,f)du \qquad [12.2]$$

where ε represents the dielectric constant of the material, $\alpha_x, \alpha_\varepsilon$ (K^{-1}) are respectively the relative dependencies of the local expansions and the electric permittivity with temperature in the material, and $\rho(z)$ (Cm^{-3}) the space charge.

12.4. Modeling of the thermal gradient

In FLIMM instrumentation, it is impossible to get an evaluation of the temperature variations inside the material in an experimental way. Some methods allow the measurement of the temperature at the sample surface (bolometric set-ups for example), but none of them can give relevant information about the local temperature distribution in a volume.

The necessity to develop a coherent mathematical model to study the local internal temperature variations was then obvious. The theoretical determination of the temperature field within an insulator under irradiation is of very great importance; the accuracy of the results will depend on it. Indeed, the extraction of the charge from the Fredholm equation [12.1] requires tricky mathematical processing, and the temperature plays a primary role. Thus, several uni-dimensional and tri-dimensional models were developed. First, a uni-dimensional model was proposed, assuming a surfacic contribution of energy, which allowed a global estimation of the charge. Taking into account that the laser diode is focused on the sample surface, this model is no longer appropriate. A new model with a "four layer" volumic contribution was then developed. The particularity of this model lies in the fact that it takes into account different media constituting the sample. Indeed, heat propagation depends on the thermal characteristics of each medium and it should be considered in order to model the physical phenomenon as finely as possible.

12.5. Mathematical deconvolution

The measurement of the pyroelectric current $I(f)$ and the modeling of the temperature distribution $T(z,f)$ lead to the determination of the charge or polarization profile by using a mathematical processing of equation [12.1].

This equation is a Fredholm integral of the first kind, in which $r(x)$ is the unknown function to evaluate. Generally, equations based on such kind of integrals are very badly conditioned (with ill-posed problems), generally due to the extreme sensitivity of the system they describe. They can therefore produce an infinite set of solutions by considering the experimental error domain.

To solve this kind of equation, several mathematical deconvolution methods have been developed, from the simplest to more complex ones.

12.5.1. *Virtual Space Charge Model*

The Virtual Space Charge Model (VSCM), a simple and original model [FRA 01], allows the average space charge in a sample to be determined. The principle of this method consists of choosing a charge distribution model which would give the same signals as those detected experimentally, induced by a real charge. This means a mathematical inversion can be avoided, by exclusively relying on a comparison between the measured induced effects and those simulated by the model.

Let us consider two constant charge distributions, ρ_1 and ρ_2, representative of each half of the sample, their sum being the average charge distribution in the sample. By giving constant values to these two distributions and knowing the thermal gradient which is propagated within the material, the associated pyroelectric current can be rebuilt. This current presents a maximum I_m at a frequency f_m, independent of the material thickness.

A general variation law can be determined: $I_m = \alpha\rho_1 + \beta\rho_2$ where α and β are constants which depend on the physical properties of the material. Consequently, the maximum current can be considered as representative of the average charge in the sample.

Nevertheless, this simple and fast model is not efficient enough for the localization of charges in the material. For this purpose, other inversion methods were implemented.

12.5.2. *The scale transformation method*

The scale transformation method, based on a simplification of the temperature equation for the highest modulation frequencies used, does not require any matrix inversion. It then lead to a simplification of the pyroelectric current equation [12.1] and an approximation of the charge function $p_a(z_r)$ in the material at thickness z_r such that [PLO 92]:

$$p_a(z_r) = \frac{kL}{D\eta j_0 S}(\Re - \Im)I(\omega = 2D / z_r^2) \qquad [12.3]$$

where $\Re$ and $\Im$ are the real and imaginary parts of the pyroelectric current, D the thermal diffusivity and $z_r = \sqrt{\dfrac{D}{\pi f}}$.

It was shown [PLO 92] that the spatial resolution decreases with the thermal gradient penetration thickness. A good approximation is then obtained with a maximal resolution near the surface of the sample. It is therefore necessary to proceed to a second measurement on the opposite side of the material in order to obtain a complete and more reliable reconstitution of the charge profile. The main advantage of this method is that it avoids mathematical treatment likely to be hidden by the noise of experimental errors. Its development is easy and immediate.

12.5.3. *The regularization method*

A more complex but more efficient method called regularization was also developed. It consists of imposing an additional constraint on the equation system to force a unique solution [PET 04a].

The fundamental Fredholm equation, taking into account a noisy environment, is given by:

$$\int_0^1 A(s,t)x(t)dt = b_0(s) + e(s) \qquad [12.4]$$

where $b_0(s) + e(s)$ is the total measured signal, $e(s)$ the noise, $A(s,t)$ the core function and $x(t)$ the solution sought.

Since the measured signal is constituted by a finite set of values s, the continuous model [12.4] can be replaced by a discrete linear equation, such as:

$$Ax = b_0 + e = b \qquad [12.5]$$

It is not easy to solve these discrete linear equations because of the great number of small singular values which tend to increase the influence of errors. A significant solution can be calculated by using the regularization methods. Their goal consists of introducing some additional information with respect to the solution sought in order to stabilize the inverse problem, then to extract a solution expected to be close to the reality. Several additional constraints can be added, but generally a minimization of the norm (order 2) of the solution is imposed.

Different regularization methods were described by Hansen [HAN 96], [HAN 94]. Our attention was focused on the Tikhonov techniques, like *Truncated Singular Value Decomposition* (TSVD) and *Piecewise Polynomial* TSVD [TIK 77]. The difference between these two lies in the way that the additional constraints are imposed to the solution in order to reduce the influence of errors.

The most famous regularization technique was developed by Tikhonov and applied in various areas, for example in image processing [AND 77] or biomedical technology [SKI 02].

The principle of this method is based on the choice of a solution x_λ which satisfies the problem:

$$\min_x \left\{ \|Ax - b\|_2^2 + \lambda \|Px\|_2^2 \right\} \qquad [12.6]$$

where P is the first order differential operator and λ the regularization parameter which controls the weight of the lateral constraint minimization $\|Px\|_2$ with respect to the residual norm minimization $\|Ax - b\|_2$.

Several methods allowing the determination of the regularization parameter λ exist; the most recent and reliable are the *L-Curve* (LC) [HAN 93] and *Self-Consistency* (SC) [HON 90].

12.6. Results

The FLIMM technique allows charge profiles to be obtained in 1-D [AGN 05], 2-D [PET 04a] or 3-D [PET 06]. The main advantage of this method is linked with its ability in getting multidimensional charge cartographies, with very good spatial resolutions: 1 µm in depth and 10 µm for lateral resolution.

The following sections present the results obtained by FLIMM for 1-D, 2-D and 3-D charge distributions.

12.6.1. *1-D study of PEN (Polyethylene Naphtalate) subjected to high fields*

The films used for these measurements were supplied by DuPont de Nemours (Luxembourg) (stabilized films with bi-axial orientation). For these 25 µm thick films, the crystallinity rate is about 44%. The samples were metallized with a 30 nm gold layer on which a thin ink layer was deposited in order to increase the optical absorption of the laser beam. These films were polarised at room temperature for 30 minutes and the applied field was chosen in a range from 12 kV/mm to 300 kV/mm. They were then placed in short circuit conditions for 30 minutes [AGN 05].

The residual field and space charge profiles obtained after 30 minutes of depolarization are presented in Figures 12.3 and 12.4. The charge profiles shown in Figure 12.4 were obtained by measurements made on both sides of the sample. Indeed, the in-depth information given by thermal methods is not sufficient if only one side of the study is considered, and thus, a second measurement is often necessary. This implies a resolution loss around the middle of the sample, and therefore all interpretations are limited close to interfaces. In these zones, an accumulation of homocharges (positive charges at the anode and negative at the cathode) can be noticed, which increase with the field. At low field, the charge density is very weak ($\sim 0.5\, C/m^3$) and can be associated with a dipolar orientation. On the other hand, a massive charge injection appears for fields greater than 225 kV/mm.

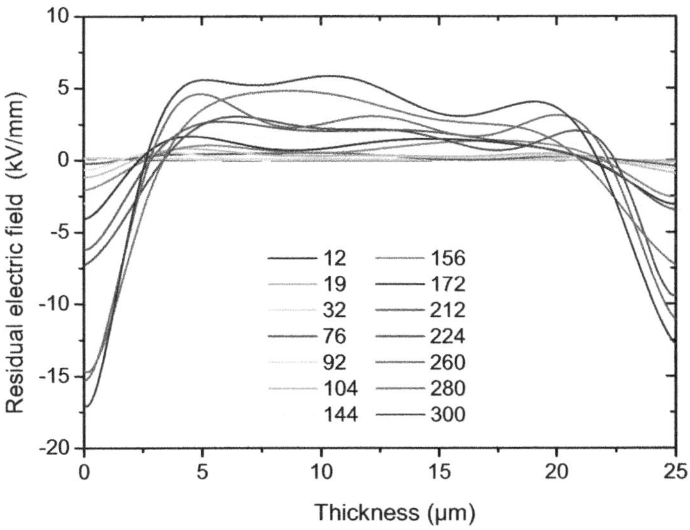

Figure 12.3. *Residual internal fields as a function of the applied voltage*

Close to the anode, the development of a positive homocharge is revealed by the increase of peak charges. It can also be noticed that the level of charges accumulated at the anode is greater than at the cathode.

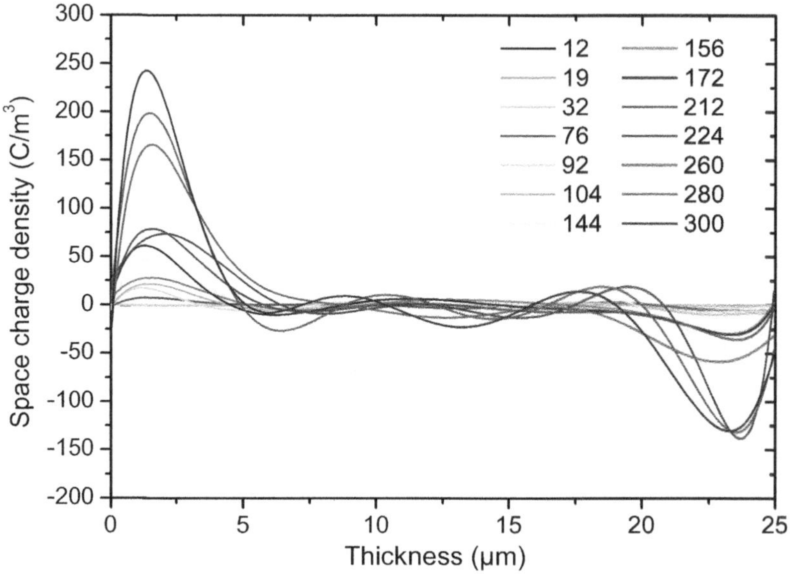

Figure 12.4. *Space charge distribution*

From the distributions in Figure 12.4, the average value of charges located in zones near the interfaces was calculated on a thickness of 5 µm and from each electrode. The values obtained for each curve are presented in Figures 12.5(a) and (b).

Three zones can be distinguished, corresponding to the zones already revealed by Augé *et al.* [AUG 00] using I(V) and electroluminescence measurements.

In the first zone, the charge density is very weak, which means that there is no injection in the range of fields from 12 kV/mm to 75 kV/mm.

Between 75 and 160 kV/mm, a charge injection is produced from the electrodes. Augé *et al.* [AUG 00] showed that in this field range, the current increases in a non-linear way, but there is still no permanent electroluminescence emission. Furthermore, by space charge measurements with the LIPP technique, an injection of homocharges from electrodes was enhanced. For higher fields (160–300 kV/mm), a massive charge injection takes place. This injection is also correlated with a significant increase of the current and electroluminescence with the field. These results are in agreement with others achieved by current and electroluminescence measurements on similar samples.

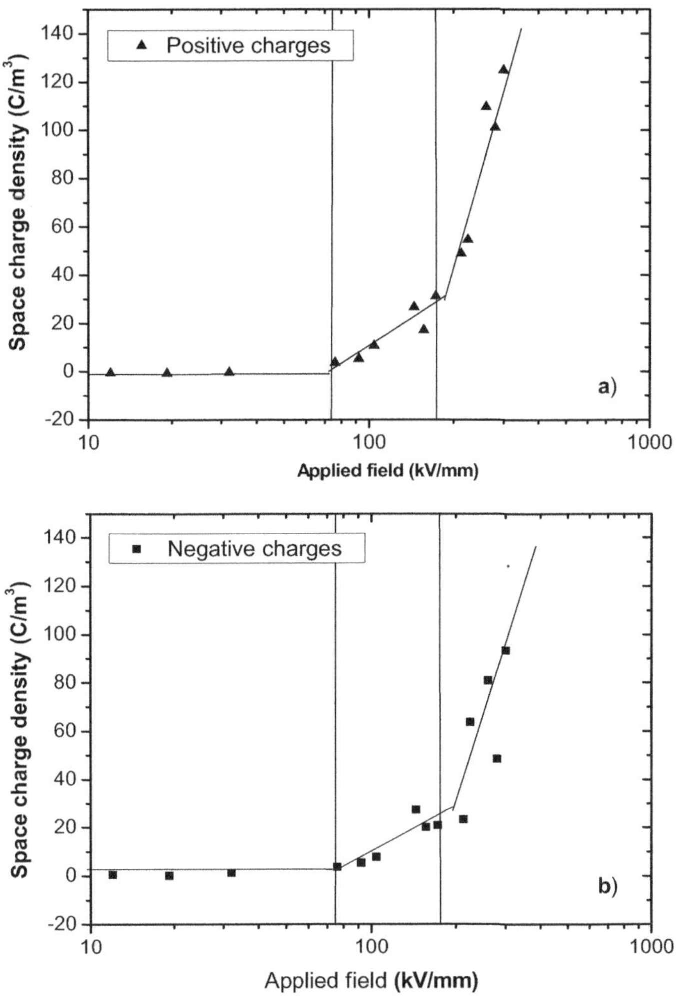

Figure 12.5. *Average space charge values as a function of the applied field: (a) positive charges near the anode; (b) negative charges near the cathode*

12.6.2. *2-D charge distribution*

The FLIMM method was also used for the study of space charges induced by ultraviolet (UV) irradiation, as a complement of electroluminescence measurements [PET 04b].

In our research, electroluminescence measurements were carried out on thin PEN films. It was noted that the emission of the electroluminescence was weaker on the irradiated zones. This phenomenon was associated with the apparition of space charges within the irradiated zones. To confirm (or invalidate) this hypothesis, space charge measurements were undertaken using FLIMM.

For this purpose, thin Poly(ethylene 2.6-naphthalene) films (PEN, DuPont de Nemours) were used. Space charge measurements were made on 25 µm thick films, and on 50 µm ones for electroluminescence studies.

The samples' ageing was induced by UV rays from a 30W insulating system used for printed circuits production. The sample subjected to UV presented a special configuration of space charge measurements, showed in Figure 12.6. During irradiation, the sample was protected by a mask, and only three zones were irradiated, for different durations of, respectively, 12, 24 and 48 hours. The irradiated zones are here represented by circles of 1 mm diameter.

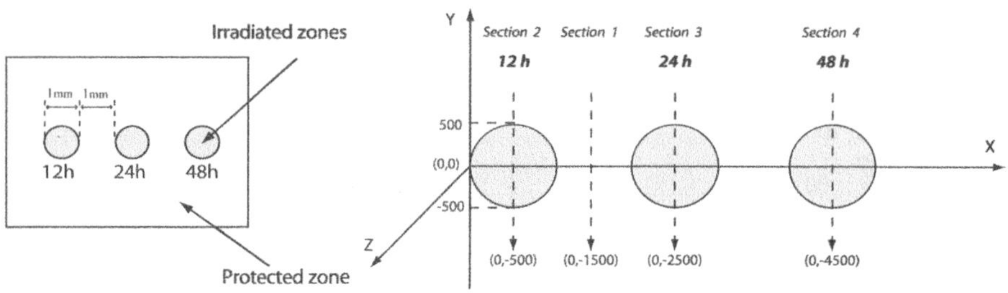

Figure 12.6. *Configuration of the sample for space charge measurements*

After the irradiation, a semi-transparent gold layer (50 nm) was sputtered on both sides of the sample. Finally, a thin carbon layer was sputtered on the irradiated zones in order to facilitate sample positioning and analysis.

The space charge measurements were made according to the X and Y axes, with a 50 µm step (Figure 12.6). The results obtained in both directions being similar, only results according to the Y axis are presented. The Z axis is the direction of the sample thickness.

Section 1 of Figure 12.7 presents the cartography of a non-irradiated zone; Sections 2, 3 and 4 show those of irradiated zones for 12, 24 and 48 hours. For each section, 22 measurement points were necessary. The acquisition of a measurement point was made in a frequency range of 100 Hz to 10 kHz, requiring 6 minutes for the total recording.

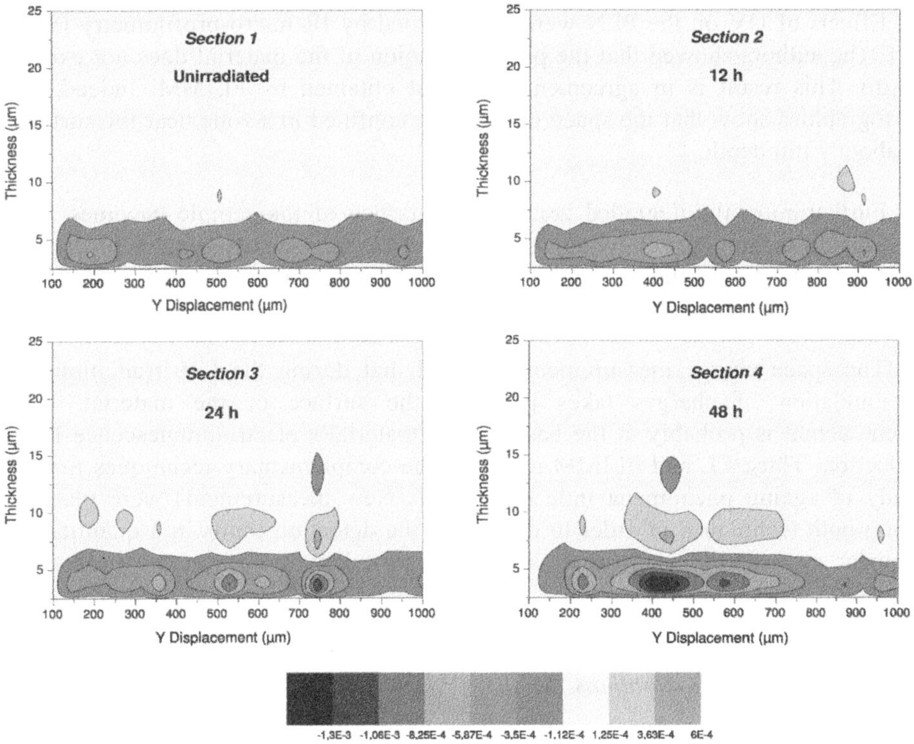

Figure 12.7. *Space charge cartographies for different irradiation durations*

Figure 12.7, Section 1, gives the charge distribution relative to Section 1 (the non-irradiated zone) and it will be considered as a reference. By visual analysis, it can be seen that there is not a great difference between Sections 1 and 2. This expresses the fact that the material is not affected by the irradiation for weak exposure durations (<12h). On the other hand, the charge level increases with exposure time. For a 24-hour irradiation, negative charges appear in a zone near the surface, in a non-homogenous way. For 48 hours, the charge level becomes important compared with the reference profile level, and the accumulation zone becomes wider.

Another important observation concerns the depth of the charge accumulation zone. This one, situated near the surface of the sample, is present in all sections studied and its spreading depends on exposure time. This result was expected, considering that the photo-oxidation induced by UV irradiation is a surface phenomenon [SCH 97].

Effects of UV on the PEN were also studied by IR micro-profilometry [SCH 97]. The authors showed that the photo-oxidation of the material does not exceed 10μm. This result is in agreement with that obtained by FLIMM. Indeed, the cartographies show that the space charges are confined in a zone near the surface, at about 7 μm depth.

Furthermore, the degraded zone at the surface of the sample becomes very absorbant, causing a yellowing of the material, thus constituting a protection screen for the rest of the material. This surface phenomenon was also revealed by photoluminescence measurements [MAR 01] taken in our laboratory.

The space charge measurements showed that during the UV-irradiation, an accumulation of charges takes place at the surface of the material. This phenomenon is probably at the basis of the material's electroluminescence level reduction. Thus, EL and FLIMM appear to be complementary techniques for the study of ageing phenomena induced by UV. New measurements were planned using both techniques in order to determine the detection limits in a quantitative manner, particularly concerning the minimal irradiation surface, the irradiation intensity or durations of exposure.

12.6.3. *3-D charge distributions*

12.6.3.1. *Sample preparation*

For three-dimensional cartographies [PET 06], the following conditions must be fulfilled:

– the material under study must trap charges for a long period of time;

– the zones where space charges are accumulated must be easy to identify.

The first condition is important because the execution of the cartography requires the recording of several measurement points according to surface scanning in X and Y directions. The total acquisition of data can last several days, hence the importance of a material possessing the ability to store charges for a long time. To that purpose, the choice was made of Teflon (PTFE) polymer, as it is well known for its charge trapping properties.

The exact knowledge of the zones likely to contain space charges is obtained by the prior pasting of a microscope grid on the sample. The grid allows a selective implantation of electrons according to its geometry. By knowing the physical characteristics of the grid used with high accuracy (i.e. the dimensions of the squares), the zones in which charges were implanted can be precisely determined.

The samples used for this study are thin (50 µm thick) PTFE films, metallized on both sides with a 30 nm gold layer. Microscope grids with different geometries were stuck on the samples, either with silver paint, or with cyanolite. The samples were irradiated by a scanning electron microscope (SEM, JEOL JSM- 6060LV) and then a very thin layer of carbon (~20 nm) was evaporated on the whole sample. This allows a print of the grid to be kept once the mask is removed.

12.6.3.2. *3-D cartographies*

Six measurement points organized according to X and Y axes, and 50 µm spaced, are depicted in Figure 12.8. The Z axis still represents the direction of the sample thickness.

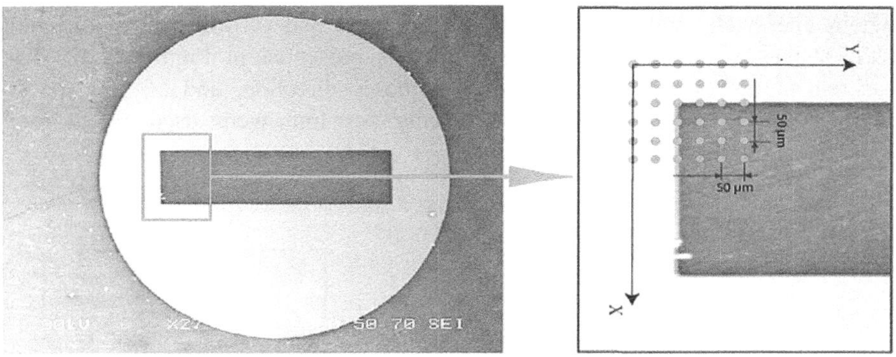

Figure 12.8. *Visualization of the analysis zone*

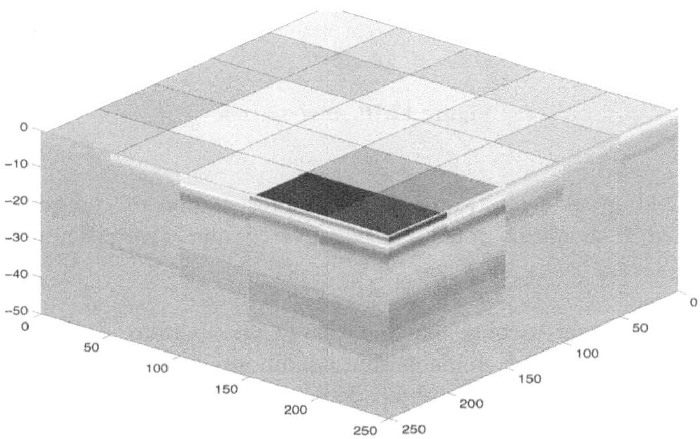

Figure 12.9. *3D Visualization of the charge distribution*

The 3-D charge representation obtained is shown in Figure 12.9.

First of all, electrons seem to remain confined in a zone near the sample surface. However, this zone is slightly smaller than the irradiated zone. This could be explained by edge effects due to a non-homogeneous implantation of electrons in the vicinity of the grid edges. The 3-D cartography is very coherent with the configuration of the sample; nevertheless, the spatial resolution can be improved.

An interesting study was to test and reach the limits of FLIMM, in terms of spatial resolution as well as sensitivity, with respect to the charge evolution with time. For this purpose, similar samples elaborated and irradiated in the same way as previous ones were studied. The grid used this time had vertical bars of different width. The chosen zone for this new study is presented in Figure 12.10. The measurement points are 10 μm spaced in the X direction, and 50 μm spaced according to the Y axis; 44 measurements per line were used for a total measurement duration of 2 days.

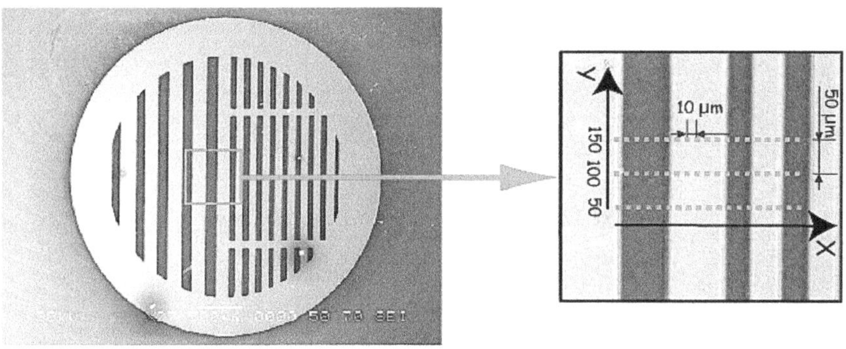

Figure 12.10. *Analysis zone*

Figure 12.11 gives a multi-dimensional representation of charges which have been implanted. It should be noted that negative charges were efficiently implanted in the non-protected zones.

It is also important to note that the charge distributions detected in FLIMM correspond exactly to the implantation conditions and according to the grid geometry. The total charge penetration depth is about 5 μm, with a peak around 2 μm.

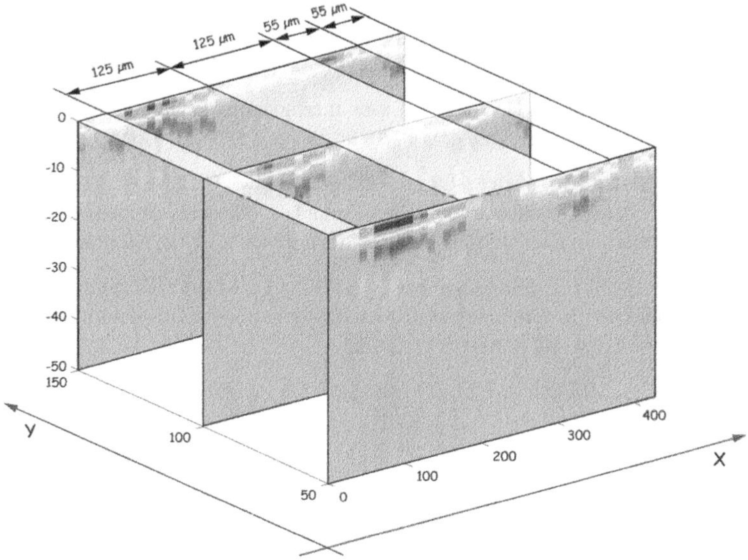

Figure 12.11. *Representation of the implanted charge with the visualization of bars*

12.7. Conclusion

In this chapter, the FLIMM method was presented from both a theoretical point of view and with experimental results. Some of its specificities for the execution of multidimensional representations of space charges in solid insulators have been shown. The results obtained show that FLIMM can be used as a diagnosis tool with a lateral spatial resolution of about 10 µm, that is to say 1,000 times lower than other more traditional methods (PEA, thermal wave, etc.).

This resolution could be improved by decreasing the spot size as well as increasing the upper modulation frequency, which will require more complex developments (which are under way).

12.8. Bibliography

[AND 77] ANDREWS H. C., HUNT B. R., *Digital Image Restoration*, Prentice-Hall, Englewood Cliffs, 1977.

[AGN 05] AGNEL S., NOTINGHER JR P., CASTELLON J., TOUREILLE A., FRANCESCHI J.L., LAURENT C., MARTY-DESSUS D., MARY D., PETRE A., TEYSSEDRE G., "Comportement diélectrique du polyéthylène téréphthalate (PET). Mesures de charges d'espace", *Revue Internationale de Génie Electrique*, vol. 8, no. 2, p. 205–220, 2005.

[AUG 00] AUGÉ J. L., TEYSSÈDRE G LAURENT C., DITCHI T., HOLÉ S., "Combined electroluminescence and charge profile measurements in poly(ethylene-2.6-naphthtalate) under a DC field", *J. Phys. D: Appl. Phys*, vol. 33, p. 3129–33138, 2000.

[FRA 97] FRANCESCHI J.L., HAAS V., "Laser thermoacoustic modulation for space charge measurement", *Appl. Phys. Letts.*, vol. 70, no. 17, p. 2236–2237, 1997.

[FRA 00] FRANCESCHI J.L., MARTY-DESSUS D., BIELLMANN C., BERQUEZ L., MOUSSEIGNE M., "Profils de charges d'espace dans les isolants polymères minces par une méthode d'excitation thermique avec laser focalisé", *Congrès SFE 2000*, Montpellier, 2000.

[FRA 01] FRANCESCHI J. L., BIELLMANN C., BERQUEZ L., MARTY-DESSUS D., "Virtual Space Charge Model for a frequency-based characterization of insulators", *Jpn. Journal of Applied Physics*, vol. 40, p. 880–890, 2001.

[HAS 97] HAAS V., MOUSSEIGNE M., FRANCESCHI J.L., *J. Electrostatics*, vol. 40–41, p. 307–312, 1999.

[HAN 93] HANSEN P.C., O'LEARY D.P, "The use of the L-curve in the regularization of discrete ill-posed problems", *SIAM J. Sci. Comput.*, vol. 14, no. 6, p. 1487–1503, 1993.

[HAN 94] HANSEN P. C., "Regularization Tools: A Matlab package for analysis and solution of discrete ill-posed problems", *Num. Algorithms*, vol. 6, p. 1–35, 1994.

[HAN 96] HANSEN P. C., MOSEGAARD K. "Piecewise polynomial solutions without *a priori* break points", *Num. Lin. Algebra with Appl.*, vol. 3, no. 6, p. 513–524, 1996.

[HON 90] HONERKAMP J., WEESE J., "Tikhonovs regularization method for ill-posed problems", *Continuum Mech. Thermodyn*, vol. 2, p. 17–30, 1990.

[LAN 81] LANG S. B., DAS-GUPTA D. K., "A technique for determining the polarization distribution in thin polymer electrets using periodic heating", *Ferroelectrics*, vol. 39, p.1249–1252, 1981.

[LAN 91] LANG S. B., "Laser intensity modulation method (LIMM): Experimental techniques, theory and solution of the integral equation", *Ferroelectrics*, vol. 118, p. 343–361, 1991.

[LAN 98] LANG S. B., "An analysis of the integral equation of the surface Laser Intensity Modulation Method using the constrained regularization method", *IEEE. Trans. on Dielec. Electr. Insul.*, vol. 5, no. 1, p. 70–76, 1998.

[MOP 82] MOPSIK F.I., DEREGGI A.S., *J.Appl. Phys.* vol. 53, 4333–4339, 1982.

[PET 04a] PETRE A., Optimisation de la méthode FLIMM pour la caractérisation en volume des charges d'espace dans les isolants polymères minces, Doctoral thesis, Paul Sabatier University, Toulouse, December 2004.

[PET 04b] PETRE A., MARY D., MARTY-DESSUS D., "Study of PEN ageing under UV irradiation by electroluminescence and space charge measurements", *J. Optoel. and Adv. Mat*, vol. 6, no. 3, p. 1049–1054, 2004.

[PET 04a] PETRE A., MARTY-DESSUS D., BERQUEZ L., FRANCESCHI J.L., "A Comparison of Different Mathematical Treatments for Solving the Inverse Problem in Focused Laser Intensity Modulation Method", *Jpn. Journal of Applied Physics*, vol. 43, no. 5A, p. 2572–2579, 2004.

[PET 06] PETRE A., MARTY-DESSUS D., BERQUEZ L., FRANCESCHI J.L., "Space Charge Cartography by FLIMM on Thin PTFE Films Irradiated by Scanning Electron Microscope", *J. Electrostatics*, vol. 64, nos. 7–9, p. 492–497, July 2006.

[PLO 92] PLOSS B., EMMERICH R., BAUER S., "Thermal wave probing of pyroelectric distributions in the surface regions of ferroelectric materials – A new method for the analysis", *J. Appl. Phys.*, vol. 72, p. 5363–5370, 1992.

[SCH 97] SCHEIRS J., GARDETTE, J.-L., "Photo-oxidation and photolysis of poly(ethylene naphthalate)", *Polym. Degr. Stab.*, vol. 56, p. 339–350, 1997.

[SKI 02] SKIPA O., NALBACH M., SACHSE F. B., DOSSEL O., "Comparison of regularization techniques for the reconstruction of transmembrane potentials in the heart", *Biomedizinische Technik*, vol. 47, p. 246–248, 2002.

[TAK 99] TAKADA T., "Acoustic and Optical Methods for Measuring Charge Distribution in Dielectrics", *Proc. IEEE CEIDP*, p. 1–14, Austin, USA, 1999.

[TIK 77] TIKHONOV A.N., ARSENIN V.Y., *Solution of Ill-posed Problems,* Wiley, New York, 1977.

Chapter 13

Space Charge Measurement by the Laser-Induced Pressure Pulse Technique

13.1. Introduction

Three non-destructive techniques currently exist to determine the space charge in a solid insulator using the propagation or production of a pressure wave. These are the electroacoustic PEA (*Pulsed Electro Acoustic*) method described in the previous chapter, a method using propagation of a pressure wave generated by a piezo: PIPS or PIPP (*Piezoelectrically Induced Pressure Step or Pulse*), and a method using propagation of a pressure wave induced by the impact of a laser beam, called the LIPP (*Laser Induced Pressure Pulse*) technique. The method of production of this pressure wave is the parameter which distinguishes these different techniques. In PIPPS, PIPP and LIPP techniques, the acoustic wave is generated in a mechanical way, either by a *piezo* transducer, or by a laser impact on a target fixed on the sample to be tested. This impulse acoustic wave propagates throughout the insulator, thus modifying the relative position of the charges it encounters. A variation of induced charges then appears at the electrodes, whose effect is the outbreak of a pulse voltage or current in the measurement circuit.

After a brief discussion of the development history of this method's principle, the equations connecting the measured external pulse current or voltage magnitude to the internal charges distribution will be described. A description of the experimental setup as well as the expected performances follows, and the

---

Chapter written by David MALEC.

possibilities offered by this technique will then be illustrated. Finally, perspectives on the use of this method will be presented.

13.2. History

The suggestion to use a short mechanical perturbation to determine the internal charge amount accumulated in a solid insulator [LAU 77] originates from the *École supérieure de physique et de chimie industrielle* in Paris. Applicable to all insulators in which the charges to be measured move with the induced deformation of the atomic lattice, this non-destructive method has given rise to numerous experimental implementations. Logically, the first was that proposed by the *Laboratoire d'électricité générale* [LAU 76], [DAR 80]. This technique used a steep-fronted pressure wave generated by a shockwave tube. Unfortunately, the spatial resolution of the measurements was strongly dependent on the position of the tube with respect to the sample and reproducibility was not guaranteed. A second implementation was achieved by creating a pressure wave with a sudden capacitor discharge in a liquid. The spatial resolution obtained with this method was mostly destined for thick samples (of a few cm).

The first use of a laser impact as a generation source for the pressure wave was made by a Russian team [ROZ 79]. The fairly long duration of the laser pulse did not permit samples of less than a few mm thick to be studied. We had to wait until the early 1980s for spatial resolutions which allowed the study of thinner samples, owing to the use of lasers emitting laser pulses of a short duration (a few ns) [ALQ 81], [SES 81], [ALQ 83]. The problems of spatial resolution, reproducibility and parallelism between the optical wave front and the sample were solved and this technique has, since then, been improved and adapted to different sample structures (whether plane or coaxial). Considering the characteristics it presents, it was adopted by numerous university research teams around the world, and also by big industrial groups working on the optimization of insulating materials.

13.3. Establishment of fundamental equations for the determination of space charge distribution

The equations (in a plane structure) which allow the connection between signals induced by the transit of the pressure wave in an homogenous insulator and the internal charges distribution $\rho(x)$ (assumed uniform on the x-axis) will be established. These internal charges induce image charges on the electrodes A and B, whose superficial densities σ_A and σ_B are given for different cases (Figure 13.1).

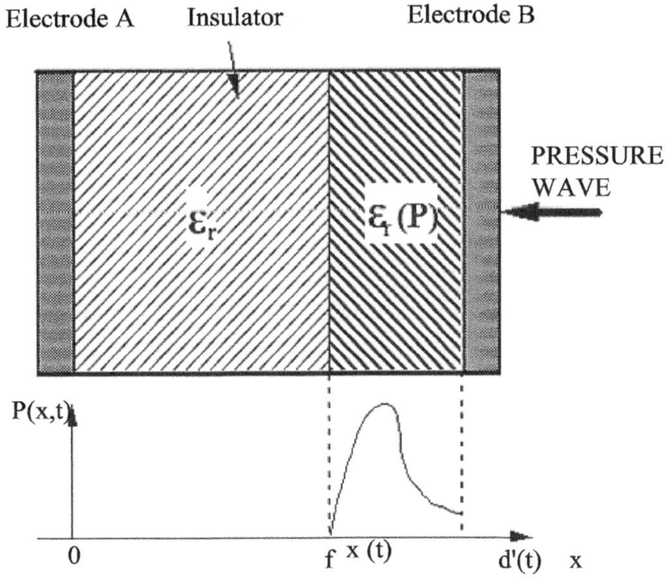

Figure 13.1. *Application of an inhomogenous pressure*

13.3.1. *Specific case: uncharged or charged and short-circuited sample (V=0)*

When the sample is uncharged, the charge density ρ(x) is null on the x-axis. After application of an electrical potential difference $V=V_B-V_A$, the charge density σ_A given by equation [13.1a] only depends on the value of this electrical potential difference. If the sample is charged and short-circuited, the charge density at the electrodes only depends on the charge densities ρ(x) accumulated in the sample and on its geometry. We can define a voltage potential difference V_0, such that σ_A is given by equation [13.1b]:

$$\text{a) } \sigma_A = -\frac{\varepsilon_0 . \varepsilon_R . V}{d} \qquad \text{b) } \sigma_A = \frac{\varepsilon_0 . \varepsilon_R . V_0}{d} \qquad [13.1]$$

where d corresponds to the thickness of the insulator, ε_0 and ε_R, respectively, to the vacuum permittivity and that of the insulator, and V_0 is the equivalent voltage of the electret [DRE 73], i.e. the voltage to apply to the sample to eliminate the surface charge σ_A.

13.3.2. *General case: charged sample submitted to an electrical potential difference*

By superposing both the previous cases, σ_A is given by:

$$\sigma_A = \frac{\varepsilon_0 . \varepsilon_R}{d} . (V_0 - V) \qquad [13.2]$$

The electrical potential difference V being fixed, calculation of the voltage V_0 is required for the determination of σ_A. The detail of this calculation can be found in [ALQ 83]:

$$V_0 = -\frac{1}{\varepsilon_0 . \varepsilon_R} . \left(d - \langle x \rangle\right) \frac{Q}{S} + \frac{1}{\varepsilon_0 . \varepsilon_R} . \int_0^d P(x).dx \qquad [13.3]$$

where P(x) represents the pressure wave spatial distribution, Q the total charge contained in a surface S and $\langle x \rangle$ the average penetration depth of the charges. The polarization (assumed uniform on the x-axis) is equivalent to a charge distribution $\rho_{pol}(x)$, defined by equation [13.4a]. For the sake of simplification, the charge density $\rho(x)$ will therefore be considered defined by equation [13.4b]:

$$\text{a) } \sigma_{pol} = -\frac{\partial P(x)}{dx} \qquad \text{b) } \rho(x) = \rho_{charge}(x) + \rho_{pol}(x) \qquad [13.4]$$

Finally, the expression for σ_A, taking into account equations [13.3] and [13.4], becomes:

$$\sigma_A = -\frac{d - \langle x \rangle}{d} . \frac{Q}{S} - \frac{\varepsilon_0 . \varepsilon_R}{d} . V \qquad [13.5]$$

The determination of σ_B is immediate since:

$$\sigma_A + \sigma_B + \frac{Q}{S} = 0 \qquad [13.6]$$

The superficial charge densities σ_A and σ_B are therefore likely to supply information on the charges accumulated within the insulator.

13.3.3. *Application of a pressure wave*

Under the effect of uniform compression, and because of the local concentration variation of dipoles and charges, the relative permittivity ε_R takes a uniform value ε'_R in the material. If we admit that the stored charges follow the atomic lattice deformation, then a basic calculation shows that the quantity $\frac{d - \langle x \rangle}{d}$ remains unchanged and that only the term related to the dielectric thickness is affected by this constraint. The application of a uniform compression therefore does not allow the spatial distribution of charges to be found.

By contrast, when the sample is subjected to an unhomogenous compression (see Figure 13.1), the value of ε_r in the compressed zone as well as the quantity $\langle x \rangle$ are modified, as are σ_A and σ_B. This variation depends on the internal charges $\rho(x)$ but also the mechanical perturbation profile P(x) the knowledge of which therefore turns out to be fundamental.

13.3.4. *Relationships between measured signals and charge distribution*

The difference in electrical potential existing at the boundaries of the sample during the propagation of the pressure wave is expressed by:

$$V(d',t) - V(0,t) = -\int_0^{d'} E(x,t).dx \qquad [13.7]$$

where d' is a function of time:

$$d'(t) = d - \Delta d(t) \text{ with } \Delta d(t) = \chi \int_{x_f(t)}^{d'(t)} P(x,t).dx \qquad [13.8]$$

In this equation, χ represents the compressibility of the insulator and $x_f(t)$ the position of the pressure wave front. The detailed calculation permitting the electric field distribution to be worked out can be found in [ALQ 83]:

$$\frac{d}{dt}\left(V(d',t)-V(0,t)\right)=-\frac{i(t).d_0}{\varepsilon_0.\varepsilon_R.S}+\chi.G(\varepsilon_R).\int_{x_f(t)}^{d} E(x_0,0).\frac{\partial P(x_0,t)}{\partial t}.dx_0 \qquad [13.9]$$

where i(t) is the current circulating in the measurement circuit and $G(\varepsilon_R)$ a function of the pressure dependence of ε_R. When the measurement circuit presents a very high impedance at high frequency, equation [13.9] becomes [13.10] (electrode A being related to the mass). If this impedance is very weak, equation [13.9] transforms into [13.11]. C is the equivalent capacity of the sample fraction submitted to the pressure wave (before compression):

$$V(d',t)=V(d,0)+\chi.G(\varepsilon_R).\int_{x_f(t)}^{d} E(x,0).P(x,t).dx \qquad [13.10]$$

$$i(t)=C.\chi.G(\varepsilon_R).\int_{x_f(t)}^{d} E(x,0).\frac{\partial P(x,t)}{\partial t}.dx \qquad [13.11]$$

In the case where the pressure wave evolves during its progression (by attenuation and dispersion of elastic waves in the material), it is not possible to simplify equations [13.10] and [13.11]. Nevertheless, if the pressure pulse is of a short duration (or less than the desired spatial resolution), then the measured voltage [13.10] is directly proportional to the electric field, or the measured current [13.11] is directly proportional to the charge quantity [ALQ 83].

The equations which have just been established show that the electric charge distribution in an electrical insulator can be deduced with no ambiguity by the measurement of the signal produced by the propagation of a pressure wave.

13.4. Experimental setup

13.4.1. *Synoptic schema of the measurement setup*

The implementation of the LIPP technique is composed of three steps: the production of a pressure wave, the measurement of the induced signal and the processing of these signals. After amplification, the collected signal (potential or current difference) is then stored in a transient recorder. The digital processing of this is then ensured by a computer, allowing the space charge distribution to be established. The synoptic outline of the experimental setup was reproduced in Figure 13.2 (current measurement). Figure 13.3 shows a final realization.

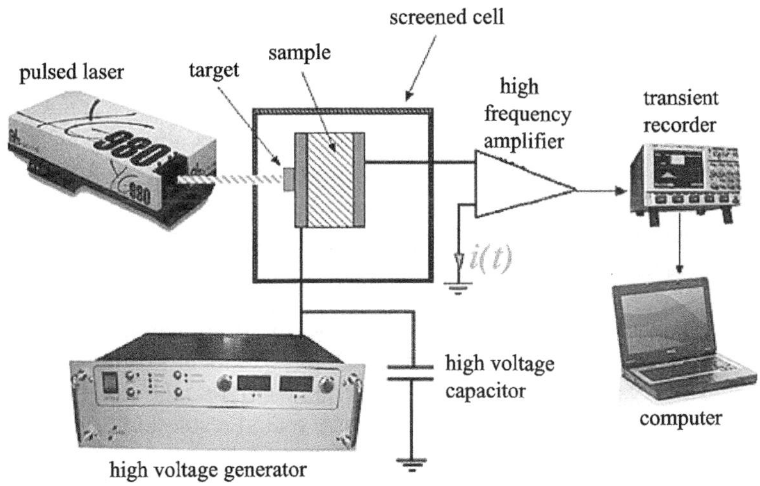

Figure 13.2. *Synoptic outline of the experimental setup*

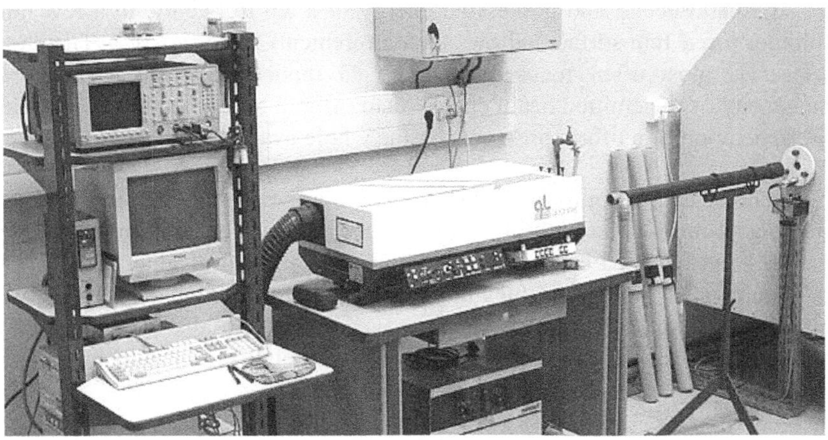

Figure 13.3. *View of an experimental setup (digital oscilloscope and signal-processing computer on the left, laser generator at the center and measurement cell on the right)*

13.4.2. *Generation of pressure*

The production of a pressure wave by the impact of a laser pulse on an absorbant surface, suggested by R.M. White in 1963 [WHI 63], was the goal of numerous research. The target absorbs part of the incident energy and transfers it to the lattice in a very brief time, producing local heating. A thermal expansion then appears and subsequently a spatial deformation gradient, i.e. a pressure which propagates in the form of a quasi-plane wave perpendicular to the irradiated surface. This absorption, which depends on both the optical properties of the material and laser beam wavelength, is attributed to electron excitation in higher energy states which then come back very rapidly to their fundamental state (typically 10^{-12} s). For metallic targets (gold, aluminum, copper, etc.), the absorption ranges in depth from 0.1 to 1 µm, but they reflect up to 90% of the laser beam energy (at λ=1,064 nm). When part of the target is vaporized (volatile target), the vapor thus created exerts an additional pressure on the still solid part.

To obtain relatively high pressures (typically from ten to twenty, to a few hundred bars) and therefore a strong signal/noise ratio, it is fundamental to have power densities varying from 10^6 to 10^8 W/cm^2. Further, the duration of the laser pulse must be fairly brief to obtain a satisfying spatial resolution (a few microns). In polymers, where the speed of sound ranges from 1,000 to 3,000m/s at 20°C, the laser pulse duration must not exceed a few nanoseconds, such to be able to process samples from a few hundred microns to a few millimeters thick.

Considering these orders of magnitude, the use of a laser emitting pulses of the order of a nanosecond and possessing energy of a ten to twenty, to a few hundred millijoules (on a 1cm surface) allows the requirements to be satisfied. The choice of the laser is most often focused on Nd-YaG models, offering a better spatial homogeneity of the emitted beam energy than offered by gas models, which permits to the generation of uniform pressures on the whole irradiated surface.

The choice and the optimization of the target are essential steps in the search for a high signal/noise ratio permitting the induced signals to be correctly processed. The target can be fixed on the sample either with the aid of glue, or with a semi-liquid couplant. In both cases, they must allow the transmission of the wave without strong attenuation. A deposit by vacuum evaporation can also be envisaged (on a metallic target); in that case the thickness of the metallization must be sufficiently high to avoid any photo-induced currents.

To increase the level of the induced pressure, and therefore the signal/noise ratio, various solutions have been found, notably the coating of the target by a substance optically transparent at the wavelength of the laser beam (water, ethanol, resin, varnish, etc.) [AND 70], [YAN 74], [FAI 74], [VON 77]. This coating allows the pressure induced in the target to be increased by using the overpressure generated in the air in the vicinity of the laser beam impact. An example of pressure obtained by use of an ethanol film left on a semi-conductive target is given in Figure 13.4. Let us note that the gain obtained on the pressure level is to the detriment of the spatial resolution (because of an increase of the pressure duration).

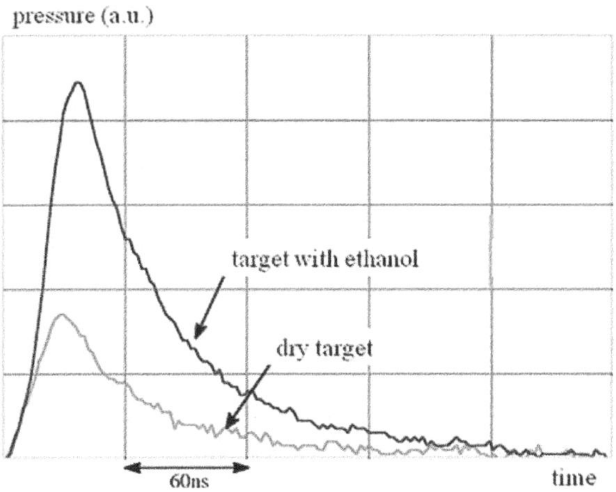

Figure 13.4. *Impact of coating a target on induced pressure level*

13.4.3. *Signal recording*

When considering the propagation speed of a pressure wave in the insulators, the current or the induced electrical potential difference will have to be recorded by a piece of equipment with a high bandwidth (a few 100MHz). In the outline of Figure 13.2, the measurement circuit (of low impedance) is made up of a high-voltage capacitor and an amplifier input impedance (50Ω). An induced current related to the internal distribution of the electric field by equation [13.11] will circulate in this measuring circuit. From an electrical point of view, the sample traversed by the pressure wave then behaves like a high-frequency current source. The non-irradiated part of the sample placed under the electrodes also behaves like a low impedance at high frequencies and consequently a part of the laser induced current circulates in this impedance. In order to reduce this lost part of the current, the dimensions of the electrodes must be, as much as possible, close to that of the laser beam. Further, a frequential analysis including all the elements of the topology of the measurement circuit is essential to check that in the "useful" spectral window, the signal collected by the amplifier is actually the image of the current generated by the sample [DER 94], [MAL 96], and that the use of filters does not truncate it [MAL 96].

13.4.4. *Calibration of the experimental setup*

The calibration operation consists of determining the evolution of the pressure wave in the insulator. This term plays a part in equation [13.10] and its time derivative in equation [13.11]. This determination can be carried out when the internal electric field is known, for example when the sample to test is empty of charge and when the applied voltage is sufficiently low to avoid any injection or dissociation of charges during the calibration measurement. If the pressure wave has a low spatial extension, the distance it covers to entirely penetrate the insulator is sufficiently weak to consider that it does not get distorted, which is mathematically expressed by:

$$P(x,t) = P(0,t\text{-}x/v) \qquad [13.12]$$

where P(x,t) represents the value of the pressure at point x and instant t, v being the wave speed and 0 the input point of the wave. If we derive P(x,t) with respect to time, we then obtain a current being, respectively, during the input and during the output of the pressure wave (in a plane insulator):

$$i_{input}(t) = v\chi G(\varepsilon_R)CE_0P(0,t)\ ;\ i_{output}(t) = -v\chi G(\varepsilon_R)CE_0P(d,t) \qquad [13.13]$$

where E_0 represents the value of the applied field, assumed uniform in the insulator. The measured signal is therefore, apart from the hypotheses, proportional to the pressure at the interface between the input electrode and the insulator (*zone a*, Figure 13.5). The wave then moves in the material (*zone b*, Figure 13.5), and is subjected to an attenuation and a dispersion to reach the interface between the insulator and the output electrode, where we can hold the same reasoning as for the input (*zone c*, Figure 13.5).

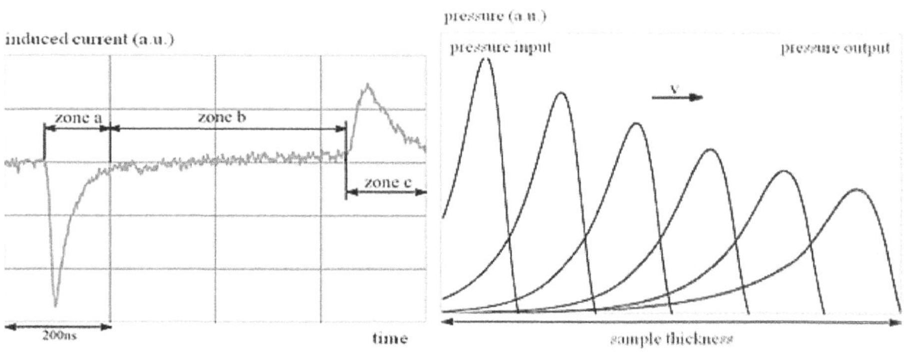

Figure 13.5. *Left: signal induced during the calibration step (Polyethylene; d=1mm, V=30kV, T=20°C). Right: Determination of the evolution of the pressure wave's profile during its transit in the sample*

From the signal induced during the calibration step, appropriate digital processing is then made to reproduce the wave propagation by taking into account both attenuation and dispersion laws [DIT 90]. The verification of the wave propagation modeling can be done in two different ways. The first, experimental [DIT 90], consists of using a quartz-crystal transducer to measure the pressure at the output of samples of a similar nature but with different thicknesses, and to compare them with the calculated pressures. The second method, mathematical [MAL 96], consists of recalculating the induced current (equation [13.11]) from the calculated pressures and comparing it with the current actually measured. An example of a pressure wave propagation calculation is given in Figure 13.5. Let us note that this kind of measurement has, rightly, been proposed for the determination of the elastic properties of insulating materials [DIT 93].

13.4.5. *Signal processing*

In the particular (and ideal) case where the wave can be assimilated to a pulse which is not subjected to any deformation during its transit in the insulator, equation [13.11] is a second-order Fredholm equation. Its resolution, known as a deconvolution operation, can be achieved using different methods [BIR 76]. In practice, the wave is subject to a deformation: equation [13.11] is then called a first-order Fredholm equation. Its resolution is not easy because the noise, even if low, together with errors due to discretization and rounding up in calculations, induce parasitic oscillations which are sometimes important in the solution [MAX 72]. The pressure wave-induced current being known in discrete form, the first step in the resolution of equation [13.11] consists of replacing the integral by a summation. In matrix form, the resolution of the equation induces strong oscillations in the solution (no longer unique), inherent to matrix inversion methods. This inversion must therefore be performed with a constraint imposed on the solution to only conserve its natural variations [BOU 87]. To validate the solution obtained, it is essential to possess *a priori* information on E(x). Two verifications are generally done on this account: the first consists of recalculating the induced signal from the found solution, E(x), and to compare it to the measured signal, whilst the second verification consists of ensuring that E(x) satisfies the equation:

$$\int_0^d \mathrm{E(x)dx} = V = V_B - V_A \qquad [13.14]$$

These conditions are directly expressible in terms of mathematical equations and therefore can be integrated into the processing software; ρ(x) is then obtained using Poisson's equation. An example of processing of the signal induced by the transit of a pressure wave in a low-density polyethylene sample is given in Figure 13.6. The laser duration being short (9 ns), the signal processing result is very close to the measured signal.

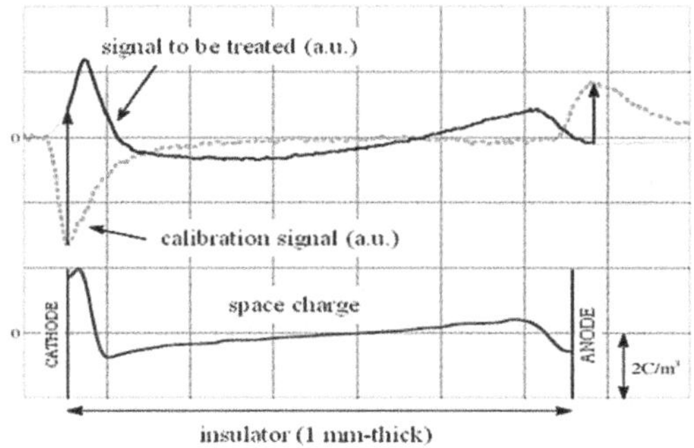

Figure 13.6. *Example of processing of a signal induced by the transit of the pressure wave in a charged sample*

13.5. Performances and limitations

13.5.1. *Performances*

Considering the high level of pressures induced by the LIPP method, measurements may be performed on thick samples (up to a few millimeters thick), allowing investigations on power cables [MAH 91]. The sensitivity of the measurements is high because the high level of induced pressures generates signals with strong signal/noise ratios. It is nevertheless quantifiably difficult because related to the noise level. However, a simplified approach can bring an estimation of its value [DIT 90]. As an example, for a polymeric insulator with a thickness of a few mm, the sensitivity of the measurement and the spatial resolution after processing of the signal are, respectively, of the order of a few mC/m^3 and a few µm (9 ns laser, 450 mJ) [MAL 96].

The duration of a measurement being short (i.e. 1 µs for a 2 mm-thick polyethylene sample at 20°C), under DC voltage, the LIPP method allows the growth and the evolution of the space charge whose dynamics is much slower to be followed in *real time*. Further, its use under industrial frequency (sine 50 Hz–60 Hz) has already been validated [BER 95], [HO 02]. This property is particularly interesting because it permits parameters essential for the study of the space charge formation in insulators to be accessed.

Laser beam focusing [QIN 99] or the use of masks [MAL 96] permits localized measurements to be made. The displacement of the focused beam (or the masks) allows the realization of 3-D space charge cartographies. It also allows the research of non-uniform distributions betraying the presence of "defects" in the insulator.

13.5.2. *Limitations*

The first limitation of the LIPP method is related to the actual principle of the measurement, i.e. the displacement of the pressure wave. Indeed, the passage from one medium to another (in the case of multilayer insulators, interfaces, electrodes, etc.) is accompanied by a reflection of all or part of the wave which depends on the acoustic mismatch between both media. If the acoustic impedances are very different, the pressure wave is entirely reflected: the method does not allow the measurement beyond the interface between both media. If the reflection is partial, signal processing is very difficult, if not impossible. It should be noted that this limitation factor can be taken advantage of to detect possible assembly defects between two layers of a similar insulating system [AIN 99].

The use of a volatile target is the second limitation factor of the LIPP method. Indeed, target ablation modifies the induced pressure profiles and successive measurements lose their synchronicity (Figure 13.7, left). The signals can no longer be processed owing to only one calibration signal, which makes the space charges study process considerably heavier. Further, after impact of the laser beam on the target, a plasma made up of extracted charged particles forms in front of it, absorbing part of the energy of the next laser shot (Figure 13.7, right). The induced pressure level is therefore attenuated after each laser shot. A system of ventilation or aspiration can be used for the evacuation of this plasma [MAL 00].

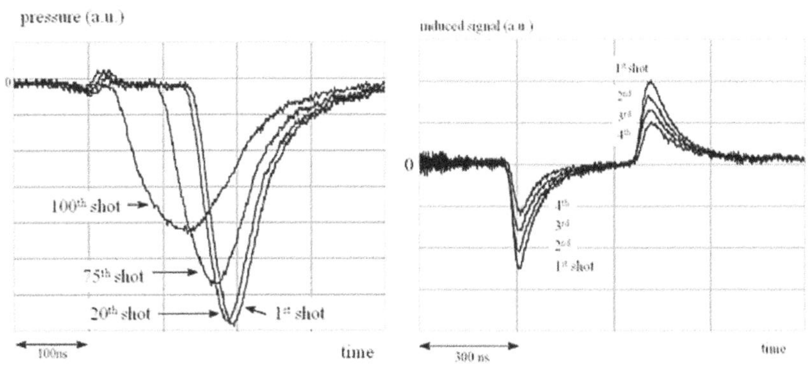

Figure 13.7. *Modification of the induced signal due to the ablation of a volatile target submitted to 1Hz laser shots (left) and due to the formation of a plasma in front of a volatile target (right) [MAL 00] (a.u. = arbitrary units)*

13.6. Examples of use of the method

Figures 13.8 and 13.9 report a few examples of measurements carried out on different types of solid insulators.

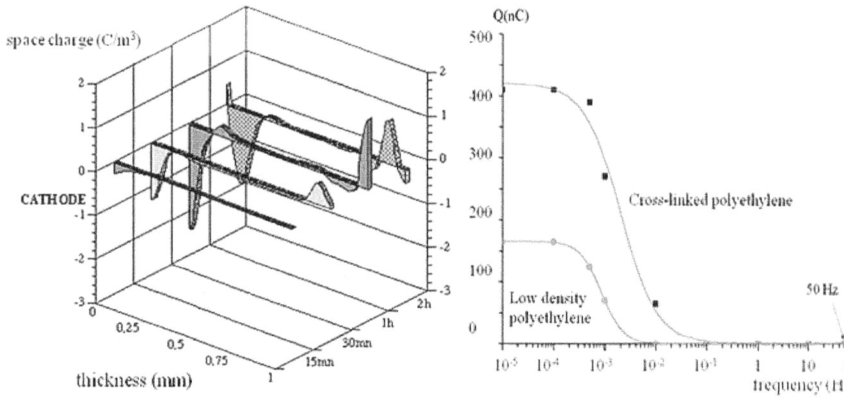

Figure 13.8. *Left: evolution of a space charge in low density polyethylene (V=50 kV, T=20°C, d=1 mm) [MAL 96]. Right: evolution of the total accumulated charge in XLPE and LDPE polyethylenes as a function of the applied voltage and frequency (20 kV peak sine voltage; T=60°C) [FAN 01]*

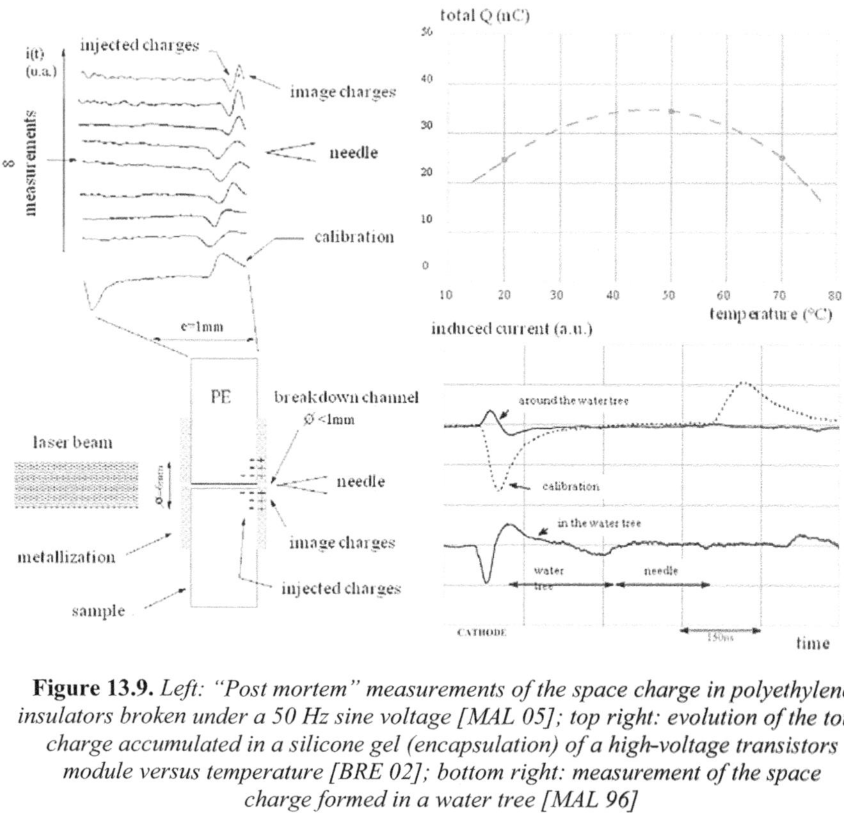

Figure 13.9. *Left: "Post mortem" measurements of the space charge in polyethylene insulators broken under a 50 Hz sine voltage [MAL 05]; top right: evolution of the total charge accumulated in a silicone gel (encapsulation) of a high-voltage transistors module versus temperature [BRE 02]; bottom right: measurement of the space charge formed in a water tree [MAL 96]*

13.7. Use of the LIPP method for surface charge measurement

In certain conditions, the LIPP measurement technique is also applicable to the determination of surface charges, as proposed by [ALQ 82], which opens its application field to the electrostatic domain [MAL 04] and the study of partial discharges [MAL 99], [MAL 02].

13.8. Perspectives

The LIPP method is currently widely used in the study of low and high-voltage insulators, electrets, piezoelectric and pyroelectric materials. In addition to these classic uses, this technique is being developed in the domain relative to surface charges (electrostatic, partial discharges, etc.). Further, it was recently validated for the study of semi-conductors [HOL 04]. Its strong potential therefore traces a future for the method just as successful as the past 20 years have been.

13.9. Bibliography

[AND 70] ANDERHOLM N.C., "Laser-generated stress waves", *Appl. Phys. Lett.,* vol. 16, no. 3, p. 113–115, 1970.

[AIN 99] AÏNOUCHE L., ALQUIÉ C., "Application of the pressure wave propagation method for adhesion defects detection and quantification in balayer structures", *J. Appl. Phys.*, vol. 86, no. 2, p. 1156–1166, 1999.

[ALQ 81] ALQUIÉ C., DREYFUS G., LEWINER J., "Stress wave probing of electric field distributions in solid dielectrics", *Phys. Rev. Lett.*, vol. 47, no. 20, p. 1483–1487, 1981.

[ALQ 82] ALQUIÉ C., CHARPAK G., LEWINER J., "Pulsed laser determination of surface electric charge distribution", *Journal de Physique-Lettres,* vol. 43, p. no. 19, 687–693, 1982.

[ALQ 83] ALQUIÉ C., Détermination de la distribution du champ électrique par perturbation locale induite par laser, PhD thesis, Pierre and Marie Curie University, Paris, 1983.

[BER 95] BERT C., HENNION C., LEWINER J., ALQUIÉ C., HAMPTON N., FREESTONE J., VERNE S., "Measurement of space charge distributions under 50Hz AC electric stresses using the LIPP method", *Proceeding of the 4th International Conference on Insulated Power Cable*, p. 95–98, SEE Edition, June 25–29, Versailles, France, 1995.

[BIR 76] BIRAUD Y.G., "Les méthodes de déconvolution et leurs limitations fondamentales", *Revue de Physique appliquée*, vol. 11, no. 2, p. 203–206, 1976.

[BOU 87] BOUILLIER G., Etude d'une méthode de mesure des distributions de charges électriques dans des structures métal-isolant-semiconducteur, PhD thesis, Pierre and Marie Curie University, Paris, 1987.

[BRE 02] BREIT F., MALEC D., LEBEY T., "Investigations on DC conductivity and space charge in silicone gel", *Proceeding of the Conference on Electrical Insulation and Dielectric Phenomena,* p. 48–51 , IEEE Edition, October 20–24, Cancun , Mexico, 2002.

[DAR 80] DARMON D., Etude d'une méthode de détermination des potentiels dans un diélectrique, PhD thesis, Pierre and Marie Curie University, Paris, 1980.

[DER 94] DE REGGI A.S., "Effects of the measuring circuit on broadband piezo and pyroelectric measurements", *IEEE Trans.Diel.Elec.Ins*., vol. 1, no. 1, p. 146–152, 1994.

[DIT 90] DITCHI T., Perfectionnement de la méthode de l'onde de pression. Application à l'étude des structures isolantes à base de polyéthylène, PhD thesis, University of Paris VI, Paris, 1990.

[DIT 93] DITCHI T., ALQUIÉ C., LEWINER J., "Broadband determination of ultrasonic attenuation and phase velocity in insulating material", *J. Acoust. Soc. Am.,* vol. 94, no. 6, p. 3061–3066, 1993.

[DRE 73] DREYFUS G., LEWINER J., "Electric fields ad currents due to excess charges and dipoles in insulators", *Phys. Rev. B,* vol.8, no. 6, p.3032–3036, 1973.

[FAIR 74] FAIRAND B.P., CLAUER A.H., JUNG R.G., WILCOX B.A., "Quantitative assessment of laser-induced stress waves generated at confined surfaces", *Appl. Phys. Lett.,* vol. 25, no. 8, p. 431–433, 1974.

[FAN 01] FANJEAU O., Etude de la formation des charges d'espace dans les isolants pour câbles d'énergie sous champ électrique alternatif 50Hz, PhD thesis, University of Paul Sabatier, Toulouse, 2001.

[HO 02] HO Y.F.F., CHEN G., DAVIES A.E., SWINGLER S.G., SUTTON S.J., HAMPTON R.N., HODBELL S., "Measurement of space charge in XLPE under a 50Hz AC stress", *IEEE Trans. Diel. Elec. Ins*., vol. 9, no. 3, p. 362–370, 2002.

[HOL 04] HOLÉ S., LEWINER J., "Direct measurement of trapped and free charges distributions in semiconductors", *Appl. Phys. Lett*., vol. 84, no. 8, p. 1308–1310 , 2004.

[LAU 76] LAURENCEAU P., BALL J., DREYFUS G., LEWINER J., "Une méthode nouvelle de détermination de la distribution spatiale des potentiels dans les diélectriques", *C.R. Acad.Sc.*, vol. 283, no. 8, p.135–138, 1977.

[LAU 77] LAURENCEAU P., DREYFUS G., LEWINER J., "New principle for the determination of potential distributions in dielectrics", *Phys.Rev.Lett.*, vol. 38, no. 1, p.46–49, 1977.

[MAH 91] MAHDAVIS., ALQUIÉ C., LEWINER R J., "Mesure directe de la charge d'espace dans les câbles à isolation synthétique par la méthode de l'onde de pression", *Proceeding of the 3rd International Conference on Insulated Power Cable*, p. 534–541, SEE Edition, June 24–28, Versailles, France, 1991.

[MAL 96] MALEC D., Etude du rôle des charges d'espace dans la rupture diélectrique des isolants solides par la méthode d'onde de pression induite par impact laser: réalisation technologique et étude théorique, PhD thesis, University of Paul Sabatier, Toulouse, 1996.

[MAL 99] MALEC D., LEBEY T., "Coupled measurements of partial discharges and space charge in a polyethylene model sample under a 50Hz AC field", *Proceeding of the Conference on Electrical Insulation and Dielectric Phenomena*, p. 235–238, IEEE Edition, October 17–20, Austin, USA, 1999.

[MAL 00] MALEC D., "Technical problems encountered with the laser induced pressure pulse method in high voltage cable insulators studies", *Meas. Sci. Technol.,* vol. 11, no. 5, p. N76–N80, 2000.

[MAL 02] MALEC D., LEBEY T., "Laser induced pulse as a tool to determine surface charges in inhomogeneous (solid–gas) dielectric", *Appl. Phys. Lett.,* vol. 80, no. 8, p. 1421–1423, 2002.

[MAL 04] MALEC D., "Study of static electricity using the laser induced pressure pulse method", *Meas. Sci. Technol.,* vol. 15, no. 1, p. N1–N5, 2004.

[MAL 05] MALEC D., MARY D., FANJEAU O., "Relationship between charge injection and dielectric breakdown of polyethylene", *Proceeding of the Conference on Electrical Insulation and Dielectric Phenomena*, p. 67–70, IEEE Edition, October 16–19, Nashville, USA, 2005.

[MAX 72] MAS J., *Traitement du signal*, Masson, Paris, 1972.

[QIN 99] QIN X., SUZUKI K., TANAKA Y., TAKADA T., "Three-dimensional space charge measurement in a dielectric using the acoustic lens and PWP method", *J. Phys. D: Appl. Phys*., vol. 32, no. 2, p. 156–160, 1999.

[ROZ 79] ROZNO A.G., GROMOV V., "Measurement of the space charge distribution in a solid dielectric", *Phys. Rev. Lett.*, vol. 5, no. 5, p. 266–267, 1979.

[SES 81] SESSLER G.M., WEST J.E., GERHARD-MULTHAUPT R., "Measurement of charge distribution in polymer electrets by a new pressure pulse method", *Polym. Bull.*, vol. 6, no. 1–2, p. 109–111, 1981.

[VON 77] VON GUTFELD R.J., MEICHER R.L., "20Mhz acoustic waves from pulsed thermoelastic expansions of constrained surfaces", *Appl. Phys. Lett*, vol. 30, no. 6, p. 257–259, 1977.

[WHI 63] WHITE R.M., "Elastic wave generation by electron bombardment or electromagnetic wave absorption", *J. Appl. Phys.,* vol. 34, no. 7, p. 2123–2124, 1963.

[YAN 74] YANG L.C.,"Stress waves generated in thin metallic films by a Q-switched ruby laser", *J. Appl. Phys.*, vo. 45, no. 6, p. 2601–2608, 1974.

Chapter 14

The Thermal Step Method for Space Charge Measurements

14.1. Introduction

The principle of non-destructive methods allowing the measurement of space charges is to detect the charges and access their distribution without modifying the electrical state of the material. The principle is therefore to make the influence charges vary at the electrodes with the aid of a non-homogeneous excitation in time and in space. The excitation can have several origins (thermal, mechanical, electrical, etc.). According to the experimentation conditions, we obtain in the external circuit linking the electrodes a signal whose analytical expression is a function of the volume charge density distribution. From this signal, the internal electric field and charge distributions can be determined after adequate processing.

Historically, a thermal stimulus was used at the *Laboratoire d'electrotechnique de Montpellier* (LEM). Indeed, the Thermal Step Method (TSM) has been developed at the LEM since 1986 and is based on the application of a temperature step to an insulating sample. In this chapter, after describing the principle of this technique, we shall present in detail its conditions of use, the numerical resolution methods allowing a space charge density distribution to be taken from the experimental signal, as well as the method's recent evolutions and perspectives related to its application.

Chapter written by Alain TOUREILLE, Serge AGNEL, Petru NOTINGHER and Jérôme CASTELLON.

14.2. Principle of the thermal step method (TSM)

This method originally developed in 1987 [TOU 87], [TOU 90] consists of the measurement of the capacitive current which appears in the external circuit after application of a positive or negative temperature step in the vicinity of the sides of a sample. If the temperature distribution across a sample is known at any instant, it is possible to go back to both the electric field and the space charge distributions. This method presents the advantage of being applicable to thin and thick samples, whether they are plane (plates) or cylindrical (cables).

14.2.1. *The TSM in short circuit conditions*

Let us consider an insulating plate of thickness d and surface S with two electrodes whose abscissae are respectively x=0 and x=d (see Figure 14.1). The material is considered homogeneous and infinitely flat ($d \ll \sqrt{S}$); thus, the electric field is considered constant in a plane parallel to the electrodes. The sample is placed in short circuit at temperature T_0.

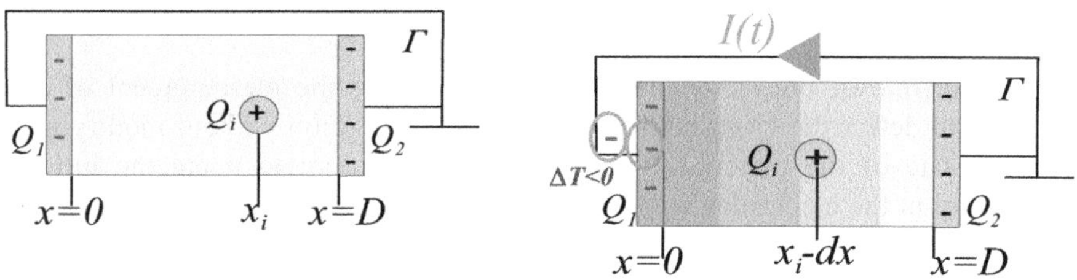

Figure 14.1. *Principle of the TSM in short circuit conditions: (a) sample at equilibrium at T_0 and (b) application of the thermal step*

Let us consider a charge Q_i situated in the insulator at a depth x_i. Since the sample/electrodes/wire system is in electrostatic equilibrium, this charged plane Q_i induces influence charges Q_1 and Q_2 at the electrodes.

The expression of the influence charges can be obtained with the aid of:

– the short circuit condition: $\int_0^d E(x)dx = 0$

– the law of conservation of the electric charge: $Q_1 + Q_2 + Q_i = 0$

– boundary conditions at the dielectric/conductor interfaces.

We deduce that:

$$Q_1 = -\frac{d - x_i}{d} Q_i \text{ , } Q_2 = -\frac{x_i}{d} Q_i \qquad [14.1]$$

If a temperature step $\Delta T = T - T_0$ is applied on one side of the sample (see Figure 14.1b), the temperature diffusion in the sample $\Delta T(x,t) = T(x,t) - T_0$ will generate local variations of the permittivity and the abscissae (dilation or contraction), which read (in first order):

$$\varepsilon = \varepsilon_{T_0}(1 + \alpha_\varepsilon \Delta T), \; dx = dx_0(1 + \alpha_d \Delta T) \qquad [14.2]$$

where α_ε is the variation coefficient of the material permittivity with the temperature and α_d its dilation coefficient.

$$\alpha_\varepsilon = \frac{1}{\varepsilon}\frac{d\varepsilon}{dT}, \; \alpha_d = \frac{1}{x}\frac{dx}{dT} \qquad [14.3]$$

The variations of the axis and the permittivity with temperature lead to a modification of the influence charges:

$$Q_1(t) = -Q_i \int_{x_i}^{d} \frac{dx}{\varepsilon(x,t)} \Bigg/ \int_0^d \frac{dx}{\varepsilon(x,t)}, \; Q_2(t) = -Q_i \int_0^{x_i} \frac{dx}{\varepsilon(x,t)} \Bigg/ \int_0^d \frac{dx}{\varepsilon(x,t)} \qquad [14.4]$$

with the previously used notations and by putting:

$$\alpha = -\frac{1}{C}\frac{dC}{dT} \approx \frac{1}{x}\frac{dx}{dT} - \frac{1}{\varepsilon}\frac{d\varepsilon}{dT} = \alpha_d - \alpha_\varepsilon, \text{ assumed as constant,} \qquad [14.5]$$

we obtain for $Q_2(t)$ the expression:

$$Q_2(t) = -Q_i \frac{x_i}{d}\left[1 + \frac{\alpha}{x_i}\int_0^{x_i} \Delta T(x,t)dx - \frac{\alpha}{d}\int_0^d \Delta T(x,t)dx\right] \qquad [14.6]$$

The variation of influence charges causes the apparition of a current in the external circuit, which we call thermal step current:

$$I(t) = -\frac{dQ_2(t)}{dt} \qquad [14.7]$$

If the value of the current and the spatial distribution of the temperature are known at every instant, the value of the charge Q_i and its position x_i can be calculated.

In general, if we define the volumic charge density by $\rho(x) = dD(x)/dx = \varepsilon\, dE(x)/dx$ on the x-axis in an infinitely flat layer of thickness dx, it can be shown by integration over the entire sample [ABO 91] [CHER 93] that the total current in the external circuit is expressed as:

$$I(t) = -\alpha C \int_0^d E(x) \frac{\partial \Delta T(x,t)}{\partial t} dx, \qquad [14.8]$$

where $E(x)$ is the electric field on the *x-axis* and C the electrical capacitance of the sample before thermal excitation.

The thermal step current can be measured with a current amplifier. Its amplitude depends on the quantity of charges stored in the sample, the charge distribution and the material parameters. In practice, the value of the current ranges from a few pico-amperes to a few micro-amperes (for high capacitances, like those of significant cable lengths).

With the expression for the current being known, to determine $E(x)$ we must process it mathematically. This operation requires a perfect knowledge of the derivative of the temperature $\partial \Delta T(x,t)/\partial t$ at every instant and every point, which is given by the heat equation in plane geometry:

$$\frac{\partial^2 \Delta T(x,t)}{\partial x^2} = \frac{1}{D} \frac{\partial \Delta T(x,t)}{\partial t} \qquad [14.9]$$

where $1/D = \mu\, c/\lambda$ is the reciprocal of the thermal diffusivity D of the material, with μ the specific mass of the insulator, c its specific heat and λ its thermal conductivity.

Further, the thermal step crosses radiator/electrode and electrode/insulator interfaces, which lead to a dampening of the temperature step which cannot be neglected during the processing of the current. To simplify the calculations, we can use an "equivalent thickness" model, i.e. we consider that the total thickness x_1, of different interfaces and average reciprocal thermal diffusivity $1/D_l$, is equivalent to a thickness x_0 with the same reciprocal thermal diffusivity $1/D_0$ as the insulator. The integral limits (0 and d) can thus be replaced by x_0 and x_0+d (the electric field between 0 and x_0 is null):

$$I(t) = -\alpha C \int_{x_0}^{x_0+d} E(x) \frac{\partial \Delta T(x,t)}{\partial t} dx \qquad [14.10]$$

In pratice (see Figure 14.2), the thermal step is created by a sudden circulation of a cool (or warm) liquid in a radiator in contact with the sample (or by a transient current which heats up a resistive material under the sample [ODI 00]). The electrodes are connected through a current amplifier. This latter is linked to a computer which records the signal and does the numerical processing. After the execution of a measurement, the sample can be brought back to room temperature by using a 25°C liquid, and the experiment can thus be repeated.

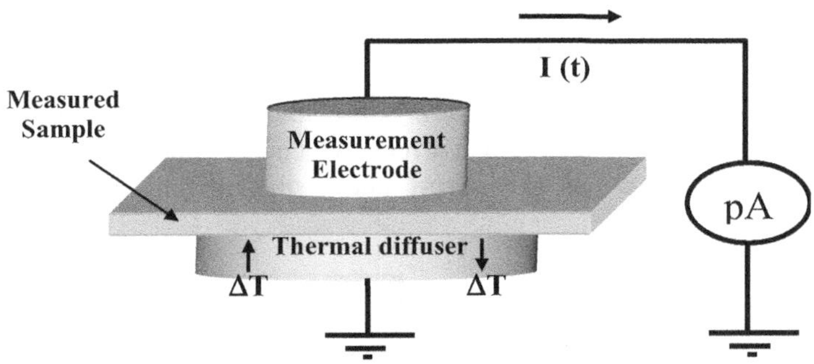

Figure 14.2. *TSM experimental assembly*

Similar equations and an analogous experimental device [SAB 91], [SAN 94] are used in the case of power cables.

The TSM in short circuit conditions reports on the residual static state of charges in a material before and after the application of the applied electric field. However, to remain close to the service conditions of insulated systems and for a better understanding of phenomena appearing when the external electric field is applied, the TSM was designed to take measurements under an applied electric field [NOT 01a]. Furthermore, the measurements under voltage can bring useful information for the calibration of the technique. This principle can also be directly used for industrial applications.

14.2.2. *Evolution of the TSM for measurements under a continuous applied electric field*

To make measurements under a continuous electric field, the device described in Figure 14.2 presents two major disadvantages:

– the current amplifier must not be in contact with the high voltage source;

– if a set up with the current amplifier placed between the sample subjected to a high voltage and the ground is used (as, for example, for conduction current measurements), the conduction and polarization currents are likely to mask the thermal step current.

A solution to these problems is to use a "compensation sample", with identical dimensions to the measurement sample, placed opposite to this latter. By connecting one side of the compensation sample to the current amplifier and the other to the measurement sample via an electrode, a "double capacitor" is created. A potential can thus be applied to the central electrode and the thermal step to the measured sample. The signal is then acquired through the compensation sample.

The use of two identical samples brings the advantage of compensating for the polarization and transient conduction currents which can appear under strong fields; the measured current is thus exclusively due to the internal electric field of the measured sample.

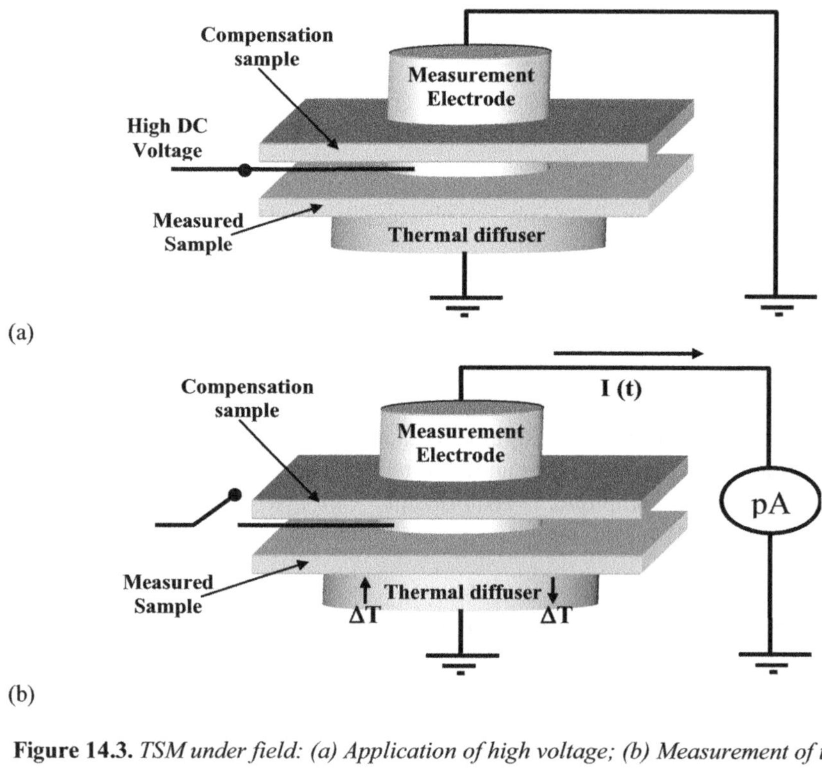

Figure 14.3. *TSM under field: (a) Application of high voltage; (b) Measurement of the signal*

The "under field" measurement is thus realized in two stages (see Figure 14.3):

– during the electrical conditioning, the potential is applied to the central electrode and the current amplifier is short circuited (the samples constitute two electrical capacitances placed in parallel with respect to the voltage source (see Figure 14.3a).

– in order to avoid the transport of influence charges to the electrodes through the high voltage source, this latter must be disconnected during the measurement (see Figure 14.3b). The signal is then measured by thermal excitation of the studied sample (in contact with the thermal diffuser), while the current amplifier is connected to the compensation sample (two samples in series with the current amplifier, in short circuit conditions).

The thermal step current in the external circuit is given by:

$$I(t) = -\alpha C_2 \int_{x_0}^{x_0+d} E(x) \frac{\partial \Delta T(x,t)}{\partial t} dx \qquad [14.11]$$

where C_2 is the capacitance "seen" by the current amplifier.

During the measurement, both samples are in series. If they are identical, $C_2=C/2$. In the absence of applied voltage, the TSM current must be half of that obtained in short circuit conditions with the thermally excited sample (where $C_2=C$).

For weak applied fields, the field can be considered constant in the material (with no injected charges): $E(x)=E_e=V/d$, where V is the applied voltage. In this case, the signal becomes directly proportional to V:

$$I(t) = -\alpha C_2 \frac{V}{d} \int_{x_0}^{x_0+d} \frac{\partial \Delta T(x,t)}{\partial t} dx \qquad [14.12]$$

This current can be used as a calibration current in the mathematical processing of the TSM signal.

If the measurement sample already contains space charges giving a residual field $E_r(x)$, and an external electric field is applied to the material ($E_e=V/d$ which does not modify the distribution of space charges), the total field in the sample becomes $E(x)=E_r(x)+E_e$.

Equation [14.11] can then be written in the form:

$$I(t) = -\alpha C_2 \int_{x_0}^{x_0+d} [E_r(x) + E_e] \frac{\partial \Delta T(x,t)}{\partial t} dx$$

$$= I_0(t) - \alpha C_2 \frac{V}{d} \int_{x_0}^{x_0+d} \frac{\partial \Delta T(x,t)}{\partial t} dx = I_0(t) + I_f(t) \quad [14.13]$$

where $I_0(t)$ is the current due to the residual space charges and $I_f(t)$ the current due to the external field.

14.2.3. *Calibration: use of measurements under low applied field for the determination of material parameters*

Let us consider an insulating sample which contains a certain quantity of space charges inducing a residual electric field $E_r(x)$. If the sample is subjected to an external field $E_e = V/d$ which does not cause any charge injection and does not modify the charge distribution in the sample, the measured TSM current, denoted as I_+, has an expression identical to equation [14.14]:

$$I_+(t) = -\alpha \frac{C}{2} \int_{x_0}^{x_0+d} [E_r(x) + E_e] \frac{\partial \Delta T(x,t)}{\partial t} dx \quad [14.14]$$

The application to the sample of an external electric field of similar value E_e but of opposite polarity gives a current I as follows:

$$I_-(t) = -\alpha \frac{C}{2} \int_{x_0}^{x_0+d} [E_r(x) - E_e] \frac{\partial \Delta T(x,t)}{\partial t} dx \quad [14.15]$$

The difference between the currents obtained by both measurements is then:

$$I_D(t) = I_+(t) - I_-(t) = -\alpha C \int_{x_0}^{x_0+d} E_e \frac{\partial \Delta T(x,t)}{\partial t} dx = -\alpha C E_e \int_{x_0}^{x_0+d} \frac{\partial \Delta T(x,t)}{\partial t} dx =$$

$$-2\alpha \frac{C}{2} E_e \int_{x_0}^{x_0+d} \frac{\partial \Delta T(x,t)}{\partial t} dx = -2\alpha \frac{C}{2} \frac{V}{d} \int_{x_0}^{x_0+d} \frac{\partial \Delta T(x,t)}{\partial t} dx = 2 I_{calibration}(t) \quad [14.16]$$

where $I_{calibration}$ (t) (calibration current), which is a function of the applied voltage and the insulating material parameters, is defined by:

$$I_{calibration}(t) = -\alpha \frac{C\,V}{2\,d} \int_{x_0}^{x_0+d} \frac{\partial \Delta T(x,t)}{\partial t} dx = -\alpha \frac{C\,V}{2\,d} K(t) \qquad [14.17]$$

The theoretical expression of the thermal stimulus is (in the case of only one thermal source):

$$\frac{\partial \Delta T(x,t)}{\partial t} = \Delta T_0 \frac{x}{t} \sqrt{\frac{1}{D4\pi t}} \exp\left(-\frac{x^2}{D4t}\right) \qquad [14.18]$$

The integral $K(t)$ is therefore:

$$K(t) = \int_{x_0}^{x_0+d} \frac{\partial \Delta T(x,t)}{\partial t} dx = \int_{x_0}^{x_0+d} \Delta T_0 \frac{x}{t} \sqrt{\frac{1}{D4\pi t}} \exp\left(-\frac{x^2}{D4t}\right) dx =$$
$$\Delta T_0 \left(\frac{\exp\left(-\frac{x_0{}^2}{D4t}\right) - \exp\left(-\frac{(x_0+d)^2}{D4t}\right)}{\sqrt{\frac{1}{D} \pi t}} \right) \qquad [14.19]$$

Thus, the temporal knowledge of $K(t)$ allows the parameters $1/D$ and x_0 to be determined, necessary for the mathematical processing of the TSM signals.

14.3. Numerical resolution methods

The TSM was one of the first techniques to give values for electric fields and space charge densities in polyethylene plates and in insulated cables. During the last few years, several calculation techniques for the field have been developed. They have followed the evolution of calculation software available on the market. In the 1980s, laborious programming (in Basic, Fortran, etc.) was necessary, whereas today performance algorithms are included in the laboratory software, making the calculations easier and noticeably improving the presentations.

The fundamental problem, as in all other indirect methods using a thermal stimulus or based on pressure, takes the form of an integral equation.

In the case of insulating plates or films with a thicknesss from a few tens of microns to several millimeters, the TSM determines the electric field $E(x)$ according to Equation [14.10]. The source of temperature – due to the "sudden" arrival of a cold or hot liquid in a small radiator – is supposed to provide a temperature step at x=0 (origin). The resolution of the heat equation, by assuming the temperature source placed near the side of the sample in contact with the radiator, gives the spatio-temporal expression for the thermal stimulus in the sample, corresponding to equation [14.18]. This stimulus, which has the appearance of a wave, propagates fairly slowly in the insulator. Its most advanced position is called $A(t)$.

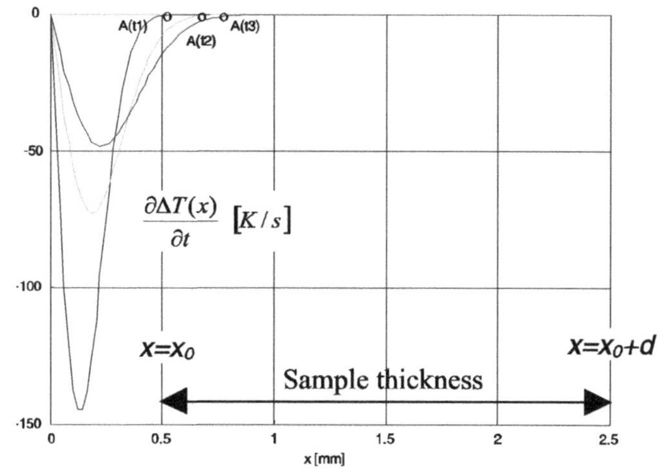

Figure 14.4. *Progression of the thermal step*

If we now consider that there is a second source of temperature at the end of the sample, corresponding to a second massive electrode situated at the x_0+d axis and therefore maintained at room temperature during the experiment, then the resolution of the heat equation gives, for the stimulus:

$$\frac{\partial \Delta T(x,t)}{\partial t} = \left(\frac{2\Delta T_0 \pi D}{L^2}\right) \sum_{n=1}^{\infty} n \exp\left(-t/\tau_n\right) \sin\left(n\pi x/L\right) \qquad [14.20]$$

with $L=x_0+d$ and $\tau_n = L^2/(D\, n^2\, \pi^2)$

This form of the stimulus leads us to express the electric field in the form of a Fourier series:

$$E(x) = \sum_{n=1}^{\infty} A_n \sin\left(n\pi x/L\right) \qquad [14.21]$$

Unfortunately, this natural form could only give the electric field in part of the sample, since $E(x)=0$ in $x=L$, which is not realistic. We have therefore preferred to use a more general form, which has turned out efficient, in the form of a Fourier pseudo-series, non-null in $x=L$.

In this processing by even and odd Fourier pseudo-series, we develop $E(x)$ in cosine and sine series, by taking the center of the sample as the origin for the total thickness:

$$E(x)=\sum_{n=0}^{\infty} A_n \cos(n\pi(x-ll)/d)+\sum_{n=0}^{\infty} B_n \sin(n\pi(x-ll)/d) \quad [14.22]$$

where $ll = x_o + d/2$.

By inserting equation [14.22] into equation [14.10], then by integrating, it is possible to obtain by fitting or by Laplace transform 7 to 14 terms of the Fourier pseudo-series.

An example of resolution is shown in Figure 14.5. Good results were also obtained on cables, with an important noise percentage (5–10%) [NOT 07].

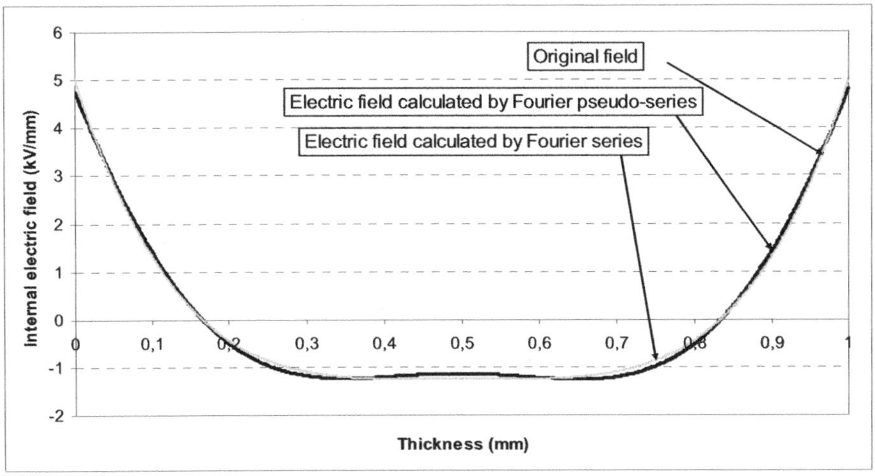

Figure 14.5. *Example of internal electric field distributions obtained for the mathematical processing by Fourier pseudo-series*

14.4. Experimental set-up

This section concerns the manner in which the TSM current measurement is made. The cases of both plate-type materials and cable-type components are addressed.

14.4.1. *Plate-type samples*

As can be deduced from equation [14.8], the measured amplitude of the TSM current principally depends on five parameters: the thermal diffusivity of material D, the relative variation of the electric capacitance with temperature α, the electrical capacitance of sample C, the internal electric field $E(x)$ and the amplitude of the thermal step ΔT_0. It is then evident that the current amplitude will get bigger as the sample capacitance and the step amplitude get bigger. The effect of the combination of these two parameters is that the thickness characterizable by the TSM is not really limited, but depends on the electrical capacitance of the sample and on the ability of the thermal source to transfer effectively the thermal step to the sample. In other words, the thicker the sample, the more powerful the thermal source must be. However, it is necessary to avoid temperature rises which could generate undesirable thermally stimulated discharges. In general, the use of negative thermal steps is preferable, because a cooling of the sample does not affect its internal electrical state (it is also necessary to check for each studied material that no phase change appears in the range of temperature used).

The measurement bench of the TSM for the characterization of plate-type samples consists of two cryo-thermostats (reservoirs filled with a mixture of water and ethanol, temperature regulated) permitting the mixture of refrigerant liquids to circulate in a radiator (for thermal step generation, then back to temperature after the measurement of the signal).

The circulation of the calo-carrier liquid in the radiator is generated by a set of electrovalves controlled by a computer. The radiator is situated in a measurement cell ensuring the role of a Faraday cage (see Figure 14.6). The measurement of current, during the application of the thermal step (generally negative, as we have previously described it, in the case of insulating samples: with the circulation of a liquid at -5°C for a few seconds), is realized with the aid of a Keithley 428 current amplifier. The thermal step amplitude must not modify the electrical characteristics of the material (amplitude of the smallest possible step, generally of the order of 20 to 30°C, in the service domain of the material).

Figure 14.7 below presents the setup used in the case of metal-oxide-semiconductor (MOS) structures for microelectronics. The MOS samples are placed on a radiator producing the thermal step. A needle (connected to the current amplifier input) is placed on the gate of the MOS structure, the rear side of the sample (below the Si substrate) being connected to the ground. The current amplifier used offers the possibility of applying a voltage to the sample during the measurement of current (bias voltage V_{BIAS}). This setup thus allows the charge evolution in the oxide to be followed during different polarization regimes of the structure. To avoid humidity issues, the applied thermal step is positive in the case of MOS structures (making sure not to modify the charge carrier density in the silicon during the measurement).

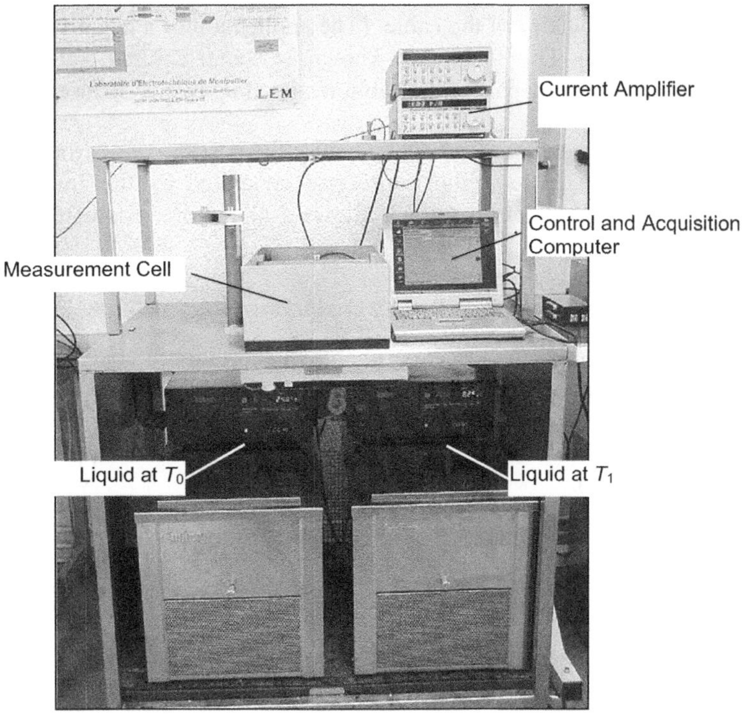

Figure 14.6. *General description of the TSM measurement bench*

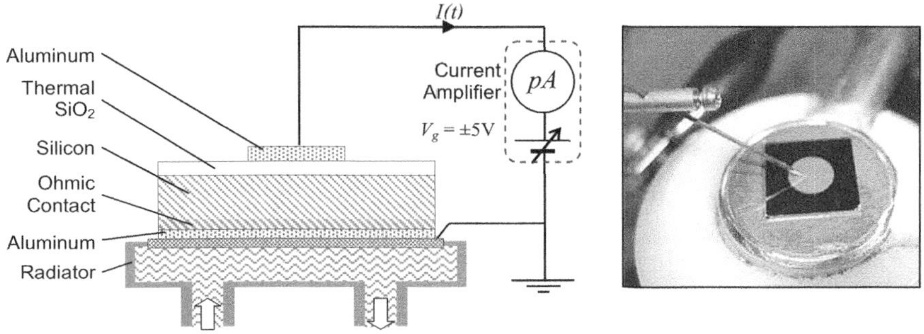

Figure 14.7. *Configuration of a thermal step measurement on a metal-oxide-semiconductor structure*

14.4.2. *Power cables*

The TSM can also be applied to power cables in short circuit conditions and under an applied DC field. In each case, the temperature step can be generated owing to the circulation of a calo-carrier or frigo-carrier liquid in a thermal diffuser adjusted for the geometry of the cable. (The application of a negative thermal step is generally preferred). This technique, known as "outer cooling", allows a localized analysis of the insulating part of the cable wrapped up by the diffuser.

The thermal step can also be generated by the passage of a strong current in the central core of the cable (temperature rise generated by the Joule effect). This technique, known as "inner heating", allows a global analysis of the insulation on the cable length.

14.4.2.1. *Measurements in short circuit conditions*

The equipment dedicated to experiments in short circuit conditions on cables by the external cooling technique is made up of a "cooling" part and of an "acquisition" part. This equipment (see Figure 14.8) is analogous to the one used for flat samples, except for the thermal diffuser, which is a cylindrical cover wrapping the cable (see Figure 14.9.a and b).

Figure 14.8. *TSM experimental bench – External cooling technique*

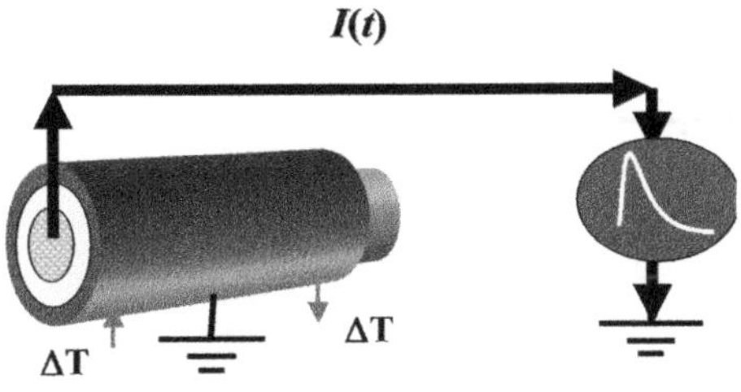

Figure 14.9a. *TSM – External cooling*

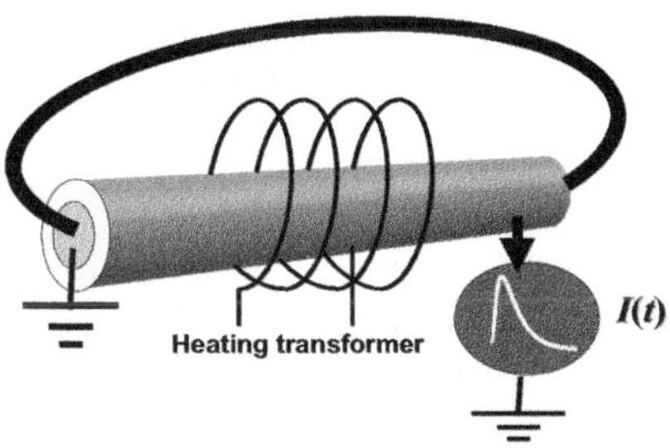

Figure 14.9b. *TSM – Inner heating*

The conductor heating technique (central core), composed of a part known as "heating" and a part known as "acquisition", uses a coil of inductive heating (4) which wraps up the cable, constituting the primary part of a heating transformer which has as its secondary part, the core of the measured cable. The heating time is controlled with the aid of two controlled rectifiers, or thyristors, mounted on opposite sides (12) (see Figure 14.11b). The heating time is calculated beforehand in order to generate a thermal step sufficiently important to reveal the electric charges present in the entire insulating thickness of the cable. The command order for the controlled rectifiers is generated remotely by software integrated in the "acquisition" part. This command system permits the heating duration of the cable core to be managed. The acquisition part is similar to that of the external cooling part, but the command software controls the thyristors, not an electrovalve system. The "acquisition" part is made up of a computer, a GPIB card (8) and an acquisition board fitted with digital and analog input/output ports (9).

14.4.2.2. *Measurements under applied electric field*

As in the case of flat samples (see section 14.2.2), the experiment under applied electric field requires the use of two identical cables (the same electrical capacitance and dimensions) whose external electrodes are insulated from one another [NOT 01a]. This "under field" measurement technique has been installed at the Nexans-France site in Calais since the year 2000.

First, the cables are subjected to an electric field via a high voltage relay (2), the current amplifier (3) fitted with a GPIB interface in short circuit (see Figure 14.10.a). The second stage consists, in the case of external cooling (see Figure 14.10.b), of measuring a current due to the application of a thermal step generated by a cryo-thermostat (5) owing to a heat exchanger (6) – a temperature diffuser – wrapping up a part of the measured cable to a length of 10 to about 100 cm.

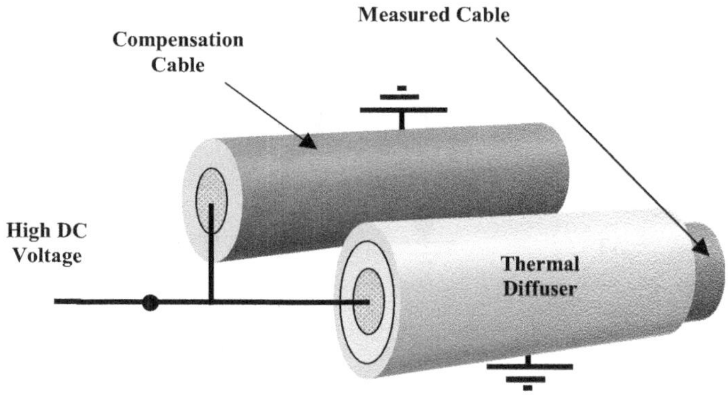

Figure 14.10a. *Cables placed under high voltage*

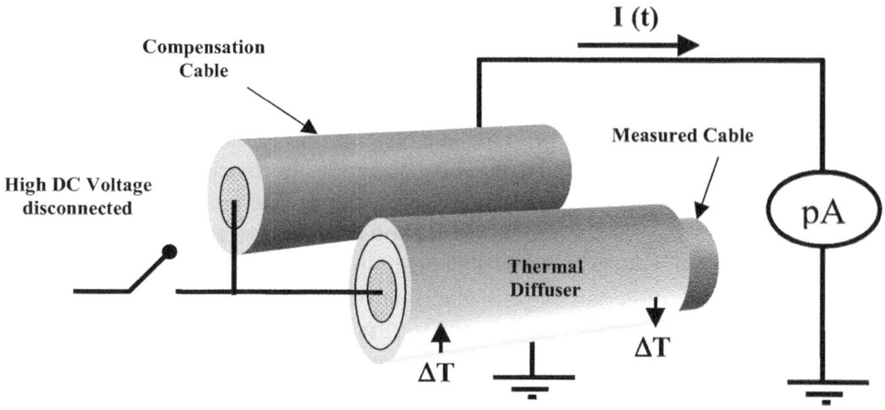

Figure 14.10b. *Measurement of TSM current*

The heat exchanger (connected to the ground) is placed on the test cable (A); a measurement is made with the aid of the current amplifier placed between the external semi-conductor of the other cable (B) and the ground.

In the case of inner heating, the thermal step is generated by the injection of a strong current to the central core of the cable to be measured; the measurement is made on the compensation cable. During the current measurement, the voltage source (1) must be disconnected from the central core of the cables because the system must imperatively remain insulated during the acquisition of the signal. This imposes the use of a remote high voltage switch (2), controllable from the command desk (10).

Both "external cooling" (see Figure 14.11.a) and "inner heating" (see Figure 14.11.b) techniques use the same equipment as for the short circuit measurements. Only the algorithms of command software are different, in order to additionally command the high voltage switch and the short circuit relay (11) of the current amplifier. The acquisition of the signals remains unchanged.

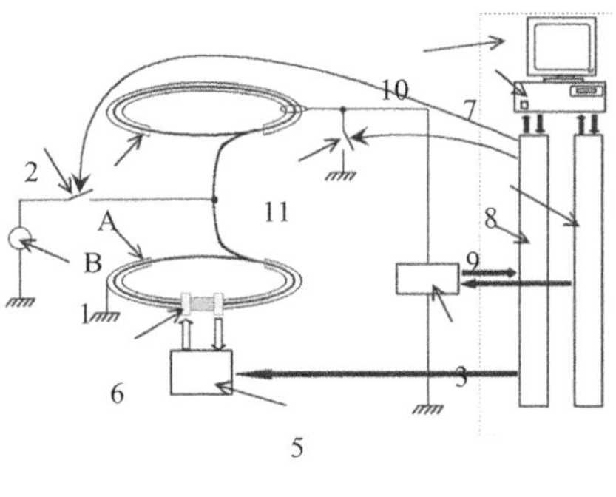

Figure 14.11a. *External cooling*

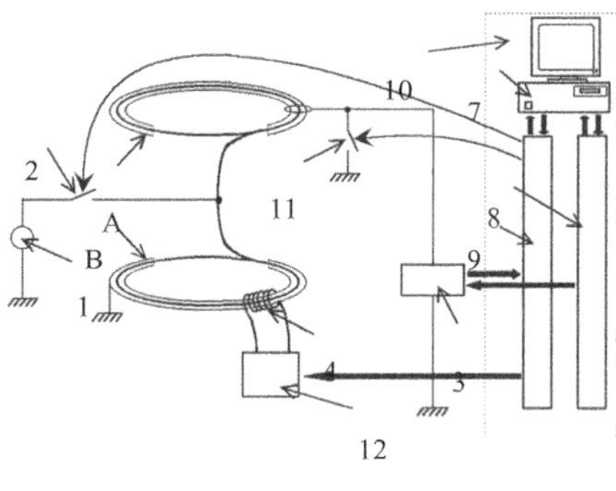

Figure 14.11b. *Conductor heating*

Figure 14.12. *Experimental equipment for measurements on cables under electric and thermal gradient installed at the Nexans-France site in Calais. A: measurement cable; B: compensation cable; 1: HV DC source; 2: H relay; 3: current amplifier; 4: heating; 5: thermal sources; 6: thermal diffuser; 7: computer; 10: command desk*

14.5. Applications

14.5.1. *Materials*

14.5.1.1. *Influence of molar weight and cooling rate on the presence of space charges in polyethylene [TOU 98]*

In order to study the appearance of space charges in polyethylene (PE) after manufacturing, several samples were made by press-moulding PE ribbons. Four samples in the form of 2mm thick plates with similar density (~960 kg/m^3) were made according to the same procedure, but with different molar weights (Mw). The characteristics of the samples are presented in Table 14.1.

Sample	Molar weight (Mw) (g/mol)	Density (kg/m^3)	Crystallinity (%)
1	100,000	958	77.7
2	71,000	965	81.2
3	59,000	962	77.7
4	48,000	959	77.6

Table 14.1. *Description of PE samples*

Figure 14.13 shows that all the space charge distributions are symmetrical and the space charge levels seem directly related to the molar weight. A strong molar weight implies a strong level of space charge. Further, the microscope observation of samples with the same density but different molar weights (see Figure 14.14) shows that a high molar weight induces small spherulites. We can infer from this that the space charge level is inversely proportional to the size of the spherulites. Indeed, this idea agrees with the theory of Ikezaki *et al.* [IKE 94], who showed that space charges are principally trapped in the core and at the periphery of spherulites, which are amorphous regions rich in defects and impurities.

During the manufacture of a PE sample, we have also made an asymmetric cooling. Thus, a side of the sample was rapidly cooled towards room temperature (cooling rate of 40°C/min, using a water cooling system), while the the other side was left to cool down naturally (at a rate of 1°C/min, imposed by thermal inertia of the moulding system). The experimental results give an asymmetric space charge distribution for asymmetric cooling (see Figure 14.15). The side which was rapidly cooled presents more space charges than that cooled more slowly. A microscope visualization of samples shows that the cooling rate also has an influence on the size of spherulites (see Figure 14.14).

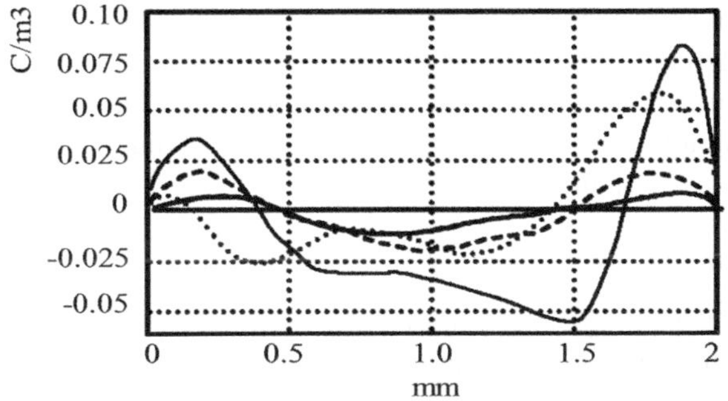

Figure 14.13. *Space charge distribution on PE samples after manufacturing (thin line: sample 1; dotted line: sample 2; dashed line: sample 3; bold line: sample 4)*

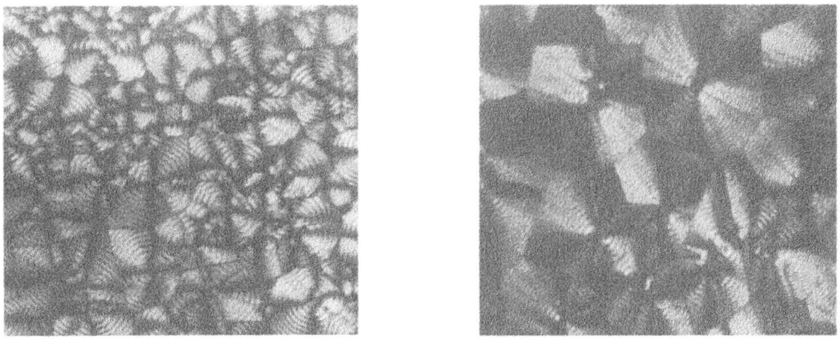

Figure 14.14. *Microscopic photograph of sample 1 (left) and sample 4 (right) (slow cooling)*

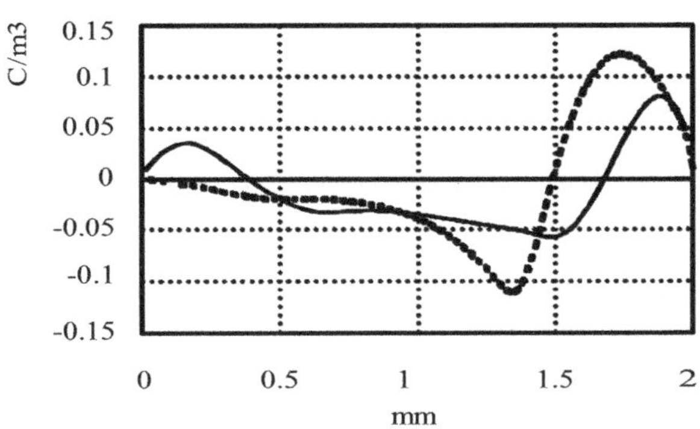

Figure 14.15. *Influence of the cooling rate on space charges in PE (continuous line: symmetrical cooling, dotted line: asymmetric cooling)*

The results presented show that press-moulded PE plates contain space charges just after manufacture. The development and the charge trapping are then produced during manufacturing whilst the material is being solidified (the thermal gradient present in the sample during cooling could cause the transit of negative charges towards the core of the sample).

14.5.1.2. *Revealing the heterogenity of composite materials (charged epoxy resin)*

Composites are heterogenous materials by definition. The aim of this study was to verify their behavior regarding the accumulation of space charges when they are subjected to a strong electrical and thermal constraint which was also heterogenous. In order to verify these heterogenities, the surface of a 1mm thick charged epoxy resin sample was divided into 40 pins of 25 mm^2, arranged in a circular shape of diameter 5cm (see Figure 14.16.a). After an electrical conditioning (10 kV/mm, 80°C, 22 hours) carried out with two electrodes of diameter 5cm placed on both sides of the sample (see Figure 14.16.b), each pin was analyzed by the TSM: the results obtained are presented in Figure 14.17, where the maximal values of TSM signals collected on the 40 pins spread out on the surface of the electrode are represented.

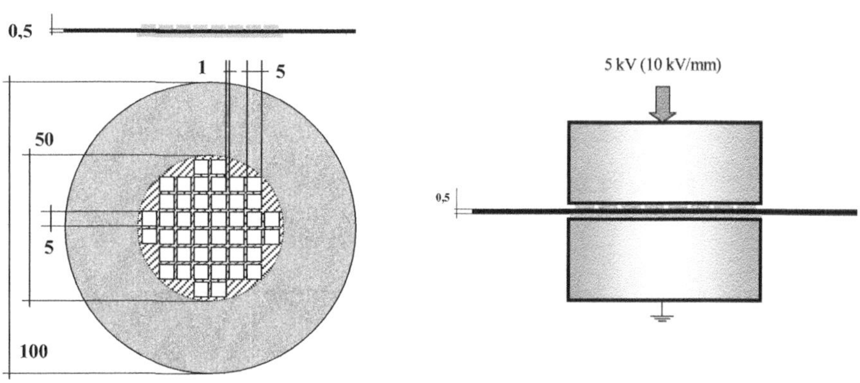

Figure 14.16. *(a) Electrode configuration; and (b) electrical conditioning of the sample*

The results of this study allow the observation of heterogenities as a function of the different analyzed zones (pins). Essentially, three zones presenting different volume densities of residual space charges can be seen: a central zone showing a rather weak average level, a peripheral zone showing a higher average level, and a zone of the order of 25% of the tested surface showing a clearly more important average level (amplitude of the TSM currents up to 6 times higher; remember that the amplitude of the TSM signals is a function of the space charge density).

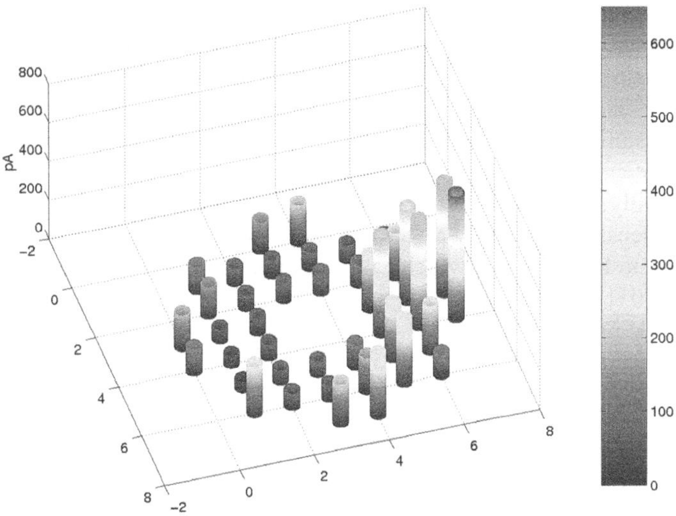

Figure 14.17. *TSM current amplitude as a function of the position of the pins – Revealing the heterogenity of an epoxy resin sample*

Analogous studies were also made on other samples and components, and showed that the heterogenities of a polymeric material strongly depend on the sensitivity of the measurement tool which is used [CAS 03].

14.5.1.3. *Evolution of space charges in materials for cables subjected to an alternative electrical constraint (50 Hz)*

If direct current is considered to be the future of high voltage power transport, the crushing majority of high and medium voltage cables with synthetic insulation are used today for AC transport (50 Hz). When we know that the required lifespan for this type of component reaches 40 years, we can easily understand the importance of the subject. If the accumulation of space charges and their role in the degradation process are clearly established in the case of a solid insulator subjected to a strong DC field, a few points on this subject are still to be cleared up when the material is subjected to an alternative constraint. Notably, the question: "space charge: cause or effect of the degradation under alternative field?" has yet to be solved.

However, whatever the envisaged degradation mechanisms, it is clear that space charge measurements give important indications concerning the state of an insulator subjected to an alternative constraint. Indeed, the gradual accumulation of charges in deep traps (greater than 1 eV [NOT 01b]) necessarily means a multiplication of these latter, which corresponds to an increase in the number of morphological defects and to chemical bond breaks. In this sense, we talk about "static" charges, since they remain trapped for a long time after the stress has been turned off, while the "dynamic" charges, weakly trapped, recombine with the inversion of the field. Revealing the static charges is quite difficult, especially if the material is not very "old", since their density is relatively weak because of the alternative nature of the field. Incidentally, most techniques cannot detect them with sufficient precision. Owing to its sensitivity, the TSM allows these accumulations to be detected and their evolution to be followed, which can be estimated as parallel to the degradation process [AGN 03].

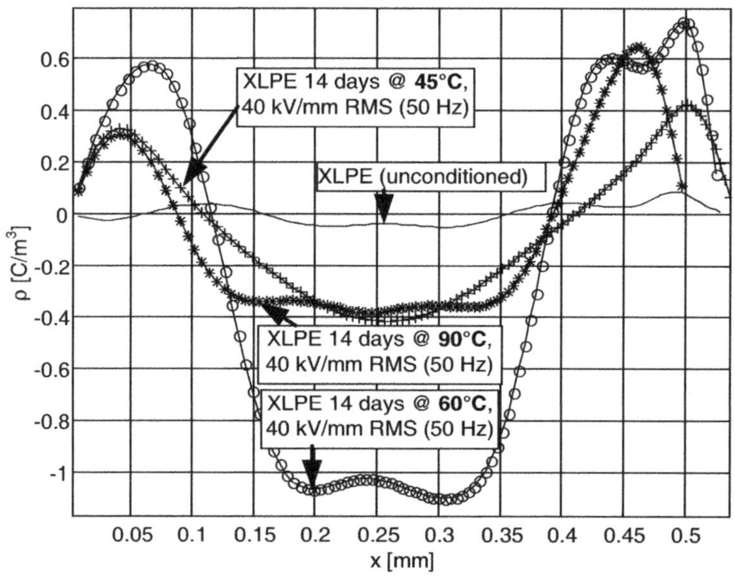

Figure 14.18. *Charge density distribution in cross-linked polyethylene (PR) subjected to a 40 kV/mm AC field (50 Hz) for 15 days at different temperatures [NOT 01b]*

Figure 14.18 shows results obtained in plane samples of cross-linked polyethylene used in high voltage cables. We see here the "static" charge distribution (deeply trapped), with a symmetric and bilateral shape (related to the alternative nature of the constraint). Figure 14.18 also shows that the accumulation of alternative charges is not an increasing function of the temperature, but can present maxima for particular values of the thermal constraint (the most important accumulation is at 60°C in the case presented). These values of the temperature depend on the structure and on the initial morphology of the material, as well as on their modification in time.

We can use the total absolute value of the deeply trapped charge $Q_{ABS} = |Q_+| + |Q_-|$ in order to follow the evolution of the insulator. This amplitude can thus be considered as the image of the concentration of deep apparent defects (filled traps) created during ageing. The evolution of the apparent defect density in high voltage cables subjected for several thousand hours to strong constraints is presented in Figure 14.19, where we observe a global upward trend in ageing over time. This increase is very strong for conditionings carried out at 145 kV and 325 kV, as well as for the conditioning at 225 kV up to 5,000 h. Because the measurements were made on sections periodically cut from loops conditioned during several thousand hours, the observed decrease at 225 kV after 5,000 h is related to the non-homogenity of one of the cable sections, as was shown in [CAS 03]. An important increase of the number of filled deep traps is also observable when the applied voltage increases, especially if we compare the cables conditioned at 145 kV with the others. The application of the stress for several thousand hours has therefore caused structural changes in the insulation, which manifest themselves by an increase in concentration of deep traps; this latter could be used as a marker for the evolution of the cable.

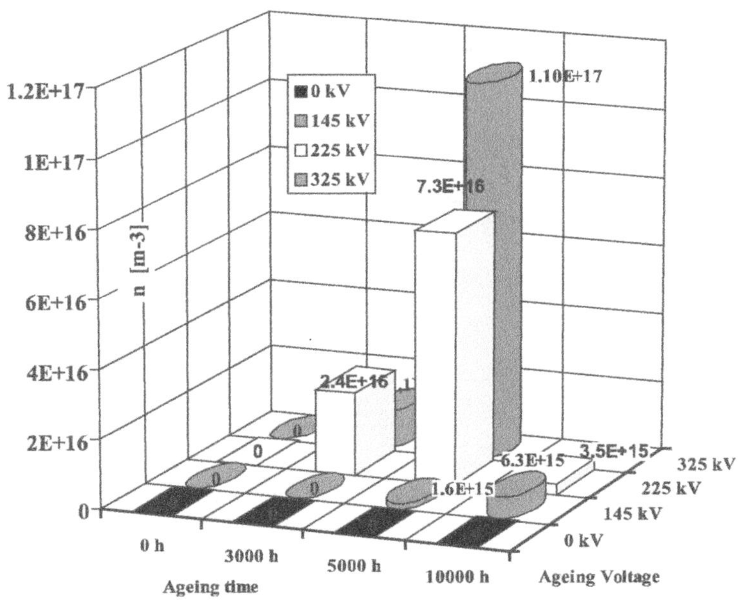

Figure 14.19. *Evolution of the apparent concentration of filled deep traps in high voltage cables (90 kV) subjected to different voltages, at 90°C [NOT 03]*

14.5.2. *Components*

14.5.2.1. *Monitoring of the internal electric field of a cable subjected to electrical and thermal stress*

Electric field monitoring in the insulation of a DC cable in service conditions can represent an essential stage for the assessment of the long-term behavior of the cable. Such monitoring was carried out on a 20 m long cable with an insulating thickness of 5.8 mm, by using the thermal step bench from the Nexans-France site in Calais.

Several test conditions were applied to the cable; the results presented in Figure 14.20 were obtained for an ageing performed under electrical and thermal gradients according to the following conditions: -85kV DC and 80°C on the central core of the cable. TSM measurements have been regularly made by using the equipment described in section 14.4.2.2 (see Figure 14.12) to observe the evolution of the internal electric field during the test period [CAS 05].

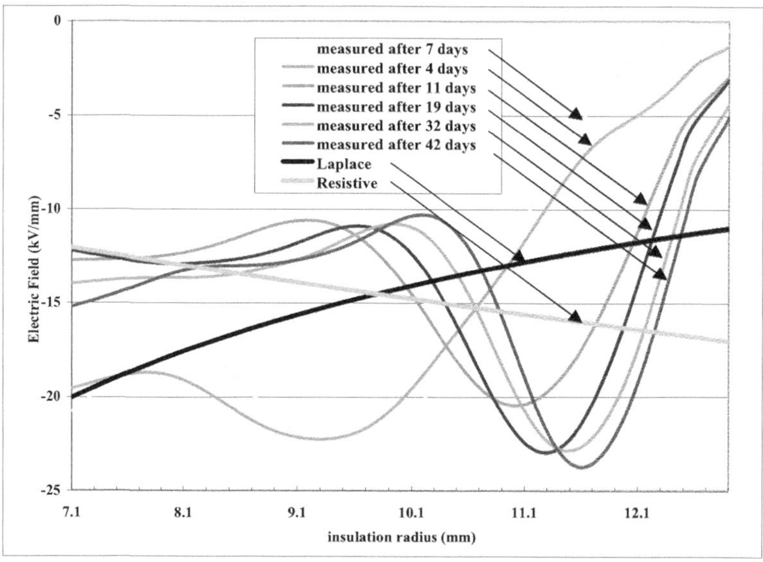

Figure 14.20. *Evolution of the internal electric field measured in the DC cable insulation (-85 kV DC, 80°C applied on the conducting core situated on the left of the figure)*

The electric field distributions presented in this figure are strongly distorted with respect both to a Laplace theoretical distribution (capacitive distribution assumed in the absence of space charge and thermal gradient) and to a "resistive" repartition (theoretical electric field in an insulation subjected to electrical stress and thermal gradient, in the absence of space charge) [COE 01]. In some areas of the cable insulation, the measured electric field is two times higher than the theoretical values because of the presence of space charges.

The trend of the field repartition to switch toward a resistive-like curve after 7 days of testing must be underlined, because such an inversion, theoretically demonstrated, had never been experimentally observed before. The internal field measured near the conductor thus tends to pass from the Laplace field to the minimum value predicted by the resistive model. Further, the field distortion is more important than predicted by the resistive model, more particularly in the vicinity of the semiconducting shield. This behavior underlines the complex dynamics of space charges, notably by taking into account the injection phenomena. Despite the field model which seems established beyond 11 days of testing, an evolution is still apparent after 42 days under constraint, thus suggesting that the equilibrium state is still not reached.

14.5.2.2. *Monitoring of the ageing of micaceous composite insulation from a power alternator winding*

This work aimed at monitoring the evolution of space charges in a 2 mm thick insulator from a stator winding of a high power AC machine. For this purpose, the sample was subjected to accelerated AC ageing under 18 kV RMS voltage (50 Hz) applied on the conductor at room temperature, i.e. twice the service voltage (9 kV/mm average electric field). Several measurements were made during ageing, until breakdown occurred [CAS 02].

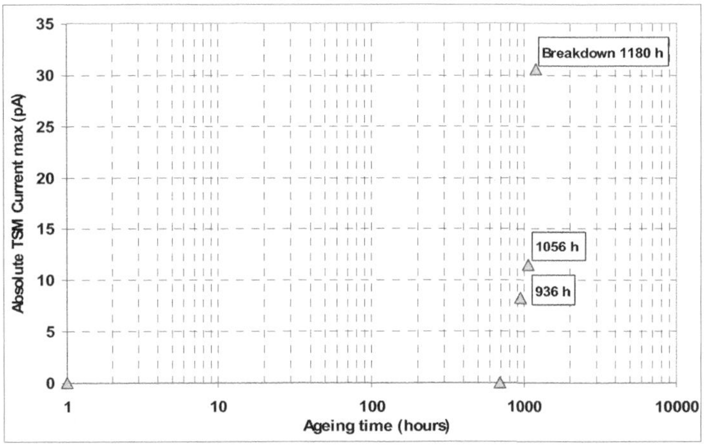

Figure 14.21. *Evolution of the TSM current amplitude in the insulation of a high power AC machine submitted to AC stress*

Figure 14.21 shows an increase of the TSM current amplitude during ageing, the breakdown occurring after 1,180 hours of ageing. In these ageing conditions, 936 hours correspond to 70% of the lifespan of the sample.

The space charge density profiles presented in Figure 14.22 show an increase of the levels with the ageing time. We also notice negative charge injection from the copper (on the right in Figure 14.22), and a migration of negative charges towards the outer electrode after breakdown (on the left in Figure 14.22).

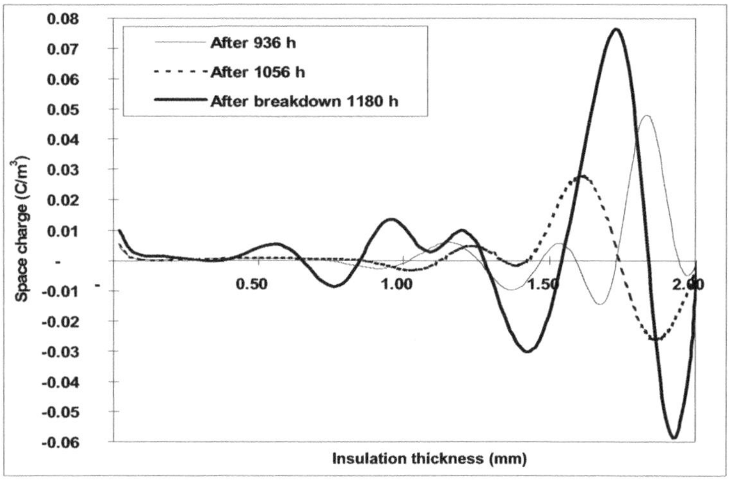

Figure 14.22. *Evolution of space charge during ageing (the copper conductor bar is on the right)*

14.5.2.3. *Characterization of Metal-Oxide-Semiconductor (MOS) structures for micro and nanoelectronics*

Providing that the sample capacitance and the thermal step amplitude are sufficient, there is in principle no restriction concerning the thickness the TSM is able to characterize. Although initially developed for the study of thick insulators, the technique could thus be adjusted for a large spectrum of thicknesses and geometries, and notably for thin layers, such those of semi-conducting components.

We recall that a MOS capacitance (the basis of most electronic components) is composed of an *n* or *p* doped silicon substrate, from which a silicon dioxide (SiO_2) layer is grown, most often by thermal oxidation (see Figure 14.23). When a variable voltage (noted further V_g or V_{gs}) is applied to the gate (electrode in contact with the SiO_2), the overall capacitance of the structure varies proportionally to the extension of a charge zone at the oxide/semi-conductor interface, which is comparable to a variable capacitance in series with the oxide. For a value of the voltage V_g called the threshold voltage, the thickness of the space charge zone in the semi-conductor becomes so high that the total capacitance of the structure collapses, allowing the switching of the component which integrates the MOS structure from the off-state to the on-state, or vice-versa (see Figure 14.23).

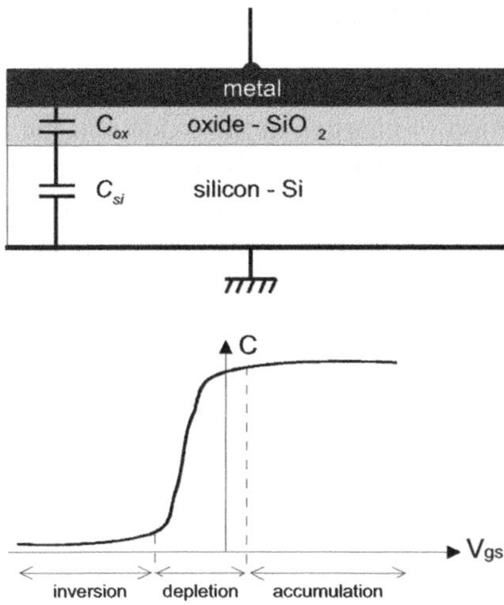

Figure 14.23. *Composition and functioning of an n-type substrate MOS structure*

To illustrate the functioning of a MOS capacitance, let us first envisage the case of an "ideal" component. For such a structure, we suppose that:

– the doping of the semi-conductor is such that the work functions of the metal and the semi-conductor are identical;

– there are no traps at the interface between the oxide and the semi-conductor;

– there are no charges in the oxide, the insulator is perfect.

As previously evoked, the capacitance of the structure is a grouping in series of the semi-conductor's capacitance C_{Si} and the insulator's capacitance C_{ox}: $1/C_{MOS} = 1/C_{ox} + 1/C_{Si}$. The oxide's capacitance is independent of the voltage V_{gs} applied to the structure: $C_{ox}=\varepsilon_0\varepsilon_{ox}S/d_{ox}$, with ε_0 the vacuum permittivity, ε_{ox} the relative permittivity of the oxide, S the surface of the gate and d_{ox} the thickness of the oxide. On the other hand, the capacitance of the Si is determined by the apparition of a space charge zone in the semi-conductor, which varies as a function of V_{gs}.

Three regimes are possible:

– accumulation of majority carriers;

– depletion of majority carriers;

– inversion (the density of minority carriers becomes greater, in the neighborhood of the SiO_2-Si interface, than the density of majority carriers).

However, in real structures, the oxides cannot be exempt from charges. It is well known that the accumulation of charges in the oxide and the increase of traps at the oxide/semi-conductor interface strongly affect the structure, first by moving its switching threshold (so leading to a bad functioning of the device), then by reducing the lifespan of the oxide. The charge measurement (especially that of interface charges) is a major issue for the improvement of the reliability of semi-conducting devices. However, because of the shallowness of the interface traps, it may be difficult to quantify them with traditional techniques.

In the following figures, thermal step currents are presented, measured for several values of the gate voltage V_g for n- and p-doped substrates 10^{15} cm^{-3}. Capacitance-voltage (C-V) measurements performed on the same samples are also shown.

The thermal step currents permit the three functioning regimes of the structures to be identified. Thus, the accumulation regime (which corresponds, in the case of ideal MOS capacitances, to a negative gate voltage for a p-substrate and a positive one for an n-substrate) is defined by a slight reduction of the thermal step current amplitude when the gate voltage increases. We also note that the currents are opposite for n-substrate and p-substrate structures: this is due to the opposite signs of charges near the Si surface in both types of structures. Thus, in accumulation, the charge at the Si-SiO_2 interface is negative for an n-type substrate and positive for a p-type substrate.

The same reasoning can be made for the inversion regime. In this case, the charge signs in the substrates, i.e. the TSM current signs, are reversed: the charge at the interface is positive for the n-substrate and negative for the p-substrate.

The depletion regime is characterized by a very fast variation of the thermal step current amplitude, while the gate voltage V_g does not vary much; this variation is related to the very rapid variation of the space charge zone width in the Si. The space charge zone in the Si induces a collapse of the total capacitance of the structure observed by C(V). This zone reaches a maximum for the threshold voltage V_{TH}. From this threshold, the minority carriers appear at the Si-SiO_2, interface causing a sudden increase of the capacitance until its accumulation value C_{ox}.

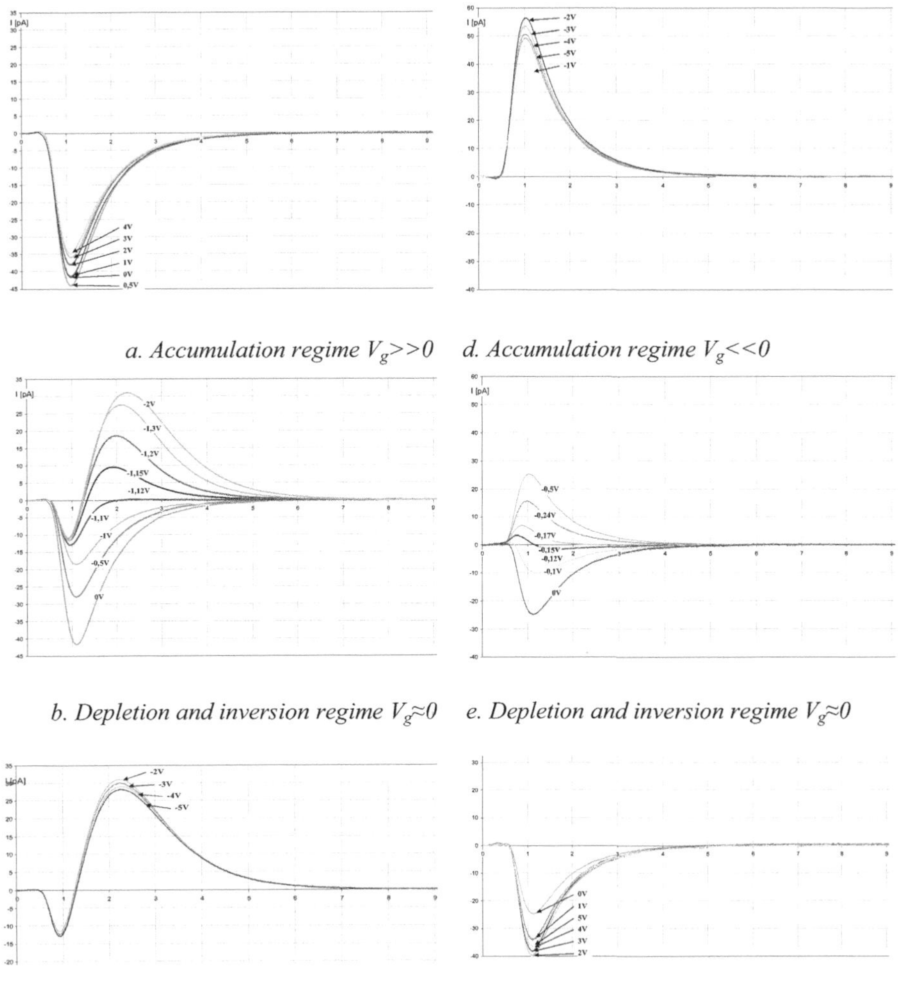

a. Accumulation regime $V_g>>0$ *d. Accumulation regime $V_g<<0$*

b. Depletion and inversion regime $V_g\approx 0$ *e. Depletion and inversion regime $V_g\approx 0$*

c. Strong inversion regime $V_g<<0$ *f. Strong inversion regime $V_g>>0$*

Figure 14.24. *Thermal step currents obtained with the TSM on both types of MOS structures with gate oxide thickness of 120 nm (a–c with n-type substrate; d–f with p-type substrate)*

The measurement of the structure capacitance made by C(V) at high frequency (1 kHz) does not permit the contribution of minority carriers to be revealed. On the other hand, this rise is observed at low frequency (20 Hz) C(V) and especially with the TSM. Indeed, the TSM currents reach, in inversion, quasi-equivalent amplitudes to those of accumulation currents (but of an opposite sign).

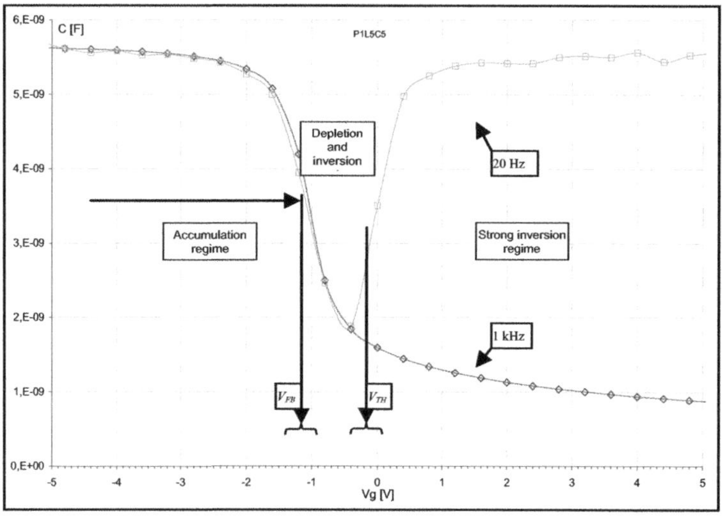

Figure 14.25. *Capacitance-voltage measurements on the p-type substrate MOS structure*

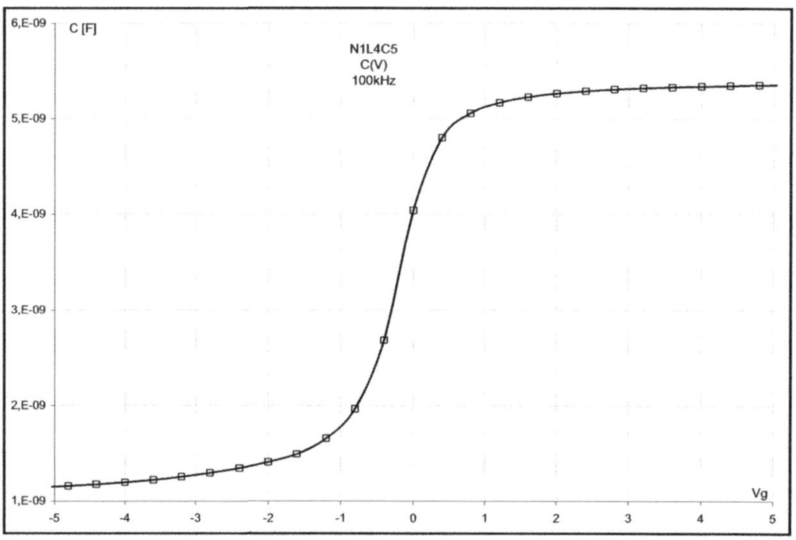

Figure 14.26. *Capacitance-voltage measurements on the n-type substrate MOS structure*

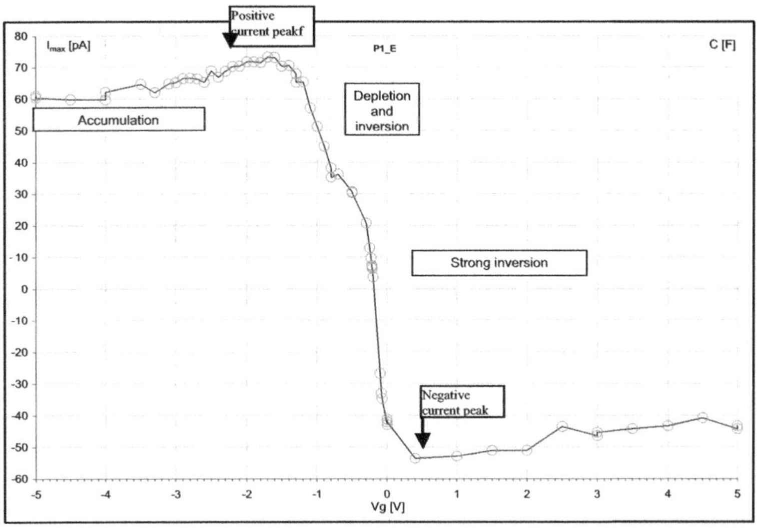

Figure 14.27. *Amplitude of thermal step currents (I_{max}) as a function of the gate voltage (V_g) on a p-type substrate MOS capacitance*

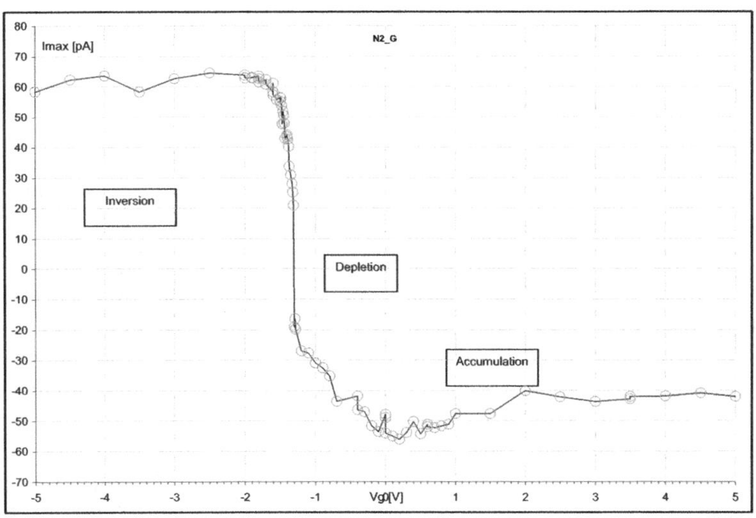

Figure 14.28. *Amplitude of thermal step currents (I_{max}) as a function of the gate voltage (V_g) on an n-type substrate MOS capacitance*

The representation of the thermal step currents amplitude (I_{max}) as a function of the gate voltage (see Figures 14.27 and 14.28) permits the current maximum (corresponding to the flatband) and its minimum (threshold voltage) to be spotted accurately, and for them to be compared with the estimations given by C(V) (see Tables 14.2 and 14.3).

We observe a good correlation between both characterization types; however, the values measured with both techniques are not identical.

	Theoretical values (without charge in the oxide)	Values determined by C(V)	Values determined with the TSM
VFB	-0.59 V	-1.2 V [CFB=4 nF]	≈ -1.7 V
VTH	+1.12 V	Between -0.5 V and +0.5 V	≈ 0.5 V

Table 14.2. *Values obtained by the C(V) and the TSM for p-type substrate*

	Theoretical values (without charges in the oxide)	Values determined by C(V)	Values determined with the TSM
V_{FB}	+0.03 V	0 V	+ 0.2 V
V_{TH}	-1.12 V	Between -0.8 V and -1.7 V	-1.6 V

Table 14.3. *Values obtained by the C(V) and the TSM for n-type substrate*

In fact, the threshold values obtained with the TSM and the C(V) could only be identical if the measurements of C(V) were made at the same frequency as that of the TSM, i.e. at a very low frequency (0.1 Hz). This means that the TSM permits to charges to be found that the C(V) cannot reveal, even at a frequency of 20 Hz. These are notably the weakly trapped interface charges, which do not appear with the C(V). The differences between both methods are stronger when the interface state density increases. We must note that the contribution of the thermal step technique is more important when the interface state density is high, and notably in the case of irradiated components, as it was shown in [FRU 06]. In this case, the use of the TSM permits all of the interface charges to be easily revealed, while the use of the C(V) requires fairly tricky calculations and may not allow all of these states to be measured.

It is therefore possible to calculate the total charge in the structure and, by measurements at different voltages, to separate the contributions of different carriers. For example, at the removal of the thermal step current (I_{max}= 0), we can calculate the charge quantity in the structures $Q_0 = -C_{ox}V_g$. For both previous structures, we obtain:

	N-type substrate	P-type substrate
Vg [IMax(Vg) ≈ 0]	-0.8 V	-0.3 V
Q0	+4.4.10-9 C	+1.65.10-9 C

Table 14.4. *Values of Q_0 calculated for V_g [$I_{Max}(V_g) \approx 0$]*

For IMax≈0, the inversion is already made in the n-type substrate MOS structure, because the charge is positive. On the other hand, for the p-type substrate MOS structure, the charge quantity at the interface has decreased compared to the flat band regime, but the inversion has not yet appeared.

Studies undertaken on gate insulator MOS capacities with even weaker thicknesses, up to 1 nm, have shown that the TSM can be applied on this type of component, allowing the different charges existing in the structure to be evaluated, and notably the interface charges [FRU 06]. This allows the use of this technique as a characterization method in the micro and nanoelectronic domain to be envisaged, in the near future.

14.6. Conclusion

The thermal step method for measuring space charges was invented in 1986. Since then, the technique has not stopped evolving and the different studies and researches led for the last few years have made it a renowned non-destructive characterization technique, both by the academic community and by industry.

The latest advances permit us to catch a glimpse of numerous studies allowing a better understanding of the phenomena of appearance and development of space charges in solid insulating materials subjected to various stress. Today, characterizations of insulators under continuous electric fields and thermal gradients are realizable, both in a laboratory and in industry, in close to operational conditions (for measurements applicable to materials and components).

The progress achieved and the industrial collaborations made have thus resulted in the technological transfer of this method to the cable industry. The use of this type of testing technique in real conditions should allow the processes leading to insulation ageing being fully understood.

14.7. Bibliography

[ABO 91] ABOU DAKKA M., Mesures de charges d'espace dans divers polymères par la méthode de l'onde thermique, Doctoral Thesis, University of Montpellier 2, 1991.

[AGN 03] AGNEL S., PLATBROOD G., TOUREILLE A., "Study of AC electrical and thermal ageing of XLPE polyethylene by space charge measurements", *Proceedings of the Jicable 2003 IEEE International Conference on Insulated Power Cables*, Versailles, p. 479–484, 2003.

[CAS 02] CASTELLON J., MALRIEU S., TOUREILLE A., "Measurement of space charge evolution in insulation of electrical machine winding: correlation with AC ageing", *Proceedings of the INSUCON 02 Conference*, p. 165–170, 2002.

[CAS 03] CASTELLON J., MALRIEU S., NOTINGHER Jr P., TOUREILLE A., BECKER J., DEJEAN P.M., JANAH H., MATALLANA J., VERITE J.C., "On site measurements on HV cable loops as part of the ARTEMIS project", *Proceedings of the Jicable 2003 IEEE International Conference on Insulated Power Cables*, p. 714–719, 2003.

[CAS 05] CASTELLON J., NOTINGHER P., AGNEL S., TOUREILLE A., MATALLANA J., JANAH H., MIREBEAU P., SY D., "Industrial Installation for Voltage-On Space Charge Measurements in HVDC Cables", *Industry Applications Conference, 40th IAS Annual Meeting, Conference Record of the 2005, Hong Kong*, p. 1112–1118, vol. 2, 2005.

[CHE 93] CHERIFI A., Contribution à l'étude des charges d'espace dans les polymères isolants utilisés en haute tension. Améliorations de la méthode de l'onde thermique, Doctoral Thesis, University of Montpellier 2, 1993.

[COE 01] COELHO R., ALADENIZE B., GUILLAUMOND F., "Charge build-up in lossy dielectrics with induced inhomogeneities", *IEEE Transactions on Dielectrics and Electrical Insulation*, vol. 4, no. 5, p. 477–486, 1997.

[FRU 06] FRUCHIER O., Etude du comportement de la charge d'espace dans les structures MOS. Vers une analyse du champ électrique interne par la méthode de l'onde thermique, Doctoral Thesis, University of Montpellier 2, 2006.

[IKE 94] IKEZAKI K., YAGISHITA A & YAMANOUCHI H., "Charge trapping in spherulitic polypropylene", *Proceedings of the 8th International Symposium on Electrets*, Paris, France, p. 428–433, 1994.

[NOT 01a] NOTINGHER P. Jr, AGNEL S., TOUREILLE A., "The Thermal Step Method for Space Charge Measurement under Applied DC Field", *IEEE TDEI*, vol. 8, no. 6, p. 985–994, 2001.

[NOT 01b] NOTINGHER P. Jr, TOUREILLE A., SANTANA J., ALBERTINI M., MARTINOTO L., "Study of Space Charge Accumulation in Polyolefins Submitted to AC Stress (50 Hz)", *IEEE TDEI*, vol. 8, no. 6, p. 972–984, 2001.

[NOT 03] NOTINGHER P. Jr, TOUREILLE P., AGNEL S., DIDON N., CASTELLON J., MALRIEU S., "Survey of space charge evolution in high voltage cables sublmitted to AC voltage as part of artemis project", *Conference Proceedings of the Jicable 2003 IEEE International Conference on Insulated Power Cables*, Versailles, France, p. 468–472, 2003.

[NOT 07] NOTINGHER P. Jr, TOUREILLE A., AGNEL S., CASTELLON J., "Determination of Electric Field and Space Charge in the Insulation of Power Cables with the Thermal Step Method using a New Mathematical Processing", *IEEE Transactions on Industry Applications,* vol. 45, no. 1, p. 67-74, 2009.

[ODI 00] ODIOT F., Mesures de charges d'espace dans le SiO_2 par la méthode de l'onde thermique: application aux technologies MOS, Doctoral Thesis, University of Montpellier 2, 2000.

[SAB 91] SABIR A., Sur une nouvelle méthode de mesure des charges d'espace dans les câbles haute tension, Doctoral Thesis, University of Montpellier 2, 1991.

[SAN 94] SANTANA J., Mesures de charges d'espace dans les câbles de transport de l'énergie électrique, Doctoral Thesis, University of Montpellier 2, 1994.

[TOU 87] TOUREILLE A., "Sur une méthode de détermination de la densité de charges d'espace dans le polyéthylène", *Proceedings of the Jicable 87 IEEE International Conference on Insulated Power Cables*, p. 89–103, 1987.

[TOU 90] TOUREILLE A., SABIR A., REBOUL J.P., BERDALA J., MERLE P., "Determination of space charge densities in a high voltage thick insulated cable using the thermal step technique", *Revue de Physique Appliquée*, vol. 25, no. 4, p. 405–408, 1990.

[TOU 98] A. TOUREILLE, P. NOTINGHER Jr, N. VELLA, J. CASTELLON, S. MALRIEU, S. AGNEL, "The Thermal Step Technique: An advanced method for studying the properties and testing the quality of polymers", *Polymer International*, vol. 46, no. 2, p. 81–92 , 1998.

Chapter 15

Physico-Chemical Characterization Techniques of Dielectrics

15.1. Introduction

The physico-chemical characterization techniques of gaseous, liquid and solid dielectrics have been used simultaneously to better understand their nature, improve their characteristics, and follow their behavior as a function of the stresses they are subjected to in the role of electrical insulating materials.

It is evident that these techniques have evolved in time, their resolution power has increased, the acquisition of data modernized and their range widened. They allow, in particular, the analysis of materials for their selection and development, to interpret their evolution and degradation, including analysis of samples extracted from operating equipment (when possible), or from damaged material to try to explain the causes.

We can therefore, for example, use the same physico-chemical characterization techniques for operating transformers or power capacitors as for a buried cable which has been damaged. For this reason, only one description will be made of all the domains where they are used, and only the information they provide will be commented on.

Chapter written by Christine MAYOUX and Christian MAYOUX.

15.2. Domains of application

15.2.1. *Transformers and power capacitors*

Electrical discharges of higher or weaker intensity generating variable current densities, as well as the local rising of temperature, can lead to the degradation of the oil and/or the paper in this type of equipment. Degradation of cellulose can be caused by hydrolysis, oxidation and thermal effect. The latter leads to a formation of water, monoxide and carbon dioxide. The products which appear are essentially in gaseous form. A small quantity of products (a few ppm) is enough for analysis by gaseous chromatography (GC) i.e. chromatography in the gaseous phase. This routine analysis has been practiced for over 40 years. The furanic by-products are analyzed by high performance liquid chromatography (HPLC) [DUV 77], [LAM 81], [DUV 82]. A good description of all the analyzes realized on transformers has been made by Vergne [VER 92] and Dhiba [DHI 95].

15.2.1.1. *Principle of chromatographic analysis and results*

The products to be analyzed in gaseous chromatography pass through an inert material, during a *stationary phase*, such as alumina or activated carbon, on which the selection is made from *retention time*. These materials with a porous character fill a stainless metal tube of diameter varying from 2 to 4 mm; described as a *column*, this tube (from 1 to 6 m long) presents itself in the form of a coil. The products injected in the column are conveyed by an inert gas such as helium, hydrogen, nitrogen or argon. The samples are sent into an *injection chamber* and led by the vector gas which puts them in contact with the inert phase. At the exit of the column we proceed, in general, to electrical detection by a heated filament mounted on a resistance bridge. The electrical equilibrium of this latter varies because of the cooling created by the organic component; a comparison is made with the resistance of the filament around which *the effluent* circulates alone. It is the area of current peaks which gives the concentration of the products, this concentration being a linear function of the areas of recorded peaks. Other types of detectors are used: flame ionization, argon ionization, electron capture. A chromatograph can be connected to a mass spectrometer or an infrared spectrophotometer.

In practice, we take, for example with the aid of a syringe, an oil sample at the bottom or, in certain cases, from the upper part of the transformer. This sample is sent to a chemical laboratory for separation of gaseous products, which are analyzed by chromatography. The techniques of sampling are variable depending on the country. The analyzed gases are: H_2, CH_4, C_2H_2, C_2H_6, CO and CO_2, whose concentrations depend on the type of defect. For example, acetylene ($CH \equiv CH$) principally comes from an arc which produces temperatures of several thousand degrees, ethylene ($CH_2 = CH_2$) comes from hot spots where temperatures vary between 150°C and 1,000°C. The partial discharges, corresponding to a cool plasma, will particularly generate hydrogen (H_2) [DUV 84], [DUV 89], [DUV 02].

Let us note that this technique can be applied to the study of the degradation of solid insulating materials used for other types of insulation. This is the case of polyethylene, for example, to which we want to know the method of degradation during exceptional thermal stress (> 250°C). Protection against the effects of dangerous toxic gases is needed, since mixtures of oxygen and low weight molecular products such as aldehydes, ketones and acids are often found. By associating the chromatograph with a mass spectrometer, 44 components have thus been identified, fatty acids representing the principal organic products with formic acid [HOF 81].

15.2.1.2. *High performance liquid chromatography (HPLC)*

This technique works on the same principles as chromatography on a column where the phenomena of sharing, absorption, ion exchange or exclusion are produced. This depends on the stationary phase, constituting the column, which can be polar (amine, nitrile) or apolar (alkyl, phenyl). The polar character of the stationary phase is aimed at a weakly mobile polar phase like a hydrocarbon.

In their furan analysis, certain researchers use steel capillary columns from 10 to 30 cm long, with an inner diameter ranging from 3 to 5 mm, filled with solid particles. For the furan analysis, these latter are composed of silicium to which carbon chains (octadecyl) are chemically bonded. The sample is injected in the column through which an organic solvent is pumped. Depending on their chemical nature, the components interact differently with the solid support. At the output, different detectors, electrochemical or thermal conductivity, can be associated. For example, a UV detector operating between 200 and 400 nm was used by Unsworth [UNS 90]. The location of the recorded peaks characterizes the component and, as previously, the peak area gives its concentration.

15.2.1.3. *Gel Permeation Chromatography (GPC)*

Sometimes called exclusion or molecular sieving chromatography, this method leads to the separation of molecules as a function of their size. It is the grains of porous gel filling the columns which play the role of molecular sieve. The smallest molecules remain included in the gel and are eluted after a delay. The diameters of pores lower than 50 nm can be used for molecules present in oils. The technique turns out to be very practical for the analysis of additives incorporated in transformer oils; we note, for example, the clear separation revealed on a chromatogram between a UV stabilizer and an antioxidant [BUR 88].

A recent report, processing the diagnostics on power equipment containing oil, traces a critical history of chromatographic analysis and proposes automatization as a means to avoid experimental errors [ARA 02].

15.2.2. *Energy transport cables and dry capacitors*

Low density polyethylene (LDPE), chemically reticulated polyethylene (XLPE), ethylene-propylene-hexadiene 1-4 terpolymer (EPR) and, for certain applications, chlorosulfonated polyethylene (HYPALON) [GUE 92], are generally used for the insulation of high voltage cables. For dry capacitors, we keep the use of terephthalate polyethylene (PET), polypropylene (PP) or polyethylene naphthalene (PEN) films.

These materials are most often examined by the same techniques which are used for their selection, or for the understanding of ageing or failure. The most current ones are microscopy, infrared spectrophotometry, calorimetric analysis, thermally stimulated currents, and electron paramagnetic resonance.

15.2.2.1. *Microscopy*

In optical microscopy, using the shortest possible waves, i.e. ultraviolet, the resolution limit is about 200 nm. Since the photon does not allow us to go further, we use another elementary particle, the electron, instead. The wavelength associated with an electron is indeed much lower than that of the ultraviolet photon and the final resolution is thus of the order of a nanometer. This wavelength is of the same order as that of atomic bonds. Thanks to electromagnetic lenses, the electron beam emitted by an electron gun is focused on the material crossing it with a bigger or smaller absorption. The image forms behind the target on a fluorescent screen.

Except for the fact that electron absorbers are heavy metals, the same revelation techniques as for microscopy in direct light can be used. There are two variants of electron microscopy: scanning electron microscopy and transmission microscopy.

Scanning electron microscopy (SEM) consists of exciting a conductive surface by an electron beam and recreating its topography by capturing the reemitted electrons. The difference in mechanical resistance between amorphous and crystalline zones of polyethylene, for example, makes a difference in relief appearing when test tubes were obtained by the sudden fracture of the polymer at a temperature lower than glass transition temperature. A gold metallic layer of about 30 nm is left on the insulating material by a cool plasma in order to avoid it from being charged during bombardment by the electron beam. We thus observe a dense spherolitic structure.

The observation of the same material by *transmission* (TEM) permits the contrast between the amorphous zone and the crystalline zone to appear. However, this latter technique is tricky in its development because the thickness of the sample must be reduced to about 0.1 µm and the contrast method is difficult to control. This method requires a coloring of the polymer first, by diffusing chlorosulfonic acid in the amorphous zones of polyethylene, then uranyl acetate which is fixed on these same zones, making them opaque to the electron beam. The contrast then appears between white crystalline lamella (C) and the dark amorphous zones (Figure 15.1).

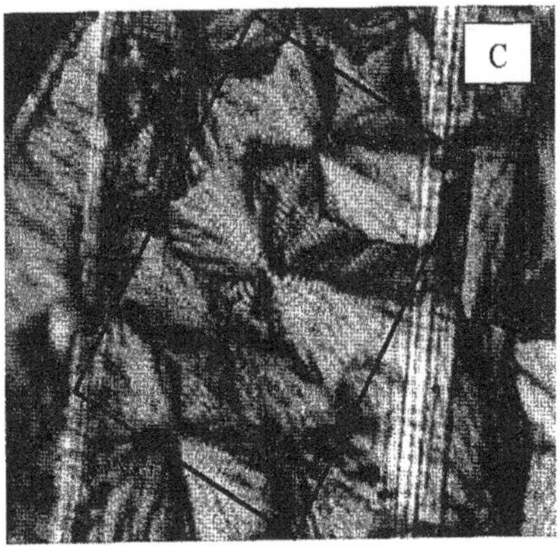

Figure 15.1. *Crystalline and amorphous zones of a LDPE [BOU 01]*

Fluorescence microscopy, derived from light microscopy, is a technique well known in the field of biology. It allows fluorescent objects or those marked by fluorescent substances to be observed.

Figure 15.2. describes the principle of the microscope. The image formed by the luminous beam is collected after reflection; this is called the epi-fluorescence mode. The fluorescent marker can be rhodamine, for example, which, when excited by blue light, will emit in red. Owing to filters, the sample is lit in blue and the red light is captured. The selection of wavelengths improves the contrast by comparison to clear field light microscopy, owing to the elimination in the final image of incident light reflection. A variant of this type of microscopy consists of replacing the beam of light by a laser beam of wavelength adjusted to the fluorescent substance and focused on the sample; this is laser scanning *confocal microscopy*. The acquisition of the image is made by scanning of the beam according to the XY focus plane and by displacement of the object according to the Z axis, using precise motors. A computing system processes the images, which allows a tri-dimensional structure to be revealed, by using successive optical sections, superposing them and reconstituting a stereographic view.

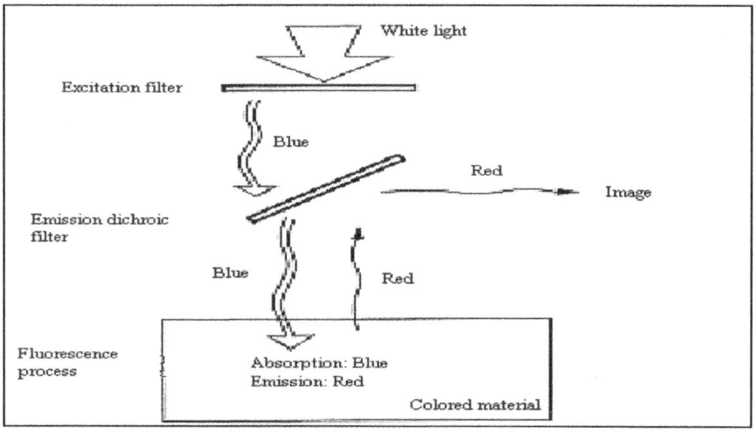

Figure 15.2. *Principle of the fluorescence microscope [MOR 92]*

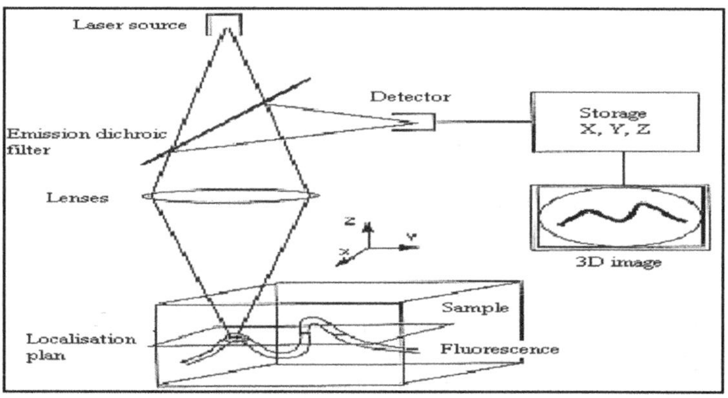

Figure 15.3. *Principle of the confocal microscope [MOR 92]*

One of the problems posed by this type of defect, water treeing, is its structure. Indeed, the characterization methods described above can lead to erroneous conclusions: semi-spherical cavities, micro-channels, etc. This latter gives different forms according to the section plane; this is due to the preparation technique for the observations by SEM (fracture) or TEM (100 nm thick sections). Also, the processing in order to improve the contrast creates artefacts [BAM 83,] [MEL 81]. We note that this purely experimental approach can support hypotheses made on the breakdown mechanism of the insulating material and perhaps, back up certain proposed models.

In practice, we most often remember that an energy transport cable can continue to ensure its function, despite the presence of defects of the type represented in Figure 15.4. Confocal microscopy has also been used to reveal microcavities which can exist in different thin (25 µm) films, amongst which certain were destined for the manufacture of capacitors: polypropylene, polyimide, polyvinylidene fluorure, etc. The obtained information concerns the surface of the film, but also a sub-layer which can reveal micro voids [SUT 92].

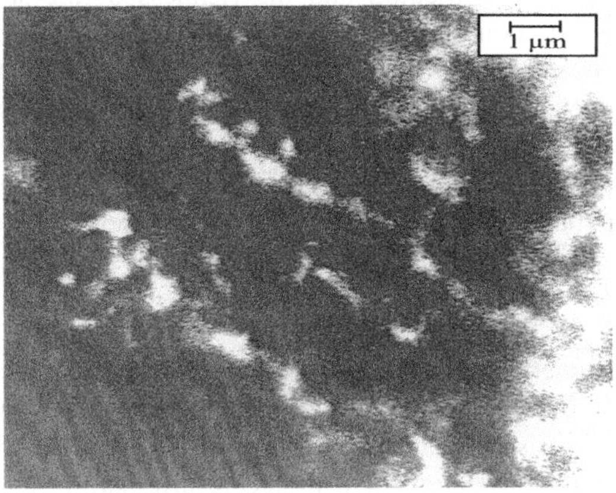

Figure 15.4. *Water treeing element stemming from a cable, observed in confocal microscopy [MOR 92]*

All these observation modes are particularly developed to compare PE and XLPE with samples where water treeing have progressed [MOR 92], [MOR 93].

The X-fluorescence induced by particles, i.e. the emission of X-rays induced by charged particles (*PIXE*), most often uses protons and allows the search for chemical species. By penetrating the material, a beam particle ejects an electron near the core of an atom which thus finds itself in an excited state. An electron in an outer orbit fills the gap and the atom then emits an X-ray expressing its energy excess, which is a proper value for each element. The device itself is composed of a 3 MeV particle accelerator, lenses which allow a focus of beams on a diameter < 1µm and an X-ray detector which realizes the counting and measurement of energy. In a few minutes, we can simultaneously take measurements from sodium (Na) to uranium (U), the weakest quantity being of the order of 1 part/10^6. For Z ranging from 20 to 30, the sensitivity is 10 ppm. The beryllium windows of the vacuum-packed detector limit its response to elements whose atomic number Z is greater than 11 (Na).

The technique has been applied to the research of impurities in cable insulation and, more particularly, to one of its defects: a *bowtie*-type of water treeing. The penetration depth in XLPE of electrons accelerated under 50 kV, for example, is 50 μm and 12 μm for protons accelerated under 1 MeV. The first measurements of the target surface being concerned with the emission of X-rays, only a few nanogram of polymer are analyzed, hence care must brought to the preparation of test tubes in order to avoid contamination. Taken on a 20 kV cable, this type of defect revealed an important quantity of chlorine comparatively to the rest of the material. The antioxidant is the source of sulfur that we also detect. The concentration of lead is decreasing, showing its conductor diffusion towards the outside, in this specific case. Despite the sensitivity of the method and the richness of the species found, it is difficult to reach a decision on the role of each one of them regarding the degradation process [HIN 88], [HOU 92]. From the sulfur measurement, the radial distribution of the concentration in the antioxidant was studied in XLPE insulated cables. As we have previously mentioned, the analysis was completed by using IR and UV spectrophotometries. We could thus notice a uniform antioxidant concentration in the core of the insulating material and decreasing towards the semi-conductor screen. Both types of antioxidants (Irganox and Santonox) have been tested here [MAN 83], [PAR 99].

Electronic paramagnetic resonance (EPR), in the area we are concerned with, is used for the detection of free radicals in old insulating materials. In the same way as free electrons in metals or semi-conductors, free radicals possess an unpaired electron. The magnetic moments of an atom or a molecule originate from the circulation of electrons around the cores, the orbital moments and the rotation of the electron on itself called the *spin* moment. When a material containing unpaired electrons is subjected to a magnetic field of frequency ν, an energy can be absorbed if a static magnetic field of value H is simultaneously applied to it. In practice, we make H increase in a continuous manner, such that the energy absorbed passes a maximum which corresponds to an energy level transition of the unpaired electron $h\nu = g\beta H$, which is called the resonance condition, H and β being constants and the factor g having a value close to 2. The resonance condition implies that, at a value of the applied magnetic field, only a frequency ν corresponds. Experimentally, the absorption lines present a certain width due to the interaction of *spin*s between themselves and their environment. Amongst the paramagnetic bodies we shall evoke here, the free radicals are not generally at a natural state but are created by a certain number of stresses, such as interactions of the insulating material with electric discharges, irradiations, or chain ruptures consecutive to electrical or mechanical stresses.

When discharges are produced in cavities within solid insulation, their noxiousness leads to the breakdown of the material over a more or less long term; when they are used to modify the surface state of a film before metallizing it and making it a capacitor, the interaction can be considered positive. In both cases, we know that ions and electrons are elements of the plasma which constitute these discharges and whose efficiency is good to know. The experiments made in particular on PP films destined for the manufacture of capacitors have shown the role played by ions and electrons accelerated between 50 and 240 eV, as well as the high efficiency when these same species work in synergy. Owing to ESR, it was particularly shown that the observed radical was of a peroxide (COO) type, the signal presenting an increasing amplitude by passing from argon ions to oxygen ions [GOM 89]. The same technique was used for the study of PEN irradiated by γ-rays [ROG 71]. The existence of fracture kinetics of polymers subjected to a mechanical stress was also demonstrated by this type of analysis. We observed in particular that the formation speed of the free radicals increased in an exponential manner as a function of the applied stress [ZHU 66]. At last used for the first time in the study of treeing phenomenon in LDPE and XLPE, this type of analysis revealed different signals according to whether it was a tree or dielectric breakdown phenomenon; however, the cores could not be identified during these preliminary experiments [BAC 78].

15.3. The materials themselves

We have just examined a few examples of current applications of physico-chemical analyzes on the ageing of systems with, for some of them, the possibility of leading to an intervention before serious damage, whilst for others leading to better understanding the system's ageing. The methods presented in the following sections are more particularly focused on the material before its use.

15.3.1. *Infrared spectrophotometry*

Infrared spectrophotometry is one of the most ancient methods at the disposal of researchers and engineers in the dielectric material field. Its use has surely been emphasized by the introduction of Fourier transform (IRTF) devices. The central organ of the Fourier transform device is a Michelson interferometer, a device destined for the analysis of multiple frequencies from a signal. Microscopes and accessories allow, in particular, surface analyzes of insulating materials which are presented in the form of films, similarly to those destined for the coils of dry capacitors. The transmission method calls on the absorption principle according to which the ratio between the transmitted light and the incident light follows a decreasing exponential law expressed by the Beer–Lambert law: $I = I_0 e^{-kx}$, where k is a function of the wavelength λ and the molecular concentration, x representing the matter crossed thickness. In the process of multiple reflections, the law remains applicable, but the coefficient k also contains parameters relative to the contact surface and the number of reflections. In the case where interest only lies on particular points of the surface, we can adjoin a beam condenser to the apparatus.

When the absorbant centers characteristic of a degradation or a simple transformation are in weak concentration and situated near the surface, it is necessary to resort to two experimental processes, *specular reflection* and *internal multiple reflections*: in the first case, the incident beam traverses the sample (tacked on a mirror) twice, the reflecting sides of a prism at 120° ensuring the transfer of the beam; in the second case, the sample is placed on both sides of a crystal, fitted for its transmission to the infrared beam which will meet the material to be analyzed n times; the value of n only depends on the crystal thickness. These two techniques increase the encounter probability of absorbant centers in a significant manner. Thus, a 2 mm thick thallium iodo-bromure crystal (KRS5) allows about 20 reflections, with a penetration depth from 0.64 to 5.1 μm in the range from 4,000 to 400 cm^{-1}. Let us note here that we express the spectral space analyzed in wave number (cm^{-1}) which is the inverse of wavelength. If we take PE for example, particular attention is often paid to the formation of carbonyl (C = O) groups around 1,720 cm^{-1}, and we then know that the penetration depth of the infrared beam for this wavelength is about 1.85 μm. An example is given in 15.5. Decomposition powders and products can be analyzed after mixture in the potassium bromide powder (KBr).

We then proceed to observation by transmission of a chip obtained by compression of the mixture. Obviously, liquid and gas cells containing transparent windows in the same range of wavelengths are also used.

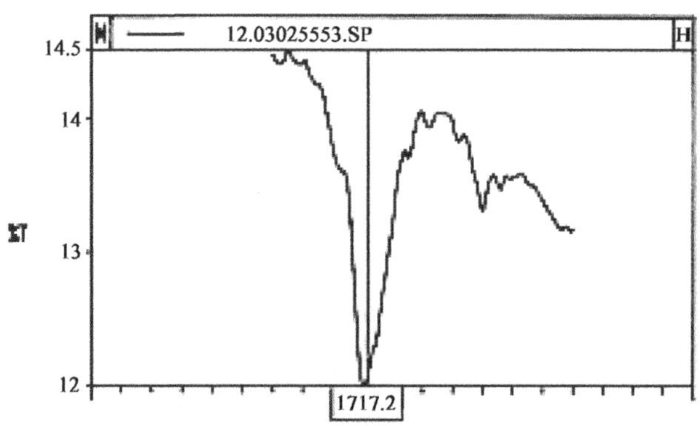

Figure 15.5. *Characteristic absorption of oxidation phenomena in PE*

The comparison between spectra obtained by direct transmission and by the multiple reflection technique (Figure 15.6) has allowed a difference between the volume (1) and the surface (2) of a 12 μm PEN film destined for the manufacture of capacitors to be shown. The manufacturing method of the film, generating its stretching, gives rise to a more crystalline surface morphology. The analysis shows that the surface morphology favors motions in the naphthalene plane, while in volume, vibrations outside the plane prevail [KRA 96].

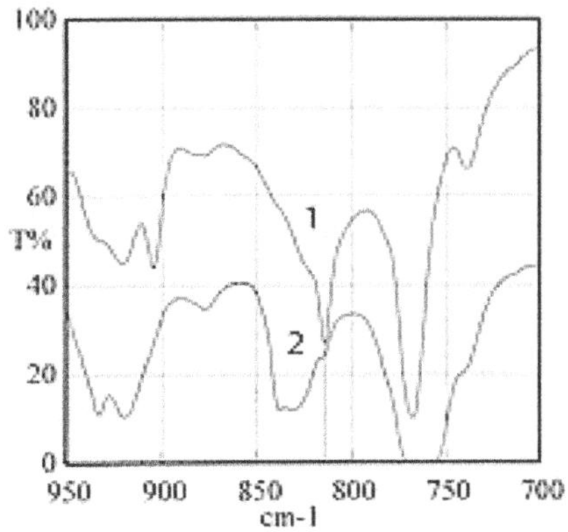

Figure 15.6. *Comparison between the volume (1) and the surface (2) of a PEN film [KRA 96]*

Even though manufacture technologies of energy transport cables have evolved since they first came out, the phenomenon of water treeing has always caught the attention of manufacturers, as well as researchers. After several years of service of 5 kV cables, XLPE samples containing antioxidants were taken and subjected to a certain number of analyzes, amongst which was the IRTF. The principal difficulty in this case is the extraction of the information of spectra to numerous differences when they are compared to the basic material. Thus, sulfate (1,130, 625 cm^{-1}) and carboxylate (1,575 cm^{-1}) anions and the classic oxidation attributed to carbonyl groups (evoked above) (~1,725 cm^{-1}) appeared. The differences between spectra, which can be realized on the apparatus, by comparing the clean and contaminated parts of the polymer, show the more important oxidation of the degraded part (in the region exhibiting treeing). This analysis thus reveals the early ageing of the insulating material, in the regions in which the antioxidant vanishing was important and, simultaneously, the catalytic effect of contaminants regarding oxidation. We can then better understand the weakening of dielectric behavior [GAR 87]. In 1994, to celebrate the 25th anniversary of the discovery of the phenomenon, an examination of theories was published, in which we underline that most researches are based on the infrared technique (micro-IRTF). An analysis of the literature published until then underlines the differences between the results concerning cable samples aged in a laboratory and those aged under an electrical field stress. The normal chemical entities given as examples are: carboxylate ions (RCOO-), ketones ($R_2C = CO$), and esters (RCOOR) which we detect in the region ranging from 1,700 to 1,730 cm^{-1} [XU 94]. The breakdown process of polyepoxide resins also seems preceded by a chemical modification of the material, revealed by IRTF, favoring the creation of electrical treeing [MIT 81]. In section 15.6.2, we emphasized the techniques permitting the control of liquid insulation; obviously IRTF contributes to the analysis of this type of insulation and what has been particularly shown through the detection of impurities is that we can identify the nature and give the proportion in the transformer oils for concentrations of the order of 30 ppm. The normal corresponding resolution of specters is 4 cm^{-1} [PER 98].

It is now common to use several techniques to study the same type of ageing of electrical insulating materials. The last two analysis methods which we just presented, ESR and IRTF, were used to study the chemical evolutions due to the impact on high density PE of ions accelerated between 50 and 150 kV; the goal being to find an analogy with the evolution of the insulating material of a cable under voltage [SCH 87].

15.3.2. *Calorimetric analysis*

The state change of a dielectric material leads to a variation of temperature, the heat being absorbed or emitted. If the comparison is made with an inert material, we can measure the necessary supply for the preservation of the equilibrium between these two systems. U being the energy of the system, this energy can evolve from a state A to a state B. The energy therefore depends on the state of the system, like pressure P and volume V. The function of state H, possessing the same properties is *the enthalpy*: $H = U + PV$. Differential thermal systems record the difference between the enthalpy changes of an analyzed material and an inert reference during simultaneous heating. When the sample and the reference are heated by the same heating source we record, as a function of time, the temperature difference between them owing to sensors placed near them both, or fixed to the bowls containing them. We then have a *Differential Thermal Analysis* (DTA). The most used technique is the *Differencial Entropic Analysis* (DEA) or *Differential Scanning Calorimetry* (DSC). The difference with the previous technique is the use of a heating element intrinsic to each part, sample and reference. The apparatus then consists of two loops, one controlling the temperature as a function of a predetermined growth speed, the other allows the input power to be adjusted as a function of the temperature difference which can appear between the reference and the sample and if, for example, a reaction is produced in this latter, whether it is exo- or endo-thermal. This process is sometimes called power compensation. It is important to mention that the head temperature measurement of the sample is maintained constant until the head temperature measurement of the reference is adjusted in a continuous manner. We record a signal proportional to the difference of heat supplied to the sample and the reference (dh/dt). This type of analysis was most often practiced for the study of a material before its use. It thus permitted the reticulation of PE to be studied, the phenomena resulting from thermal processing of this material destined for the manufacture of cables [CAI 72]. Associated with infrared spectrophotometry, it completes the analysis of oxidation phenomena of PE when it is in contact with metals with greater or lesser, according to whether we are dealing with Cu, Pb, Ag and Zn or Al and Au, etc. Oxidation is once again characterized by the optical density of the absorption at 1,720 cm^{-1} which corresponds to the vibrations of the carbonyl (C = O) groups. With metals such as copper and lead, whose chemical activity is important, the formed components diffuse easier in the polymer when the temperature of the material is high. We remember that these two metals can be an essential part of a cable, the first as a conductor, the second in the role of a screen [EGO 75]. DEA, associated with thermally stimulated depolarization currents, dielectrical spectroscopic analysis and dynamic mechanical analysis, allow a comparative study of an amorphous and a partially crystalline PEN. The interest of the study was on the effect of the degree of crystallinity on the electrical properties in particular, knowing that this dielectric is used for the manufacture of capacitors and that, comparatively to PET, its fusion temperature (268°C) is greater by about 10°C and its glass transition temperature T_g is also higher than 50°C [CAN 00]. A similar approach had been realized for the study of the formation of aggregates due to water and which we know can lead to the breakdown of cable insulations [JOH 80]. By assuming the nature of the physical properties of polyepoxides from biphenol-A or cyclo-aliphatics to be unstable at temperatures less than the glass transition temperature T_g, this technique used at the same time as dielectric measurements shows the interest of annealing [LI 94].

More recently, the study of total and partial isothermal crystallization by DEA permitted the existence of a negligible proportion of rigid amorphous phases in PEN to be quantified. The complement of analysis brought by IRTF led to a quantitative study from certain absorption bands (813 and 839 cm^{-1}), showing that the crystallization process induced an increase of *trans conformations*. All the results show which semi-crystalline structure we should tend towards for good use of PEN as a capacitor dielectric [ZOU 02].

A better knowledge of the morphology of an organic insulating material in particular can help to understand the conduction and breakdown phenomena of this type of material. When LDPE is used for the manufacture of energy transport cables, we can take an interest in the gradual evolution of the material, from the basic resin, as a function of a temperature rise, then as a function of its transformation such as a reticulation from dicumyle peroxide and, finally, the adjunction of antioxidants already mentioned. Fusion specters obtained in DEA for samples annealed at different temperatures are represented in Figure 15.7. Three zones of temperature could then be defined: $T_f < 45°C$, $45°C < T_f < 100°C$ and $T_f > 100°C$. The same technique used to analyze the material with its antioxidant reveals an increase of the fusion peak which is interpreted as a reduction of the number of defects inside crystallites. A correlation was found with electrical behavior [BOU 01].

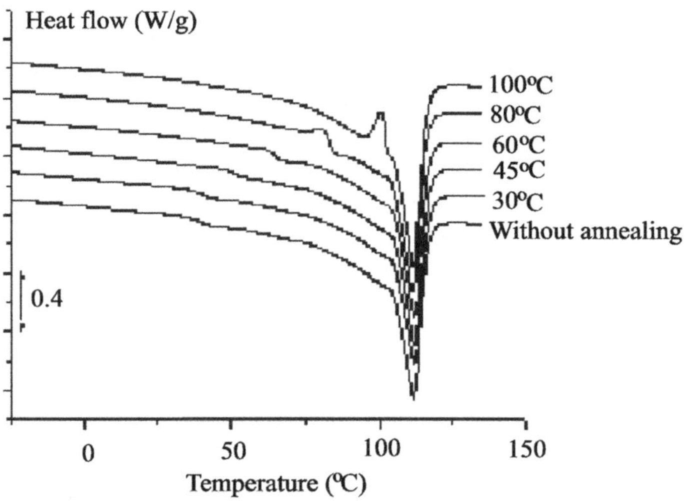

Figure 15.7. *Influence of the annealing temperature on an LDPE and its antioxidant analyzed by DEA [BOU 01]*

15.3.3. *Thermostimulated currents*

This technique, proposed 40 years ago [BUC 66] to characterize the punctual defects in alkali halogen crystals has since been widely applied to other materials, like those used in electrical insulating [LAV 93] shows the schematic representation of an experimental device which is composed of a measurement cell (1), placed in an enclosure regulated between the temperature of liquid nitrogen and 300°C. The cryostat (2) is constituted of a triple wall permitting it to be insulated by a secondary vacuum (3); the liquid nitrogen (4) is destined to cool the cell. The thermal exchange between the cell and the walls of the nitrogen reservoir is ensured by the gaseous helium under weak pressure (20 mbar) (5). A heating resistance (6) surrounding a removable stainless steel cylinder (7) is connected to a power supply controlled by a regulator. A probe placed in the vicinity of the sample measures its temperature (8). The electrodes (9), between which the sample is placed (10), are connected to a rotating relay allowing the application of a static electric field and the measurement of the depolarization current, as a function of its position. The detection of the depolarization current is recorded by an electrometer sensitive to 10^{-16} A.

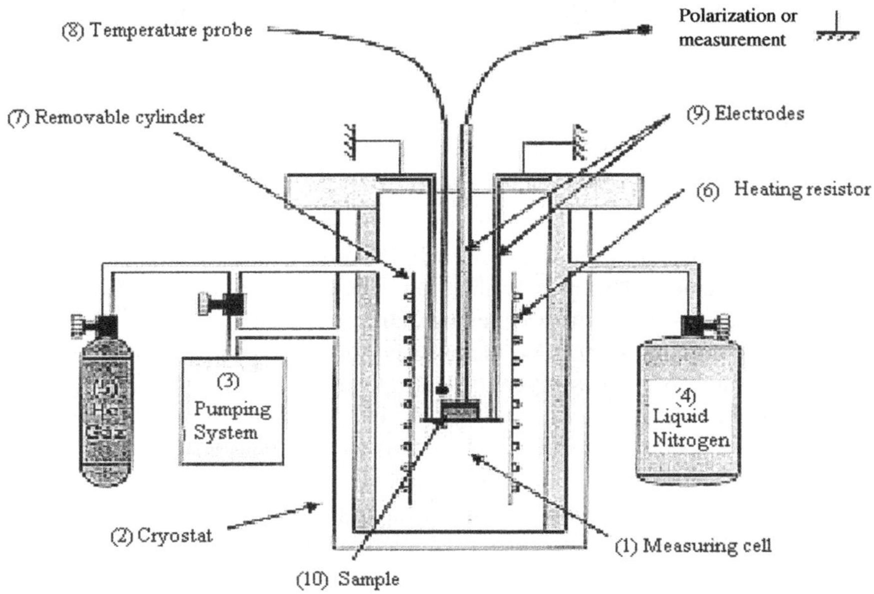

Figure 15.8. *Experimental device for measurement of thermally stimulated currents [MAY 00]*

At given polarization temperature T_p, the samples are subjected to a static electric field F_p for about 2 mn to allow the orientation of mobile dipoles at this temperature, according to the field F_p; the equilibrium polarization thus obtained, we proceed to a decrease of the temperature until $T_0 << T_p$ in order to fix the dipoles. After suppression of the field, we place the test tube in short circuit, which allows the flow of surface charges. A linear growth of the temperature at the speed of a few degrees/mn induces the gradual relaxation of dipolar entities. We then record a depolarization current. The recording of this current as a function of the temperature constitutes a global or complex spectrum of the dielectric relaxation modes. Each peak being associated with a relaxation mode, certain criteria ensure their dipolar nature: these are reproducibility, proportionality between their intensity and F_p, and invariance with different polarization temperatures. There is a relaxation time distribution which makes the exploitation of this type of spectra difficult; also another procedure is applied, that of fractional polarizations. Here, the temperature is decreased under field until a value $T_d = T_p - \Delta T$. After the suppression of the field, the temperature is maintained constant for a time t_d allowing the return to the equilibrium of entities whose relaxation time $\tau(T_d)$ is smaller than t_d. The spectrum is then recorded as in the previous method. By displacing the polarization window

[T_p, T_d] along the axis of temperatures, we obtain a set of elementary spectra corresponding to the decomposition of a complex spectrum.

The electric stress can be replaced by a mechanical cutting one σ; we then realize the thermo-stimulated flow (TSF) which allows the molecular mobility to be studied. The mechanical stress is applied for a time t_σ of the order of 2 mn at temperature T_σ, which allows the mobile entities of the polymer to get oriented. This configuration out of equilibrium is frozen by a rapid quench which prevents any molecular displacement. The stress is then removed and the return to equilibrium of the sample is generated by a rise of temperature at constant speed (~7 K/mn); the mobile entities can then reorient themselves. In the ageing domain of cable insulation affected by water treeing, this technique brought certain physical indications on the role of water and electrical field. Thus, the decrease of activation enthalpy appears as a consequence of water absorption which transforms the structure under the influence of an electrical field [MAR 92].

The few references which we have given here simultaneously take into account the most fundamental aspects of results for this type of analysis, and also the aspects purely applied in the world of insulation, knowing that this selection is not exhaustive and that further examples can be found in the references supplied.

The effect of the thermal expansion of elastomers on TSC was revealed on both sides of T_g in this type of polymer [VAN 87]. In spite of wide use in the insulation domain, the complexity of the poly-epoxy structure which we have already evoked makes it difficult to understand the conduction mechanism, in particular because of the ionic or electronic character that can be attributed to it as a function of the applied electrical field range. Photo-currents and TSC can come in helpful for understanding this [KAW 88]. The interest in simultaneously using stimulated thermoluminescence and TSC has been shown in several insulating materials such as XLPE, PET, and EPR [FLE 89]. More recently, by choosing three poly-epoxides whose architecture is well defined, the distinction was made between the processes of localized and non-localized relaxation, owing to the study of the influence of growth speed in temperature [HEN 00].

15.4. Conclusion

We have attempted to inspect a few analysis techniques practiced in the domain of electrical insulating materials. As we have underlined, the apparatus and the investigations have a history; with progress being due to developments in electronics and computer science. We have also pointed out that several techniques can be used simultaneously, sometimes in complement to electrical measurements, to try to gain a better understanding of the initial structure and behavior of materials subjected to device stresses. Published papers contain more and more results obtained by this type of approach, even if it is costly. Among other analyzes, we could have cited for example: UV spectrophotometry, ESCA, XPS, the interaction of weak energy electrons with the surface of solid dielectrics (LEET), and ESD, described in a synthesis article by Sanche [SAN 93].

We have insisted on using a more commonly practiced method because, to our mind, beyond the results it already brings, it could help towards the validation of certain models for which outlines have already been conceived and which rely on the existence of free volumes. This method belongs to the series of spectroscopic methods using γ-rays, PASCA (*Positron Annihilation Spectroscopy for Chemical Analysis*). The positron, known for about half a century, is the antimatter of the electron carrying a positive charge of the same order as the negative charge of the electron, and with properties identical to those of electron. The most common source is ^{22}Na. This new probe (PASCA) is well described by Stevens [STE 90] and Jean [JEA 90] and a demonstration of its use was made, for example, in the ageing study of cables insulated by cross-linked polyethylene [BER 92].

15.5. Bibliography

[ARA 02] ARAKELIAN V.G., "Effective diagnostics for oil-filled equipment", *IEEE Electrical Insulation Magazine*, vol. 18, no. 6, p. 26–38, 2002.

[BAC 78] BACQUET G., DIB J., WU C.Y., DENSLEY J.R., BOGGS S.A., "ESR study of free radicals in electrical trees in polyethylene", *IEEE Trans. Electr. Insul.*, vol. 13, no. 3, p. 157–163, 1978.

[BAM 83] BAMJI S.S., BULINSKI A., DENSLEY J., GARTON A., "Etching and the morphology of cross-linked PE cable insulation", *IEEE Trans. Electr. Insul.*, vol. 18, no. 1, p. 32–41, 1983.

[BER 92] BERNSTEIN B.S., STINIVAS N., "Accelerated aging of crosslinked polyethylene cable insulation; electric Positron Annihilation Spectroscopy study", *Conference Record of the 1992 IEEE Intern. Symp. Electr. Insul. Dielectr. Phenom.*, 92CH3150-0, p. 104–109, 1992.

[BOU 01] BOUDOU L., Influence des paramètres de mise en œuvre, sur la morphologie et la conductivité électrique d'un polyéthylène soumis destiné à la fabrication des câbles, Doctoral Thesis, University of Paul Sabatier, Toulouse, 2001.

[BUC 66] BUCCI C., FIESCHI R., GUILDI G., "Ionic thermocurrents in dielectrics", *Phys. Rev.*, vol. 148, no. 2, p. 816–823, 1966.

[BUR 88] BURTON M.J., CARBALLEIRA M., DUVAL M., FULLER C.W., GRAHAM J., DE PABLO A., SAMAT J., SPICAR E., "Applications de la chromatographie liquide à l'analyse des matériaux isolants électriques", *Proceedings of the International Council on Large Electric Systems Conference* (CIGRE), 15–03, 1988.

[CAI 72] CAILLOT C., AUDOUX C., AUCLAIR H., "Application de l'analyse enthalpique differentielle à l'étude du polyéthylène haute pression pour câbles", *Rev. Gén. Ele.*, vol. 81, no. 11, p. 740–745, 1972.

[DHI 95] DHIBA D., Etude du vieillissement de l'isolation papier/huile dans les transformateurs de puissance. Influence des inhibiteurs d'oxydation, Doctoral Thesis, University of Paul Sabatier, Toulouse, 1995.

[DUV 77] DUVAL M., LAMARRE C., "The characterisation of electrical insulating oil by HPLC", *IEEE Trans. Electr. Insul*, vol. 12, no. 5, p. 340–344, 1977.

[DUV 82] DUVAL M., GIGUÈRE Y., "Simultaneous determination of the antioxidant, the crosslinking-agent and decomposition products in polyethylene by reverse-phase HPLC", *J. Liquid Chromatography*, vol. 5, no. 10, p. 1847–1857, 1982.

[DUV 84] DUVAL M., LAMARRE C., GIGUERE Y., "Reversed-phase high-performance liquid chromatographic analysis of polar oxidation products in transformer oils", *J. Chromatography*, vol. 284, no. 1, p. 273–280, 1984.

[DUV 89] DUVAL M., "Dissolved gas analysis: it can save your transformer", *IEEE Electrical Insulation Magazine*, vol. 5, no. 6, p. 22–23, 1989.

[DUV 02] DUVAL M., "A review of faults detectables by gas-in oil analysis in transformers", *IEEE Electrical Insulation Magazine*, vol. 18, no. 3, p. 8–17, 2002.

[EGO 75] EGORENKOV D.G., LIN D.G., BELY V.A., "Effect of metals on melt oxidation of polyethylene", *J. Polym. Sc.*, vol. 13, no. 7, p. 1493–1498, 1975.

[FLE 89] FLEMING R.J., "Thermally-stimulated conductivity and luminescence in organic polymers", *IEEE Trans. Electr. Insul.*, vol. 24, no. 3, p. 523–531, 1989.

[GAR 87] GARTON M., BAMJI S., BULINSKI A., DENSLEY J., "Oxidation and water tree formation in service-aged cable insulation", *IEEE Trans. Electr. Insul.*, vol. 22, no. 4, p. 405–412, 1987.

[GOM 89] GOMÈS DE LIMA P., LOPEZ J., DESPAX B., MAYOUX C., "Sur l'efficacité des ions et des électrons dans un plasma froid venant en contact avec un polymère", *Rev.Phys. Appl.*, vol. 24, no. 3, p. 331–335, 1989.

[GUE 92] GUEGUEN V., Vieillissement d'élastomères utilisés comme isolants électriques en ambiance nucléaires, Thesis, Ecole Normale Supérieure d'Arts et Métiers, Paris, 1992.

[HEN 00] HENN F., GLUNTINI J.C., HALARI J.L., VANDERSCHUEREN J., MURACCIOLE J.M., "TSDC in epoxy networks of well-controlled architecture. Experimental evidence for local and non-local relaxation process", *IEEE Trans. Electr. Insul.*, vol. 7, no. 4, p. 551–555, 2000.

[HIN 88] HINRICHSEN P.F., HOUDAYER A., BELHADFA A., CRINE J.P., PELISSOU S., CHOLEWA M., "A localized trace element analysis of water trees in XLPE cable insulation by micro-PIXE and EDX", *IEEE Trans. Electr. Insul.*, vol. 23, no. 6, p. 971–978, 1988.

[HOU 92] HOUDAYER A.J., HINRICHSEN P.F., KAJRIS G., PARPAL J.L., FOURMIGUÉ J.M., CRINE J.P., "μ-PIXE analysis of impurity migration in polyethylene", *Proceedings of the 4th International Conference on Conduction and Breakdown in Solid Dielectrics*, Catalog 92CH3094-6, p. 433–439, 1992.

[HOF 81] HOFF A, JACOBSSON S., "Thermo-oxidative degradation of low-density polyethylene close to industrial processing conditions", *J. Appl. Polymer Sci.,* vol. 26, no. 10, p. 3409–3423, 1981.

[JEA 90] JEAN Y.C., "Positron Annihilation Spectroscopy for Chemical Analysis: A new probe for microstructural analysis of polymers", *Microchem.*, vol. 42, no. 1, p. 72–102, 1990.

[JOH 80] JOHNSON G.E., BAIR H.E., MATSUOKA S., ANDERSON E.W., SCOTT J.E., "Water sorption and its effect on a polymer's dielectric behaviour", *Water in Polymers*, 27, p. 451–468, American Chemical Society, 1980.

[KAW 88] KAWAMOTO A., SUZUOKI. Y., IKERJIRI T., MIZUTANI T., IEDA M., "Photocurrents and thermally stimulated currents in epoxy resin. Effects of mechanical stress", *IEEE Trans. Electr. Insul.*, vol. 23, no. 2, p. 201–208, 1988.

[KRA 96] KRAUS E., Comportement électrique et physicochimique de films de poly (éthylène-naphtalène 2,6 dicarboxylate) soumis à des contraintes électriques et thermiques, Thesis, No. 2288, Toulouse, 1996.

[LAM 81] LAMARRE C., DUVAL M., GAUTHIER J., "Dosage par CLHP du DBPC dans les huiles de transformateurs neuves oxydées", *J. Chromatography.*, vol. 213, no. 3, p. 481–490, 1981.

[LAV 93] LAVERGNE C., LACABANNE C., "A review of thermo-stimulated current", *IEEE Electrical Insulation Magazine*, vol. 9, no. 2, p. 5–21, 1993.

[LI 94] LI Y., UNSWORTH J., "Effect of physical aging on dielectric, thermal and mechanical properties of cast-epoxy insulators", *IEEE Trans. Dielec. Electr. Insul.*, vol. 1, no. 1, p. 9–17, 1994.

[MAN 83] MANGARAJ D., MONDRON P., EPSTEIN M.M., "Analysis of antioxidant in aged cables", 1983 Annual Report of the *Conference on Electrical Insulation and Dielectric Phenomena*, catalog 83CH1902-6, p. 286–293, 1983.

[MAR 92] MARTINEZ J.J., BOULOUZ M., MAYOUX C., LACABANNE C., "Thermally stimulated creep for the study of ageing in insulating polymers under electric field", *Conference Record of the 1992 IEEE International Symposium on Electrical Insulation and Dielectric Phenomena*, Catalog 92CH3150-0, p. 62–65, 1992.

[MAY 00] MAYOUX CH., Polymérisation radicalaire en suspension inverse d'un poly (acide acrylique) réticulé. Etude de la dynamique de chaîne, Thesis, No.3802, Toulouse 2000.

[MEL 81] MELTON C.W., MANGARAJ D., EPSTEIN M.M., "Morphology of thermoplastic and cross-linked polyethylene cable insulation", *Conference Record of the 1981 IEEE Conference on Electrical Insulation and Dielectric Phenomena*, catalog 81CH1668-3, p. 299–305, 1981.

[MIT 81] MITSUI H., YOSHIMITSU T., MIZUTANI Y., UMEMOTO K., "Electrical failure properties of cast epoxy resins", *IEEE Trans. Electr. Insul.*, vol. 16, no. 6, p. 533–541, 1981.

[MOR 92] MOREAU E., Un phénomène de vieillissement du polyéthylène sous champ électrique: les arborescences d'eau, leur structure fine et leur caractéristique électrique, Thesis, No. 1294, Toulouse, 1992.

[MOR 93] MOREAU E., BOUDET A., MAYOUX C., LAURENT C., WRIGHT M., "Fine structure of defects in polyethylene used for power cable insulation observed by fluorescence microscopy", *J. Mater. Sc.*, vol. 28, no. 1, p. 161–169, 1993.

[PAR 99] PARPAL J.L., GUDDEMI C., HINRICHSEN P.F., "Antioxydant concentration distribution measurement in XLPE cable insulation by PIXE analysis, FTIR and UV spectroscopy", *Proceedings of the 5th International Conference on Insulated Power Cables*, p. 324–328, 1999.

[PER 98] PERCHERANCIER J.P., VUARCHEIX P.J., "Fourier Transform Infrared (FT-IR) spectrometry to detect additives and contaminants in insulating oils", *IEEE Electrical Insulation Magazine*, vol. 14, no. 3, p. 23–29, 1998.

[ROG 71] ROGOWSKI R.S., PEZDIRTZ G.F., "Electron spin resonance of γ-irradiated poly(ethylene-2,6-naphtalene dicarboxylate)", *J. Polym. Sc.* A2, vol. 9, p. 2111–2117, 1971.

[SAN 93] SANCHE L., "Electronic aging. Electron interactions in thin-film dielectrics", *IEEE Trans. Electr. Insul.*, vol. 28, no. 5, p. 789–819, 1993.

[SCH 87] SCHAIBLE M., HAYDEN H., TANAKA, "Chemical changes created by high energy ions in polyethylene", *IEEE Trans. Electr. Insul.*, vol. 22, no. 6, p. 699–708, 1987.

[SCH 88] SCHUPPE W.H., SAURE M., ANDRESS H., MÖLLER K., MEURER D., KRAUSE K., "Méthodes analytiques et physiques sensibles pour diagnostiquer les modifications d'état dans des plaques ou des isolations de câbles", *Proceedings of the International Council on Large Electric Systems Conference (CIGRE)*, Papier 15–03, 1988.

[STE 90] STEVENS J.R., "Probe and label techniques", *Methods of experimental physics*, vol. 16A, Ch. 5, Academic Press, p. 371–403, 1990.

[SUT 02] SUTAR J.L., LAGHARI J.R., CHENG P.C., "LSCM: A non-destructive diagnostic tool for examining the microstructure of polymer dielectric films", *IEEE Electrical Insulation Magazine*, vol. 8, no. 4, p. 20–24, 1992.

[UNS 90] UNSWORTH J., MITCHELL F., "Degradation of electrical insulating paper monitored with high performance liquid chromatography", *IEEE Trans. Electr. Insul.*, vol. 25, no. 4, p. 737–746, 1990.

[VAN 87] VANDERSHUEREN J., LADANG M., NIEZETTE J., CORAPCI M. "The effect of thermal expansion on thermally stimulated currents in polymers; 1. Elastomeric materials changes", *IEEE Trans. Electr. Insul.*, vol. 22, no. 1, p. 19–22, 1987.

[VER 92] VERGNE J., Etude du vieillissement physico-chimique du complexe papier-huile dans les transformateurs électriques, Thesis no. 522, Toulouse, 1992.

[XU 94] XU J.J., BOGGS S.A., "The chemical nature of water treeing: Theories and evidence", *IEEE Electrical Insulation Magazine*, vol.10, no. 5, p. 29–37, 1994.

[ZHU 66] ZHURKOV S.N., TOMASHEVSKI E.E., "An investigation of fracture process of polymers by the Electon Spin Resonance method", *Proc. Conf. Physical basis of yield and fracture*, Oxford, p.200–208, 1966.

[ZOU 02] Zouzou N., Etude de la mobilité moléculaire du PEN destiné à la fabrication des condensateurs: influence de la microstructure, Thesis, University of Paul Sabatier, Toulouse, 2002.

Chapter 16

Insulating Oils for Transformers

16.1. Introduction

Insulating oils are used in electrical engineering as impregnators of cellulose insulations or as a filling product for various electrical equipment: transformers (power, distribution, traction, furnace, instruments, etc.), reactors, capacitors, cables, crossings, circuit-breakers, tap-changers, etc. [BER 97]. The volumes of oil which are used vary from a few liters for capacitors to several tens of thousand liters for power transformers. The main role of this oil is to remove the air and other gases so as to improve the dielectric behavior of the solid insulation. However, in many applications, they are used to ensure both the electrical insulation and the heat transfer of a component or a system, as in the case of transformers.

Oils can also be used for their capability of extinguishing electric arc (in the arcing chambers of switches and some circuit-breakers), thanks to their lubricating power for materials containing pieces in motion (plug selectors, immersion pumps) and in certain cases to improve resistance to fire, as is the case for distribution transformers near populations. Finally, they are used to slow down direct oxidation of cellulose insulation.

Mineral oil is the most commonly used liquid in power transformers, both for its physico-chemical properties and for its low cost [ROU 98]. Originally, the electric insulation of transformers was ensured by non-impregnated natural products in direct contact with air, which is a rapid source of oxidation. In 1854, the immersion of coils in turpentine [ABD 68] enabled their dielectric performance and their

---

Chapter written by Abderrahmane BEROUAL, Christophe PERRIER, Jean-Luc BESSEDE.

lifespan to be increased, and showed that we can remove solid insulation from direct oxidation. In 1891, petrol oil was used experimentally in the insulation of a transformer [BER 02] to replace air. It was only at the beginning of the 20th Century, with the multiplication of electric energy distribution networks and the increase of installed power, that air-insulated transformers (known as "dry") became enormous in size. Since then, they have been substituted little by little for transformers filled with mineral oil (known as "immersed"); the mineral oil used as an electric insulator and fluid calocarrier ensuring heat transfer. One of the disadvantages of mineral oil lies in the fact that its resistance to fire is weak. To solve this problem, synthetic liquids such as polychlorobiphenyls (PCB) were developed in the 1930s. In the 1970s, the use of PCB was forbidden because of their toxicity; indeed they have been recognized as dangerous for populations, animals and the environment, notably by the fact that they could emit dioxins during incomplete combustion. Thus, new synthetic liquids of low flammability, such as silicone oils or synthetic esters were perfected to replace the PCB. However, the use of these latter remains limited to small transformers (distribution and traction) because of their higher cost. Nowadays, in the face of the increasing demand for the use of friendly products in industry, distribution transformers are more and more likely to be filled with vegetable oils instead of synthetic products.

This chapter gives a description of different insulator liquids used in transformers. After a general presentation of diverse insulating liquids destined for the electrical engineering industry, we describe in a more precise manner the five types of insulating liquids which can be found in transformers, i.e. mineral oil, synthetic ester oil, silicone oil, PCB and vegetable oil (natural ester). Finally, we give security instructions related to their use.

16.2. Generalities

There are three types of insulating liquids (Table 16.1): (i) vegetable oils (natural products), (ii) mineral oils (based on refined petroleum products), (iii) synthetic liquids.

This listing is related to their chronological order of their appearance in electrical engineering applications. Indeed, vegetable oils were the first used, then mineral oils were developed, then synthetic liquids were manufactured to solve the shortcomings of mineral oils in certain applications [BER 02].

Nowadays, with concerns related to the environment, the trend is oriented again towards vegetable oils (based on soya, colza, sunflower, etc.). Even though their use is restricted for the moment to distribution transformers, numerous ongoing projects try to extend their application to power transformers.

	Category	Type
(i)	Vegetable oils	Castor oil, then soya, rapeseed, …
(ii)	Mineral oils	Naphthenics
		Paraffinics
		High molecular weight
(iii)	Unsaturated linear synthetic hydrocarbons	Polybutenes
	Synthetic aromatic hydrocarbons	Alkylbenzenes
		Alkylnaphtalenes
		Alkylbiphenyls
	Chlorinated linear hydrocarbons	Perchlorinatedethylene
	Chlorinated aromatic hydrocarbons	Trichlorobenzenes
		Polychlorobiphenyls (PCB)
		Polychlorophenylalkanes
	Chlorofluoridated linear hydrocarbons	Trichlorotrifluoroethane
	Ethylenic aromatic hydrocarbons	Alkyldiarylethene
	Organic esters	Benzyl neocaprate
		Diotylphalate
		Phosphate esters
		Pentaerythritol esters
	Ether-oxides	Diolylether
	Silicone liquids	Polydimethylsiloxanes (PDMS)
		Polymethylphenylsiloxanes

Table 16.1. *Different types of insulating liquids [VUA 98A]*

Table 16.2 presents the principal applications of different insulating liquids in electrical equipment. Three types of products are distinguished for their application in transformers:

– mineral oil: the most used insulating liquid in electrical equipment for its good dielectric and heat transfer properties, its good compatibility with cellulose insulators and its low cost; power transformers can contain between 40,000 and 80,000 liters of oil! It is principally for this technico-economical reason that this oil is the most commonly used in power transformers;

– *synthetic oils*: used every time some specific properties are searched for. This is notably the case when improving resistance to fire of transformers in the vicinity of populations (distribution and traction). These synthetic liquids are all obtained from different petrochemical processes. There are three principal types destined for transformers:

- pentaerythritol ester oils, still called synthetic esters or organic esters (as opposed to natural esters or vegetable oils),

- silicone oils,

- chlorinated hydrocarbons, such as PCB (now forbidden),

– *vegetable oils*: following the craze for the environment, have been and still are subject to numerous research [OOM 02] [DAR 07] [SAN 87] [MCS 02] [BER 04] to be developed in transformer-type applications. They are now commercially available and largely used in applications where fire resistance and biodegradable products are required.

Application	Power transformer and reactor	Distribution transformer	Traction transformer	Instrument transformer
Mineral oil	x	x	x	x
Silicone oil		x	x	
Synthetic ester	O	x	x	
Aromatic hydrocarbons				x
Natural esters	O	x		

Table 16.2. *Use of insulating liquids in transformers.*
Key: *x: widely used; o: less commonly used*

In a transformer, the oil ensures several functions:

a) Insulation function

In the active part of a transformer, the insulation between the elements at different electric potentials is ensured:

– by the liquid dielectric only, when dealing with an insulation between two bare metallic pieces (switch contacts for example);

– by a solid layer (paper or pressboard) impregnated with liquid dielectric; this is the case, for example, for the insulation between two conductors neighboring the same coil;

– by a mixed insulation: combination of liquid dielectric films and barriers of solid insulators; this is the case for an insulation between two concentric coils at different voltages.

The principal properties which permit the aptitude of an oil to fulfill its function as a dielectric to be judged are the dielectric strength, the dissipation factor or the loss factor (tan δ), the dielectric contant and the resistivity. These four characteristics, which depend on the temperature and the frequency at a given voltage, are essential for the design of a transformer.

b) Heat transfer function

Although its efficiency is exceptionally high (99 to 99.9%), a transformer cannot escape the energy loss which accompanies any conversion. The energy lost gets dissipated in the form of heat, thus requiring the cooling of the apparatus. Because of the losses, the temperatures of each element rise until establishment of an equilibrium between the production and evacuation speeds of the heat.

The heat to evacuate is conveyed by natural or forced circulation (streaming) of the oil, towards external radiators (Figure 16.1). A well-proportioned cooling device enables the formation of hot spots to be avoided, owing to important and well regulated circulation.

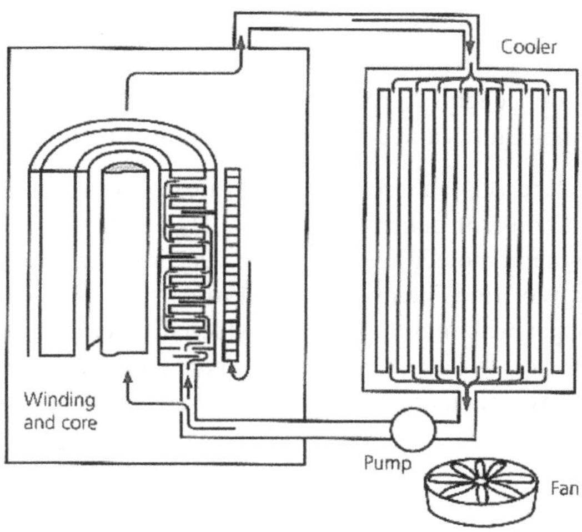

Figure 16.1. *Schematic of heat transfer in a transformer [NYN 04]*

The cooling is made by convection and thermal conduction [PER 87], and the principal characteristics which allow the ability of an oil to evacuate heat to be judged are the viscosity, the thermal conductivity and the specific heat.

c) Other functions

Oils can also be used for their power to extinguish electric arcs (in cutting chambers of charge switches and certain circuit-breakers) because of their lubricating power for materials containing pieces in motion (plug selectors, immersion pumps), and in certain cases to improve the resistance to fire, as it is the case for distribution transformers near populations. Finally, we must not forget the insulating function of solid materials in the strict sense of the term, i.e. which is that of slowing down the direct oxidation of cellulose insulation.

16.3. Mineral oils

16.3.1. *Composition*

Mineral oils are obtained by refining of crude petroleum and are essentially constituted of carbon and hydrogen atoms. In each mineral oil, carbon is found (Figures 16.2 and 16.3) [NYN 04] in:

– paraffinic structure (C_P): the molecules of this group are also known as saturated hydrocarbons in strait chain or alkanes, which can be linear (n-alkanes) or ramified (iso-alkanes) and whose general formula is C_nH_{2n+2}. The n-alkanes are also known as paraffins or waxes, and have bad flow properties at low temperatures;

– naphthenic structure (C_N): the molecules of this group are also known as saturated cyclics or cycloalkane hydrocarbons. They correspond to closed carbonated chains of general formula C_nH_{2n}. The cycloalkanes present better properties at low temperatures and a better solvent power than n-alkanes;

– aromatic structure (C_A): these molecules are cyclic components, also called unsaturated hydrocarbons, of general formula C_nH_{2n-6}. They are totally different to paraffinic and naphthenic molecules, but play a very important role in the properties of a mineral oil. We find them in two forms: the monoaromatics and the polyaromatics amongst which certain are considered carcinogenic. These aromatic components allow the oil to have a good performance with oxidation (production of phenols which destroy the radicals) and good gas properties (strong absorption capacity of the gases).

According to the original crude oil and the percentage of carbon in each structure, we distinguish two types of mineral oils: the naphthenic oils and the paraffinic oils. An infrared measurement permits the type of mineral oil to be defined with respect to percentage [NYN 04]:

$C_P < 50\%$ gives a naphthenic oil,

$C_P \approx 50–56\%$ gives an intermediary oil,

$C_P \geq 56\%$ gives a paraffin oil.

Mineral oils also contain a weak percentage of hydrocarbon molecules containing in their structure other elements such as nitrogen, sulfur and oxygen.

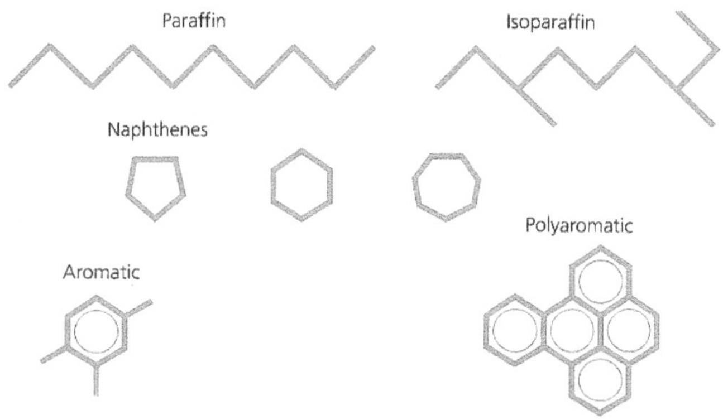

Figure 16.2. *Basic hydrocarbon structure in a mineral oil [NYN 04]*

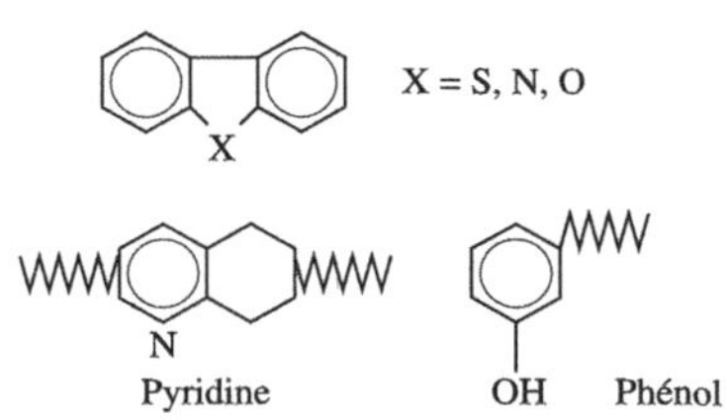

Figure 16.3. *Heteroatoms in mineral oil [NYN 04]*

In most cases, these elements called heteroatoms (Figure 16.3), are related to the aromatic structures and we can find them, for example, in pyridine (C_5H_5N) or phenol (C_6H_5OH).

It is important to note that sulfur plays a natural inhibiting role to oxidation which destroys the peroxides, whereas the synthetic inhibitors which can be added (such as DBPC) destroy the radicals.

16.3.2. *Implementation*

Mineral oils are obtained according to a process which usually includes a distillation followed by a deparaffining operation, an extraction by solvent of undesirable molecules and a catalytic hydrogenation.

Distillation is the process during which a splitting is realized, i.e. the separation of the crude oil into distillates, having different boiling intervals, fitted to the dielectric application. The maximal splitting temperature is about 350°C; above this temperature, a thermal decomposition (cracking) of the oil starts to occur.

The naphtenic crudes practically contain no n-alkanes and do not require deparaffining. However, for paraffinic crudes, the deparaffining is necessary to obtain properties at acceptable low temperature. This process consists of mixing the oil with a solvent, then cooling down the whole. The n-alkanes then crystallize in long needles and the paraffin is taken out by filtering. After filtering, the solvent is removed from the oil by distillation.

The extraction process permits the removal of the undesirable distillate molecules in order to improve a desirable property such as stability for oxidation and gassing. These undesirable constituents include unsaturated hydrocarbons, nitrogen and sulfur components.

The hydrogenation process is realized to complete the solvent extraction. Indeed, this latter does not allow all undesirable components to be removed. Thus, the distillates undergo a hydrogen catalytic processing at high pressure and temperature (30 to 100 bars, 200 to 300°C) [DIM 69], which transforms the aromatic hydrocarbons into saturated hydrocarbons and notably decreases the rate of carcinogenic polyaromatic hydrocarbons (HPA), and the organosulfur and nitrogen in neutral chemical species. However, the strong decrease of the aromatic rate causes a degradation of the oxidation performance and the gassing.

16.3.3. *Characteristics*

The classes and specifications of mineral oils of electrical use are defined in the publication IEC 60296 [STA 03] (Table 16.3). There is also a maintenance guide of mineral oils used in electrical apparatuses (IEC 60422 [STA 05B]).

Mineral oils are the only insulating liquids used in power transformers for their properties (which have never stopped improving since their first use at the beginning of the 20th Century), their availability and their low price. Even if, nowadays, their performances reach satisfactory levels, manufacturers of electrical devices still seek oils having:

– a better stability for oxidation, to prolong the lifespan of transformers and notably that of cellulose insulators;

– a better dielectric performance, to reduce the size of transformers or increase the security coefficients.

Property	Test method	Limits	
		Transformer oil	Low temperature switchgear oil
1 – Function			
Viscosity at 40 °C	ISO 3104	Max. 12 mm^2/s	Max. 3,5 mm^2/s
Viscosity at –30 °C [a]	ISO 3104	Max. 1 800 mm^2/s	--
Viscosity at –40 °C [b]	IEC 61868	–	Max. 400 mm^2/s
Pour point [a]	ISO 3016	Max. –40 °C	Max. –60 °C
Water content	IEC 60814	Max. 30 mg/kg [c]/ 40 mg/kg [d]	
Breakdown voltage	IEC 60156	Min. 30 kV / 70 kV [e]	
Density at 20 °C	ISO 3675 or ISO 12185	Max. 0,895 g/ml	
DDF at 90°C	IEC 60247 or IEC 61620	Max. 0,005	
2 – Refining/stability			
Appearance	–	Clear, free from sediment and suspended matter	
Acidity	IEC 62021-1	Max. 0,01 mg KOH/g	
Interfacial tension	ISO 6295	No general requirement [f]	
Total sulfur content	BS 2000 Part 373 or ISO 14596	No general requirement	
Corrosive sulfur	DIN 51353	Not corrosive	
Antioxidant additive	IEC 60666	(U) uninhibited oil: not detectable (T) trace inhibited oil: max. 0,08 % (I) inhibited oils: 0,08 – 0,40 %	
2-Furfural content	IEC 61198	Max. 0,1 mg/kg	
3 – Performance			
Oxidation stability [1]	IEC 61125 (method C) Test duration: (U) Uninhibited oil: 164 h (T) Trace inhibited oil:332 h (I) Inhibited oil: 500 h		
- Total acidity		Max. 1,2 mg KOH/g[1]	
- Sludge		Max. 0,8 % [1]	
DDF at 90 °C	IEC 60247	Max. 0,500 [1]	
Gassing	IEC 60628, A	No general requirement	
4 – Health, safety and environment (HSE)			
Flash point	ISO 2719	Min. 135 °C	Min. 100 °C
PCA content	BS 2000 Part 346	max. 3 %	
PCB content	IEC 61619	Not detectable	

[a] This is the standard LCSET for an transformer oil (see 5.1) and can be modified depending on the climatic condition of each country. Pour point should be minimum 10 K below LCSET.

[b] Standard LCSET for low temperature switch gear oil.

[c] For bulk supply.

[d] For delivery in drums and IBC.

[e] After laboratory treatment (see 6.4).

[f] Where it is used as a general requirement, a limit of minimum 40 mN/m is recommended.

Table 16.3. *Specifications of mineral oils according to the standard IEC 60296 [STA 03]*

One of the advantages of mineral oils and, more particularly, naphthenic mineral oils, lies in the fact that they have a weak viscosity compared with other insulating oils for transformers (Figure 16.4). This allows a good evacuation of the heat and a correct cool start as well as a good impregnation of cellulose insulators.

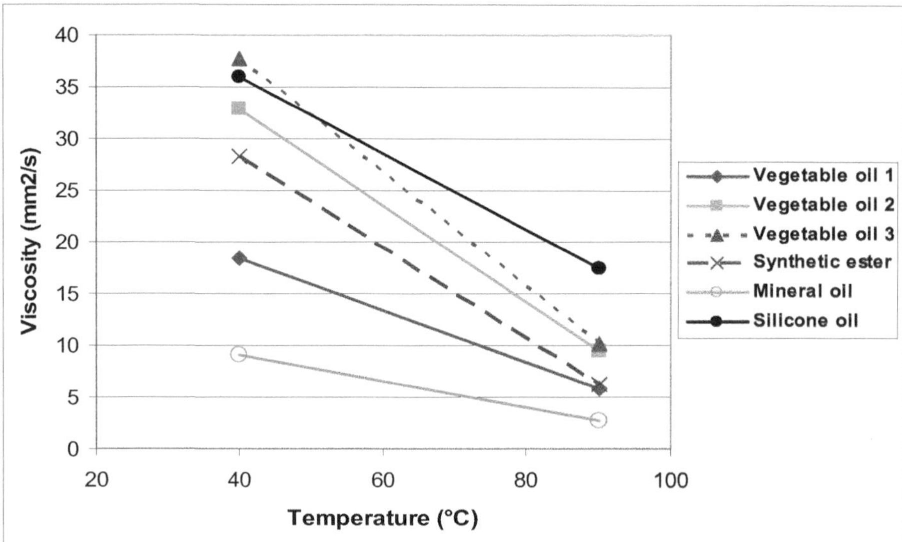

Figure 16.4. *Viscosity of different liquid dielectrics as a function of temperature [DAR 07]*

One of the disadvantages of mineral oils is their flammability. These oils indeed possess a relatively low flash point (between 140 and 150°C), which poses a problem for apparatus installed near populations, such as traction or distribution transformers. This problem is less important for big power transformers which are generally installed far from the population. However, it still remains serious because of the fact that these latter can cause important damages to the factories they supply (nuclear power stations and other industries).

The principal problems regarding toxicity are related to the fact that certain polyaromatic molecules contained in mineral oils are listed as carcinogenic. Regarding ecotoxicity, mineral oils can be harmful to the environment when there are leaks at the transformer level, where they are potentially accumulable (through bioaccumulation, with possible contamination of the supply chain).

16.4. Synthetic esters or pentaerythritol ester

16.4.1. *Composition and implementation*

Ester oils for transformers are known as pentaerythritol ester or organic esters (as opposed to natural esters). They were developed following the banishment of PCB for the impregnation of fire resistant transformers. These esters are also known as tetraesters because they are obtained from a tetraalcohol (the pentaerythritol) and a mixture of monocarboxylic acids containing from 7 to 9 carbons (Figure 16.5). The esterification is brought to 100°C and followed by neutralization and distillation operations and a processing on activated ground.

pentaerythritol + monocarboxilyc acids ⇒ pentaerythritol ester + water

(tetra-alcohol) *(tetraester)*

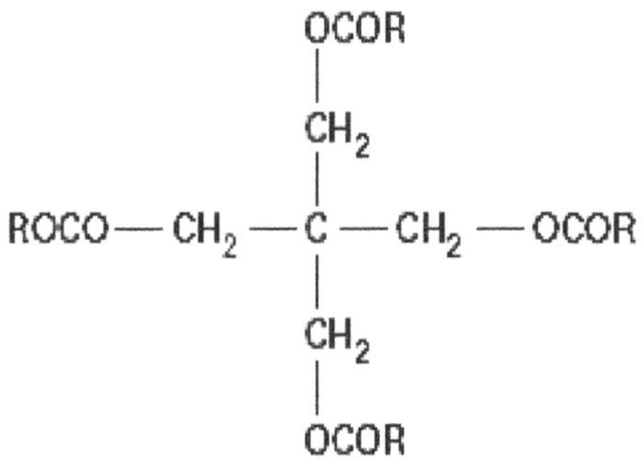

Figure 16.5. *Structure of the pentaerythritol ester, also called tetraester*

16.4.2. *Characteristics*

The specifications of organic ester oils with electrical use are defined in IEC 61099 [STA 92A] (Table 16.4). There is also a maintenance guide of synthetic esters for transformers (IEC 61203 [STA 92B]).

One of the principal assets of this type of oil is the high concentration of water it can contain compared with other insulating oils. Indeed, they can contain up to 2,000 ppm of water at 20°C against 55 ppm for mineral oils (Figure 16.6, at $1/T \approx 3.4.10^{-4}$). For this reason, humidity becomes a factor less limiting for dielectric properties. Numerous studies were made of the pentaerythritol esters, and notably on Midel 7131 oil [FOF 01] [BOR 91] [GOC 02] [DUM 95] [FOF 02A] [FOF 02B], principally stressing its high ability to absorb water (high solubility of water) and notably the assets which we can benefit from by mixing with mineral oil at the level of electrical properties.

Property	Test method (Subclause)	Permissible values
Physical		
- Colour	9.1	Max. 200
- Appearance	9.2	Clear, free from suspended matter and sediment
- Density at 20 °C (kg/dm³)	9.3	Max. 1,000
- Kinematic viscosity at 40 °C (mm²/s) at –20 °C (mm²/s)	9.4	Max. 35,00 Max. 3 000
- Flash-point (°C)	9.5	Min. 250
- Fire-point (°C)	9.6	Higher than 300
- Refractive index at 20 °C	9.7	(1)
- Pour-point (°C)	9.8	Max. –45
- Crystallization	9.9	No crystals
Chemical		
- Water content (mg/kg)	9.10	Max. 200 (2)
- Neutralization value (mg KOH/g)	9.11	Max. 0,03
- Oxidation stability	9.12	
Total acidity (mg KOH/g) Total sludge (% mass)		Max. 0,3 Max. 0,01
Electrical		
- Breakdown voltage (kV)	9.13	Min. 45 (2)
- Dielectric dissipation factor, tan δ at 90 °C and 50 Hz	9.14	Max. 0,03 (2)(3)
- D.C. resistivity at 90 °C (GΩ·m)	9.14	Min. 2

Table 16.4. *Specifications of synthetic ester oils according to standard CEI 61099 [STA 92A]*

The Midel 7131 ester oil allows good breakdown properties to be kept despite a significant water content (Figure 16.7) by comparison with other oils, and notably with mineral oil. It also allows the degradation of paper at the electrical insulation level to be reduced.

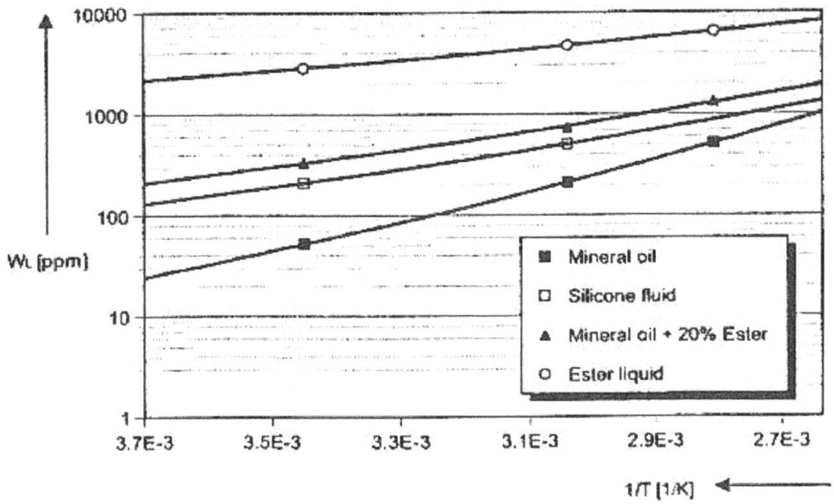

Figure 16.6. *Saturation limit in water (W_L) as a function of temperature for different insulating liquids [FOF 01]*

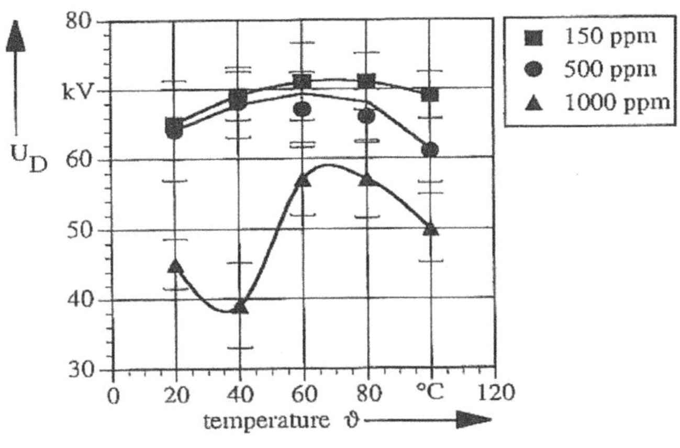

Figure 16.7. *Breakdown voltage of the Midel 7131 ester oil as a function of temperature for different water contents (ppm) [DUM 95]*

By studying the variation Δ of the relative dissipation factor, which is in fact the ratio of tan δ of aged paper on tan δ of non-aged paper, it was shown [FOF 02B] that the electrical insulation of paper was less altered in the presence of ester and notably in mixture with mineral oil (Figure 16.8).

The ester oils also allowed the degradation of paper at the mechanical level to be reduced. This is related to the fact that highly soluble oils, such as esters, absorb the humidity of the paper (Figure 16.9), thus reducing its alteration (Figure 16.10).

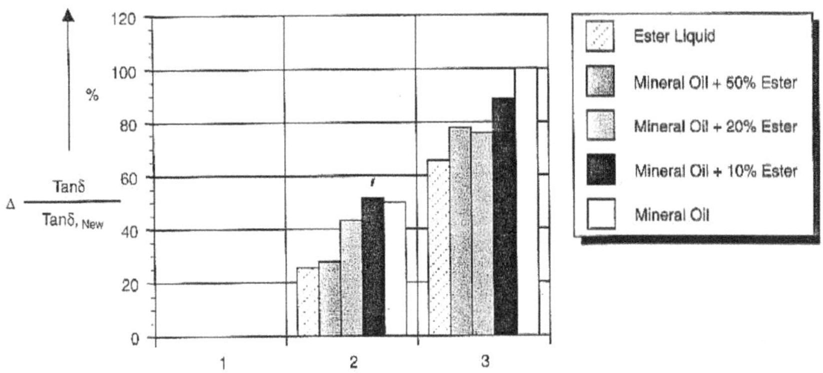

Figure 16.8. *Evolution of the relative dissipation factor of a cellulose paper impregnated with different insulating liquids. 1: dried and impregnated paper (non-aged), 2: impregnated paper (non-aged), 3: dried and impregnated paper (aged) [FOF 02B]*

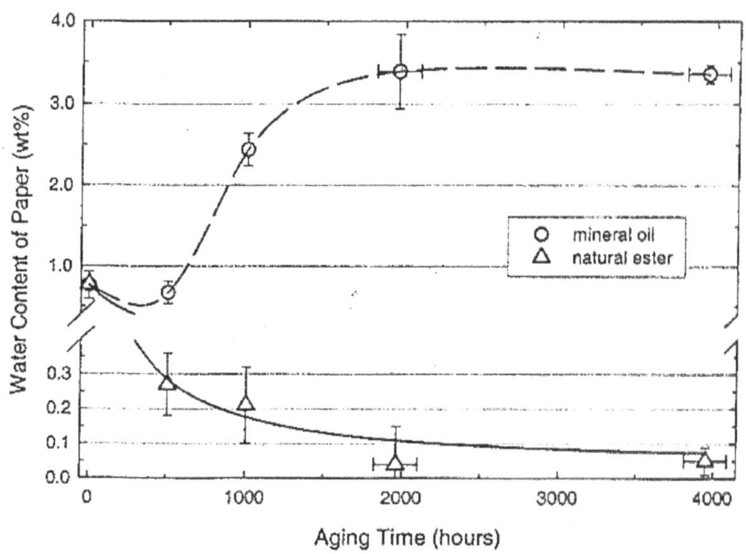

Figure 16.9. *Evolution of the water content as a function of time, for Kraft paper aged in an ester oil and a mineral oil [MCS 02]*

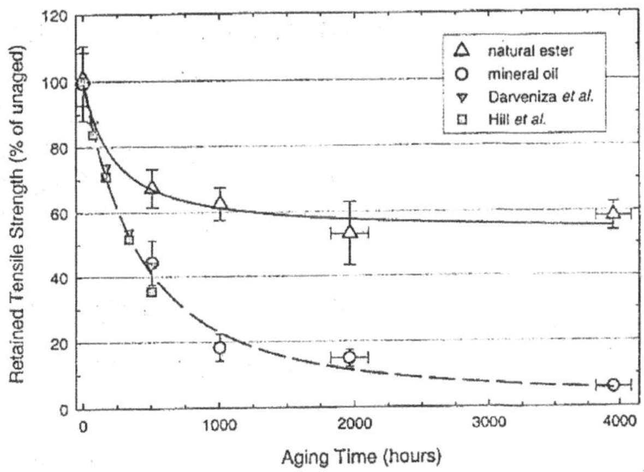

Figure 16.10. *Evolution of the mechanical traction performance of paper, as a function of time, for Kraft paper aged in an ester oil and a mineral oil [MCS 02]*

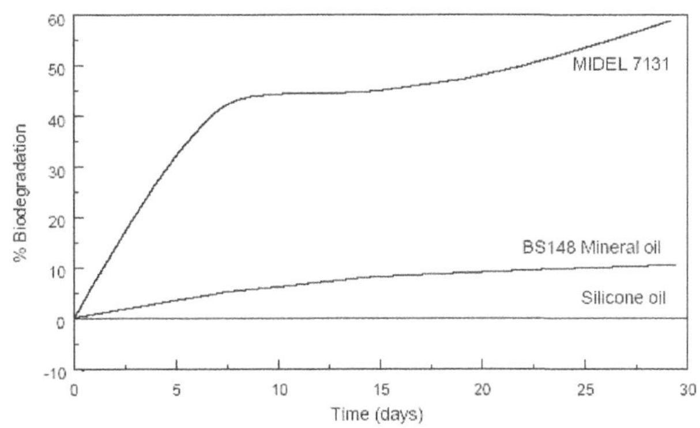

Figure 16.11. *Biodegradability test (test OECD 301D) on a mineral oil, a silicone oil and a synthetic ester oil (7131 Midel) [MID XX]*

Another asset of organic ester oil is its high resistance to fire with a fire temperature higher than 300°C. Its final asset is its biodegradability compared with mineral or silicone oil (Figure 16.11).

Its weak points are situated at the viscosity level which is a bit high and becomes very significant at low temperatures. This point constitutes one of the harmful effects which ester oil can have by mixing with mineral oil, as indicated in Figure 16.12.

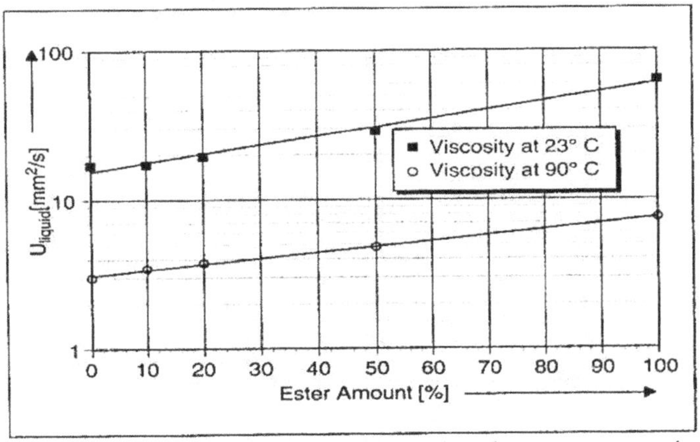

Figure 16.12. *Viscosity evolution of an ester/mineral oil mixture according to the percentage of ester, at different temperatures [FOF 02A]*

The other weak point concerns their instability for gassing. Indeed, a study made by DUMKE *et al.* [DUM 96] revealed their bad stability with a strong production of gas in comparison with other types of oil (Figure 16.13).

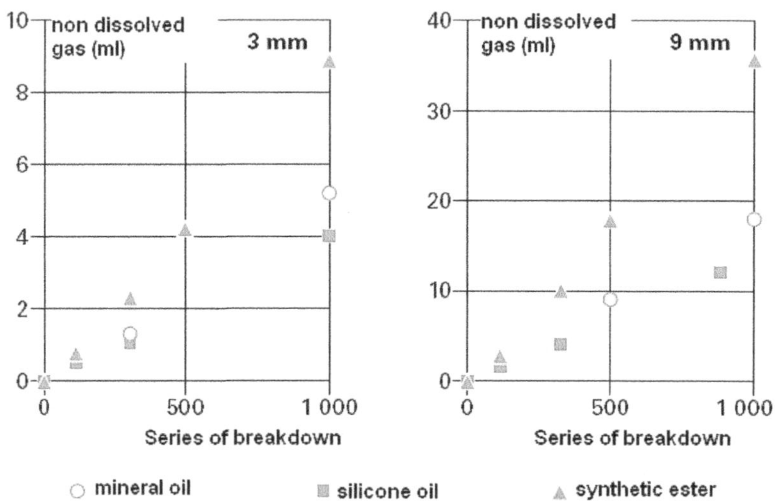

Figure 16.13. *Volume of gas produced during successive breakdowns for two inter-electrode distances [DUM 96]*

16.4.3. *Application*

Tetraesters are used for the filling of "fire resistant" distribution transformers. Their high fire point (> 300°C) with respect to mineral oils constitutes the primordial characteristic of these products. We thus find them in devices near populations, in buildings, tunnels, etc. However, their use being relatively new (only about 15 years), there is not yet enough hindsight to have a complete idea of their long-term behavior, knowing that the normal lifespan of a transformer is at least 25 years.

Principally because of their higher cost (about 4 to 8 times more expensive than mineral oils), tetraesters are not well developed in power transformers. The most famous synthetic ester, destined for transformers and available on the market, is *Midel 7131*, manufactured by M&I Materials.

16.5. Silicone oils or PDMS

16.5.1. *Composition and implementation*

Silicone oils, like synthetic esters, were proposed as replacement liquids for PCB in transformers. The most commonly used product is the polydimethylsiloxane or PDMS. It is obtained from dimethylchlorosilane, which hydrolyse in silanol, an unstable component giving (by polycondensation) the chains:

O – Si – O – Si – O – … .

The polymerization reaction is followed by a filtering and a passage in a vacuum-packed devolatilization column. The general formula is the following (Figure 16.14).

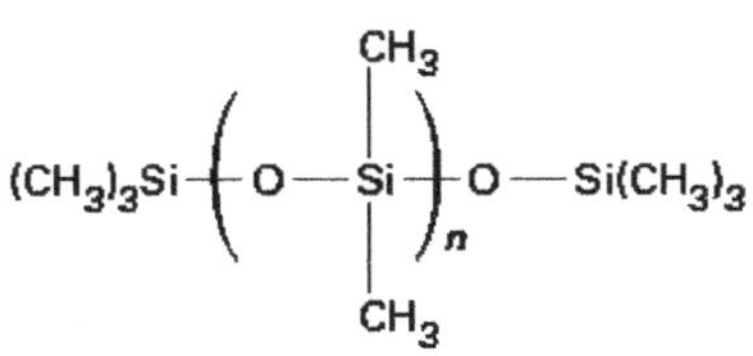

Figure 16.14. *Structure of silicone oils or PDMS*

PDMS corresponds to chains where *n* ranges from 40 to 50 in order to obtain an acceptable viscosity and a good resistance to fire.

16.5.2. *Characteristics*

The specifications of silicone oils with electrical use are defined in the standard IEC 60836 [STA 05A] (Table 16.5). There is also a maintenance guide (IEC 60944 [STA 88]).

Property	Test method (clause or subclause)	Permissible values	Notes
Colour	8.1.1	Max.35	
Appearance	8.1.2	Clear, free from suspended matter and sediment	
Density at 20 °C (kg/dm^3)	8.2	0,955 to 0,970	
Kinematic viscosity at 40 °C (mm^2/s)	8.3	40 ± 4	
Flash point (°C)	8.4	Min. 240	
Fire point (°C)	8.5	Min. 340	
Refractive index at 20 °C	8.6	1,404 ± 0,002	
Pour point (°C)	8.7	Minus 50 or below	
Water content (mg/kg)	8.8	Max. 50	1
Neutralization value (mg KOH/g)	8.9	Max. 0,01	1
Breakdown voltage (kV)	8.10	Min. 40	1
Dielectric dissipation factor (DDF) at 90 °C and 50 Hz	8.11	Max. 0,001	1,2
Permittivity at 90 °C	8.11	2,55 ± 0,05	1
DC resistivity at 90 °C (GΩ × m)	8.11	Min. 100	1
NOTE 1 For untreated liquid, as received.			
NOTE 2 For frequencies *f* (Hz) in the range of 40 Hz to 60 Hz, convert values as follows:			
DDF(50 Hz) = DDF *f* (Hz)/ 50.			

Table 16.5. *Specifications of silicone oils or PDMS according to the standard IEC 60836 [STA 05A]*

Owing to their very weak volatility, their chemical stability and their resistance to oxidation, silicone oils are capable of conserving their insulation properties during extended periods at high temperature, unlike mineral oils [BUR 74]. Below 150°C and in the presence of air, the oxidation of these oils is negligible [WIL 80] [CAN 81]. A study realized by Dow Corning shows that the dissipation factor of silicone oils stays relatively stable during ageing tests (Figure 16.15).

The other strong assets which characterize PDMS are its weak viscosity at low temperatures, its high viscosity index (weak viscosity variation with temperature), and its good performance with fire (high fire point).

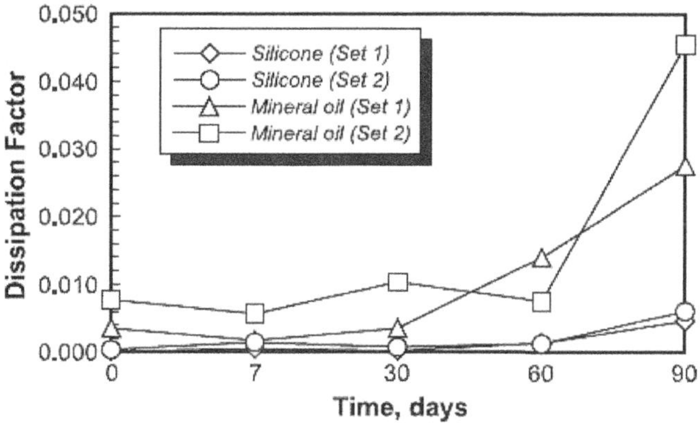

Figure 16.15. *Comparison of the dissipation factor evolution (tan δ) for silicone and mineral oils as a function of time, during an ageing at temperature [GOU 97]*

In a recent study, tetraesters, silicone oils and mineral oils were compared with regard to their aptitude to produce gas under the effect of breakdowns or partial discharges [DUM 96]. It appears that tetraesters produce more gas than other oils, while silicone oils produce less gas (Figure 16.13).

On the other hand, these oils present two disadvantages:

– having a good viscosity index, silicone oils certainly present a weak viscosity at low temperatures, but a high viscosity at positive temperatures in comparison with other types of oil (Figure 16.16), and more particularly at the functioning temperatures of a transformer (90°C);

– having a very high stability for oxidation, silicone oils are not very biodegradable (Figure 16.11).

16.5.2.1. *Use*

PDMS is essentially used for the filling of distribution and traction transformers, where a certain resistance to fire is sought. Their use is less widespread than those of organic esters because of their very high cost (8 times more expensive than mineral oil) and the difficulty to remove them after use because they are not at all biodegradable. Nevertheless, they are very used for traction transformers (embedded in trains) because they have a good thermal stability by supporting temperatures from 125 to 150°C.

The most famous silicone oils are *Rhodorsil 604V50* manufactured by the Rhodia company, and *561* from Dow Corning.

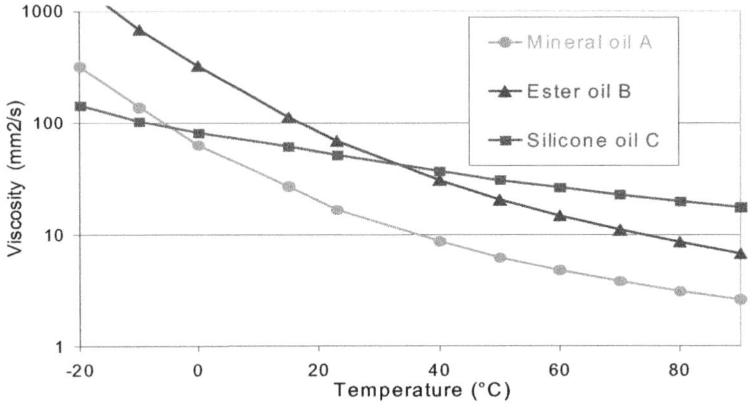

Figure 16.16. *Evolution of viscosity as a function of temperature for different oil types [PER 04]*

16.6. Halogenated hydrocarbons or PCB

16.6.1. *Composition and implementation*

In the past, the use of chlorinated products (halogenated) by the electrical engineering industry was justified by the fact that these latter presented the advantage of not giving off flammable or explosive gases under the effect of temperature, partial discharges or during an electrical breakdown. The principle then amounts to replace part of the hydrogen atoms of the molecule by chlorine atoms which, under ionization or electric arc, form HCl molecules instead of gaseous hydrogen (Figure 16.17). Polychlorobiphenyls (PCB), still known as askarels (in Greek: fire resistant) or in France as pyralene, were widely used until their use was limited, then forbidden in 1985 because of their persistence in the environment and their toxicity (notably by emission of dioxins when they burn).

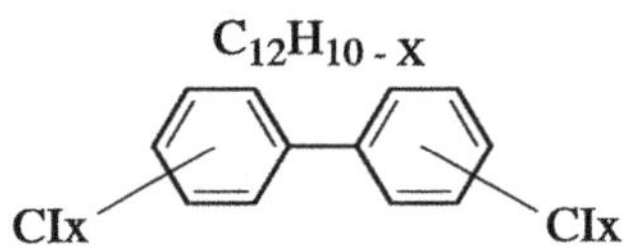

Figure 16.17. *General structure of PCB*

The big problem lies in the fact that numerous pieces of equipment are still filled with this liquid or contain traces of it. By 2010, all apparatus containing over 500 ppm of PCB should be removed. Apparatus containing between 50 ppm and 500 ppm will be declared slightly polluted in the European Union and will be removed at the end of their normal use. Finally, those containing less than 50 ppm will be declared un-contaminated.

16.6.2. *Characteristics*

The specifications of PCB are given in the IEC 60588-3 standard specification [STA 77], while the IEC 60588-5 [STA 79A] and 60588-6 [STA 79B] standard specifications describe the compatibility test with solid materials.

The volumic mass of PCB varies enormously with chlorine content and can range from 1,200 kg/m^3 (dichlorinated) to 1,600 kg/m^3 (hexachlorinated). Under partial discharges and electric arcs, PCB liberates chlorhydric acid, which can attack the cellulose. They must then contain acid fixative additives. The dielectric strength of PCB, at industrial frequency or lightning impacts, is less than that of mineral oils [VUA 98B].

16.6.3. *Retro-filling*

Considering the costs of transformers and to solve the banning of PCB, users of materials immersed in these latter wished to keep their transformers and replace the PCB by biodegradable liquids, at weak flammability and non-toxic risk. For this purpose, they realized a retro-filling operation which, however, poses a certain number of problems [0SB 84]:

– the efficiency of the retro-filling does not permit the limit value of 50 ppm to be reached in one go (notably because of residual liquid which remains in the cellulose);

– the replacement liquids do not present the same listing regarding fire, and apparatus must be delisted;

– the replacement liquids have lower electrical qualities and the apparatus cannot supply the same power.

16.7. Natural esters or vegetable oils

16.7.1. *Composition and implementation*

Vegetable oils were part of the first liquid insulators used for the manufacture of electrical apparatus [CLA 62]. They are essentially composed of triglycerides, which is in fact a triester (the formula of an ester is R-COO-R’), as represented in Figure

16.18. Vegetable oils are also known as natural or vegetal esters (as opposed to synthetic or organic esters) because they are naturally synthesized by any living organism and can stem from different seeds such as the castor oil plant, colza, sunflower, etc.

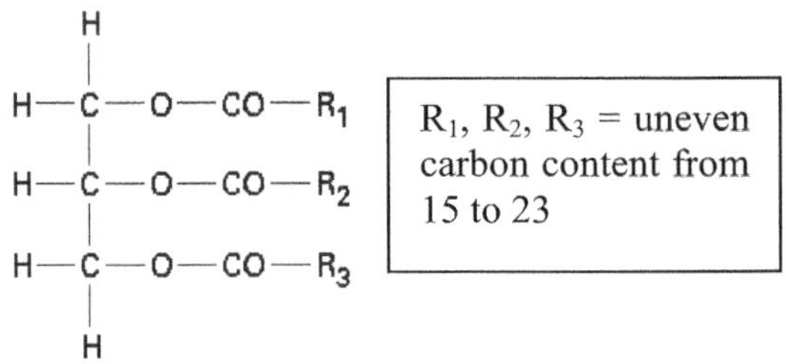

Figure 16.18. *Chemical structure of a vegetable oil, also called triester (triglycerides)*

Vegetable oils are naturally synthesized by all living organisms and are obtained by esterification of a simple tri-alcohol, the glycerol, with three fatty acids. These acids are monocarboxylics (of formula R-COO-H), with unplugged linear chains containing an even number of carbon atoms (between 8 and 24). They can be saturated (without double bonding) or unsaturated and sometimes hydroxyled. The esterification reaction of a vegetable oil is the following:

Glycerol	+	3 fatty acids	$\Rightarrow$	Triglycerides	+	water
(trialcohol)		*(acid)*		*(triester)*		

The tryglycerides are obtained by trituration (grinding and pressure) of seeds (soya, corn, colza, sunflower, castor oil plant, etc.).

16.7.2. *Characteristics*

There are no specifications defined by the IEC today for vegetable oils, and it is therefore difficult to use this type of oil at the industrial level in transformers. Nevertheless, Technical Committee 10 (insulating fluids) from the IEC is actively working on the topic.

One of the big assets of these oils is their excellent biodegradability, which unfortunately confers a high sensitivity for oxidation on them [OOM 02] [DAR 07] [SAN 87] [MCS 02] [BER 04] with a higher increase in the dielectric dissipation factor, acidity and viscosity.

For this reason, their use is limited to sealed electrical equipment (i.e. without any communication with the atmosphere), where this sensitivity is compensated by antioxidants whose environmental properties can be harmful ("non-green" products).

Their other assets are that they are not very flammable and have a high water solubility (like synthetic esters), compared with mineral oils.

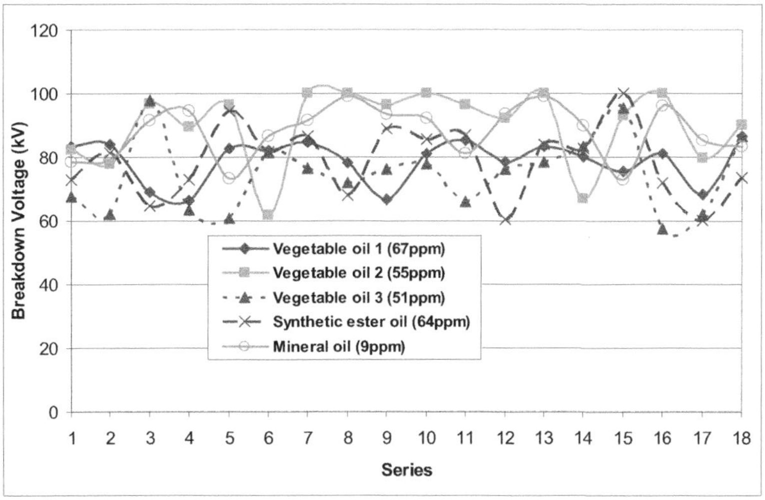

Figure 16.19. *Breakdown voltage of different insulating liquids in accordance with IEC 60156 test (60 Hz and 2.5 mm gap) [DAR 07]*

Apart from these two advantages, this type of oil presents a higher viscosity than mineral oils (Figure 16.4) and notably a fairly high pour point (in the neighborhood of –20°C), which limits heat transfer and restrains their use in countries where the climate is not too cold.

Oils presenting a high percentage of unsaturated fatty acids present lower viscosities and lower pour points, whereas oils in which saturated fatty acids are in the majority turn out more viscous and are known for having a better resistance to oxidation [BER 04].

Finally, new vegetable oils have electrical properties near those of mineral oils for small gaps, in accordance with IEC 60156 [STA 95] test (Figure 16.19); the gap indicated by IEC 60156 specifications is equal to 2.5 mm. Several studies are running to define vegetable oil's behavior for larger gaps (of tens of mm). Their resistivity is lower (and, inversely, their dissipation factor is higher).

16.7.3. *Use*

As indicated in Table 16.2, this type of oil is mainly used in small transformers. However, and notably with the recent interest in biodegradable products, numerous power transformers were made or retrofilled in order to replace mineral oils by vegetable oils for impregnation [DAR 07] [STA 07]. We can thus find *Envirotemp FR3* marketed by the Cooper Power Systems company, *Midel eN* manufactured by M&I, or *Biotemp* oil marketed by ABB.

16.8. Security of employment of insulating oils

The security of employment of insulating liquid concerns the risks of fire and explosion which they could be the cause of, as well as the influence they can have on the health of individuals or on the environment. The fire hazard lies in the flammability characteristics (flash point, fire point, auto-inflammation temperature) and certain combustion characteristics (oxygen index, emitted heat quantity) [FOF 01] [MIL 78].

The explosion hazard is related to the nature of gas produced by decomposition of liquids under electric discharges and hot spots. Hydrogen, methane and acetylene are the principal gases concerned.

The dangers concerning health and the environment, respectively related to toxicological and eco-toxicological characteristics, are subject to public organization research, such as that undertaken by the National Institute of Scientific Research, and are published in the form of biochemical studies, toxicity profiles, etc. [EEC 92].

16.8.1. *Characteristics related to fire*

In certain applications, where the risk in case of a fire is considered unacceptable, we use fire resistant liquids. The fire hazard is estimated from flammability characteristics such as flash point, fire point, and auto-inflammation temperature as well as certain combustion characteristics such as the oxygen index or amount of emitted heat.

PCBs (polychlorobiphenyls) were considered non-flammable, i.e. they could not be burned without an energy supply from an external fire. Their banishment and the banning of the use of chlorinated products no longer allows such flammable liquids to be used. We currently use less flammable liquids which are defined as having a fire point greater than 300°C.

16.8.1.1. *Flash point, fire point and auto-inflammation*

The gradual heating of a liquid causes the emission of vapors. Once the concentration of these vapors in the atmosphere surmounting the liquid becomes sufficient to form a flammable mixture, they burn near a flame. The corresponding temperature is called the flash point of the liquid.

There are two types of methods to determine the flash point:

– the open-cut method, such as the Cleveland method (ISO 2592 [STA 00A]);

– the closed-cut method, such as the Pensky–Martens method (ISO 2719 [STA 02B]).

For the open-cut method, as indicated by its name, the liquid is in contact with free air and the vapors can freely mix with the surrounding air.

In the other method, the volume above the cupel is closed and the vapors remain in the neighborhood of the liquid surface. Closed-cut methods generally give weaker values than open-cut methods.

If we carry on heating, experiments show that a permanent combustion settles in the presence of a flame, from a certain temperature corresponding to the fire point of the liquid. Non-flammable liquids such as PCB do not have a fire point.

We generally consider that, for the inflammation hazard to become negligible, the temperature must remain 25 to 30°C cooler than the flash point [BER 02].

The auto-inflammation temperature of a liquid is the minimal temperature at which an instantaneous combustion occurs. This temperature is much higher than that of the flash point and therefore is of less interest. It can be determined by the IEC 60079-4 method [STA 75], which consists of injecting a weak volume of liquid in an enclosure containing air and determining the temperature at which a flame appears in a fixed period of time.

The auto-inflammation temperature corresponds to the behavior of a liquid in a fire, while the fire point characterizes thermal limits of service.

16.8.1.2. *Combustion characteristics*

Since the banning of PCBs, there are no more non-flammable liquids (without fire point) for sale. These products were replaced by liquids less flammable than mineral oils, with a fire point greater than 300°C.

However, they still burn just as much and their behavior in a fire poses a problem. To address this issue, combustion characteristics such as the oxygen index and the lower calorific power (LCP) are studied.

The oxygen index corresponds to the minimal oxygen content in an air/nitrogen mixture, which allows the combustion to be maintained in defined conditions (ASTM D 2863 [STA 00B]).

The LCP is the net heat quantity liberated by the total liquid combustion; it is determined by the ASTM D 240 [STA 02A] method.

16.8.2. *Toxicology and ecotoxicology*

16.8.2.1. *The toxicological properties*

These properties concern all tests relative to the harmful effects which an oil can have on mankind. These tests are realized in a laboratory on animals (rats, rabbits, guinea-pigs).

We find [EEC 92]:

– the acute toxicity estimated from the DL_{50} (lethal dose) which is the dose or concentration which causes the death of 50% of individuals in 24 to 48 hours;

– the acute irritating aspect for skin or eyes;

– the sensitization power, which permits the allergic reactions that the product can cause to be estimated;

– the genotoxicity which indicates whether the liquid can alter genetic material, but also indicates the carcinogenic potential of the substance;

– the subacute toxicity, which is determined by administering substance doses for 28 days. The highest concentration which does not lead to any effect is defined as the dose without effect.

16.8.2.2. *The ecotoxicological properties*

These properties concern all tests relative to the harmful effects which an oil can have on the environment. We find [EEC 92]:

– the acute toxicity, is determined from tests on fish, daphnia and seaweed. According to the species, we establish respectively the CL_{50} (lethal atmospheric concentration for 50% of individuals) to know if a product is very toxic, the CE_{50} (atmospheric concentration having an effect on 50% of individuals) to define a product as toxic, and the Cl_{50} to know if the product is more or less harmful.

– for substances not very soluble in water (such as insulating oils), the tests are realized at the limit of solubility.

– the biodegradability permits the potential persistence of the product in the environment to be estimated. The normalized tests were originally developed for products very soluble in water.

– these tests then allow “easily biodegradable” products to be distinguished from “not easily biodegradable” products.

– other tests were developed, notably for not very soluble products like insulating oils. They then allow “potentially biodegradable” products to be distinguished from “non-biodegradable” products.

– the accumulative potential permits the risk of accumulation in the food chain to be estimated. The water-sharing coefficient (W) / octanol (O), called lg P_{ow}, indicates the solubility difference of the substance in water and fat. A product is considered potentially accumulable if $3 < \lg P_{ow} < 6$.

– the bioconcentration factor (BCF) permits the accumulative potential of a product to be estimated by tests, most often on fish. The products having a BCF > 100 are generally recognized as potentially accumulable.

– the long-term toxicity is established on fish or daphnia by 21–day tests. Almost insoluble products are tested at the limit of their solubility.

16.9. Conclusion and perspectives

It emerges from the observations made in this chapter that ester oils distinguish themselves with their strong ability to absorb humidity (high water solubility). The difference between synthetic esters and natural esters lies at the level of their stability for oxidation and their pour point. The natural esters, which are easily biodegradable, have a weak performance to oxidation and get fixed at much higher temperatures. Silicone oils have a very good stability for oxidation but are not at all biodegradable.

Mineral oil, which has been used for over a century, still remains the reference insulating liquid in the transformer industry for:

– its good physico-chemical characteristics;

– its good compatibility with cellulose insulators;

– its low viscosity;

– its relative low cost.

Nevertheless, numerous recent problems, related to the corrosivity of mineral oils, challenge its reign and little by little drives transformers manufacturers to choose esters, which in addition to their good properties, do not have this problem and are less harmful for the environment and to humans.

Growing interest in vegetable oil-based dielectric fluids is also due to (1) their excellent fire safety characteristics which ensure better safety in operation, handling, storage and transportation and thus the operational safety of transformers using such liquids; and (2) the alarming predictions concerning the shortage of petroleum oils by the middle of this century; one can expect a serious crisis for petroleum oils and very significant and rapid increases in their price.

16.10. Bibliography

[ABD 68] ABDY C.A., "The early years of the power transformer", *Electronics and Power*, vol. 14, no. 8, p. 335–338, August 1968.

[BER 97] BERGER N., RANDOUX M., OTTMANN G., VUARCHEX P., "Revue des isolants liquides", *Electra (f)*, no. 171, p. 32–57, 2002.

[BER 02] BERGER N., "Liquides isolants en électrotechnique – Présentation générale", *Techniques de l'ingénieur*, Techniques de l'ingénieur, Génie Electrique, ISSN 0992-5449, vol. D3, no. D2470, pp. 2470.1–2470.40, 2002.

[BER 04] BERTRAND Y., HOANG L.C., "Vegetable oils as substitute for mineral insulating oils in medium-voltage equipments", *CIGRE, Paper D1-202*, 2004.

[BOR 91] BORSI H., "Dielectric behavior of silicone and ester fluids for use in distribution transformers", *IEEE Transactions on Electrical Insulation*, vol. 26, no. 4, p. 755–762, August 1991.

[BUR 74] BUROW R.F., VINCENT G.A., *IEEE PES Winter Meeting*, New York, N.Y., January 27–February 1, 1974.

[CAN 81] CANDILLON C., BOSS P., "Utilisation des fluides silicones dans les transformateurs de distribution, en vue du remplacement des askarels", *Bull. ASE / UCS*, p. 142–146, 7 February 1981.

[CLA 62] CLARK F. M., *Insulating materials for design and engineering practice*, John Wiley & Sons, 1962.

[DAR 07] DARWIN A., PERRIER C., FOLIOT P., "The use of natural ester fluids in transformers", *3rd European Conference on HV and MV Substation Equipment: Evolution, Trends and Perspectives*, Lyon, France, Paper 0036, 15–16 November, 2007.

[DIM 69] DIMELER G. R., MILLS J. W., MELCHIOR J. J., "The scope of hydrogenation as a refining tool for manufacture of transformer oils", *IEEE Trans. Elect. Insul.*, EI-4, no. 1, p. 7–12, March 1969.

[DUM 95] DUMKE K., BORSI H., GOCKENBACH E., "A synthetic insulating liquid for applications in transformers", *9th ISH, Graz*, subject 1, p. 1025, August–September 1995.

[DUM 96] DUMKE K., BORSI H., GOCKENBACH E., "Experimental investigation on the behaviour of decomposition gases in insulating liquids caused by partial discharge and breakdown", Conference Record of the ICDL'96, *12th Intern. Conf. Conduction and Breakdown in Dielectric Liquids*, Rome, Italy, p. 300–303, July 15–19, 1996.

[EEC 92] "Directive 92/69/CEE de la Commission, du 31 juillet 1992, portant dix-septième adaptation au progrès technique de la directive 67/548/CEE du Conseil concernant le rapprochement des dispositions législatives, réglementaires et administratives relatives à la classification, l'emballage et l'étiquetage des substances dangereuses", *Journal officiel no. L 383A*, p. 235, 29 December 1992.

[FOF 01] FOFANA I., BORSI H., GOCKENBACH E., "Fundamental investigations on some transformer under liquids under various outdoor conditions", *IEEE Transactions on Dielectric and Electrical Insulation*, vol. 8, no. 6, p. 1040–1047, December 2001.

[FOF 02A] FOFANA I., BORSI H., GOCKENBACH E., WASSERBERG V., "Challenge of mixed insulating liquids for use in high-voltage transformers, Part 1: Investigation of mixed liquids", *IEEE Electrical Insulation Magazine*, vol. 18, no. 3, p. 18–31, May–June 2002.

[FOF 02B] FOFANA I., BORSI H., GOCKENBACH E., WASSERBERG V., "Challenge of mixed insulating liquids for use in high-voltage transformers, Part 2: Investigation of mixed liquids impregnated paper insulation", *IEEE Electrical Insulation Magazine*, vol. 18, no. 4, p. 5–16, July–August 2002.

[GOC 02] GOCKENBACH E., BORSI H., "Performance and new application of ester liquids", *Proceedings of 14th ICDL, Graz* (Austria), p. 203–206, 7–12 July 2002.

[GOU 97] GOUDIE J.L., "Silicones materials in new high temperature liquid transformer designs", *Electrical Insulation Conference and Electrical Manufacturing and Coil Winding*, vol. 23, p. 455–458, September 1997

[MCS 02] MCSHANE C.P., RAPP K.J., GAUGER G.A., LUKSICH J., CORKRAN J.L., "Aging of Kraft paper in natural ester dielectric fluid", *IEEE Proceedings of 14th International Conference on Dielectric Liquids (ICDL)*, Graz (Austria), p. 173–177, 7–12 July 2002.

[MID XX] http://www.midel.com/English/PDFFiles/EnvironmentalBehaviour.pdf

[MIL 78] MILLER D.B., "Tests and standards to evaluate the fire safety of electrical insulating fluids", *IEEE Transactions on Electrical Insulation, EI-B*, no.5, p. 378–381, October 1978.

[NYN 04] *Transformer oil handbook*, Copyright Nynas Naphtenics AB, Sweden, 2004.

[OOM 02] OOMMEN T.V., "Vegetable oils for liquid-filled transformers", *IEEE Magazine*, vol. 18, no. 1, p. 6–11, January–February 2002.

[0SB 84] OSBORN S. W., "PCB removal from transformers", *EPRI Report EL-3345, Project 2028–2 final report*, May 1984.

[PER 87] PERRET J., PARIS M., "Les huiles silicones pour transformateurs", *E.D.F Bulletin des etudes et recherché, Série B, Réseaux électriques, matériels électriques*, no. 2, 1987, p. 5–13, 1987.

[PER 04] PERRIER C., BEROUAL A., BESSEDE J-L., "Experimental investigations on different insulating liquids and mixtures for power transformers", *IEEE International Symposium on Electrical Insulation (ISEI 2004)*, paper 11–3, Indianapolis, Indiana, USA, 19-22 September (2004).

[ROU 98] ROUSE T.O., "Mineral insulating oil in transformers", *IEEE Electrical Insulation Magazine*, vol. 14, no. 3, p. 6–16, May–June 1998.

[SAN 87] SANKARALINGAM S., KRISHNASWAMY K.R., "New dielectric liquids from vegetable origin – A feasibility study on indian rape seed oil", *CIGRE Symposium*, Section 5, paper 500–01, Vienna, 1987.

[STA 75] Standard IEC 60079-4 (Ed. 2), *Matériel électrique pour atmosphère explosives – Partie 4: Méthode d'essai pour la détermination de la température d'inflammation*, January 1975.

[STA 77] Standard IEC 60588-3 (Ed. 1), *Askarels pour transformateurs et condensateurs – Partie 3: Spécifications pour askarels neufs*, January 1977.

[STA 79A] Standard IEC 60588–5 (Ed. 1), *Askarels pour transformateurs et condensateurs – Partie 5: Essai éliminatoire pour déterminer la compatibilité des matériaux avec les askarels pour transformateurs*, January 1979.

[STA 79B] Standard IEC 60588-6 (Ed. 1), *Askarels pour transformateurs et condensateurs – Partie 5: Essai éliminatoire pour déterminer les effets des matériaux sur les askarels*, January 1979.

[STA 88] Standard IEC 60944 (Ed. 1), *Guide de maintenance des liquides silicones pour transformateurs*, September 1988.

[STA 92A] Standard IEC 61099 (Ed. 1), *Spécifications pour esters organiques de synthèse neufs à usages électriques*, May 1992.

[STA 92B] Standard IEC 61203 (Ed. 1), *Esters organiques de synthèse à usages électriques – Guide de maintenance des esters pour transformateurs dans les matériels*, December 1992.

[STA 95] Standard IEC 60156, *Insulating liquids – Determination of breakdown voltage at power frequency – Test method*, 1995.

[STA 00A] Standard ISO 2592, *Détermination des points d'éclair et de feu – Méthode Cleveland en vase ouvert*, September 2000.

[STA 00B] Standard ASTM D 2863, *Standard test method for measuring the minimum oxygen concentration to support candle-like combustion of plastics (oxygen index)*, 2000.

[STA 02A] Standard ASTM D 240, *Standard test method for heat of combustion of liquid hydrocarbon fuels by bob calorimeter*, 2002.

[STA 02B] Standard ISO 2719, *Détermination du point d'éclair – Méthode Pensky-Martens en vase clos*, November 2002.

[STA 03] Standard IEC 60296 (Ed. 3), *Fluides pour applications électrotechniques – Huiles minérales isolantes neuves pour transformateurs et appareillages de connexion*, November 2003.

[STA 05A] Standard IEC 60836 (Ed. 2), *Spécifications pour liquides isolants silicones neufs pour usages électriques*, May 2005.

[STA 05B] Standard IEC 60422 (Ed. 3), *Huiles minérales isolantes dans les matériels électriques – Lignes directrices pour la maintenance et la surveillance*, October 2005.

[STA 07] Technical report no. B900-04062, *E-FR3 Large and medium Power transformer users list (Retrofill and new installations)*, Cooper Power Systems, December 2007.

[VUA 98A] VUARCHEX P.J., "Caractéristiques des diélectriques liquides", *Techniques de l'ingénieur, Traité K 714*, p. 3, January 1998.

[VUA 98B] VUARCHEX P.J., "Les isolants liquides dans les condensateurs et les transformateurs : emploi, nature et recherche", *Journée électrotechnique, 10–11*, p.21, March 1998.

[WIL 80] WILSON A.C.M., *Insulating liquids: Their uses, manufacture and properties*, Peter Peregrinus, Stevenage, UK, p. 124, 1980.

Chapter 17

Electrorheological Fluids

17.1. Introduction

The study of setting dielectric fluids (liquids, suspensions, etc.) in motion and their flow properties under electric fields and mechanical stress constitutes an important section of scientific, technological and practical investigation which has simultaneously caught the interest of researchers and manufacturers. The physical phenomena originating these laws of hydrodynamic behavior can be grouped into three categories: electrokinetic effects, electroviscous effects and electrorheological effects.

17.1.1. *Electrokinetic effects*

Electrokinetic effects are produced in the following situations: (i) the presence of a porous material immersed in a liquid; and (ii) when colloidal particles are placed in suspension in a liquid [HUN 81]. The electrical double layer, compact and diffuse (Stern and Gouy-Chapman layers), appearing at the solid/liquid interfaces of the initially neutral medium, reacts to the external solicitation (electric field or mechanical constraint) by implementing a reversible "electrokinetic conversion". Thus, the application of an electric field causes a motion in the fluid (displacement of the liquid or the particles): an electrical-mechanical conversion (motor operating) or the setting in motion of the fluid polarizes the electrode, which then takes a voltage U: the mechanical-electrical conversion (generator operating). By taking the previous examples of dielectric media (i) and (ii), we obtain the following associated

---

Chapter written by Jean-Numa FOULC.

electrokinetic effects: for motor operating, (i) a electro-osmosis or (ii) a electrophoresis, and for generator operating, (i) a flow potential or (ii) a sedimentation potential, also called Dorn effect.

17.1.2. *Electroviscous effects*

When an isotropic liquid of weak conductivity placed between two metallic electrodes is subjected to a sufficiently high electric field, an agitation appears followed by the liquid being set in motion. This phenomenon is due to a reaction at the electrode/liquid interface whose effect is the creation of charge carriers (injection phenomenon), the establishment of a space charge, then Coulomb forces exerted on the space charge under the action of the electric field. From a certain applied voltage (critical voltage Vc ~ 10 V for a polar liquid and ~ 100 V for a non-polar liquid), an instability regime appears and the liquid, led by the displacement of the charge carriers, is set in motion (electroconvection phenomenon). In the case where the liquid, subjected to a pressure gradient or shear, flows longitudinally between parallel electrodes supporting an electrical potential V, a secondary transversal flow perturbs the principal flow due to the electroconvection and increases the flow losses [ATT 82]. The apparent viscosity of the liquid can thus be controlled by the potential difference applied between the electrodes. For a strongly polar liquid, the increase in viscosity can be very important (in a ratio > 20).

Another situation containing a monophasic homogeneous fluid based on liquid crystal solutions can be encountered. These fluids, whose anisotropic behavior confers a privileged orientation on them during the application of an electric field, cause a strong increase (in a ratio > 30) of the fluid viscosity [YAN 92]. The principal interest of these fluids is their good stability (no sedimentation) but they also have a few disadvantages: their viscosity without a field is quite high (i.e. high viscous losses) and their operating temperature range is relatively limited.

It is important to note that in electroviscous effects, the fluid always remains in the liquid state (only the apparent viscosity of the fluid is modified).

17.1.3. *Electrorheological effects*

Electrorheology presents a more general character and is applied to suspensions, generally concentrated, called electrorheological (ER) fluids, of thin solid particles in a dielectric liquid [JOR 89]. ER fluids have the remarkable property of being able to "solidify" in the presence of an electric field, then to get their initial flow property back when the field is removed. Another particularity of ER fluids is that they are able to continuously control the flow resistance of the fluid, i.e. its apparent viscosity, by an electric field. The obtained variation of viscosity is very important (in the "solid" phase, the fluid viscosity is infinite). We can distinguish two kinds of ER effect: the positive ER effect, which presents most interest, characterized by an increase of the apparent viscosity of the fluid (with possibility of "solidification") in the presence of an electric field [CON 92], [TAO 93] and the negative ER effect, which leads to a decrease in fluid viscosity under the action of the electric field[1] [BOI 95], [LOB 99].

17.2. Electrorheology

17.2.1. *Electrorheological effect*

The implementation of the ER effect is relatively easy. Numerous powders (starch, cellulose, etc.) can constitute the solid phase of an ER fluid, and a standard insulating liquid can be used as the liquid phase [BLO 88]. The most spectacular manifestation of the ER effect is the reversible change of liquid/solid state of the fluid under the action of an electric field (Figure 17.1).

Electrorheology is an interdisciplinary domain of study which catches the attention of numerous specialists: physicists (for modeling of interactions between particles, formation and behavior of organized and distortable media), chemists (particle synthesis with specific characteristics, elaboration of suspensions), electricians (conduction properties under the high electric field of the dielectric materials), fluid mechanics (rheological study of complex fluids), engineers (conception of new applications). In this chapter, after a presentation of the main characteristics and the composition of ER fluids (choice of solid and liquid constituents), we tackle the study of the basic mechanisms of the positive ER effect in more detail by showing the importance of the conduction properties of solid and liquid dielectric materials in the modeling of interactions between particles.

---

1. The decrease in fluid viscosity (in the case of a suspension whose particles are more insulating than the carrier liquid) is due to a particular configuration of the solid phase. Under the action of the electric field, the particles acquire a charge (the diffuse layer is driven away by the field) and migrate towards an electrode: the suspension then becomes a medium with two separate phases (liquid and solid). Further, when the suspension is set in motion, the electric field can cause a rotation of the particles according to the vorticity axis (Quincke rotation), whose effect is an increase in the flow speed and therefore a decrease in the apparent viscosity of the suspension.

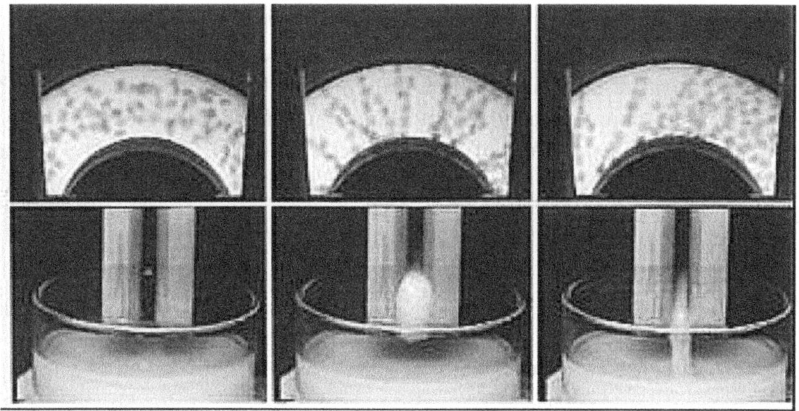

Figure 17.1. *Illustration of the reversibility of the electrorheological effect [BOI 96a].Top row: suspension of polymer grains (size ~ 2mm) in an insulating liquid; bottom row: suspension of cellulose particles (size ~ 20–30µm) in a hydrocarbon (insulating liquid for transformer). (a) The ER fluid is liquid;(b) for sufficient E (E > 2 kV/mm) the particles (or the grains) line up with the field and the fluid becomes a "solid" gel; (c) the fluid becomes liquid again*

17.2.2. *Characterization of electrorheological fluids*

17.2.2.1. *Rheological characteristics*

The basic feature of an ER fluid (rheogram) is given by the variation of the mechanical stress (shear stress) as a function of the velocity gradient (shear rate) for various applied electric fields [LEM 92]. The measurement is generally performed in Couette flow and in cylindrical geometry. Subject to special dimensional conditions, this configuration can reduce to the simple case of a shear between two parallel plane plates between which the ER fluid is applied. These plates, which constitute the electrodes subjected to an electric potential difference V, create a constant electric field E = V/d (d, distance between the electrodes) in the fluid. The typical characteristics obtained show that the rheological behavior of an ER fluid is similar to that of a Bingham body (Figure 17.2):

$$\tau(E, \dot{\gamma}) = \tau_S(E) + \eta_{S0}\dot{\gamma} \qquad [17.1]$$

where τ: shear stress, τ_S: threshold stress or yield stress, $\dot{\gamma}$: shear rate, η_{S0}: dynamic viscosity of the suspension without electric field, E: applied electric field (the field usually ranges from 2 to 5 kV/mm).

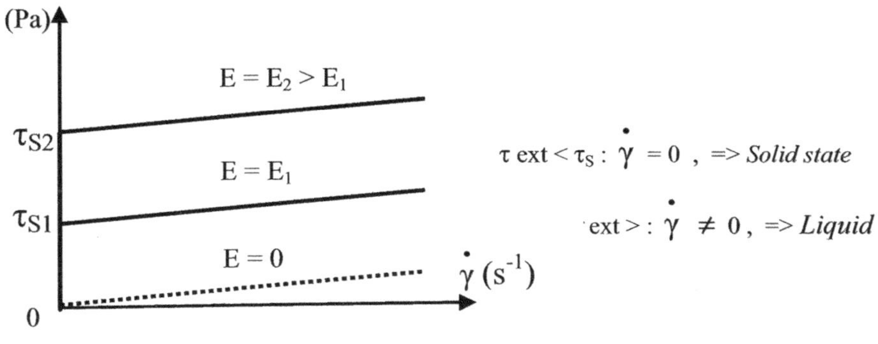

Figure 17.2. *Typical rheogram of an ER fluid. The behavior of the fluid is of Bingham-body for E ≠ 0 and Newtonian-body for E = 0. τ ext: external applied shear stress, τ_{S1}, τ_{S2}: threshold stress of the fluid flow*

The value of the yield stress τ_S is the most important characteristics of an ER fluid. In the case of ER fluids based on micrometer-size particles, the maximum value of τ_S is of the order of 5 to 8 kPa (for an applied electric field about 5 kV/mm). Its dependence regarding the electric field E and numerous other parameters (constituent characteristics, temperature, volume fraction, nature and geometry of the particles, fluid structure, experimental conditions etc.) is not established in a precise manner. Nevertheless, an approximate expression allows a few elements of modeling to be brought out:

$$\tau_S \propto \varepsilon_L \, \phi^m \, E^n \qquad [17.2]$$

where ε_L: permittivity of the liquid carrier, ϕ: volume fraction of the particles ($m \sim 1$ for $\phi < 10\%$, $m \sim 2$ to 3 for $\phi > 20\%$), E: applied electric field ($n \sim 2$ for $E < 0.5$ kV/mm, $n \sim 1$ for $E > 2$ kV/mm).

Further, each solid or liquid state of the fluid possesses specific characterization parameters:

a) With ER fluid in solid state: under oscillatory shear flow of strain amplitude γ and pulsation ω, we obtain [JOR 92], [MCL 91]:

$$\tau(\omega) = [G'(\omega) + i\,G''(\omega)]\,\gamma(\omega) \qquad [17.3]$$

where G': storage modulus (elasticity), G'': loss modulus (viscosity).

b) With ER fluid in liquid state: under steady flow with shear stress τ and shear rate $\dot{\gamma}$, the associated apparent viscosity η_{app} is defined by:

$$\eta_{app} = \tau / \dot{\gamma} = \eta_0 (1 + K/Mn), \qquad [17.4]$$

with Mn: Mason number and K a constant:

$$Mn = \frac{\text{Hydrodynamic forces}}{\text{Electric forces}} \cong \eta \dot{\gamma} / 2 \varepsilon_L E^2 \qquad [17.5]$$

The Mason number, the ratio between the hydrodynamic and electric forces, expresses the structure level of an ER fluid [MAR 89]. The electric forces, induced by the field, have a structuring effect (formation of fibers and columns) and tend to "solidify" the fluid. Conversely, hydrodynamic forces, due to the flow, have a destructuring effect (break of fibers, separation of particles) and cause the flow of the fluid. It is therefore the competition between these two antagonist forces which determines the actual state (solid or liquid behavior) of the ER fluid.

17.2.2.2. *Electrical characteristics*

Concerning the electrical characteristics of the ER fluids, the most significant parameter for the applications is the average current density ($j_{av} = I/S$), where I and S are the current and the cell surface. A magnitude of current density greater than a few tens of µA/cm2 must be avoided. Indeed, in the contact zone of the particles, where the current densities are strengthened, a degradation of the materials resulting from local heating is possible.

The dielectric spectroscopy analysis of the suspensions (at weak or high electric field and with or without flow) leads to a correlation of the electrical properties of the materials (conductivity and permittivity) to the rheological behavior of the suspension in static or dynamic flow.

17.2.2.3. *Energy assessment*

In DC voltage, it is easy to express the density of the control losses P_C (power control consumption) and then, by associating the transmitted power density P_T, to determine the power gain G_P and the efficiency η of the system.

For example, in the case of a device composed of two parallel electrodes of surface S, separated by a distance d (input electrode, index I and output electrode, index O), subjected to a mechanical stress (speed V and force F) and an electrical stress (voltage U et current I) by inserting an ER fluid, we can define the following mechanical and electrical values:

– Speed gradients: of the ER fluid: $\dot{\gamma} = (V_O - V_I)/d$, of the output electrode: $\dot{\gamma}_O = V_O / d$,

– Mechanical parameters: $\tau_O = F_O / S$, $\tau_I = F_I / S$, viscous losses: $P_V = (\tau_I - \tau_O)\dot{\gamma}$,

– Applied electric field: $E = U / d$, average current density: $j_{av} = I / S$,

– Control loss density: $P_C = E.j_{moy}$, transmitted power density: $P_T = \tau_O \dot{\gamma}_O$,

We obtain:

$$G_P = P_T / P_C \qquad [17.6]$$

$$\eta = P_T / (P_T + P_C + P_V) \qquad [17.7]$$

NOTE: In steady shear flow, we have:

– for ER fluid in solid state, $V_O = V_I$ and $\tau_O = \tau_I$ $(P_V = 0)$

– for ER fluid in the liquid state, $V_O < V_I$ and $\tau_O < \tau_I$ $(P_V \neq 0)$.

17.2.3. *Composition of electrorheological fluids*

An ER fluid is a suspension of solid particles (dispersed phase) immersed in a dielectric liquid (continuous phase). The volume fraction ϕ of the solid phase ($\phi = V_S / (V_S + V_L)$) is about 20 to 40%. In order to avoid the sedimentation of the particles, their density must be as close as possible to that of the liquid[2].

The choice and the characteristics of the materials making up the suspensions condition the ER fluid performances (value of the yield stress, current density, etc.) widely [BLO 90]. A great number of liquid and solid constituents produce an ER effect but until recent years, all experimental results published in this field led to a

2. This restraint disappears in the case of much reduced-size particles. Indeed, the sedimentation speed is proportional to the square of the radius of the particles and the presence of the Brownian motion can inhibit the gravitational effect resulting from the difference in solid and liquid material densities.

relatively low efficiency for the fluids tested (especially for the yield stress strength).

a) Continuous phase: the choice of a liquid is not the most important issue when making up an ER fluid. The liquid must nevertheless be a good electric insulator, even at high temperature and under high electric field, and if possible, have a weak viscosity ($\eta_L < 20$ mPas) and a moderate permittivity ($\varepsilon_r \sim 3$). The commonly used liquids can be listed in three categories: mineral oils with "standard" properties, silicon oils (good temperature stability but relatively poor dielectric performance) and chlorinated liquids ("high" density and permittivity but moderate insulating power) [VUA 01].

b) Dispersed phase: this is the dispersed solid phase which conditions the performances of an ER fluid, almost on its own. The knowledge of their required characteristics leading to production of a strong ER effect was ignored for a long time. It is only over the past 15 years that research has shown that the electrical conduction properties of the materials were crucial in the implementation of the ER effect[3] [AND 92], [DAV 92], [FOU 92].

We list below the various types of solid materials (the particles) used, in three categories:

i) The insulating homogeneous materials (silica, starch, cellulose, etc.) requiring the contribution of an external activator (water, surfactant). These materials are not thermally stable in the temperature range of about 0–100°C [IKA 98], [OTS 92];

ii) The homogeneous materials with intrinsic, ionic or electronic conductivity (aluminosilicates, zeolites, semi-conducting polymers, polyaniline, etc.) [GOW 90], [ISH 95];

iii) The inhomogeneous (multilayer) materials with functionalized surface (silica covered with polyaniline, aluminum covered with an insulating layer, etc.) [AKH 99], [WU 98a].

NOTE:

– The materials cited above, which possess a suitable electric conductivity, produce an ER effect in DC voltage but practically no effect in AC voltage. However, some ferroelectric materials (barium titanate, etc.) produce an ER effect under high frequency AC voltage;

---

3. This finally justifies the observations made for decades on the necessity of using slightly hydrated solid particles to produce an ER effect (anhydrous or hydrophobic particles were without effect) and that particles too hydrated reduced the effect (it is actually water adsorbed at the surface of the particles which supplies sufficient conductivity).

– The composition and geometry of the solid phase (mono or polydispersed particles, particle aspect ratio, etc.) also have an impact on the rheological behavior of ER fluids [QI 02]. However, the size of the particles and their surface functionalization are the most important factors which contribute to producing a very strong ER effect [SHI 94], [WEN 04], [WU 05];

– The dependence of the ER effect on temperature is explained by its role on liquid viscosity, the conductivity of the materials and the surface polarization of the particles. Experiments show that the ER effect (value of the yield stress) passes through a maximum for temperatures ranging approximately between 60 to 80°C [WU 98b].

17.2.4. *Applications of electrorheological fluids*

The applications of ER fluids turn to good account the ability to control the structure of suspensions by an electric field E. An ER suspension can thus pass from a homogeneous and isotropic dispersion (where $E = 0$) to an inhomogeneous medium having a strong orientation anisotropy (where $E \neq 0$). The two behaviors of the fluid can operated: with a reversible change in liquid/solid state (switching-mode) or a continuous control of the flow (modulation-mode). The first applications appeared in the world of mechanics (for shock absorbers, clutches, brakes, etc.) but the performance of ER fluids and the powers brought into play make the implementation of such applications difficult [FUR 99], [HAR 91], [MON 97], [STA 92]. Other types of application require little power and turn to good account the easiness of miniaturization of ER devices include micro-wave transmitters, microfluid displays, etc. [FAN 02], [NIU 05], [WEN 05].

17.3. Mechanisms and modeling of the electrorheological effect

Discovered in the 1940s by W. Winslow [WIN 49], the ER effect then gave rise to numerous projects and led to successive advancements concerning the elaboration and characterization of new fluids and new approaches to ER effect mechanisms [BLO 88], [HAV 95], [KLA 67], [UEJ 72], [WEN 03]. However, because of the insufficiency of ER fluid performances (yield stress and temperature range too weak) and a relative lack of knowledge about basic ER effect mechanisms, no decisive advancement regarding application could be realized in the past. The recent discovery of the giant ER effect (see section 17.5.) strongly revived interest in these fluids and permits us now to envisage the implementation of operational ER devices in the near future.

17.3.1. *Forces exerted on and between the particles*

In the absence of an electric field, the particles immersed in a liquid are subjected to several types of forces: a gravitational force due to the difference of densities in solid and liquid phases (this effect goes hand in hand with the size of the particles) and could lead to their sedimentation; Brownian motion (not very important for particles of size > 1µm and for temperatures T < 50°C); and hydrodynamic force (generated by the liquid flow). Further, two types of interaction between particles can also be taken into consideration: the van der Waals forces play a part at nanometric distances which have an actual effect only for particles of very small size (< 0.1 µm), and electrostatic repulsion forces due to diffuse layers surrounding the particles.

In the presence of an electric field, a dielectrophoretic force (dipole subjected to a field gradient) [POH 78] resulting from the polarization of particles and an electrostatic interaction force (interfacial charges) appears between the particles. Finally, an electrophoretic force (Coulomb force) exists if the particles are charged (with an intrinsic charge or solid/liquid interface effect due to the electrical double layer). In the case of particles more insulating than the liquid, the application of a DC electric field induces electrophoretic forces, which drain the particles towards an electrode: the separation of the solid (particles) and liquid phases leads to a decrease in the apparent viscosity of the suspension (negative ER effect).

NOTE: For ER suspensions composed of micrometer-size particles (common ER fluids), only two force categories must be taken into consideration (the other forces are negligible) to explain the behavior of the fluid: electric (dielectrophoretic and electrostatic) and hydrodynamic forces.

17.3.2. *Mechanisms of the electrorheological effect*

An ER fluid not subjected to an electric field is a concentrated suspension of solid particles randomly spread out in a liquid and relatively near one another (for a volume fraction of 20%, the average distance between mono-dispersed spherical particles is about 0.7 times the radius of the particles).

The mechanisms leading to the (positive) ER effect are as follows:

1 – From the application of an electric field, the particles get polarized then attract or repel one another, according to their relative position.

2 – The particles which attract each other group together to form a multitude of chains, or elementary fibers, parallel to the field lines (response time of a few milliseconds). The electric current which is established throughout the fibers ensures their preservation and provides them some rigidity. The ER fluid then presents itself as a solid gel.

3 – The application of an external force (shear or squeeze force) distorts the fibers and induces a reaction force from the fluid depending on the distortion level of the fibers and the value of the electric field. For example, in the case of a plane shear, the reaction force increases with the inclination θ of the fibers and reaches its maximal value (yield stress) for $\theta \sim 20°$ [CON 92].

17.3.2.1. *Polarization and bringing the particles closer*

A non-fibrated ER suspension can be perceived as a sequence of solid (particle) and liquid materials. The electrical behavior of such a medium then comes down to that of two dielectric layers in series (a solid layer, of index S and a liquid layer, of index L).

Reasoning for a comparable system but with a simpler geometry: the presence of two plane layers with the same surface, constant thicknesses d_S and d_L, conductivities σ_S and σ_L and permittivities ε_S and ε_L, the response of the system to a voltage V(t) shows the presence of a density of charge q_S at the solid/liquid interface (interfacial Maxwell–Wagner polarization) and defines the distribution of the electric field in the medium.

Thus, for example, the response to an applied voltage of magnitude V leads to:

– The initial state ($t = 0^+$): $q_S = 0$, the electric field distribution depends on the value of the permitivities: $\varepsilon_S E_S = \varepsilon_L E_L$, avec $d_S E_S + d_L E_L = V$;

– The final state ($t \rightarrow \infty$): $q_S = (\varepsilon_S E_S - \varepsilon_L E_L)$, the electric field distribution depends on the value of the conductivities: $\sigma_S E_S = \sigma_L E_L$.

NOTE: The time constant value τ of the system which corresponds to the temporal limit between the initial state and the final state is:

$$\tau = \frac{\varepsilon_S d_L + \varepsilon_L d_S}{\sigma_S d_L + \sigma_L d_S} \qquad [17.8]$$

For $t \ll \tau$ (case of high frequency AC voltage): the polarization mechanism is dominant and the field distribution depends on the permittivities of the materials.

For $t >> \tau$ (case of steady state in DC voltage): the conduction mechanism is dominant and the field distribution depends on the conductivities of the materials.

Let us now consider a spherical particle of radius R, with conductivity σ_S and permittivity ε_S, immersed in a liquid of conductivity σ_L and permittivity ε_L. The application of a uniform electric field E_o polarizes the particle and induces an equivalent dipole p:

$$p = 4\pi\, \varepsilon_L\, R^3 \beta\, E_o \qquad [17.9]$$

β depends on the difference in the electrical characteristic values of the materials.

By reasoning in the AC and DC conditions previously defined, we respectively obtain $\beta = \beta(\varepsilon)$ for the AC regime at high frequency and $\beta = \beta(\sigma)$ for the DC regime at steady state, with:

$$\beta(\varepsilon) = \frac{\varepsilon_S - \varepsilon_L}{\varepsilon_S + 2\,\varepsilon_L} \quad \text{and} \quad \beta(\sigma) = \frac{\sigma_S - \sigma_L}{\sigma_S + 2\,\sigma_L} \qquad [17.10]$$

The introduction of a second particle, identical to the previous one, at a distance s from the first particle ($s >> R$), the line Δ passing through the centers of the particles and the direction of the electric field forming an angle θ, the component F_Δ of the dipolar interaction force between the particles is expressed by:

$$F_\Delta(s, \theta) = 12\pi R^2 \varepsilon_L \beta^2\, E_o^2 \left(\frac{R}{2R + s}\right)^4 (3\cos^2\theta - 1) \qquad [17.11]$$

Equation 17.11 (point-dipole approximation) [AND 94], [KLI 91a] shows that for $\theta < 55°$ the particles attract each other and they repel one another for $\theta > 55°$. Consequently, during the application of an electric field to an ER suspension, the nearby particles will either attract (those which are localized in an attraction cone at angle $\theta \cong 55°$), or repel, and form rows of chains or fibers separated by a liquid[4] zone.

For weaker separation distances s ($s < R$) the previous dipolar model is no longer valid (the polarization charges concentrate in the zone near the particles). The multipolar approach, which allows this situation to be taken into account, leads to a

4. The local and microscopic behavior between the particles (attraction or repulsion) generates a diphasic macroscopic structure in the fluid (the solid fibers composed of particles alternate with the liquid layers).

reinforcement of the interaction between the particles. Then the replacement of the two isolated particles by a set of particles structured in fiber (multi-body system) allows the electrostatic interaction force of a chain of particles to be calculated [CLE 93], [HAL 90], [KLI 91b]. However, this approach which only takes into account the polarization effect of the materials (polarization model of the ER effect) and which is based on the hypothesis that the solid and liquid dielectrics used are perfect insulators (conductivity is zero) leads to theoretical values of the interaction force between particles and the yield stress much less than those experimentally obtained. Furthermore, when the particles get in contact (formation of chains), the calculations show that the electric field in the liquid near the contact becomes infinite.

The consideration of the actual characteristics of the materials and of the experimental conditions (DC or AC voltage) therefore turns out to be very necessary. As we have previously underlined, in a DC electric field and steady state, the distribution of the electric field within the suspension (diphasic dielectric medium) is determined by the conduction properties of the materials.

17.3.2.2. *Formation of chains and conduction current*

Considerations of a general nature show that under a DC field, the attraction of particles and the formation of stable chains leading to the electrorheological effect require two conditions [BOI 96a]:

a) Conductivity of particles greater than that of the liquid: $\Gamma = \dfrac{\sigma_S}{\sigma_L} > 1$

NOTE: The induced dipole p (equation [17.9]), with $\beta = \beta(\sigma)$, must be in the direction of the electric field. In the opposite case, the particles get polarized, come closer then turn back simultaneously and separate.

b) Conductivity of particles limited: $\sigma_{S\,max} \cong 10^{-7}$ S/m

NOTE: The contact time of the particles (t_{mc}) must be less than the relaxation time of the charges between two particles (t_{rl}). In this case, chains form between the electrodes and the current which flows through them, holds the superficial charges and induces an electrostatic force of attraction between the particles. It is this force which determines the rigidity of the chains (or the fibers) and leads to the value of the yield stress.

$$t_{mc} \cong \frac{\eta_L}{\varepsilon_L E_0^2}, \quad t_{rl} \cong 10\,\frac{\varepsilon_L}{\sigma_S}, \text{ with } \eta_L\text{: dynamic viscosity of the liquid.}$$

NOTE: This double condition on the conductivity value of the particles justifies the necessity of using properly hydrated powders to produce an ER effect. It is the presence of water molecules at the surface of the particles which confers on these latter an appropriate conductivity.

The almost instantaneous structure formation of the solid phase in a chain network evolves and transforms little by little into thick fibers or columns composed of a group of chains (characteristic time of a few seconds to a few minutes) [GIN 95]. In this case, chains inside the columns most often form a body-centered-tetragonal lattice. This structure in columns, which only appears with chains or very extended elementary fibers (with a size of the particles much smaller than the interelectrode distance) reinforces the rigidity of the ER fluid in the "solid" state [GUL 93], [KLI 93], [TAO 91].

17.3.2.3. *Flow resistance*

The state of the structure of an ER fluid simultaneously subjected to an electric field and a mechanical stress (represented by the Mason number Mn) determines the rheological behavior of the fluid. Three situations can be envisaged:

i) Mn $\rightarrow$ 0 (high E and $\tau_{ext} \cong 0$): the fluid behaves like a solid gel, it can transmit a force (or torque) without speed slipping (synchronous transmission);

ii) Mn $\rightarrow \infty$ (E $\cong$ 0 and high τ_{ext}): the fluid is in a liquid state and its viscosity, constant, is near that of the suspension in the absence of electric field. The force is transmitted with speed slipping (asynchronous transmission);

iii) Intermediate Mn (moderate E and τ_{ext}): the fluid is in the liquid state and its viscosity is variable, it depends on E and τ_{ext} (controlled asynchronous transmission).

NOTE: Measurements performed on a model ER fluid (suspensions of cellulose powder in mineral oil) showed the direct connection between the structure level of the fluid and, the rheological (apparent viscosity) and electrical (current density) responses of the fluid [FOU 96].

17.4. The conduction model

The conduction model, valid in a DC field or in low frequency AC field, is based on the consideration of the electrical conduction properties of the solid and liquid constituents and the liquid permittivity of an ER fluid to determine a variation law of the attraction force between nearby particles or particles in contact, as a function of the applied electric field and the characteristics of the materials [ATT 94], [ATT 02], [BOI 96b]. This approach, which relies on the results of the electrostatics and electrodynamics of the continuous media, sheds a new light on the understanding of the electrorheological effect and was experimentally verified in a large scale set-up (with an interaction between two millimeter-sized hemispheres) [FOU 94].

17.4.1. *The bases of the conduction model*

Let us consider the case of a chain formed by spherical particles in contact, immersed in a liquid and subjected to a DC electric field. The cohesive force of the chain is directly related to the attraction force between the particles themselves and, therefore, for symmetrical reasons, to that of two half-spheres in contact (Figure 17.3). Particles are assumed to be more conductive than the liquid ($\sigma_S > \sigma_L$).

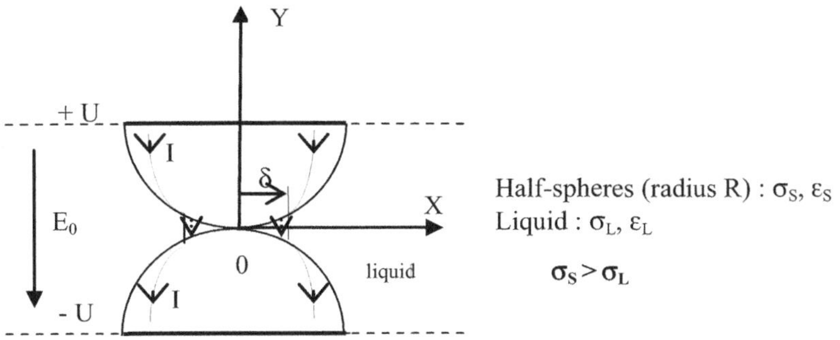

Figure 17.3. *Half-spheres in contact immersed in a dielectric liquid and brought to a voltage +U and –U, respectively. The current I passes through the half-spheres for $x > \delta$ and through the liquid for $x < \delta$*

The distribution of the current between the solid (half-sphere) and the liquid depends on the values of their respective conductances C_S and C_L ($C = \sigma$ S/l), σ, S and l representing the conductivity, the surface and the length of each element, respectively. Figure 17.3 shows that the passage of the current between the positive electrode and the negative electrode (diametral planes of half-spheres brought to the voltage +U and - U) distinguishes two zones:

Zone 1: $x > \delta$. The conductance of the solid is high (high values of σ_S and S_S) and that of the liquid is weak (weak value of σ_L and high l_L). All of the current passes through the solid. The conductance of the solid being high, the decrease in voltage V(x) due to the passage of current is negligible: the half-sphere can therefore be considered as equipotential: V(x) = U. The electric field $E_1(x)$ in the liquid is in this case equal to:

$$E_1(x) = 2U/y = 2RE_0/y \cong 2R^2E_0/x^2 \quad [17.12]$$

Zone 2: $x < \delta$. The conductance of the solid is weak (high σ_S but very weak S_S) and that of the liquid is high (weak σ_L but very weak l_L). All of the current passes through the liquid.

In this zone the voltage of the half-sphere decreases and passes from U (for x = δ) to 0 (for x = 0) and the distance y also decreases, passing from x^2/R (approached value) to 0.

The electric field $E_2(x)$ in the liquid is considered constant and equal to:

$$E_\delta = 2RU/\delta^2 = 2R^2E_0/\delta^2 \qquad [17.13]$$

17.4.2. *Attraction force between half-spheres*

The calculation of the attraction force between the half-spheres is obtained by integrating the electrostatic pressure in the electric contact zone (0-δ) and the external zone (δ-R):

$$F = \int_0^R \frac{1}{2}\varepsilon_L E^2 dx = 4\pi R^2 \varepsilon_L E_0^2 \left(\frac{R}{\delta}\right)^2 \qquad [17.14]$$

For x = δ the current changes of medium (passage from solid to liquid) and the conductance of the liquid C_L (in the external zone, x>δ) is equal to the conductance of the solid C_S (in the electric contact zone, x<δ). δ is deduced from the relation $C_S = C_L$.

The conductance C_S of the solid spheres can be approximated by assuming that the current passes through disks of radius δ (the electrical contact zone). This result can be obtained from the classic capacitance expression, c = 8ε.δ of a conducting disk of radius δ isolated in an infinite medium of permittivity ε. This gives c' = c/2 for the capacitance of one face of the disk relative to the corresponding half-sphere of infinite radius. The resistance of the medium lying between a disk face and the corresponding half-sphere of infinite radius is r such that the relaxation time is r.c' = ε/σ. We thus obtain the equivalent conductance C_S = 1/2r of the two half-sphere in series.

$C_S \cong 2\sigma_S\delta$ (conductance of an electric contact between two solids) [17.15]

The conductance of the liquid C_L is obtained by integration of the conductance dC_L of the liquid layer between cylinders of radii x and x+dx: $dC_L = 2\pi\sigma_L x.dx/y \cong 2\pi R\sigma_L dx/x$.

$$C_L = 2\pi R \int_\delta^R \sigma_L \frac{dx}{x} \qquad [17.16]$$

The conductance C_L depends on the value of the liquid conductivity [TOB 96]. Further, Equation [17.12] reveals a significant enhancement of the electric field in the liquid for values of x near δ (contact zone). In practice, the field E_δ can reaches several tens of kV/mm. For such fields, the liquid conductivity is strongly increased and depends enormously on the value of the field [ONS 34].

Consequently, the calculation of the attraction force between two particles, given by the conduction model (a two zone model), leads to two expressions:

i) In a low applied field ($E_0 < 0.1$ kV/mm)

The field in the contact zone remains limited ($E_\delta < \sim 3$–4 kV/mm), the liquid conductivity is constant (liquid conductivity in low field $\sigma_{L0} \sim 10^{-12}$ S/m) and δ is independent of E_0.

$$F = 4\pi R^2 \varepsilon_L K_\Gamma^2 \Gamma^2 E_0^2 \text{, with } K_\Gamma = [\pi \ln(R/\delta)]^{-1} \qquad [17.17]$$

This law shows that the force depends on the square of the applied electric field E_0 and the square of the conductivities ratio Γ.

ii) In a high applied field ($E_0 > 1$ kV/mm)

The field in the contact zone is very high ($E_l > \sim 8$–10 kV/mm), the conductivity of the liquid is increased and the electric contact zone gets wider (δ increases with E_0).

$$F = 2\pi R^2 \varepsilon_L E_c E_0 [\ln[(10\, \Gamma/\pi)(2\, E_0 / E_c)^{1/2}]]^2 \qquad [17.18]$$

The force varies quasi-linearly with the applied electric field and does not depend much on the conductivities ratio Γ.

The conduction model also permits the attraction force F between particles to be expressed as a function of the current I flowing through the fluid and the applied voltage U to the fluid. By using the relation $I = C_S U$ and removing δ between Equations [17.14] and [17.15], we obtain:

$$F = 4\pi\varepsilon_L (\sigma_S R)^2 U^4 / I^2 \qquad [17.19]$$

This new expression, which appears in the form of a "universal law", independent of the value of the applied electric field was experimentally verified in a wide range of the conductivities ratio (variation of Γ from 3 to 1,500) [ATT 94]. This law also reveals the correlation between the force (and the yield stress) and the current passing through the ER fluid, and confirms the validation of the conduction model of the ER effect in DC voltage.

NOTE: The conduction model presented above only considers the volumic conduction properties of the liquid. Furthermore, the high field enhanced conduction phenomenon (Onsager effect), due to the dissociation of neutral species, assumes the existence of a "neutral species reservoir" sufficiently important to permanently supply the dissociation. This hypothesis is verified in the case of a large-scale system (e.g. millimetric half-spheres) or for ER fluids with particles of medium size (> 50 μm) but cannot be kept in the other cases. For ER fluids based on micrometric, and, particularly, nanometric particles, the conduction process of the liquid under high electric fields can be explained by the presence of charge injection phenomena at the particle/liquid[5] interface [ATT 02].

17.5. Giant electrorheological effect

The giant ER effect discovered in 2003 by researchers from the University of Hong Kong [WEN 03] is characterized by a very strong mechanical response. Measurements carried out on these fluids showed that a yield stress τ_S of the order of 150 kPa could be reached for a 5 kV/mm DC electric field. Such an advance in performance allows operational applications in the domain of mechanical transmissions with quite higher power levels to be envisaged. Other work then confirmed the presence of an intense ER effect (τ_S ~ 100 kPa) with nano-suspensions [SHE 05a], [YIN 05]. We can note that this giant ER effect appears with fluids based on nanoparticles (size ~ 50 to 100 nm) functionalized on the surface. In the case of the fluid elaborated by Wen *et al.* [WEN 03], the suspension is composed of barium titanyl oxalate nanoparticles (size ~ 60nm, volume fraction ϕ = 30%) coated with 5 nm of urea, immersed in a silicone oil. The other characteristic features of these fluids are: an high current consumption (j ~ 50 to 100 μA/cm^2 for $E_{(DC)}$ ~ 4 kV/mm), a fairly high intrinsic viscosity (without electric field): $\eta_0 > 1$ Pa.s, and a quasi-linear dependence of the yield stress on the electric field.

The nanometric dimension of the particles and their functionalization (with the presence of a strong surface polarization) can explain the existence of the giant ER effect. At this reduced scale, the Van der Waals forces appearing between the particles in contact [KRU 52], [SHE 05b] on the one hand, and the very high electric field prevailing in the liquid near the contact zone (on the other hand) probably contribute to the strengthening of the attraction force between the particles and as a consequence of the yield stress of the fluid. Indeed, it is well-known that a dielectric liquid can support an electric field even higher when its volume (surface and

5. Charge injection phenomena in the presence of metallic electrodes are fairly well-known [DEN 82], but this is not the case for electrochemical processes relative to organic or inorganic material interfaces (the case of particles in ER suspensions) and insulating liquids.

thickness) is restricted; the Paschen law for gases is the most famous example. We therefore expect a similar result for a liquid [LES 02] but with different behavior laws (existence or absence of charge carriers in the case of extreme liquid confinements, injecting ability of the electrodes, etc.). At the experimental level, the effect of the size of the particles on the rheology of ER nano-suspensions is confirmed: the viscosity increases when the size of the particles decreases [GUE 06], [WEN 04]. This can also explain the high value of the intrinsic viscosity of nano-suspensions.

17.6. Conclusion

The study of the ER effect has seen a renewed interest over the past few years, owing to spectacular advancements with regard to the performance of new fluids based on nanoparticules and functionalized dielectric materials. Operational applications turning the possibility of controlling the structure level of an ER suspension from an electric field to good account are now conceivable. The mechanical sector (for couplers, shock absorbers, etc.) is the most active, but other areas of applications have already been explored: electromagnetic wave transmitters, display devices, etc. The nanometric size of particles allows tremendous yield stress under electric fields to be obtained but also causes, in the absence of a field, strong interactions between particles, whose effects are a high viscosity of the suspension limiting the use of these fluids for applications which only bring into play weak motion speeds. Other advancements which lead to the elaboration of better ER fluids, characterized by an intrinsically weak viscosity, a strong rigidity of the suspension in solid state and a limited current consumption, therefore still remains to be realized but the important point is that the issue resulting from the yield stress weakness of the old ER fluids ($\tau_S < 10$ kPa) is now solved.

17.7. Bibliography

[AND 92] ANDERSON R.A., "Effects of finite conductivity in electrorheological fluids", *Proceedings of the Third ICERF*, TAO, R. (ed), World Scientific, Singapore, p. 81–90, 1992.

[AND 94] ANDERSON R.A., "Electrostatic forces in an ideal spherical-particle electrorheological fluid", *Langmuir*, vol. 10, p. 2917–2928, 1994.

[AKH 99] AKHAVAN J., SLACK K., WISE V., BLOCK H., "Coating of polyaniline with an insulting polymer to improve the power efficiency of electrorheological fluids", *International Journal of Modern Physics B*, vol. 13, nos 14–16, p. 1931–1939, 1999.

[ATT 82] ATTEN P., HONDA T., "The electroviscous effect and its explanation I- The electrohydrodynamic origin: study under unipolar D.C. injection", *Journal of Electrostatics*, vol. 11, p. 225–245, 1982.

[ATT 94] ATTEN P., FOULC J.-N., FELICI N., "A conduction model of the electrorheological effect", *International Journal of Modern Physics B*, vol. 8, nos 20 & 21, p. 2731–2745, 1994.

[ATT 02] ATTEN P., FOULC J.-N., GONON P., "Role and nature of high field conduction of the suspending liquid in electrorheological fluids", *International Journal of Modern Physics B*, vol. 16, nos 17 & 18, p. 2662–2668, 2002.

[BLO 88] BLOCK H., KELLY J.P., "Electrorheology", *Journal of Physics D: Applied Physics*, vol. 21, p. 1661–1677, 1988.

[BLO 90] BLOCK H., KELLY J.P., QIN A., WATSON T., "Materials and mechanisms in electrorheology", *Langmuir*, vol. 6, p. 6–14, 1990.

[BOI 95] BOISSY C., ATTEN P., FOULC J.N., "On a negative electrorheological effect", *Journal of Electrostatics*, vol.35, p. 13–20, 1995.

[BOI 96a] BOISSY C., Etude des suspensions électrorhéologiques: rôle de la conduction des phases solide et liquide dans l'interaction entre les particules, Doctoral Thesis, Joseph-Fourier University, Grenoble, 1996.

[BOI 96b] BOISSY C., ATTEN P., FOULC J.N., "The conduction model of electrorheological effect revisited", *International Journal of Modern Physics B*, vol. 10, nos 23 & 24, p. 2991–3000, 1996.

[CLE 93] CLERCX H.J.H, BOSSIS G., "Many-body electrostatic interactions in electrorheological fluids", *Physical Review E*, vol. 48, no. 4, p. 2721–2738, 1993.

[CON 92] CONRAD H., CHEN Y. SPRECHER F., "The strength of electrorheological (ER) fluids", *International Journal of Modern Physics B*, vol. 6, p. 2575–2594, 1992.

[DAV 92] DAVIS L.C., "Polarisation forces and conductivity effects in electrorheological fluids", *Journal of Applied Physics*, vol.72, no. 4, p. 1334–1340, 1992.

[DEN 82] DENAT A., Etude de la conduction électrique dans les solvants non polaires, Physical Sciences Doctoral Thesis, Université Scientifique et Médicale de Grenoble, 1982.

[FAN 02] FAN J., ZHAO X., GAO X., CAO C., "Electric field regulating behaviour of microwave propagation in ER fluids", *Journal of Physics D: Applied Physics*, vol. 35, p. 88–94, 2002.

[FOU 92] FOULC J.-N., FELICI N, ATTEN P., "Interprétation de l'effet électrorhéologique", *Comptes Rendus de l'Académie des Sciences de Paris*, t. 314, Série II, p. 1279–1283, 1992.

[FOU 94] FOULC J.-N., ATTEN P., FELICI N., "Macroscopic model of interaction between particles in an electrorheological fluid", *Journal of Electrostatics*, vol. 33, p. 103–112, 1994.

[FOU 96] FOULC J.-N., ATTEN P., BOISSY C., "Correlation between electrical and rheological properties of electrorheological fluids", *Journal of Intelligent Material Systems and Structures*, vol. 7, no. 5, p. 579–582, 1996.

[FUR 99] FURUSHO J., SAKAGUCHI M., "New actuators using ER fluid and their applications to force display devices in virtual reality and medical treatments", *International Journal of Modern Physics B*, vol. 13, nos 14, 15 & 16, p. 2151–2159, 1999.

[GIN 95] GINDER J.M., CECCIO S.L., "The effect of electrical transients on the shear stresses in electrorheological fluids", *Journal of Rheology*, vol. 39, no. 1, p. 211–234, 1995.

[GUE 06] GUEGAN Q, CHEVALIER J, FOULC J.-N., AYELA F, TILLEMENT O., "Progrès récents en électrorhéologie: Les fluides électrorhéologiques à base de nanoparticules", *Actes de la 5ème Conférence de la Société Française d'Electrostatique*, ATTEN, P. and DENAT, A. (eds), Grenoble, p. 149–154, August 2006.

[GUL 93] GULLEY G.L., TAO R., "Static shear stress of electrorheological fluids", *Physical Review E*, vol. 48, no. 4, p. 2744–2751, 1993.

[GOW 90] GOW C.J., ZUKOSKI C.F., "The electrorheological properties of polyaniline suspensions", *Journal of Colloid and Interface Science*, vol. 136, no. 1, p. 175–188, 1990.

[HAL 90] HALSEY T.C., TOOR W., "Structure of electrorheological fluids", *Physical Review Letters*, vol. 65, no. 22, p. 2820–2823, 1990.

[HAR 91] HARTSOCK D. L., NOVAK R.F. , CHAUNDY G.J., "ER fluid requirements for automotive devices", *Journal of Rheology*, vol. 35, no. 7, p. 1305–1326, 1991.

[HAV 95] HAVELKA K.O., FILISKO F.E., *Progress in Electrorheology*, New York, Plenum Press, 1995.

[HUN 81] HUNTER R.J., *Zeta Potential in Colloïd Science*, Academic Press, London, 1981.

[IKA 98] IKAZAKI F., KAWAI A., UCHIDA K., KAWAKAMI T., EDAMURA K., SAKURAI K., ANZAI H., ASAKO Y., "Mechanisms of electrorheology: The effect of the dielectric property", *Journal of Physics D: Applied Physics*, vol. 31, p. 336–347, 1998.

[ISH 95] ISHINO Y., MARUYAMA T., OHSAKI T., ENDO S., SAITO T., GOSHIMA N., "Anhydrous electrorheological fluid using carbonaceous particulate as dispersed phase", in *Progress in Electrorheology*, HAVELKA K.O. and FILISKO F.E. (eds), Plenum Press, New York, 1995.

[JOR 89] JORDAN T.C., SHAW M.T., "Electrorheology", *IEEE Transactions on Electrical Insulation*, vol. 24, no. 5, p. 849–878, 1989.

[JOR 92] JORDAN T.C., SHAW M.T., MCLEISH T.C.B., "Viscoelastic response of electrorheological fluids. II. Field strength and strain dependence", *Journal of Rheology*, vol. 36, no. 3, p. 441–463, 1992.

[KLA 67] KLASS D.L., MARTINEK T, "Electroviscous fluids. I. Rheological properties, II. Electrical properties", *Journal of Applied Physics*, vol. 38, no. 1, p. 67–80, 1967.

[KLI 91a] KLINGENBERG D.J., SWOL F.V., ZUKOSKI C.F., "The small shear rate response of electrorheological suspensions. I. Simulation in the point-dipole limit", *Journal of Chemical Physics*, vol. 94, no. 9, p. 6161–6169, 1991.

[KLI 91b] KLINGENBERG D.J., SWOL F.V., ZUKOSKI C.F., "The small shear rate response of electrorheological suspensions. II. Extension beyond the point-dipole limit", *Journal of Chemical Physics*, vol. 94, no. 9, p. 6170–6178, 1991.

[KLI 93] KLINGENBERG D.J., ZUKOSKI C.F., HILL J.C., "Kinetics of structure formation in electrorheological suspensions", *Journal of Applied Physics*, vol. 73, no. 9, p. 4644–4648, 1993.

[KRU 52] KRUYT H.R., *Colloïd Science*, New York, Elsevier Publishing Co, 1952.

[LEM 92] LEMAIRE E., BOSSIS G., GRASSELLI Y., "Rheological behavior of electrorheological fluids", *Langmuir*, vol. 8, p. 2957–2961, 1992.

[LES 02] LESAINT O., TOP T.V., "Streamer initiation in mineral oil. Part I: Electrode surface effect under impulse voltage", *IEEE Transactions on Dielectrics and Electrical Insulation*, vol. 9, no. 1, p. 84–91, 2002.

[LOB 99] LOBRY L., LEMAIRE E., "Viscosity decrease induced by a DC electric field in a suspension", *Journal of Electrostatics*, vol.47, p. 61–69, 1999.

[MAR 89] MARSHALL L., ZUKOSKI C.F., GOODWIN J.W., "Effects of electric fields on the rheology of non-aqueous concentrated suspensions", *Journal of Chemical Society, Faraday Transactions I*, vol. 89, no. 9, p. 2785–2795, 1989.

[MCL 91] MCLEISH T.C.B, JORDAN T.C., SHAW M.T., "Viscoelastic response of electrorheological fluids. I. frequency dependence", *Journal of Rheology*, vol. 35, no. 3, p. 427–448, 1991.

[MON 97] MONKMAN G.J., "Exploitation of compressive stress in electrorheological coupling", *Mechatronics*, vol. 7, no. 1, p. 27–36, 1997.

[NIU 05] NIU X., WEN W., LEE Y.K., "Electrorheological-fluid-based microvalves", *Applied Physics Letters*, vol. 87, no. 1, 243501, 2005.

[ONS 34] ONSAGER L., "Deviations from Ohm's law in weak electrolytes", *Journal of Chemical Physics*, vol. 2, no. 9, p. 599–615, 1934.

[OTS 92] OTSUBO Y., SEKINE M., KATAYAMA S., "Electrorheological properties of silica suspensions", *Journal of Rheology*, vol. 36, no. 3, p. 479–496, 1992.

[POH 78] POHL H.A., *Dielectrophoresis*, Cambridge University Press, Cambridge, 1978.

[QI 02] QI Y., WEN W., "Influences of geometry of particles on electrorheological fluids", *Journal of Physics D: Applied Physics*, vol. 35, p. 2231–2235, 2002.

[SHE 05a] SHEN R., WANG X, WEN W, LU K., "TiO_2 based electrorheological fluid with high yield stress", *International Journal of Modern Physics B*, vol. 19, nos 7, 8 & 9, part 1, p. 1104–1109, 2005.

[SHE 05b] SHEN M., CAO J.G., ZHU J.T., XUE H.T., ZHOU W., "Van der Waals interaction in colloïdal giant electrorheological systems", *International Journal of Modern Physics B*, vol. 19, nos 7, 8 & 9, part 1, p. 1170–1176, 2005.

[SHI 94] SHIH Y.H., CONRAD H., "Influence of particle size on the dynamic strength of electrorheological fluids", *International Journal of Modern Physics B*, vol. 8, nos. 20 & 21, p. 2835–2853, 1994.

[STA 92] STANWAY R., SPROSTRON J.L., PRENDERGAST M.J., CASE J.R., WILNE C.E., "ER fluids in squeeze-flow mode: An application to vibration isolation", *Journal of Electrostatics*, vol. 28, p. 89–94, 1992.

[TAO 91] TAO R., SUN J.M., "Three-dimensional structure of induced electrorheological solid", *Physical Review Letters*, vol. 67, no.3, p. 398–401, 1991.

[TAO 93] TAO R., "Electric-field-induced phase transition in electrorheological fluids", *Physical Review E*, vol. 47, no. 1, p. 423–426, 1993.

[TOB 96] TOBAZEON R., "Conduction électrique dans les liquides", *Techniques de l'Ingénieur*, traité de Génie Electrique, vol. D2, nos. 430 & 431, 1996.

[UEJ 72] UEJIMA H., "Dielectric mechanism and rheological properties of electro-fluids", *Japanese Journal of Applied Physics*, vol. 11, no. 3, p. 319–326, 1972.

[VUA 01] VUARCHEX P.J., "Huiles et liquides isolants", *Techniques de l'Ingénieur*, traité de Génie Electrique, vol. D, nos. 230, 231 & 232, 2001.

[WEN 03] WEN W., HUANG, X., YANG S, LU K, SHENG P., "The giant electrorheological effect in suspensions of nanoparticules", *Nature Materials*, vol. 2, p. 727–730, 2003.

[WEN 04] WEN W., HUANG X, SHENG P., "Particle size scaling of the giant electrorheological effect", *Applied Physics Letters*, vol. 85, no. 2, p. 299–301, 2004.

[WEN 05] WEN W., WEISBUCH C., PHUONG D.M., LU G., GE W., CHAN C.T., SHENG P., "Neutral nanoparticle-based display", *Nanotechnology*, vol. 16, p. 598–601, 2005.

[WIN 49] WINSLOW W.M., "Induced fibration of suspensions", *Journal of Applied Physics*, vol. 20, p. 1137–1140, 1949.

[WU 98a] WU C.W., CONRAD H., "Multi-coated spheres: recommended electrorheological particles", *Journal of Physics D: Applied Physics*, vol. 31, p. 3312–3315, 1998.

[WU 98b] WU C.W., CONRAD H., "The temperature dependence of the conductivity of silicone oil and related electrorheology of suspensions containing zeolite particles", *Journal of Physics D: Applied Physics*, vol. 31, p. 3403–3409, 1998.

[WU 05] WU Q., ZHAO B.Y., FANG C., HU K.A., "An enhanced polarization mechanism for the metal cations modified amorphous TiO_2 based electrorheolocal materials", *The European Physical Journal E*, vol. 17, p. 63–67, 2005.

[YAN 92] YANG I.K., SHINE A.D., "Electrorheology of a nematic poly(n-hexyl isocyanate) solution", *Journal of Rheology*, vol. 36, no. 6, p. 1079–1104, 1992.

[YIN 05] YIN J, ZHAO X., "Large enhancement in electrorheological activity of mesoporous cerium-doped TiO_2 from high surface area and robust pore walls", *International Journal of Modern Physics B*, vol. 19, nos 7–9, part 1, p. 1071–1076, 2005.

Chapter 18

Electrolytic Capacitors

18.1. Introduction

Considering their ability to store energy in the form of electrostatic charges, capacitors are indispensable passive components in electronic and electronic power circuits. The physical and geometrical properties of the dielectric which constitutes them have a major impact on their characteristics.

Electrolytic capacitors are amongst the most commonly used capacitors because they associate a strong capacity with unit volume, an operating voltage which can reach several hundred volts and an attractive price. Their main fault is their relatively high rate of failure compared with the other constituents of electronic circuits. The presentation of different constituents, their foils and the technology used for the manufacture of these capacitors allows us to understand and deduce their main characteristics. Owing to the elaboration of equivalent electrical diagrams related to the different elements making up the capacitor, the behavior of this latter and the influences of electrical parameters, temperature or ageing can be analyzed.

This chapter starts with a presentation of parameters influencing the characteristics of capacitors, which permits an inventory to be made of their different families. The most used electrolytic capacitors are those with liquid electrolyte aluminum or solid electrolyte tantalum. The principle, the constituents and the foils of these two components are detailed. The models and the characteristics for the capacitors are then presented and the variations of parameters analyzed as a function of voltage and temperature for both types of electrolytic

---

Chapter written by Pascal VENET.

capacitors presented. The failures of these components are then listed and explained. To conclude, a glimpse of the current and future evolution of electrolytic capacitors is given.

18.2. Generalities

18.2.1. *Characteristic parameters*

18.2.1.1. *Presentation*

A capacitor is a device permitting energy to be stored in the form of electrostatic charges. It is composed of two conductive electrodes (the anode and the cathode) separated by an insulating material (the dielectric).

When a potential difference U is applied between both electrodes, an electric charge Q proportional to the applied voltage U and the capacity C_{AK} is accumulated in the capacitor.

18.2.1.2. *Energy and capacity*

For a capacitor under a voltage U of capacitance C_{AK}, the stored energy W is given by the following relationship [18.1]:

$$W = \frac{1}{2}Q.U = \frac{1}{2}C_{AK}.U^2 \qquad [18.1]$$

In the case of an ideal plane capacitor (see Figure 18.1), the value of the capacitance C_{AK} is proportional to the surface S of the electrodes and inversely proportional to the thickness of the dielectric e (see equation [18.2]):

$$C_{AK} = \varepsilon_0 \varepsilon_r \frac{S}{e} \qquad [18.2]$$

where:

– ε_0 is the absolute vacuum permittivity ($\varepsilon_0 = 8{,}854.10^{-12}$ F/m);

– ε_r is the relative permittivity or dielectric constant of the dielectric material;

– S is the surface of the electrodes (m²);

– e is the thickness of the dielectric (m).

Considering equations [18.1] and [18.2], the stored energy W in the capacitors is therefore a function of four parameters: U, ε_r, S and e. To increase the stored energy, we must therefore increase the voltage U, the relative permittivity ε_r, the surface S and decrease the thickness e.

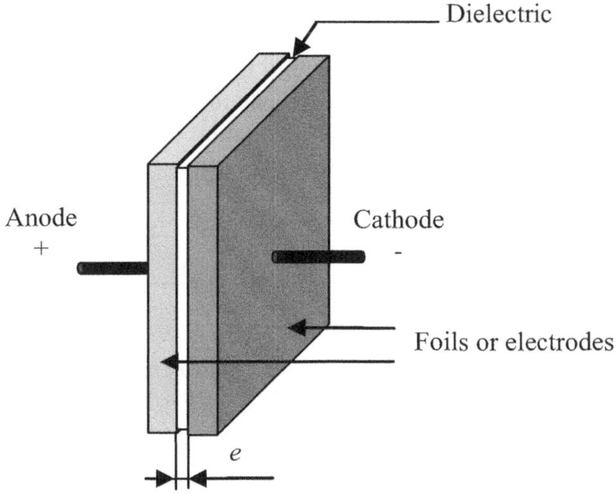

Figure 18.1. *View of an ideal plane capacitor*

18.2.1.3. *Dielectric constant and rigidity*

The parameters U, ε_r and e are interdependent of one another. Indeed, one of the essential characteristics of the dielectric is its performance in voltage as a function of its thickness.

Table 18.1 gives the value of the dielectric constant ε_r and the dielectric rigidity e_v of different materials used for the manufacture of capacitors [PER 03], [BES 90], [LAG 96]. These materials define the type of capacitor. For certain materials, the dielectric rigidity is not linear depending on their thickness [MEN 97]. The technological stresses due to materials and to the manufacture of capacitors therefore impose restrictions on the dielectric thickness and its voltage performance.

The use of a dielectric given under an appropriate voltage allows a strong capacitance. The associated dielectric constant is therefore a key factor of this capacitance. Let us note that ceramic capacitors possess a very strong relative permittivity value but, unfortunately, the minimal thickness of the dielectric is relatively large (see below).

Type of capacitor	Dielectric material	Dielectric constant ε_r	Dielectric rigidity e_v (kV/cm)
Air	Air	1.00059	21
Film	Plastic	2.1 to 6	600 to 4,500
	Paper	2 to 6	300 to 600
Electrolytic	Aluminum oxide (AL_2O_3) (aluminum electrolytic capacitor)	8 to 10	6,600 to 7,700
	Tantalum oxide (ta_2O_5) (tantalum electrolytic capacitor)	10 to 27	6,600 to 10,000
Glass, Mica, Ceramic	Glass	4.8 to 9.9	100 to 400
	Mica	5.4 to 8.7	600 to 1,800
	Ceramic	10 to 20,000	200 to 400

Table 18.1. *Dielectric constant and rigidity of the main materials used for the manufacture of capacitors*

This leads us to tackle the second dimension parameter of capacitance, i.e. the dielectric thickness.

18.2.1.4. *Thickness of the dielectric*

The minimal dielectric thickness e_{min} is a function of dielectric rigidity e_v and the rated voltage U_n of the component according to the following relationship:

$$e_{min} = \frac{U_n}{e_v} \quad [18.3]$$

Table 18.2 gives the minimal thickness of the dielectric e_{min} ([BES 90], [LAG 96]) for the different families of capacitors and the order of magnitude of the rated voltage range. For electrolytic capacitors, this thickness, which constitutes that of the oxide formed by the electrolyze process, is a function of the applied voltage during this process.

The thickness of the dielectric is a determining element for the value of the capacitance. Indeed, we see in Table 18.2 that electrolytic capacitors, which can have a minimal thickness of dielectric e_{min} up to 1,000 times less than the other components, possess a very strong capacitance in comparison with these latter.

Type of capacitor	Dielectric material	Minimal thickness of the dielectric e_{min}	Order of magnitude of rated voltages (V)
Film	Plastic	0.9 to 6 μm	10 to 10^4
	Paper	5 to 14 μm	10^2 to 10^5
Electrolytic	Aluminum oxide (AL_2O_3) (aluminum electrolytic capacitor)	1.4 nm/V	10 to 500
	Tantalum oxide (ta_2O_5) (tantalum electrolytic capacitor)	1.4 nm/V	10 to 500
Glass, Mica, Ceramic	Glass	10 μm	10^2 to 10^3
	Mica	20 μm	10^2 to 10^4
	Ceramic	20 μm	10 to 10^4

Table 18.2. *Minimal thickness of the main dielectric materials used for the manufacture of capacitors and associated voltage ranges*

18.2.1.5. *Surface of electrodes*

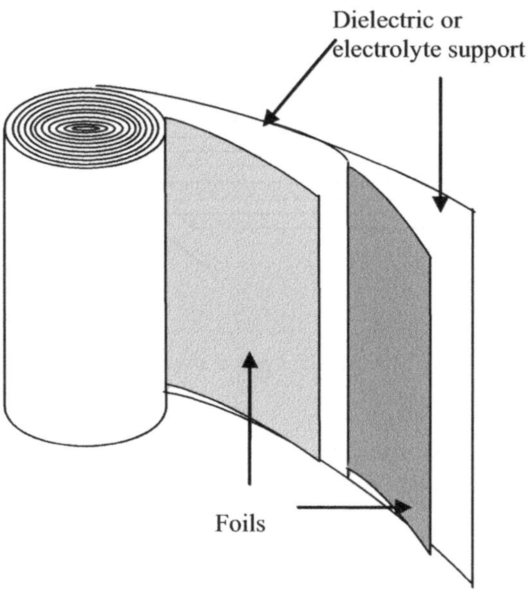

Figure 18.2. *Coiled capacitor*

The last important parameter permitting the value of the capacitance to be increased is the surface of the electrodes *S*. To obtain a large surface of electrodes for a given package volume, two processes are used:

– the dielectric material (or electrolyte support for aluminum electrolytic capacitors (see section 18.4)) is coiled with the electrodes (see Figure 18.2). The package used is most often cylindrical or possibly parallelepipede-shaped if the coil is flattened;

– the dielectric material (or electrolyte support for aluminum electrolytic capacitors (see section 18.4)) is piled up with electrodes on different layers for the insertion in a generally parallelepipedic package (see Figure 18.3). According to the type and the technology of capacitors, there are several bonding possibilities of different layers of the same nature.

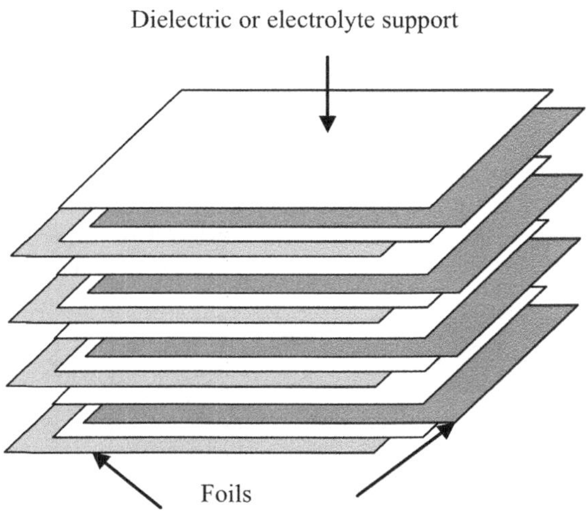

Figure 18.3. *Parallelepiped-shaped capacitor*

Let us note that the capacitance C_{AK} of a cylindrical capacitor with only one layer having a length l, of electrodes with radius r_1 and r_2 (see Figure 18.4) is given by the following expression:

$$C_{AK} = \frac{2\pi\varepsilon_0\varepsilon_r l}{ln\frac{r_2}{r_1}} \qquad [18.4]$$

In general, the difference r_2-r_1, corresponding to the thickness e of the dielectric, is small compared to r_1. The expression can therefore be written:

$$C_{AK} = \frac{2\pi\varepsilon_0\varepsilon_r l}{ln\left(1+\frac{r_2 - r_1}{r_1}\right)} \approx \frac{2\pi\varepsilon_0\varepsilon_r l.r_1}{r_2 - r_1} = \frac{\varepsilon_0\varepsilon_r S}{e} \qquad [18.5]$$

where $e = r_2$-r_1 and $S = 2\pi\, r_1.l$ corresponding to the surface of the electrodes.

The capacitance C_{AK} of a cylindrical capacitor is therefore comparable to that of a flat capacitor.

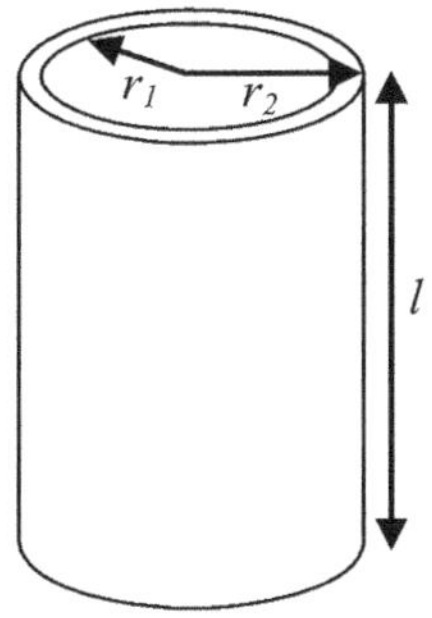

Figure 18.4. *Cylindrical capacitor with one layer*

For aluminum electrolytic capacitors (see section 18.4), the electrode surface on which the oxide is left by electrolysis is engraved, which allows it to have an effective surface larger than the apparent surface. The multiplicative factor between the apparent surface and the real surface S can vary from 20 to 100 according to the voltage range of capacitors. Let us note that the quality of the engraving has an important influence on the capacitance of the capacitor and it determines a large part of its tolerance range.

18.2.2. *Conclusions on the different families of capacitors*

The different parameters detailed above allow both main parameters to be taken into account for the choice of a type of capacitors, i.e. capacitance C_{AK} and use of the voltage U of the component. Figure 18.5 represents the orders of magnitude of capacitance and voltage ranges for the different families of capacitors on the market [ELE 04], [SAR 98], [WIL 92].

The components presenting the strongest volume capacitance value are electrolytic capacitors. For example, the theoretical volume v of a capacitor 4,700 µF / 500 V as a function of the used technology is as follows [PER 03]:

– for an aluminum electrolytic capacitor: $v \approx 0.4\ dm^3$;

– for a polypropylene film capacitor: $v \approx 70\ dm^3$;

– for a mica capacitor: $v \approx 500\ dm^3$.

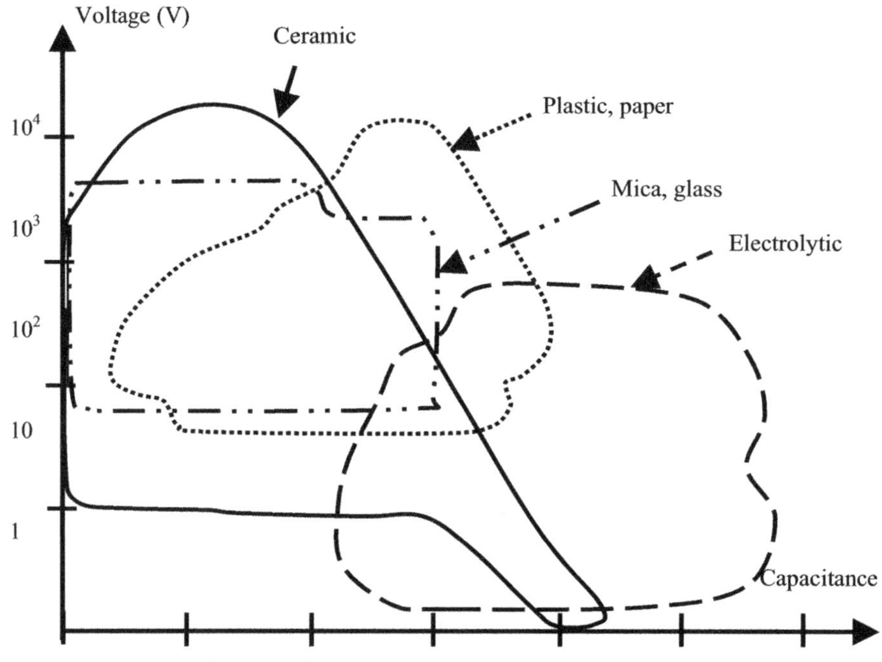

Figure 18.5. *Capacitance and voltage ranges for the different types of capacitors*

18.3. Electrolytic capacitors

The next part of this chapter deals with electrolytic capacitors. Owing to their strong capacitance volume associated with a rated voltage which can reach several hundred volts, these components are widely used in the electronic and power electronic domains. The frequential and thermal electrical stresses present in electronic circuits have a direct influence on the behavior of capacitors. The main disadvantage of electrolytic capacitors is their relatively weak reliability.

There are two large families of electrolytic capacitors: liquid electrolyte or solid electrolyte.

Aluminum electrolytic capacitors dominate the market of capacitors with strong capacitance. They are therefore much used in power electronics. They represent over a third of the world market of capacitors [LAG 96], [NIS 96]; 99% of these are liquid electrolyte [NIS 96]. This is mainly due to the relatively high cost of solid electrolyte capacitors and the technological difficulty of uniformly fitting the electrodes of these components with the aid of a solid electrolyte.

In the domain of electronics, the most used solid electrolyte capacitors are tantalum electrolytic capacitors which, at a reasonable price, allow a high degree of miniaturization necessary in computer circuits, cameras, flat screens, telephones, etc. The capacitance obtained for a reduced-volume capacitor is important owing to, amongst other things, a dielectric constant practically three times higher than that of aluminum electrolytic capacitors. These components represent over 10% of the world market of capacitors [LAG 96], [NIS 96].

The vast majority of electrolytic capacitors present in power and electronic circuits are therefore liquid electrolyte aluminum capacitors and solid electrolyte tantalum capacitors. It is therefore these which will be presented below. Most analyzes, interpretations or conclusions will nevertheless be transposable to other electrolytic capacitors.

18.4. Aluminum liquid electrolytic capacitors

18.4.1. *Principles and composition [PER 03], [ALV 95]*

Aluminum electrolytic capacitors are composed of two aluminum foils, an electrolyte support (also called a separator) made up of paper foils impregnated by the electrolyte (generally boric acid dissolved in a solvent) and an aluminum oxide (AL_2O_3) layer constituting the dielectric and formed on the surface of the anode foil by electrolysis. The composition of such a capacitor is represented in Figure 18.6.

As we saw above, the thin oxide layer (a few nm to a few hundred nm) and the engraving of the aluminum foil constituting the anode means large capacitances can be achieved.

This type of capacitor is polarized and can only support a very weak inverse voltage (of a few volts). This performance with inverse voltage is due to a thin oxide layer which is naturally created on the aluminum cathode of the component (see Figure 18.6).

The dielectric, i.e. the aluminum oxide layer, is formed by electrolysis on the anode during the manufacturing process of the component. The electrolyte which impregnates the paper has two main roles:

– to ensure the best possible electrical conduction between the aluminum foil brought to the negative potential and the aluminum oxide (the real cathode is therefore made up of the impregnated paper);

– to regenerate the integrity of the aluminum oxide in case of defect.

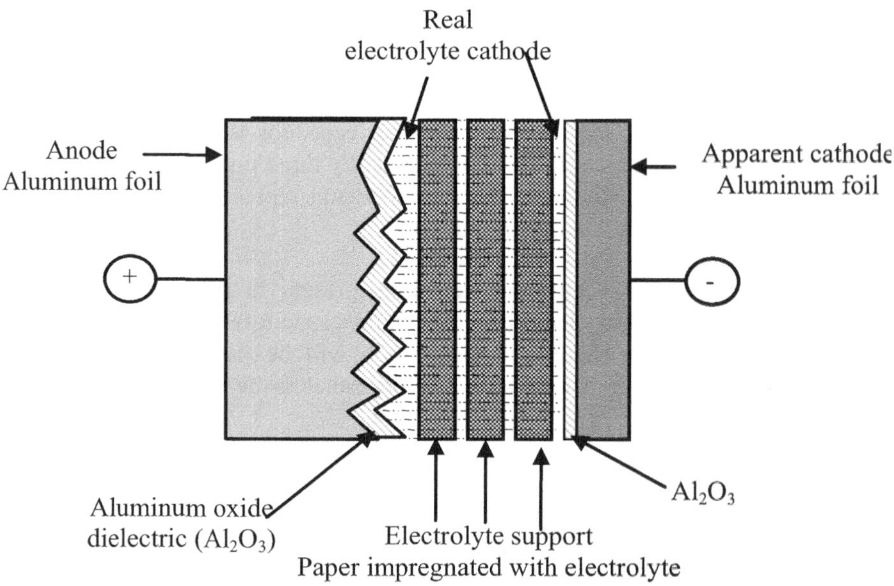

Figure 18.6. *Composition of an aluminum electrolytic capacitor*

These latter defects are related to the quality and the greater or lesser thickness of the aluminum oxide layer. In thinner zones which can appear, the leakage current I_L can increase locally. The current thus created transforms the water of the electrolyte into oxygen and hydrogen. This electrochemical phenomenon permits an oxidation of the anode, which eliminates the possible weaknesses of the aluminum oxide layer. This phenomenon is called self-regeneration of the oxide.

18.4.2. *Assembly and connections [PER 03]*

Tabs link the coil of the capacitor to the terminals on the component package. These tabs are set on the anode and on the cathode. Figure 18.7 represents the connections realized on the coil for a capacitor with radial outputs. There are several coil loops by tab. The path of the current entering the anode is therefore axial in the tabs and mainly orthoradial in the electrodes.

Knowledge of the direction and distribution of the currents in the different elements of the capacitor are essential to determine the magnetic field within and near the component, the associated parasitic inductance, the losses and the heating generated [JOU 96], [PER 05].

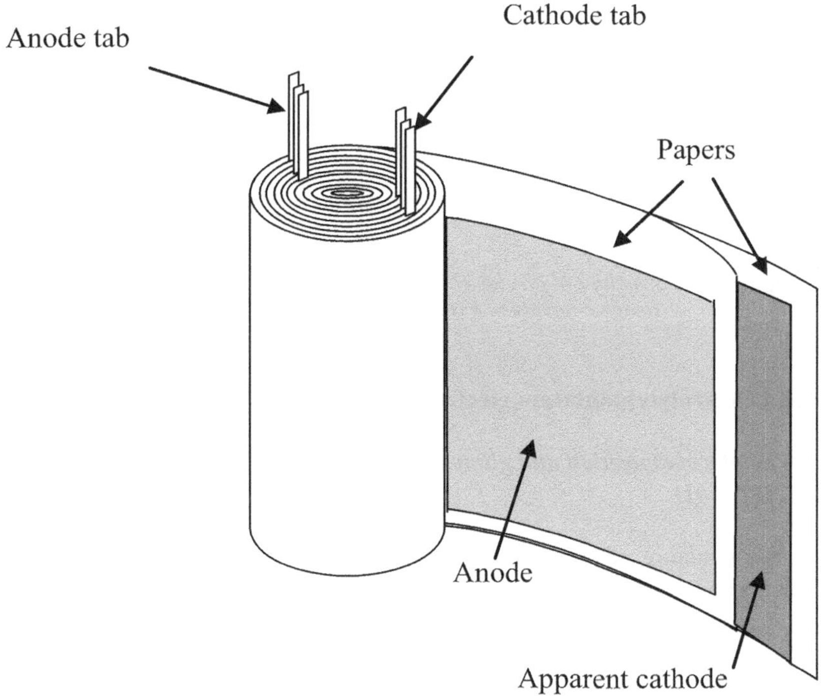

Figure 18.7. *Coil connnections*

The coil is placed in a waterproof aluminum package. For example, Figure 18.8 represents the composition of a large electrolytic capacitor possessing a connection with screwed terminals. Depending on the package, a rupture point or a safety valve (see Figure 18.8) allows the evacuation of possible over pressure in case of evaporation of the electrolyte due to internal heating (see section 18.7.3.1).

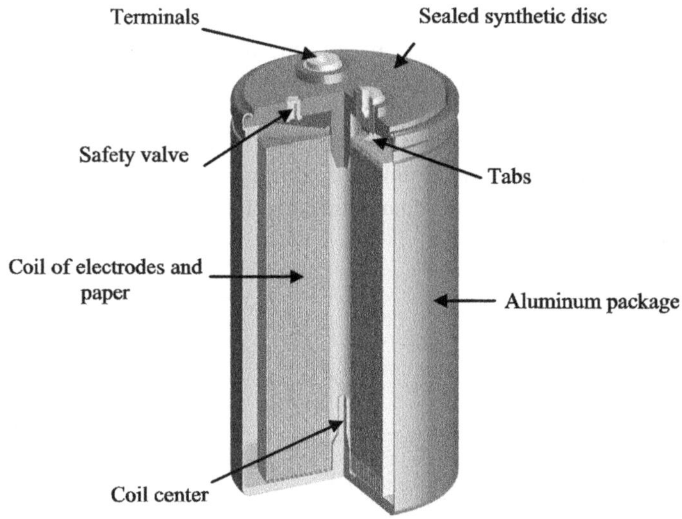

Figure 18.8. *Structure of an electrolytic capacitor with screwed terminals. (Image courtesy of Vishay Intertechnology, Inc.)*

18.5. (Solid electrolyte) tantalum electrolytic capacitors

18.5.1. *Principle, composition and glimpse of the manufacture [BES 90], [KEM 01], [LAG 96], [PRY 01]*

The basic structure of a tantalum electrolytic capacitor is represented in Figure 18.9. The role and the manufacture of the different elements are explained below.

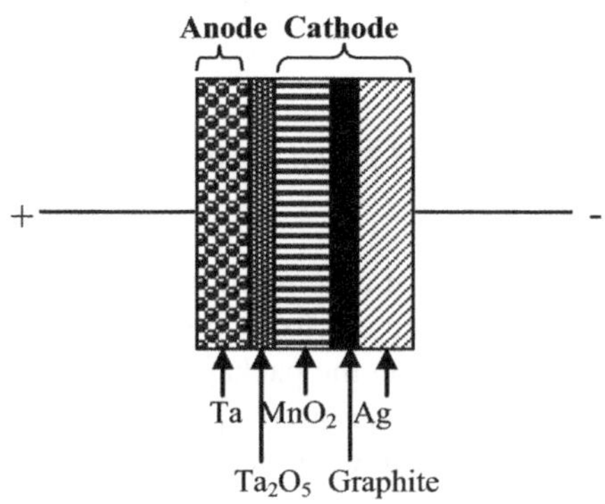

Figure 18.9. *Composition of an electrolytic tantalum capacitor*

Tantalum (Ta) is a rare metal. To obtain the anode, some tantalum powder mixed with a binding organic is pressed with a tantalum connection wire. In general, the conglomerated powder (pellet) has a cylindrical shape for traversing components (drop-shaped or cylindrical) and parallelepipedic for the SMC (Surface Mounted Components). A tantalum connection wire is placed in the pellet axis. The strengthening of the pellet is made by placing the device under vacuum at very high temperature (generally between 1,200°C and 1,800°C). This latter process is called fritting.

The formation of the dielectric (Ta_2O_5) is realized by immersing the pellet in an acid bath and by subjecting it to a continuous voltage (3 and 4 times higher than the rated voltage of the capacitor). This operation oxidizes the tantalum, and a thin layer of tantalum pentoxide (Ta_2O_5) (with thickness of about 1.4 to 2 nm/V) is formed at its surface. Owing to the use of the porous structure made up of tantalum powder, the surface affected by the oxidation can be up to 100 times greater than the apparent surface of the pellet. The thinness of the dielectric combined with this large surface and the relatively strong dielectric constant allows strong capacitance values to be obtained.

The cathode is made by dipping the pellet in a manganese nitrate solution which penetrates the porous structure and is left on the dielectric. By heating the whole, the nitrate is decomposed in manganese dioxide (MnO_2). The manganese dioxide can be replaced by a polymer, which allows the equivalent series resistance of the capacitor to be improved [PRY 01]. It is quite tricky to link the manganese dioxide (MnO_2) connection wire. For this purpose, the pellet is first dipped into a graphite solution, dried, and then a silver (Ag) layer is left on the whole. The connection wire of the cathode is welded on the silver layer or is placed in contact with the silver via the the component package.

As for the electrolyte of the aluminum electrolytic capacitor, the manganese dioxide (MnO_2) of the tantalum capacitor has two main roles:

– to ensure the best possible electrical conduction between the negative terminal of the component and the tantalum oxide (ta_2O_5);

– to insulate the possible defects of the tantalum oxide (by a self-healing phenomenon).

These latter defects are related to the quality and the presence of possible impurities in the tantalum oxide (Ta_2O_5) layer. If a defect zone is present, the leakage current I_L increases locally in this latter and in the manganese dioxide adjacent to this zone. The current thus created locally heats the manganese dioxide (MnO_2), which by liberating oxygen is decomposed into manganese sesquioxide (Mn_2O_3) of much higher resistivity. If this mechanism is successful, a zone of very high resistivity adjacent to the defect decreases the current and prevents a short circuit (see Figure 18.10) [PRY 01]. This phenomenon is called self-healing. It is only possible if the defect zone is small and if the energy dissipation is not too high. In the opposite case, local heating near the defect zone can lead to an extension of this zone and therefore make self-healing impossible.

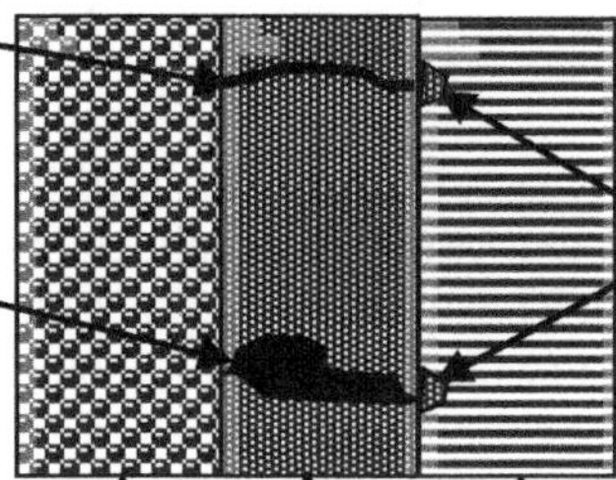

Figure 18.10. *Self-healing mechanism of tantalum capacitors*

18.5.2. *Assembly and connections*

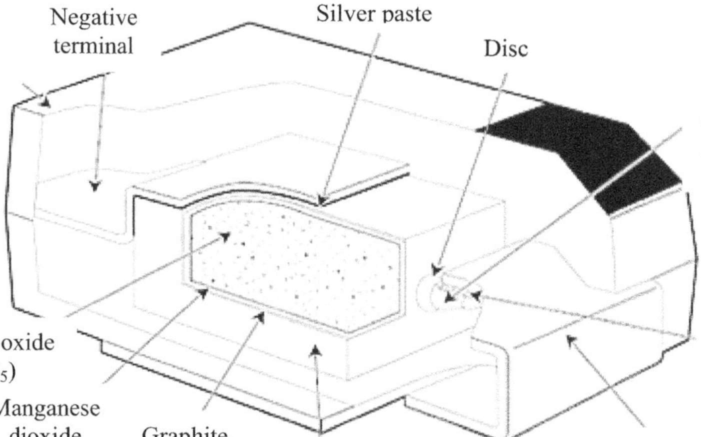

Figure 18.11. *Assembly of an SMC tantalum capacitor [KEM 06]*

The assembly described above is then coated with epoxy resin, moulded or placed in a plastic or metallic package. The traversing components can have different shapes (cylindrical, parallelepipedic, "drop") whereas the SMC (Surface Mounted Components) are generally shaped like a cobblestone; for example, Figure 18.11 represents the structure of a SMC tantalum capacitor [KEM 06].

18.6. Models and characteristics

18.6.1. *Representative electrical diagram*

There are several diagrams representing the frequential behavior of capacitors. The most used model, combining simplicity and relatively good precision, is represented in Figure 18.12, in which:

– C_{AK} represents the ideal capacitance between anode and cathode;

– R_p is parallel resistance, representing the losses in the dielectric and leakage between both electrodes (considering the defects of the dielectric). This element representing the insulating resistance of the capacitor induces a leakage current I_L. The orders of magnitude of this current are variable according to the type of dielectric and the technology of the capacitor;

– R_l is the series resistance of the connections and the electrodes (including the paper impregnated with electrolyte for aluminum electrolytic capacitors);

– L is the equivalent series inductance of the connections and coils.

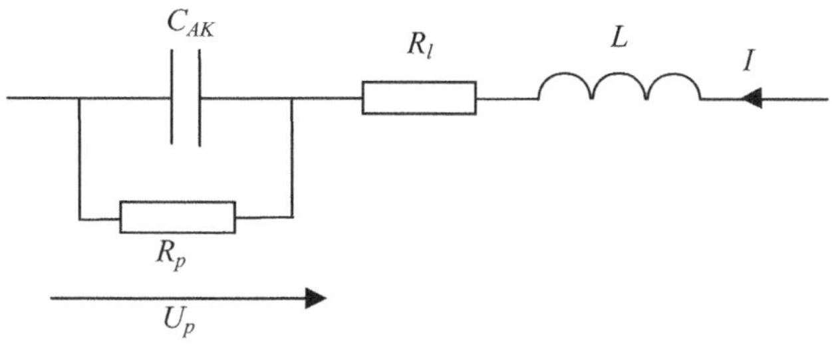

Figure 18.12. *Representative diagram of a capacitor*

The capacitance C_{AK} is a function of the relative permittivity (see equation [18.2]) which mainly depends on temperature, applied voltage and a lower frequency measurement [ROB 79].

The resistance R_p depends on the applied voltage and the temperature.

The resistance R_l is a function of the temperature and the frequency (this latter variation, visible in high frequency, is due to the skin effect and the non-homogenous distribution of currents in the electrodes [JOU 96]).

These different influences on the elements of the diagram will be detailed in later sections.

We can simplify this diagram according to the normalized representation of Figure 18.13. It is composed of elements function of the frequency which are:

– C, the capacitance;

– ESR, the Equivalent Series Resistance, representing all the losses in the capacitor;

– ESL, the Equivalent Series Inductance.

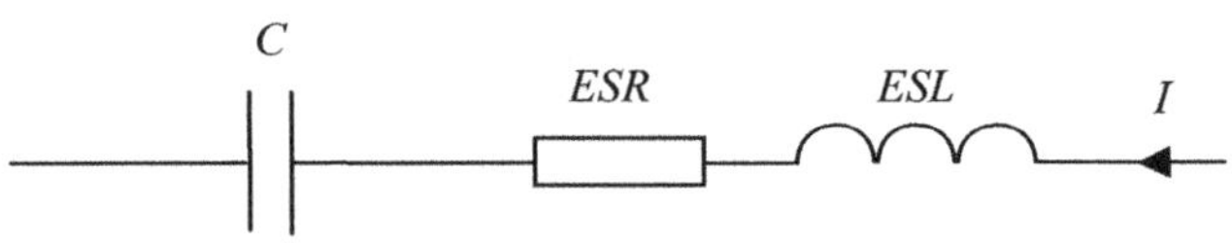

Figure 18.13. *Normalized equivalent diagram of a capacitor*

From the identity of impedances of the circuit represented in Figure 18.12 and that represented in Figure 18.13, we deduce the following relationships:

$$C = C_{AK}\left(1 + \frac{1}{R_p^2 C_{AK}^2 \omega^2}\right) \quad [18.6]$$

$$ESR = R_l + \frac{R_p}{1 + R_p^2 C_{AK}^2 \omega^2} \quad [18.7]$$

$$ESL = L \quad [18.8]$$

where ω is the electric pulsation.

This representation is important because it can be deduced quasi directly from the characterization of the capacitor during frequential measurements of the impedance Z of the capacitor. Indeed, the *ESR* resistance represents the real part of the impedance while the imaginary part is comparable to reactances $1/(C.\omega)$ at low frequency and $ESL.\omega$ at high frequency. The resonance frequency f_r is thus expressed:

$$f_r = \frac{1}{2\pi\sqrt{ESL.C}} \qquad [18.9]$$

The modulus variation of the impedance Z with the frequency for an aluminum electrolytic real capacitor 4,700 µF / 500 V is drawn in Figure 18.14 on a log-log scale. The normalized equivalent diagram represents its behavior precisely.

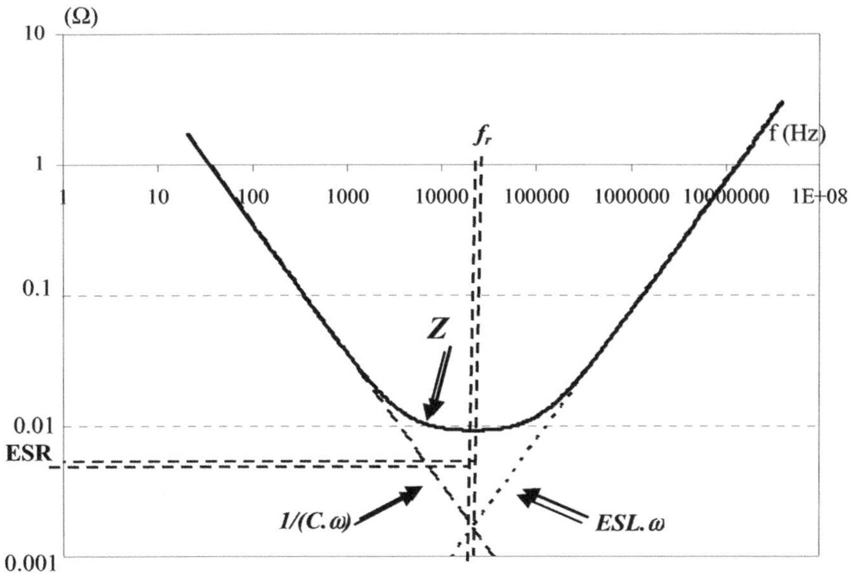

Figure 18.14. *Modulus of the impedance Z as a function of the frequency for a capacitor 4,700 µF. Values deduced from the measurement: C = 4,620 µF, ESR = 9,2 mΩ, ESL = 12 nH*

18.6.2. *Loss factors, loss angles*

The elements R_p and C_{AK} of the equivalent circuit in Figure 18.12 are directly related to the dielectric. By considering the voltage U_p to the terminal of R_p, the losses in the dielectric are written:

$$P_d = \frac{U_p^2}{R_p} \qquad [18.10]$$

These are null if the resistance R_p tends to infinity, which corresponds to a dephasing φ_p of the current in the capacitor I on the voltage U_p at -π/2. In the opposite case, the dephasing φ_p is greater than -π/2 and the loss angle of the dielectric δ_d is then defined as being the complement of the phase angle φ_p with respect to -π/2. This angle is visualizable on the complex plane by representing the complex admittance Y_p made up by R_p in parallel with C_{AK} (see Figure 18.15).

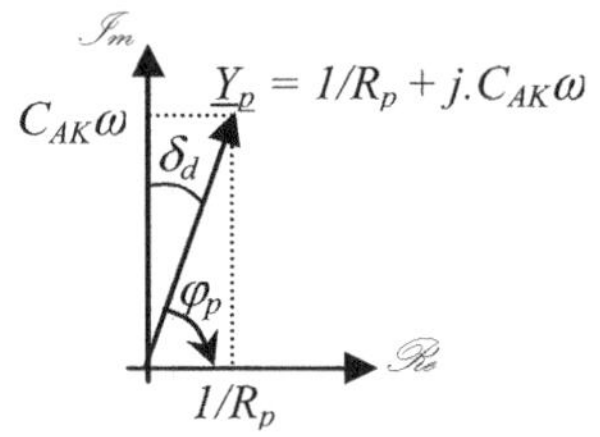

Figure 18.15. *Representation in the complex plane of Y_p and δ_d*

The loss (or dissipation) factor of the dielectric is defined by expressing the loss angle tangent of the dielectric, which amounts to expressing the ratio of the current in R_p with the current in C_{AK} (see equation [18.11]).

$$tan\,\delta_d = \frac{1}{R_p C_{AK}\omega} \qquad [18.11]$$

Thus, by considering equation [18.10], the losses in the dielectric P_d can be expressed as a function of this factor and the reactive power dissipated in the dielectric Q_d:

$$P_d = U_p^2 C_{AK}\omega.tan\,\delta_d = Q_d.tan\,\delta_d \qquad [18.12]$$

The total losses in the capacitor P are the sum of the losses in the dielectric P_d and the losses by Joule effect due to the resistance R_l of the connections and the electrodes. By considering I, the current in the capacitor, the equivalent diagrams in Figure 18.12 and Figure 18.13, they are therefore written:

$$P = P_d + R_l.I^2 = ESR.I^2 \qquad [18.13]$$

For a perfect capacitor, there are no losses P, the resistance ESR is null and the dephasing φ of the current in the capacitor I on the voltage at the terminals of the component is at -π/2. For a real capacitor, the dephasing φ is greater than -π/2 and the loss angle δ is then defined as being the complement of the phase angle φ with respect to -π/2. By neglecting ESL, the loss angle δ is visualizable on the complex plane by representing the complex equivalent impedance Z of the circuit in Figure 18.13 (see Figure 18.16).

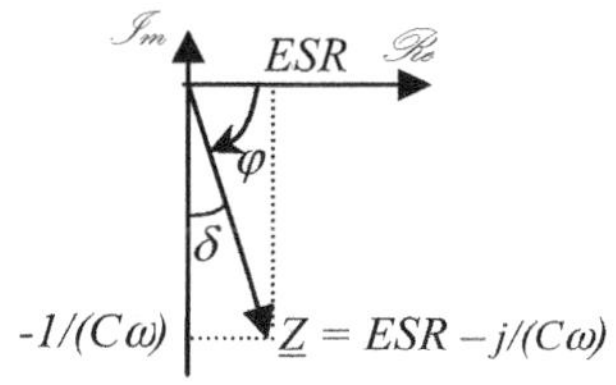

Figure 18.16. *Representation in the complex plane of Z and δ*

The loss (or dissipation) factor of the capacitor is defined by expressing the loss angle tangent:

$$tan\,\delta = ESR.C.\omega \qquad [18.14]$$

It is this factor entitled *DF* (Dissipation Factor) which is given by the manufacturers of capacitors. It permits the total losses P in the component as a function of the total reactive power Q considering equations [18.13] and [18.14] to be determined:

$$P = tan\,\delta.\frac{I^2}{C\omega} = Q.tan\,\delta \qquad [18.15]$$

The loss factor of the capacitor $tan\,\delta$ must not be confused with the loss factor of the dielectric $tan\,\delta_d$. The procedure to obtain the relationship relating these two factors is given below.

By neglecting ESL, the expression for the equality of reactive powers of equivalent circuits in Figure 18.12 and Figure 18.13 gives a relationship between the current I and the voltage U_p:

$$Q = Q_d \Leftrightarrow \frac{I^2}{C\omega} = U_p^2 C_{AK}\omega \qquad [18.16]$$

By expressing the total losses in the capacitor with the aid of equations [18.13] and [18.15], and by considering equations [18.10], [18.11] and [18.16], we find the relationship between both loss factors:

$$tan\,\delta = tan\,\delta_d + R_l.C.\omega \qquad [18.17]$$

By only taking into account the circuit elements in Figure 18.12 and Equation [18.6], this equality can thus also be written:

$$tan\,\delta = tan\,\delta_d + R_l.C_{AK}\left(1+\frac{1}{R_p^2C_{AK}^2\omega^2}\right).\omega \qquad [18.18]$$

There is equality between both factors only if the losses by Joule effect due to the resistance R_l are neglected.

18.6.3. *Variation as a function of the voltage*

The polarization voltage U (continuous voltage) applied to the terminals of the capacitor has a minor influence (1 or 2% at the maximum) on the value of the capacitance C_{AK} and the series resistance of the connections and the electrodes R_l [KEM 06], [PER 03]. On the other hand, its influence is important on the leakage current and therefore on the resistance R_p which represents, among others, these leaks between both electrodes.

This voltage sensitivity is easily understood knowing that the stronger the electric field is, the more the electrolyte (the manganese dioxide for tantalum capacitors) is able to emit electrons which migrate towards the anode via the dielectric. The leakage current is therefore mainly a function of this electric field applied between the anode and the cathode, as well as the thickness and the quality of the oxide. It is therefore higher for capacitors with a strong capacitance value, because statisticly the larger the surface of the dielectric is, the higher the number of possible defects (impurities or irregularities in the layer). Let us note that the leakage current is also variable as a function of the polarization time and depends on the previous state for liquid electrolyte capacitors. Indeed, the aluminum oxide layer of a stored aluminum electrolytic capacitor is degraded. The leakage current when brought into service can be large. It then decreases in time owing to the self-healing phenomenon.

For an aluminum electrolytic capacitor, this leakage current I_L can be written as a function of the polarization voltage U in the following form [PER 03], [RIF 95]:

$$I_L(U)=\frac{U}{R}+I_{Ln}.exp[(U-U_n).\xi]-I_{Ln}.exp(-U_n.\xi) \qquad [18.19]$$

where U_n is the rated voltage of the capacitor, R is the resistance of the dielectric considered constant for weak voltages (less than $U_n/2$), I_{Ln} is the leakage current at U_n, and ξ a coefficient depending on the component.

Figure 18.17 shows the leakage current I_L versus applied polarization voltage U, measured 5 minutes after the beginning of the charge for an aluminum electrolytic capacitor of 4,700 μF / 500 V [PER 03]. This curve was deduced from equation [18.19].

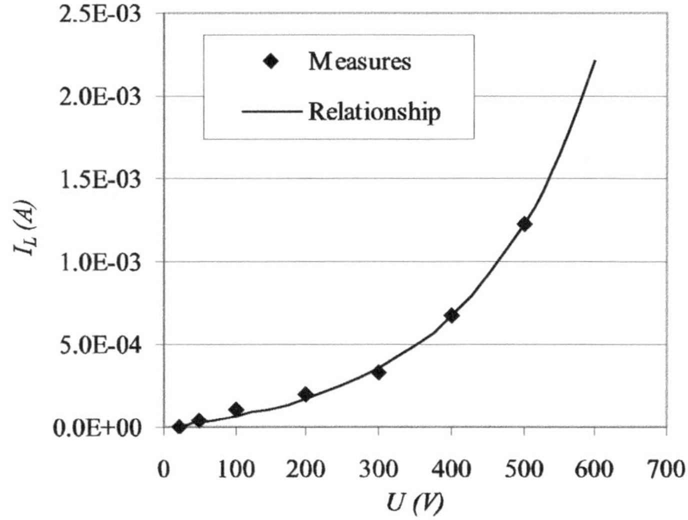

Figure 18.17. *Leakage current at 5mn as a function of the polarization voltage for a capacitor 4,700 μF / 500 V at 25°C*

The resistance R_p decreases exponentially with the voltage U because this latter can thus be expressed (knowing that the resistance R_l is negligible compared to R_p):

$$R_p(U) = \frac{U}{I_L(U)} \qquad [18.20]$$

The variations with the voltage for tantalum capacitors are near those of aluminum electrolytic capacitors.

18.6.4. *Variation as a function of the ambient temperature*

18.6.4.1. *Electrolyte liquid aluminum electrolytic capacitors*

Aluminum electrolytic capacitors are very sensitive to the variations of temperatures because of the property changes of the electrolyte. For example, Figure 18.18 represents the modulus of the impedance Z as a function of the frequency for different ambient temperatures for a capacitor 4,700 μF/500 V. When the temperature increases, we note a strong decrease of the equivalent series resistance

(ESR) (corresponding to the value of Z at the resonance frequency f_r i.e. the minimum of Z as a function of f), a shift (decrease) of the resonance frequency due to an increase of the capacity C (see equation [18.9]).

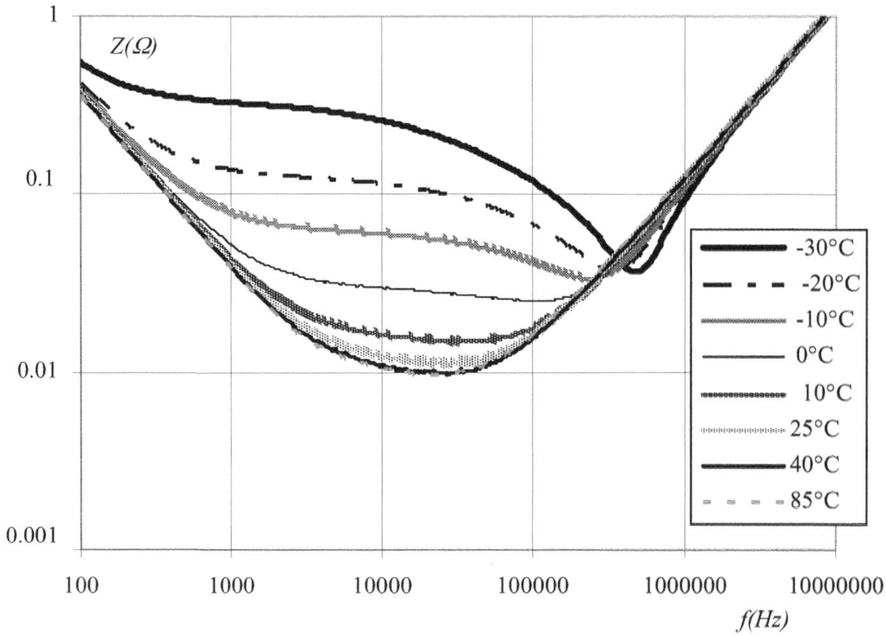

Figure 18.18. *Modulus of the impedance Z as a function of the frequency and defined by the ambient temperature for a capacitor 4,700 μF/500 V*

These variations are above all due to the change in viscosity of the electrolyte when the temperature decreases, which decreases its conductivity. Further, the temperature variations can generate a dilation of electrodes which also modifies the capacitance value.

ESR decreases exponentially with temperature T according to the formula:

$$ESR(T) = \alpha + \beta . exp(-T / \gamma) \qquad [18.21]$$

α, β *and* γ are coefficients which depend on the component. ESR decreases (a few tens per cent) between a temperature of 0°C and the maximal temperature acceptable by the component, but it can increase by a factor of more than 10 if the temperature decreases to negative values.

To a first approximation, the capacitance C increases linearly a few tens per cent with temperature.

The leakage current increases with the temperature T according to an exponential law in the form:

$$I_L(T) = \chi + \lambda.exp(T/\theta) \qquad [18.22]$$

χ, λ, θ are coefficients which depend on the component. The increase between the negative minimal and the positive maximal temperatures is important since it can reach a factor near 100.

The elements of Figure 18.12 can be determined as a function of temperature [PER 03]. The inductance L can be considered constant. The variations of the other parameters are given below, knowing that the different coefficients (Greek letters) depend on the component:

$$R_l(T) = \alpha_l + \beta_l.exp(-T/\gamma_l) \qquad [18.23]$$

$$C_{AK}(T) = \kappa + \nu.T \qquad [18.24]$$

$$R_p(T) = \frac{U}{\frac{U}{\rho exp(-T/\tau)} + I_{Ln}(T)exp[(U-U_n)\xi] - I_{Ln}(T)exp(-U_n\xi)} \qquad [18.25]$$

$I_{Ln}(t)$ is determined using equation [18.22].

The model in Figure 18.12 is valid as long as the temperature is positive. For negative temperatures, Figure 18.18 shows that the modulus curve of the impedance Z as a function of the frequency presents an inflection point. The basic model in Figure 18.12 is then no longer satisfactory and its order must be increased by adjunction of elements in parallel or in series. Possible models, representing correctly the functioning at a negative temperature, are given in [PER 06].

18.6.4.2. *Solid electrolyte tantalum capacitors*

For tantalum capacitors, variations with temperature are of a similar form to those of aluminum electrolytic capacitors. However, their amplitudes are low because the electrolyte is solid, so there is no modification related to the state change of the electrolyte. This aspect constitutes one of the advantages of solid electrolyte capacitors.

18.7. Failures of electrolytic capacitors

In comparison to other electronic components, electrolytic capacitors present relatively large failure rates [LAH 98], [USM 95] hence the importance of this section.

18.7.1. *Modes and failure rates of components*

The failure rate $\lambda(t)$ is the conditional probability of failure by unit time. It gives a measurement of the risks for a device to break down during a time interval $]t, t+\Delta t]$ when Δt tends to zero, knowing that this device lasted until time t. The failure rate $\lambda(t)$ is expressed in 10^{-6}/hour or 10^{-9}/hour (implied: failure/10^6 h or failure/10^9 h) or in FIT (*Failure In Time* with 1 FIT = 1 failure/10^9 h). Thus, if 1,000 capacitors are used for 1 million hours (114 years), 1 failure corresponds to 1 FIT. For electrolytic capacitors, according to the stress and the components, the failure rates can range from a few FIT to a few thousand (even a few tens of thousand) FIT (for reasonable ambient temperatures).

The failure rate λ(t) follows for many components the bathtub curve represented in Figure 18.19.

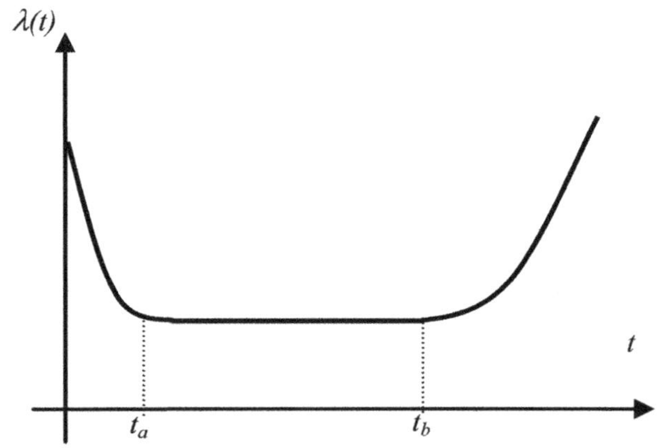

Figure 18.19. *Failure rate $\lambda(t)$ as a function of time*

This curve is split in three parts:

– the early failure period ($t \leq t_a$), due to the youth defects of capacitors, where the failure rate $\lambda(t)$ decreases;

– the intermediary period (t $\in\,]t_a, t_b[$), where the failure rate is approximately constant, corresponding to the normal lifetime period;

– the wear period (t $\geq t_b$) where the failure rate increases. In this period, the failures take a systematic character. Generally, these are failures by degradation (drifts).

There are two main failure modes, by degradation and by catastrophy:

– failures by degradation are defined by a change of characteristics of the component which drifts out of its specific tolerances. The most current degradations are an increase of the equivalent series resistance (*ESR*) or of the loss factor $tan\delta$, a decrease in the capacitance *C*, an increase in the leakage current I_L;

– catalectic failures (sometimes called catastrophic by manufacturers) are sudden failures corresponding to the disappearance of the component function. They are most often characterized by a short circuit or an open circuit of the capacitor.

18.7.2. *Influence of temperature*

Temperature is a major stress with a direct influence on failure rate $\lambda(t)$. The temperature of a component is a function of the ambient temperature and of the dissipation of power *P* in the component. The current *I* which traverses the capacitor and the equivalent series resistance *ESR* (see equation [18.13]) therefore causes an increase in the temperature of the component package and influences the lifetime. It is recognized that the breakdown process is equivalent to a chemical reaction. Now, the speed constant of chemical reaction k_v depends on the absolute temperature *T*, according to the Arrhenius Law:

$$k_v \text{ is proportional to } exp\left(-\frac{E_a}{kT}\right) \qquad [18.26]$$

E_a is the activation energy of the reaction (which depends on the type of failure) expressed in electronvolts (eV) and k the Boltzmann constant $(k = 8.62.10^{-5}\ eV/K)$.

The time of apparition of a failure is inversely proportional to the speed of apparition of this failure, so this time t_1 can therefore be written for a temperature T_1 considering equation [18.26]:

$$t_1 = A.exp\left(\frac{E_a}{kT_1}\right) \qquad [18.27]$$

A is a coefficient which depends on the component. The time t_1 until the failure can be accelerated (until time t_2) by increasing the temperature (of value T_1 to T_2) by the multiplicative acceleration factor of the failure A_{CC}. This factor can thus be written:

$$A_{cc} = \frac{t_1}{t_2} = exp\left(\frac{E_a}{k}\left(\frac{1}{T_1} - \frac{1}{T_2}\right)\right) \qquad [18.28]$$

This formula is much used to extrapolate the duration of apparition of a failure at any temperature following an ageing accelerated under maximal temperature.

18.7.3. *Failures of liquid electrolyte aluminum electrolytic capacitors*

18.7.3.1. *Ageing of liquid electrolyte aluminum capacitors*

The most probable failure of liquid electrolyte capacitors is related to its ageing. Like all components based on liquid electrolyte, the aluminum electrolytic capacitors present a wear period ($t \geq t_b$ in Figure 18.19), their failure then becoming inevitable.

This is the main disadvantage of this component. The life end of the capacitors is caused by an internal failure phenomenon, evaporation of the electrolyte which manifest itself by the following degradation of electric parameters: the capacitance C decreases (a few tens per cent) and the equivalent series resistance *ESR* increases (by more than 100%).

The decreasing rate in the capacitance C is lower than the increasing rate of *ESR*, so the loss factor $tan\delta$ defined by equation [18.14] increases.

The equivalent series resistance *ESR* is the parameter which is the most rapidly degraded with the component ageing. It therefore constitutes an indicator of ageing preferred for these components.

The equivalent series resistance mainly depends on the resistance of the electrolyte which impregnates the papers.

Since the electrolyte is evaporated, the equivalent surface of this latter decreases, leading to the increase of *ESR*. The increase in this resistance as a function of time depends on the type of electrolyte, the component package and, more particularly, its watertightness. These leaks are not always visible on the component. They are on the capacitor represented in Figure 18.20.

Figure 18.20. *View from above of a capacitor presenting electrolyte leakage*

Several models describe its evolution as a function of time. A simple linear model of the evolution as a function of the ageing time of ESR in 1/ESR was developed by [RHO 84]. Other more precise models show an increase of the ESR according to the exponential law [PER 03].

The variable elements of Figure 18.12 as a function of time t_1 for a temperature T_1 are represented by the following equations [PER 03]:

$$R_l(t_1) = R_l(0) + a_R.t_1 \exp(b_R.t_1) \quad [18.29]$$

$$C_{AK}(t_1) = C_{AK}(0) - a_C.t_1 \quad [18.30]$$

$R_l(0)$ and $C_{AK}(0)$ are the values of R_l and C_{AK} for a healthy capacitor. The different coefficients a_R, b_R and a_C are functions of the component. They depend on the geometry of this latter, of its technological characteristics and of the electrolyte used. Considering equation [18.28], we can deduce the evolution of the parameters of this model as a function of time for a temperature T_2. The activation energy E_a of this equation was estimated at 0.405 eV by [RHO 84] for this failure mode.

Since the drifts of the different parameters are known and can be expressed as equations, predictive maintenance systems of electrolytic capacitors can be elaborated [LAH 98]. Some propose an individual supervision of the capacitors [BES 03], [VEN 02] by measuring their impedance near resonance frequency, the equivalent series resistance *ESR* (see Figure 18.14) can be determined whilst functioning and compared to that of a healthy component.

18.7.3.2. *Other failures of liquid electrolyte aluminum capacitors*

A large number of catalectic failures related to the manufacture defect of the capacitor or its use can come occur. Figure 18.21 summarizes the main modes, mechanisms and causes of failure [PER 03], [ABD 08].

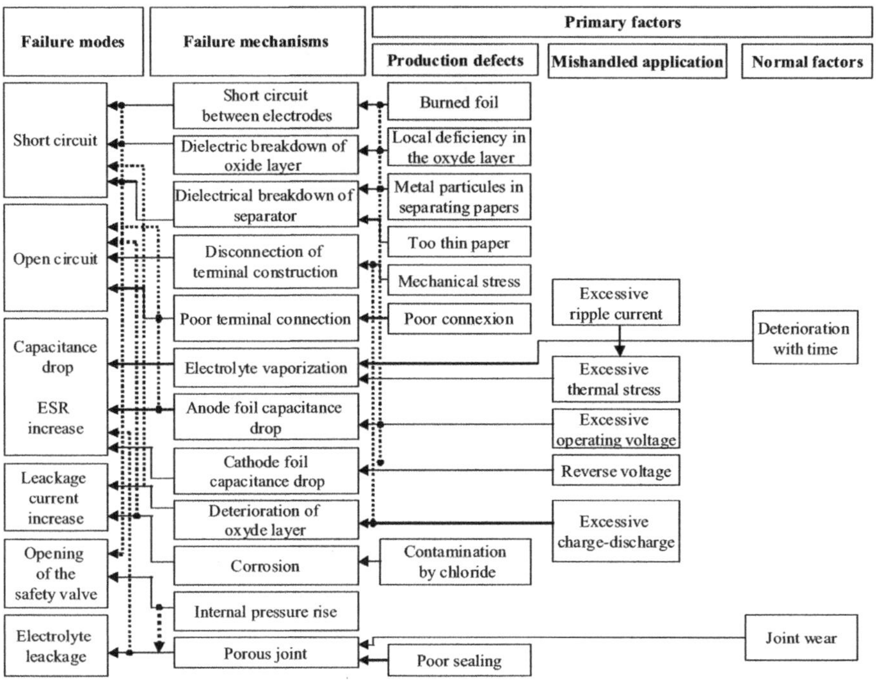

Figure 18.21. *Summary of the failures and their causes*

18.7.4. *Failures of solid electrolyte tantalum capacitors*

Since the electrolyte of the tantalum capacitors considered is solid, there is no electrolyte leakage risk and therefore these components do not present any wear period (see Figure 18.19). Their failure rate therefore does not increase in time, which is one of their advantages.

The majority of defects of these components are related to their thinness and to the quality of the tantalum oxide layer (Ta_2O_5); 90% of the failures induce an increase of the leakage current I_L or short circuits, which can lead to the flammability of the component.

The stresses revealing these failures are temperature, polarization voltage and current.

These failures are related to the absence of the self-healing phenomenon (see section 18.5.1). As for the component, this latter is only possible if:

– the manganese dioxide (MnO_2) thickness is sufficient and regular;

– the tantalum (Ta) pellets do not present any impurities;

– silver or graphite do not migrate to the tantalum oxide (Ta_2O_5) dielectric film through a possible crack of the manganese dioxide (see Figure 18.9). In these penetration zones, the self-healing effect of the capacitor is held, leading to a deterioration of the dielectric film, then a breakdown of the dielectric [POZ 98].

Moreover, for the self-healing to take place, the energy dissipated in the capacitor needs to remain at a decent level. The charge/discharge resistance of the capacitor must be higher than a few Ω/V for a good self-healing process. By default, locally, a very high temperature can appear, which associated with the electric field can lead to a crystallization of the tantalum pentoxide (Ta_2O_5) and lead to an increase of the leakage current or to a short circuit.

These different defects are inherent to manufacturing problems, storage under high humidity or to a poor use of the component.

18.8. Conclusion and perspectives

Electronic devices must be more and more compact. Since the complexity of electronic circuits increases, the power demanded for their supplies is therefore higher. The current devices consequently require the use of not too large electrolytic capacitors capable of supplying a relatively high power. This is the case for the market of SMC-type electrolytic capacitors, which is today boosted by the increase of IT and popular electronics (flat screens, DVD players and recorders, etc.).

To answer these criteria, several routes are conceivable for tantalum electrolytic capacitors. Some of them are cited below:

– the component miniaturization first needs an increase of its volume capacity. The use of tantalum nanopowder allows this need to be answered. The use of new dielectrics to replace tantalum pentoxide (Ta_2O_5) is also considered. Niobium pentoxide (Nb_2O_5), for example, presents a larger dielectric constant than that of tantalum pentoxide (41 for Nb_2O_5 against 27 for Ta_2O_5);

– the use of polymers to replace manganese dioxide (MnO_2) will allow components with weaker *ESR* to be made, therefore with less losses and a higher energy density;

– the manufacture of components including several tantalum pellets in parallel (multi-anodes) allows the *ESR* to be reduced, and capacitance *C* to be increased.

Liquid electrolyte aluminum capacitors are also subject to miniaturization. To reduce their *ESR*, weak resistivity electrolytes are studied. To improve their reliability, devices to monitor their healthy state are conceived. Finally, to combine the advantages of solid and liquid electrolyte capacitors, hybrid capacitors are investigated.

We cannot finish this chapter, and particularly this section, on the perspectives of electrolytic capacitors without tackling the new component in the domain of energy storage, i.e. the supercapacitor. This component is particular as its principle does not rely on the presence of a dielectric but on that of the double electric layer which is created at the interface between a solid electrode and a liquid electrolyte in the presence of an electric field. It is the thickness of the double electric layer of a few nanometers (related to the diameters of the solvent molecules) which defines the component capacitance. This thinness, associated with a very large surface of electrodes owing to its use of porous materials such as activated carbon, allows very high capacitances (of several thousand farads for the largest components) to be obtained. By principle, the voltage performance of this component is limited to a few volts, which corresponds to the electrolyte decomposition. The capacitance of this component is very rapidly degraded as a function of the frequency, which reserves it to the exclusive domain of energy storage. The supercapacitor can therefore only challenge electrolytic capacitors in this specific domain.

Electrolytic capacitors are therefore essential today in most applications such as for filtering, decoupling, etc.

18.9. Bibliography

[ABD 08] ABDENNADHER K., VENET P., ROJAT G., RETIF J.M., ROSSET C., "A real time predictive maintenance system of Aluminium Electrolytic Capacitors used in Uninterrupted Power Supplies", *IEEE Industry Applications Society Annual Meeting*, Edmonton, Canada, 2008.

[ALV 95] ALVSTEN B., *Electrolytic capacitors theory and application*, Rifa Electrolytic Capacitors, Sweden, 1995.

[BES 90] BESSON R., *Technologie des composants électroniques*, vol. 1, Editions Radio, Paris, 1990.

[BES 03] BESSEYRE D., *Développement d'un condensateur intelligent*, Mémoire d'ingénieur CNAM, Lyon, 2003.

[ELE 04] ELECTRONIQUE INTERNATIONAL HEBDO, *Les condensateurs évoluent à leur rythme*, 2004.

[JOU 96] JOUBERT C., Etude des phénomènes électromagnétiques dans les condensateurs à films métallisés - Nouvelle génération de condensateurs, Doctoral Thesis, Ecole Centrale de Lyon, December 1996.

[KEM 01] General Notes of the KEMET capacitance company, "*What is a capacitor?*", 2001.

[KEM 06] General Notes of the KEMET capacitance company, "*Surface Mount Capacitors*", 2006.

[LAG 96] LAGRANGE A., "Condensateurs", *Techniques de l'Ingénieur*, traité Electronique, E 2 060, Editions TI, 1996.

[LAH 98] LAHYANI A., VENET P., GRELLET G., VIVERGE P.-J. "Failure Prediction of Electrolytic Capacitors During Operation of a Switchmode Power Supply", *IEEE Transactions on Power Electronics*, vol. 13, no. 6, november 1998.

[MEN 97] MENGUY C., "Mesure des caractéristiques des matériaux isolants solides", *Techniques de l'Ingénieur*, traité Génie Electrique D 2 310, Editions TI, 1997.

[NIS 96] NISHINO A., "Capacitors: operating principles, current market and technical trends", *Journal of Power Sources*, vol. 60, no. 2, June 1996.

[PER 03] PERISSE F., Etude et analyse des modes de défaillances des condensateurs électrolytiques à l'aluminum et des thyristors, appliquées au système de protection du LHC (Large Hadron Collider), Doctoral Thesis, Claude Bernard University, Lyon 1, July 2003.

[PER 05] PERRET R., *Mise en œuvre des composants électroniques de puissance*, Hermes, 2005.

[PER 06] PERISSE F., VENET P., RETIF J.M., ROJAT G., "Study of an Electrolytic Capacitor Model as a function of Temperature", *EPE Journal*, vol. 16, no. 2, 2006.

[POZ 98] POZDEEV Y., "Reliability Comparison of Tantalum and Niobium Solid Electrolytic Capacitors", *Quality and Reliability Engineering International*, vol. 14, no. 2, 1998.

[PRY 01] PRYMAK J., "Improvements with Polymer Cathodes in Aluminum and Tantalum Capacitors", *Applied Power Electronics Conference and Exposition*, APEC 2001. Sixteenth Annual IEEE Volume 2, 2001.

[RHO 84] RHOADES G.E., SMITH A.W.H., "Expected Life of Capacitor with Non-solid Electrolyte", *34th Component Conference Proc*, 1984.

[RIF 95] RIFA, *Electrolytic Capacitors Theory and Application*, 1995.

[ROB 79] ROBERT P, *Matériaux de l'électrotechnique*, Traité d'électricité, St Saphorin, Editions Georgi, vol II, 1979.

[SAR 98] SARJEANT W.J., ZIRNHELD J., MACDOUGALL F.W., "Capacitors", *IEEE Transactions on Plasma Science*, vol. 26, no. 5, October 1998.

[USM 95] U.S. MIL SPECS, *Military Handbook Reliability Prediction of Electronic Equipment, MIL-HDBK-217F*, notice 1: July 1992, notice 2: February 1995.

[VEN 02] VENET P., PERISSE F., EL-HUSSEINI M.H., ROJAT G., "Realization of a Smart Electrolytic Capacitor Circuit", *IEEE Industrial Applications Magazine*, vol. 8, no. 1, January/February 2002.

[WIL 92] WILLIAMS B.W., *Power Electronics: Devices, Drivers, Applications and Passive Components*, 2nd Edition, Hong Kong, Editions ELBS, 1992.

Chapter 19

Ion Exchange Membranes for Low Temperature Fuel Cells

19.1. Introduction

Since the 1960s, when Loebe in California first showed the application of cellulose membranes in the desalination of water under high pressure gradients, transport studies in membranes have become a flourishing field of research. The process of desalination under a pressure gradient across membranes is called reverse osmosis. The development of membranes from polymers (polyimides, polysulfones, etc.) with good chemical, mechanical and thermal properties promoted the desalination of water by reverse osmosis at competitive prices, for use in households and industries, overtaking the traditionally used flash distillation process [PUS 82], [BAU 62], [CAB 85].

On the other hand, the synthesis of polyelectrolytes centered the interest of numerous researchers because of the potential of these materials, in the form of ion-exchange resins and ion-exchange membranes, offer for ionic separations [SAT 02]. One of the applications which raised great interest was the use of ion-exchange membranes for electrical generation systems.

For the last forty years, polymeric proton exchange membranes (PEM) were proposed to be used as solid polyelectrolytes for fuel cells. Amongst their

---

Chapter written by Vicente COMPAÑ MORENO and Evaristo RIANDE GARCIA.

advantages (with respect to inorganic membranes), the ease of production, flexibility and good mechanical properties stand out [GRE 61]. The development of fuel cells in the last twenty years as an economically interesting alternative for the production of electricity, and their potential use in electric cars caused a burst of interest for ion-exchange membranes. At present, great efforts are being focused on the development of ionic membranes that combine two at first sight contradictory properties: high proton conductivity and high selectivity or permselectivity. Because the two types of fuel cells, *Polyelectrolyte Membrane Fuel Cells* (PEMFC) and *Direct Methanol Fuel Cells* (DMFC), are used at temperatures near 100°C, membranes used in these devices must also have high chemical stability in unfriendly working conditions. Cation-exchange membranes can also be used as polyelectrolytes in dry cell batteries.

In general, a polymeric ion exchange membrane is made up of crosslinked polymer chains with ionic groups of acid or base type covalently bonded to the chains. Membranes containing acid groups are called cation-exchange membranes, or negative membranes. In the presence of water the acid group dissociates into an anionic component covalently bonded to the membrane and a free cation. Membranes with fixed base groups are called anion-exchange membranes, or positive membranes. The positive part of the fixed group is covalently bonded to the membrane and the anion becomes mobile. While cation- and anion-exchanged membranes are used in numerous ionic separation processes, in the case of fuel cells (PEMFC and DMFC), only cation-exchange membranes in the acid form are used. This is due to the fact that acid membranes act as electrolyte paths through which protons produced in the oxidation of the fuel in the anode travel to the cathode where water is produced. For this reason, in the following, we particularly focus our attention on cation exchange membranes or negative membranes.

Depending on the ionic nature of the acid groups covalently bonded to the polymer chains, called fixed groups, cation-exchange membranes are listed as strong or weak. Strong acid membranes have, in most cases, fixed sulfonic acid ($-SO_3H$) groups whose apparent pK varies between 0 and 1. However, strong proton exchange membranes can be prepared with fixed phosphonic $-PO_3H_2$ groups; these membranes exhibit $pK_1 = 2$–3 and $pK_2 = 7$–8. Acid membranes with much lower pK contain $-CF_2SOOH$ fixed groups in their structure. In this case, the high electronegativity of fluorine atoms considerably increases the acidity of the fixed group, the pK reaching the value of -6. Weak cation-exchange membranes contain weak acid fixed groups of type–COOH whose pK lies in the range 4–6. We must point out that the $-CF_2COOH$ group is strong and its apparent pK can reach the value of 2 because of the high electronegativity of the fluorine atoms bonded to the carbon in the α position with respect to the acid residue.

From a practical point of view, cation-exchange membranes used as polyelectrolytes in fuel cells must combine the following characteristics:

1) high ion exchange capacity (IEC); this property is measured by titration of the protons released, for example, in the ion-exchange reaction:

$$R\text{-}SO_3H + Na^+ \rightarrow R\text{-}SO_3Na + H^+$$

2) high proton conductivity; this is one of the principal properties for membranes used as solid electrolytes in fuel cells;

3) high permselectivity; an ideally permselective membrane is one in which a Faraday of current (96.480C) flowing through the membrane transports the equivalent of one proton. From now on, we shall call co-ions and counter-ions, respectively, the mobile ions of the same and opposite sign to the ions covalently bonded to the membranes;

4) low electro-osmosis; protons (and cations in general) moving from the anode to the cathode drag water, drying the membrane and decreasing its conductivity;

5) weak free diffusion of electrolytes in the membrane; this property, very important in separation processes, is not so important in fuel cells;

6) high chemical stability; ion exchange membranes are used in chemically unfriendly environments;

7) high mechanical resistance and good dimensional stability;

8) high thermal stability; this characteristic is very important in fuel cells;

9) low fuel crossover.

A large number of the properties mentioned above, and which optimize the performance of membranes, are mutually incompatible. For example, a high IEC increases the conductance of membranes but may decrease the mechanical properties and dimensional stability. Consequently, the preparation of membranes requires the reaching, in some cases, of a suitable balance among incompatible properties [KOT 02], [WIH 02], [KOT 01].

For reasons related to osmotic phenomena, which will be discussed later, ion-exchange membranes swell in an aqueous medium, a process that, if excessive, decreases the mechanical properties. To decrease membrane swelling, the polyelectrolyte chains of ion-exchange membranes must be crosslinked with crosslinking agents. The crosslinking process reduces the length of the elastic chains of the polyelectrolyte, and consequently, the membrane swelling. Although crosslinking reduces the conductance of polyelectrolytes, this process does not always negatively impact on the permselectivity of membranes because the size of the pores is also reduced. In other words, the optimal density of crosslinking points is the result of an optimal balance between ohmic resistance acceptable to ion transport, permselectivity and mechanical stability. In conclusion, good performance membranes must exhibit excellent selectivity and low resistance to ion transport, combined with high mechanical resistance and a long average lifetime in working conditions.

19.2. Homogenous cation-exchange membranes

The first ion-exchange membranes were developed in the 1930s by the dispersion of ion exchange resins in polymers which act as binders. These membranes are of a heterogenous type. Homogenous membranes are characterized by a homogenous distribution, at the microscopic scale, of ionic groups in the membranes phase. Although the interest for heterogenous membranes is currently growing, most research in cation-exchange membranes has been carried out in homogenous membranes.

Owing to their thermal stability, the most interesting materials for the preparation of homogenous membranes usable in fuel cells utilize perfluorinated polymers [MAU 04], [HICK 04]. The general structure of Nafion, a representative of this type of polymers, is represented in Figure 19.1. The membranes based on these materials contain hydrophobic ($-CF_2-CF_2-$) and hydrophilic ($-SO_3H$) regions. The fact that the values of x, y and z in Figure 19.1 can be adjusted allows the properties of the ion exchange membranes to be varied at will. These membranes present good electrochemical properties combined with high chemical stability, but have high cost, poor thermal properties and are environmentally unfriendly. The perfluorinated membranes are only useful in fuel cells which produce electricity from hydrogen. They are easily permeable to fuel reformers, such as methanol. Methanol crossover causes secondary reactions in the cathode that reduce fuel-cell performance.

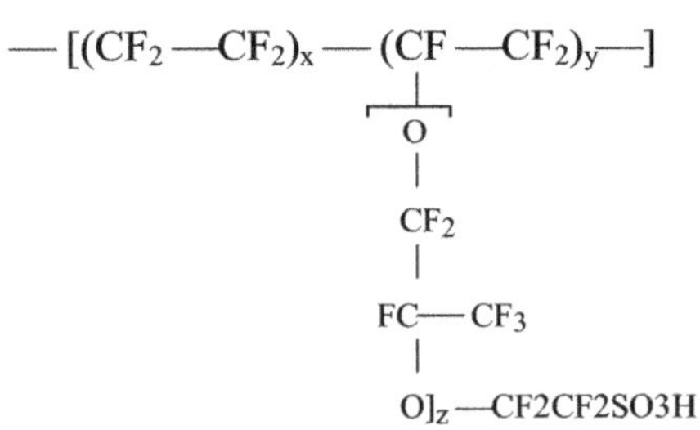

Figure 19.1. *General structure of Nafion*

The use of non-fluorinated polymers such as aromatic polyesthers, polybenzimidazolees, polyimides, polysulfones, poly(aryl sulfone ether) and polyketones containing acid functional groups are alternative materials that awakened the interest of researchers in membranes [HIC 04], [R&D 03], [LI2 03], [MAU 04], [JAN 03]. Good performance membranes can be prepared from polyarylene-ether-sulfone [ZHA 01]. Grafting of polystyrene onto polystyrene sulfonate ionomers allows the preparation of membranes with acceptable electrochemical properties for their use in fuel cells or as polyelectrolytes in dry cell batteries [DIN 01]. Descriptions of the grafting of polystyrene onto partially fluorinated polymers, such as polyvinylidene fluoride, using γ radiation, have been reported. The grafted polystyrene was subsequently sulfonated and used to prepare cation-exchange membranes [ENN 99]. The synthesis of the acid 9.9b (4-aminophenyl) fluorene-2,7-disulfonic has been recently described. The reaction of this product with 1,4,5,8- tetracarboxil naphtalene dianhydre led to the formation of sulfonated polyimides which allow the preparation of high conductivity cation-exchange membranes [GEN 01], [COR 03].

Polymers used in membranes should have high glass transition temperature, good chemical stability and good mechanical properties. The introduction of acid fixed groups in the membrane can be made by sulfonation with concentrated sulfuric acid SO3, chlorosulfonic acid, acetyl sulfate or other methods depending on the membrane [GUO 99], [HAR 03], [GEN 01], [WIL 02]. Whenever possible, it is preferable to use polymers obtained from sulfonated monomers or sulfonated pre-polymers. In this way, membranes with homogenous distribution of ionic fixed groups in their structure can be obtained.

19.3. Heterogenous ion exchange membranes

The development of heterogenous ion exchange membranes has recently awakened great interest. Membranes of this type were prepared by dispersing commercial ion exchange resins, of a sulfonated polystyrene/divinylbenzene type, into perfluoride polymers, such as vinylidene polyfluoride, which act as a binder [LÓP 01]. Mixtures of polyelectrolytes and semicrystalline polymers were also prepared, for example, by blending sulfonated EPDM with semicrystalline polypropylene.

The main obstacle for the large-scale commercialization of fuel cells containing organic membranes as solid electrolytes are the low protonic conductivity of low hydrated cation-exchange membranes, as well as the high methanol permeability and poor mechanical properties of these materials above 130°C. These drawbacks could be overcome by developing composite membranes consisting of fillers in the shape of small-size particles dispersed in a polymer matrix. The fillers used are silica, heteropolyacids, layered metallic phosphates and phosphonates. In this sense, the preparation of membranes containing high conductivity zirconium phosphonate as filler particles has been reported [YAN 04], [ULB 06], [HOG 05], [ALB 03]. In general, all PEMFC composites are obtained by dispersing organic or organo-inorganic particles (with or without ionic groups) in polyelectrolytes in the acid form On the other hand, conductive membranes can also be obtained from porous non-conductive polymers blended with ionomers or high conductivity particles [HIC 04]. Maurits *et al.* [MAU 00], [MAU 02], created a wide variety of nanocomposites formed by organic polymers and inorganic oxides in order to increase the quantity of water absorption in the polymer matrix, to reduce the methanol permeability and to increase the mechanical properties of the composite membranes [MAU 00, 02], [MIY 01].

Nano and micro fillers (depending on the particle size) are used in the development of composite membranes. There are two methods to accomplish this task:

a) dispersion of nano or micro particles in a polyelectrolyte solution, followed by the elimination of the solvent by evaporation;

b) growth *in situ* of particles in a membrane or in a polyelectrolyte solution.

We shall present below a few examples of materials used to obtain composite cation-exchange membranes.

Metal oxides have been used as fillers. Polymeric membranes with silica or zirconia have been prepared in different ways, either by dispersion of particles in a polyelectrolyte solution or by depositing *in situ* the solid in the polymeric matrix. The conductivities of the composite membranes were higher than those of the pristine polyelectrolytes and other properties around 100°C, such as methanol crossover and water permeability, improved [KRE 98, 01], [DIM 02], [MIY 01], [HOG 01, 05].

In the group of layered phosphates, phosphate-antimonic acids present swelling properties similar to smectites as well as strong acid character and high proton conductivity. The incorporation of these materials in a polymer matrix, for example in a sulfonated polysulfone, improves the conductivity and reduces both the gas crossover and membrane swelling. In this group of materials, metal phosphates (IV) such as α-ZrP are included. This phosphate does not modify the conductivity mechanism of the hydrated membranes. In the vicnity of 130°C, the performance of the pristine membrane is better than that of the composite membrane containing the phosphate owing to a better retention of water; in the vicinity of 130°C the pristine membrane exhibits better retention of water and the deterioration of the membrane decreases [HOG 01, 05].

Finally, metal phosphonates with layered structures exhibit high conductivity [ALB 97]. Up-to-date results indicating the positive effects of these materials as fillers have not been reported, and in some cases the conductivities are lower than those of the pristine membranes. These materials have also been employed as fillers of porous membranes, though their use is limited by the mechanical properties, flexibility and mechanical resistance of the membranes obtained. If this limitation is surpassed, the preparation of composite membranes using these fillers would be interesting owing to the wide variety of composite acid membranes that can be prepared with conductivities of the same order as Nafion in the temperature range 130–160°C, but at lower cost [HOG 01, 05].

In the efforts that are being made to develop membranes with optimal properties for fuel cells (PEMFC and DMFC), our research group is working on the preparation of composite membranes based on Nafion and styrene-b-(ethylene-1-butadiene)-b-styrene (SEBS). As fillers, sulfonated phenyl silica and sulfonated phenyl-sepiolite are being used [FER 07], [del RIO 05].

19.4. Polymer/acid membranes

Membranes formed by polymers containing groups of the basic type (ether, amine or imine groups) were developed with characteristics good enough to be used in fuel cells. The membranes can be doped with strong acids, such as H_3PO_4. A polymer of this type is polybenzimidazole (PBI) (pK = 5.5). Membranes obtained by immersing PBI in phosphoric acid noticeably increase their conductivity and thermal stability.

Moreover, the doped polymer presents good mechanical resistance up to 200°C [BUC 88]. Considering that the modulus of PBI is three times that of Nafion, polybenzimidazole membranes with low hydrogen permeability and thinner than Nafion can be prepared. It has ben shown that membranes of this type can be used at temperatures greater than 150°C [WAI 95]. Therefore, in these working conditions, fuel cells having doped PBI as solid electrolyte could operate using hydrogen fuel containing about 1% (v/v) of carbon monoxide. It is possible that the dopant phosphoric acid does no shrink from the polymeric matrix as a result of the strong phosphoric-PBI interactions. Moreover, the nitrogen protonation of the imine functions leads to the formation of hydrogen-bonded networks. The lack of data on durability does not permit us to know if the phosphoric acid is mantained for a long time in the membrane. If that were not the case, the conductivity of the membrane would be considerably reduced.

19.4.1. *Membranes prepared from polyme blends*

In the same way as PBI/H_3PO_4 membranes, polyelectrolytes based on blends of acid and basic ionomers have been prepared. A representation of how ionomers react is shown in Figure 19.2.

Figure 19.2. *A representation of how ionomers react*

The acid component of the blend can be sulfonated poly(ether-ether-cetone) or polyether sulfone or *ortho*-sulfone-sulfonated poly(ethersulfone) whereas the basic component can be polybenzimidazole. The membranes present excellent thermal properties, since their decomposition temperature ranges in the interval 270–350°C, as well as good proton conductivities. However, only limited durability of the order of 300h has been demonstrated in direct hydrogen fuel cells at 70°C [WAL 99].

Another blend which is subjected to investigations is that composed of sulfonated phenylene polyoxide and poly(vinylidene fluoride). In this case, the non-ionic component noticeably improves the proton conductivity of the blend. It should be noted that the performances of these blends in direct hydrogen fuel cells, at 45°C, are reported to be higher than that of the Nafion 112 membrane [WAL 99].

19.5. Characterization of membranes

A series of characteristics of membranes involving morphology, chemical structure, concentration of ionic fixed groups, as well as the chemical nature of the filler dispersed in the polymeric matrix in the case of composite membranes, affect the physico-chemical phenomena observed in ion exchange membranes. Among them, the following phenomena can be mentioned: cationic or anionic permselectivity, generation of an electromotive force when the membrane is flanked by solutions of the same nature but different concentration, ionic conductivity, electrolyte diffusion, osmotic and electro-osmotic processes, etc. These phenomena, which characterize the electrochemical behavior of ion exchange membranes, are addressed below.

19.5.1. *Nernst–Planck flux equation*

In numerous applications, ions travel across ion exchange membranes under combined driving forces of chemical and electrical potential gradients. Let us first consider the flux of an ion *i* under the action of a unidimensional electrochemical potential gradient $d\mu/dx$. The flux of the ionic particle *i*, *Ji*, in a point in the membrane depends on the velocity *ui* and the concentration *ci* of the particle in that point such that *Ji=ciui*. On the other hand, in steady state conditions, the acceleration of the particle caused by the chemical gradient driving force is dissipated by the friction of the particle with its surroundings. In this case, $d\mu_i / dx + \chi_i u_i = 0$, where χi is the friction coefficient, and the particle flux is expressed by [SCH 62], [YAM 64]:

$$J_{i\mu} = c_i u_i = -\frac{c_i D_i}{RT}\frac{d\mu_i}{dx} \qquad [19.1]$$

where the Einstein equation $\chi_i = RT / D_i$ which relates χi with the ionic diffusion D was used. In isobaric and isothermal conditions, $d\mu_i = RT\, d\ln(c_i\gamma_i)$, where γi is the activity coefficient of the particle *i*. Then, $Ji\mu$ can be written:

$$J_{i\mu} = -D_i\left(\frac{dc_i}{dx} + c_i\frac{d\ln\gamma_i}{dx}\right) \qquad [19.2]$$

If the driving force responsible for the flux is the electric potential gradient $d\psi/dx$, the electrical charge transported by mol of the particle *i* is z_iF where *F* and *zi* are, respectively, Faraday's constant and valence of the ionic species *i*. Consequently, the impulsive force by mol of the particle *i* is ziF ψ/dx. In steady-state conditions $z_iFd\psi / dx + \chi_i u_i = 0$ and the flux of the particle caused by the electric potential gradient is given by:

$$J_{i\psi} = c_i u_i = -z_i c_i \frac{D_i F}{RT}\frac{d\psi}{dx} = -z_i c_i \bar{u}_i \frac{d\psi}{dx} \qquad [19.3]$$

In this equation, the quantity $\bar{u}_i = \frac{FD_i}{RT}$ represents the ionic mobility of the particle *i*, which is generally expressed in cm^2/V.s units. The total flux of the particle *i* under the combined action of chemical and electrical potential gradients can be written as: $J_i = J_{i\mu} + J_{i\psi}$, or:

$$J_i = -D_i \left(\frac{d c_i}{dx} + c_i z_i \frac{F}{RT} \frac{d\psi}{dx} + c_i \frac{d \ln \gamma_i}{dx} \right) \quad [19.4]$$

where Ji is generally expressed in mol/(cm^2 s). Ion-exchange membranes contain fixed ionic groups covalently bonded to the polymer chains, counter-ions, co-ions and water. Moreover, the electroneutrality principle indicates that the sum of all charges in the membrane phase is zero. Consequently, the concentration of counter-ions is higher than that of co-ions. Then counter-ions impart more moment than co-ions to the liquid located in the pores of the membrane. The result is that under the electrical potential driving force, the center of gravity of the liquid pore moves with velocity v with respect to the membrane and its flux, called convention flux, is $Jconv = ci\,v$. Taken the membrane as reference frame, the flux of the particle i is:

$$J_i = -D_i \left(\frac{d c_i}{dx} + c_i z_i \frac{F}{RT} \frac{d\psi}{dx} + c_i \frac{d \ln \gamma_i}{dx} \right) + c_i v \quad [19.5]$$

This expression, named generalized Nernst–Planck equation, describes the absolute flux of the specie i through an ion exchange membrane taking the membrane as reference frame. The convention velocity can be estimated [SCH 62] by assuming that the concentration of ionic charges in the liquid of the membrane is FX/ε, where ε is the volume fraction of liquid pore in the membrane. X represents the concentration of fixed ions in the membrane in mol/cm^3. As the charge of the liquid pore is $\omega FX/\varepsilon$, where by the electroneutrality principle ω is -1 for cation exchange membranes and +1 for anion exchange membranes, the force acting on the liquid pore is $F = (\omega FX/\varepsilon)\,(d\psi/dx)$. In steady-state conditions, $F + \rho_0 v = 0$, where ρ_0 is the specific resistance of friction of the charged liquid pore against the pore walls. In this case, the pertinent expression for v is:

$$v = -\frac{\omega FX}{\rho_0 \varepsilon} \left(\frac{d\psi}{dx} \right) = -\,\omega \bar{u}_0 \frac{d\psi}{dx} \quad [19.6]$$

where ρ_0 has g/(s.cm^3 of liquid pore) units and $\bar{u}_0$ is the liquid pore mobility which can be expressed as:

$$\bar{u}_0 = \frac{FX}{\rho_0 \varepsilon} \quad [19.7]$$

In the absence of chemical potential gradients, i.e. when the ionic solutions separated by the membrane have the same concentration, the flux equation for the species i can be written as:

$$J_i = -z_i c_i \frac{D_i F}{RT}\frac{d\psi}{dx} + c_i v = -\left(\bar{u}_i c_i z_i + \omega c_i \bar{u}_0\right)\frac{d\psi}{dx} \qquad [19.8]$$

The liquid pore mobility is generally expressed in $cm^2/V{\cdot}s$ units.

19.5.2. *Osmotic phenomena and electric potential*

Owing to the presence of fixed ions in their structure, ion exchange membranes swell in water until the chemical potentials of this solvent in the membrane phase and in pure water are equal. If the membrane is immersed into an electrolytic solution, water and electrolyte ions of the solution diffuse inside the membrane, verifying at any point the electroneutrality principle:

$$\cdot\sum_i z_i c_i + \omega X = 0 \qquad [19.9]$$

In isothermal conditions, the chemical potentials of water in the solution which surrounds the membrane, μ_w, and in the membrane phase, $\bar{\mu}_w$, are:

$$\mu_w = \mu_0 + V_w p + RT \ln a_w \qquad [19.10]$$

$$\bar{\mu}_w = \mu_0 + V_w \bar{p} + RT \ln \bar{a}_w \qquad [19.11]$$

where μ_0 is the chemical potential in standard pressure and temperature conditions, VW is the partial molar volume of the solvent, a_w and $\bar{a}_w$ represent, respectively, the activity of water in the solution and in the membrane. When equilibrium is reached, $\mu_w = \bar{\mu}_w$ and equations [19.10] and [19.11] lead to the following expression:

$$\pi = p - \bar{p} = \frac{RT}{V_w}\ln\frac{\bar{a}_w}{a_w} \qquad [19.12]$$

where π is the osmotic pressure.

Consequently, the immersion of a dry ion exchange membrane in water swells the membrane and can even dissolve it, unless the cohesive forces of the molecular chains are able to support the osmotic phenomena. To avoid this process, membranes are crosslinked with crosslinking agents, thus increasing the elastic modulus and reducing swelling.

In principle, when a membrane separates two ionic solutions of different concentration, the solvent flows from the lower- to the upper concentration solution to balance the difference of chemical potential of the solvent in both solutions. In the case of ion exchange membranes surrounded by electrolytic solutions of the same nature but different concentration, fluxes of solvent larger than those predicted by the chemical potential difference (positive anomolus osmosis) may occur. In some cases, even fluxes of solvent in the opposite direction to that imposed by the chemical potential difference, i.e. from the diluted to the concentrated solution (negative anomolous osmosis) may take place. In general, these anomalous phenomena are observed in very porous membranes with high ion-exchange capacity. Anomolous osmosis is explained by the action of the two forces responsible for the flux of water: the pressure gradient due to the osmotic pressure which causes the solvent flow from the less to the more concentrated solution and the electric potential created in the membrane phase. Actually, the liquid pores of the membrane contain mobile ions which balance the fixed charges covalently bonded to the polymeric chains of the membrane. If the difference of mobility between the counter-ions and co-ions is high, the ion with higher mobility diffuses rapidly through the membrane producing a space charge, with the same sign as the ion, on the side of the membrane in contact with the less concentrated solution; consequently, the space charge on the side of the membrane in contact with the concentrated solution has an opposite sign. In this way, a diffusion potential is created.

To get a deeper insight into this matter, let us assume that the mobility of counter-ions in a cation exchange membrane is higher than that of co-ions. The counter-ions will move rapidly from the membrane phase in contact with the concentrated solution towards the phase in contact with the dilute solution creating a higher electrical potential on the side of the membrane in contact with the diluted solution than on the other one. Since the mobile liquid pore in the membrane phase (water, counter-ions and co-ions) contains on average a net positive charge, the potential created promotes the displacement of the liquid pore from the side of the membrane in contact with the diluted solution (higher potential) to the side in contact with the concentrated solution (lower potential). The liquid transport caused by the electrical potential adds to the osmotic transport arising from the chemical potential; consequently, it is responsible for the positive anomolous osmosis. On the other hand, if co-ions have higher mobility than counterions, a similar reasoning shows that the anomalous osmosis is of negative type whenever the flow arisen from the electric field caused by the mobility of co-ions is larger than that caused by the chemical potential. Negative abnormal osmosis can occur, for example, in anion-exchange membranes (positive membranes) with protons as co-ions. The high mobility of protons leads to the development of a potential which causes the transport of the negatively charged liquid pore from the concentrated to the diluted solution [SAT 02], [YAM 64].

19.5.3. *Ionic diffusion in ion exchange membranes*

For membranes used in ionic separation processes, free diffusion of electrolyte decreases membranes performance. From the Nernst–Planck equation, the flux of cations, *J*+, and *f* anions, *J*-, can be expressed as follows:

$$J_+ = -\overline{D}_+ \left(\frac{d\overline{c}_+}{dx} + \overline{c}_+ \frac{d \ln \overline{\gamma}_+}{dx} + z_+ \overline{c}_+ \frac{F}{RT} \frac{d\psi}{dx} \right) \qquad [19.13]$$

$$J_- = -\overline{D}_- \left(\frac{d\overline{c}_-}{dx} + c_- \frac{d \ln \overline{\gamma}_-}{dx} + z_- \overline{c}_- \frac{F}{RT} \frac{d\psi}{dx} \right) \qquad [19.14]$$

where $z+$ and $z-$ are the cation and anion valences, respectively, while $c+$ and $c-$ represent the respective concentrations. If the solutions are sufficiently diluted such that the activity coefficients ($\gamma+$, $\gamma-$) are equal to 1, the flux of cations is given by

$$J_+ = -\overline{D}_+ \overline{D}_- \frac{\overline{c}_+ z_+^2 + \overline{c}_- z_-^2}{\overline{c}_+ \overline{D}_+ z_+^2 + \overline{c}_- \overline{D}_- z_-^2} \frac{d\overline{c}_+}{dx} \qquad [19.15]$$

In the development of this equation, the electroneutrality principle which requires that $z_+ J_+ + z_- J_- = 0$ and $z_+ \overline{c}_+ + z_- \overline{c}_- = X$, where X is the concentration of fixed ions, was considered. By comparing this equation with Fick's first law, a global diffusion coefficient can be defined as:

$$\overline{D}_{+-} = -\overline{D}_+ \overline{D}_- \frac{\overline{c}_+ z_+^2 + \overline{c}_- z_-^2}{\overline{c}_+ \overline{D}_+ z_+^2 + \overline{c}_- \overline{D}_- z_-^2} \qquad [19.16]$$

If $z_+ = z_-$ equation [19.16] leads to:

$$\frac{1}{\overline{D}_{+-}} = \frac{\overline{x}_-}{\overline{D}_+} + \frac{\overline{x}_+}{\overline{D}_-} \qquad [19.17]$$

where $\overline{x}_+$ and $\overline{x}_-$ represent, respectively, the molar fractions of cations and anions in the membrane. Obviously, $\overline{D}_{+-} \cong \overline{D}_+$ if $\overline{x}_+ << \overline{x}_-$ and, in this case, the less concentrated ionic species in the membrane phase controls the diffusion [LAK 69], [SAT 02]. Consequently, the flux of cations in cation exchange membranes is given by:

$$J_+ = -\overline{D}_- \frac{d\overline{c}_+}{dx} \qquad [19.18]$$

In the case of anion-exchange membranes, the equivalent expression is:

$$J_+ = -\overline{D}_+ \frac{d\overline{c}_+}{dx} \qquad [19.19]$$

19.5.4. *Electromotive force of concentration cells and transport number*

The interphase separating a bulk solution from a membrane gives rise to a discontinuity in the concentration of electrolyte in the solution-membrane interphase as a consequence of the rejection of co-ions by the fixed ions of the membrane. However, the chemical potential is continuous in this region.

Let us consider an ion exchange membrane separating two solutions of the same nature, but with different concentration. Let us assume further that the electromotive force of the cell is measured with calomel electrodes, using the following configuration: Hg | Hg_2Cl_2 | bridge of saturated KCl | bridge of the saturated electrolyte | concentration of electrolyte (c_1) | cation exchange membrane | concentration of the electrolyte (c_2) | bridge of the saturated electrolyte | bridge of saturated KCl | Hg_2Cl_2 | Hg. This configuration corresponds to a concentration cell. Owing to the saturation of bridge electrolytes, the electrodes do not contribute to the electromotive force E of the concentration cell. In reversible conditions, ionic transport gives rise to a variation of free energy related to the electromotive force of the cell by:

$$dG = -F\,dE \qquad [19.20]$$

where F is Faraday's constant. The transport number of the mobile ionic species i, t_i, represents the number of equivalents of the ion i transported by Faraday of current, its maximum value being the unit. The fact that the current in the membrane is transported by counter-ions and co-ions leads to:

$$t_+ + t_- = 1 \qquad [19.21]$$

The variation of free energy is related to the chemical potentials of the ionic species by:

$$\Delta G=\sum_i \frac{t_i}{z_i} d\,\mu_i = RT\sum_i \frac{t_i}{z_i} d\ln a_i \qquad [19.22]$$

where z_i is the valence of the ionic species *i*, $\mu_i = \mu_{0i} + RT\ln a_i$ and t/z_i represents the moles of *i* transported by Faraday of current. Equations [19.20] and [19.22] lead to:

$$E=-\frac{RT}{F}\int_1^2 \sum_i \frac{t_i}{z_i} d\ln a_i \qquad [19.23]$$

To integrate equation [19.23] it is necessary to know the dependence of t_i on the concentration of the species *i* in the membrane. However, for low concentrations, transport numbers can be assumed to be constant, and equation [19.31] becomes:

$$E=-\frac{t_+}{z_+}\frac{RT}{F}\ln\frac{(a_+)_2}{(a_+)_1}+\frac{t_-}{z_-}\frac{RT}{F}\ln\frac{(a_-)_2}{(a_-)_1} \qquad [19.24]$$

where a+ and a- are, respectively, the activities of cations and anions in the solutions flanking the membrane. For a 1:1 electrolyte solution (zi = 1) it is assumed (a+)2/(a+)1 = (a-)2/(a-)1 = (a)2/(a)1. For an ideal selective membrane, t+ = 1 and the electromotive force for the monovalent electrolyte reaches its maximum value given by:

$$E_{max}=-\frac{RT}{F}\ln\frac{a_2}{a_1} \qquad [19.25]$$

Hence the transport number of the monovalent cation is related to the electromotive force of the concentration cell by:

$$t_+=0.5+\frac{E}{2E_{max}} \qquad [19.26]$$

19.5.5. *Conductivity*

Though proton diffusion in cation exchange membranes is still a poorly understood phenomenon, the process probably involves the dissociation of the proton from the acid site, subsequent transfer of the proton to the aqueous medium, screening by water of the hydrated proton from the conjugate base (i.e the anion of the fixed group) and, finally, diffusion of the proton in the water enclosed in the confined water in the polymeric matrix [PAD 03]. Consequently, the relationship between proton mobility and the quantity of water contained in the membrane is a critical factor in the conductive process. In fuel cells, good performance membranes must remain hydrated in working conditions. At the molecular level, the transport mechanism of protons in hydrated membranes is generally described either by the vehicle mechanism in which the proton diffuses through the media together with the vehicle (i.e. water, with water as H_3O^+) or by proton hopping (proton transferred from one water molecule vehicle to other) or Grotthuss transport which usually involves the reorientation of solvent dipoles, an inherent part of establishing the proton diffusion pathway [POU 92], [KRE 98, 01], [PAD 03]. This ambiguity poses a problem for the development of models predicting proton transport in new cation-exchange materials [TUC 97]. There are, however, a few empirical rules to predict protons transport in membranes.

As indicated above, the water content of cation-exchange membranes in the acid form governs proton transport in these materials. Water allows the segregation of hydrophilic moieties of the polyelectroyte chains from hydrophobic ones giving rise to the formation of hydrophobic and hydrophilic domains. In high conductive membranes, hydrophilic domains form percolation paths through which protons travel from the anode to the cathode of fuel cells.

It is important to develop new polyelectrolyte materials which combine a good segregation of hydrophilic domains from hydrophobic ones with acceptable water content. For example, the proton conductivity of Nafion at room temperature may vary between between 0.01 and 0.1 S/cm, depending on the humidity of the membrane, but it drops several orders of magnitude when the water content comes near to about two moles of water/moles of fixed anionic groups in the membrane [YAN 04], [ALB 03], [ATK 03]. The transport of protons through membranes is especially related to the diffusion mechanisms of these particles in water that involve transport of the protonic charge between two stables forms of the hydrated proton called the Zundel ion ($H_5O_2^+$) and the Eigen ion ($H_9O_2^+$), [PAD 03]. Acid contamination may increase the apparent conductivity of membranes.

In the case of composite membranes, recent studies have shown the difficulty involved in the development of models that explain proton transport in this kind of material. However, some research focused on the understanding of proton transport and the mechanisms involved in this process in nanocomposite membranes have been reported. Interest has been focused on the increase of the membrane wettability by adding hydrophilic fillers, such as organo-organic solids properly functionalized with sulfonic acid groups (for example: silica, sepiolite, SBA-15, among others). These fillers are hygroscopic because they keep water trapped in their cavities at temperatures near 250°C [FER 07].

There are other types of composite membranes obtained by dispersing conductive fillers in the polymeric matrix (for example: α-ZrP). The fillers decrease the permeability of water and methanol in the membranes as well as the molecular migration of other species. If the humidity of the composite membrane decreases a loss of conductivity is oberved.

A third generation of membranes have been prepared from composites obtained by dispersing proton conducting fillers, such as heteropolyacids and acid doped polybenzimidazole, in polymer matrices [HOG 05]. In this case, an acid of high conductivity is immobilized in the polymeric matrix so that the protonic conductivity is independent of the humidity of the membrane, and the electro-osmotic flux is reduced. In the same way, we can mention the use of acid solids such as zirconium phosphates and their by-products in PEMFCs [ALB 92, 95, 97].

From a phenomenological point of view, the electrical current flux, j, through an ion exchange membrane in equilibrium with an electrolyte solution of concentration c, promoted by an electrical potential gradient $d\psi/dx$, is given by:

$$j = F\sum_i z_i J_i \qquad [19.27]$$

Equations [19.8] and [19.27] lead to:

$$j = -F\sum_i \left(\bar{u}_i c_i z_i^2 + \omega z_i c_i \bar{u}_0\right)\frac{d\psi}{dx} \qquad [19.28]$$

Given that in the liquid pore $\sum_i c_i z_i = X$, equation [19.28] takes the form:

$$j = -F\sum_i \left(\bar{u}_i c_i z_i^2 + \omega X \bar{u}_0\right)\frac{d\psi}{dx} \qquad [19.29]$$

On the other hand, the specific conductance is given by: $\kappa = -\frac{j}{d\psi / dx}$ and, consequently, this quantity can be expressed in terms of the ionic and pore liquid mobilities by:

$$\kappa = F \sum_i \left(\bar{u}_i c_i z_i^2 + \omega X \bar{u}_0 \right) \qquad [19.30]$$

19.5.6. *Electro-osmosis*

An electric current flowing through a membrane which separates two solutions of the same electrolyte, of concentration c, causes a net transport of water from one side to the other side of the membrane. For a cation-exchange membrane, the current is transported by cations from the anodic compartment to the cathodic one. Cations not only carry the hydration water associated with them, but also drag water in their motion in such a way that a large part of the flux of water promoted by the current corresponds to dragged water. The term electro-osmosis describes the total flux of water, independently of the transport mechanism. Electro-osmosis plays a negative role in fuel cells because protons traveling from the anode to the cathode compartments across a cation-exchange membrane in the acid form drag water in the membrane phase, drying it. This phenomenon decreases the conductivity of the membrane. It must be pointed out, however, that the reaction of oxi-ions with protons in the cathodic compartment forms water that back-diffuses across the membrane, thus contributing to its humidification.

Since the sign of the liquid pore in the membrane phase is the same as that of counter-ions, equation [19.6] indicates that the velocity of the pore caused by the electrical potential gradient $d\psi/dx$ is given by:

$$v = \omega \bar{u}_0 \frac{d\psi}{dx} \qquad [19.31]$$

Obviously, the volume of solvent flowing per time unit is:

$$\frac{dV}{dt} = \varepsilon v = \omega \varepsilon \bar{u}_0 \frac{d\psi}{dx} = \omega \frac{FX}{\rho_0} \frac{d\psi}{dx} \qquad [19.32]$$

As indicated in section 19.5.1, ε is the volume fraction of liquid in the membrane and ρ_0 the specific friction coefficient.

The electro-osmotic permeability coefficient is defined as:

$$D_j = \frac{dV / dt}{j} \quad [19.33]$$

where j is the current density. From equations [19.29], [19.30], [19.32] and [19.33] the electro-osmotic permeability is expressed in terms of the characteristics of the membrane by:

$$D_j = \frac{dV / dt}{j} = \frac{\omega \varepsilon \bar{u}_0}{F\left(\sum_i z_i^2 \bar{u}_i c_i + \bar{u}_0 X\right)} = \frac{\omega F X}{\rho_0 \sigma} \quad [19.34]$$

This equation suggests that D_j is directly proportional to the ion-exchange capacity of the membrane and inversely proportional to the specific coefficient times the conductivity. A balance of matter in the electrosomotic process leads to

$$V_w t_w + \sum_i t_i V_i + \omega D_j F = 0 \quad [19.35]$$

The transport number of water, t_w, is therefore:

$$t_w = -\frac{\omega D_j F + \sum_i t_i V_i}{V_w} \quad [19.36]$$

where V_w and V_i represent, respectively, the partial molar volumes of water and and the ionic species i.

19.5.7. *Thermodynamics of irreversible processes and transport numbers*

Ion exchange membranes are non-ideal barriers through which the solvent and the solution can flow under the action of different forces, such as gradients of concentration, pressure, temperature and electrical potential. In isobaric and isothermal conditions, the forces are reduced to concentration and electrical potential gradients. The model subsequently described expresses the electromotive force of a concentration cell as a function of observable variables only [GAR 92, 94]. The model can also be used in the case of multicomponent systems [GAR 97], [COM 94], but this section will be restricted to the case of a single electrolyte solution.

Let us consider the configuration Ag | AgCl ionic aqueous solution (c_L) | cation exchange membrane | ionic aqueous solution (c_R) AgCl | Ag, in which the electrolyte is dissociated in its cationic and anionic components $C_{\nu_+}A_{\nu_-} = \nu_+ C^{z_+} + \nu^- A^{z_-}$. From the thermodynamics of irreversible processes, the dissipation of local energy in each point of the membrane Ψ is given by:

$$-T\dot{s} = \Psi = j_w \frac{d\mu_w}{dx} + j_+ \frac{d\tilde{\mu}_+}{dx} + j_- \frac{d\tilde{\mu}_-}{dx} \qquad [19.37]$$

where, i = +, is the electrochemical potential of the ion i, μ_w and j_w are respectively the chemical potential and the flux of the solvent (water), respectively, and j_i, (i=+,-), represents the flux of ionic species using the membrane as reference frame.

Let us now assume that the electrodes are reversible to the anion of the electrolyte. The electric potential gradient measured at equilibrium in the interphases solution-electrodes, $d\psi/dx$, is related to the electrochemical potential gradient of the anion by [GAR 00]:

$$\frac{d\psi}{dx} = \frac{1}{z_- F} \frac{d\tilde{\mu}_-}{dx} \qquad [19.38]$$

where F represents the Faraday constant and z- is the valence of the anion. From equations [19.37] and [19.38] we obtain:

$$\Psi = j_w \frac{d\mu_w}{dx} + \frac{1}{\nu_+} j_+ \frac{d\mu_2}{dx} + j \frac{d\psi}{dx} \qquad [19.39]$$

where $j = F(z_+ j_+ + z_- j_-)$ is the electric current density and $\mu_2 = \nu_+ \tilde{\mu}_+ + \nu_- \tilde{\mu}_-$ is the chemical potential of the electrolyte. From the electroneutrality principle, $\nu_+ z_+ + \nu_- z_- = 0$ and the Gibbs–Duhem equation $c_w d\mu_w + c_2 \nu_- d\mu_- + c_2 \nu_+ d\mu_+ = 0$, where c_2 and c_1 are the concentrations of the solution and water, respectively, the dissipation function becomes:

$$-\Psi = \frac{1}{\nu_+} j_+^w \frac{d\mu_2}{dx} + j \frac{d\psi}{dx} \qquad [19.40]$$

where:

$$j_+^w = j_+ - \frac{\nu_+ c_2}{c_1} j_w \qquad [19.41]$$

is the matter flux density of the cation, taking the center of gravity of the solvent as frame of reference.

According to the thermodynamics of irreversible processes, fluxes are linear functions of forces and therefore equation [19.40] allows the following transport equations to be stated:

$$\frac{1}{\nu_+} j_+^w = -\lambda_{11} \frac{d\mu_2}{dx} - \lambda_{12} \frac{d\psi}{dx} \qquad [19.42]$$

$$j = -\lambda_{21} \frac{d\mu_2}{dx} - \lambda_{22} \frac{d\psi}{dx} \qquad [19.43a]$$

Since the Onsager coefficient matrix is symmetrical ($\lambda_{12} = \lambda_{21}$), these equations can be transformed into [GAR 97]:

$$\begin{pmatrix} \frac{1}{\nu_+} j_+^w = j \\ -\frac{d\psi}{dx} \end{pmatrix} = \begin{bmatrix} -D_2^w & \frac{t_+^w}{\nu_+ z_+ F} \\ \frac{t_+^w}{\nu_+ z_+ F} \frac{d\mu_2}{dc_2} & \frac{1}{\kappa} \end{bmatrix} \cdot \begin{pmatrix} \frac{dc_2}{dx} \\ j \end{pmatrix} \qquad [19.43b]$$

where D_2^w is the diffusion coefficient of the electrolyte, t_+^w is the transport number of cations using the solvent as reference frame, and κ is the electrical conductivity.

If the current density is zero, the previous equation leads to:

$$\Delta\psi = E = -\frac{1}{\nu_+ z_+ F} \int_0^l t_+^w(x, c_2) \frac{d\mu_2}{dx} dx = \frac{1}{\nu_+ F} \int_0^l \tau_+(x, c_2) \frac{d\mu_2}{dx} dx \qquad [19.44]$$

where E is the electromotive force of the concentration cell, $\tau_+ = t_+^w / z_+$ is the reduced apparent transport number which expresses the number of moles of counter-ions transported by a Faraday current and l is the membrane thickness.

Let us analyze the relationship between t_+^w and the transport number of counterions taking the membrane as a frame of reference, represented by t_+. The parameter t_+^w is given by:

$$t_+^w = \frac{F z_+ j_+^w}{j} \qquad [19.45]$$

By considering the relationship between the flux of cations with respect to the liquid in the membrane, j^w, and with respect to the membrane, j_+, equation [19.41] leads to:

$$t_+ = t_+^w + \frac{\nu_+ c_2}{c_w} t_w \qquad [19.46]$$

where $t_+ = Fz_+ \frac{j_+}{j}$ and t_w is the transport number of water associated with the electro-osmotic processes. Accordingly, $t_+^w \leq t_+$ and the difference between these parameters increases as the electro-osmotic permeability increases.

According to equation [19.44], the electromotive force of concentration cells is a complex function of the external concentration profile of salt in the membrane. For homogenous membranes, the electromotive force is simplified to:

$$E = -\frac{1}{\nu_+ F} \int_0^l \tau_+(c_2) \frac{d\mu_2}{dx} dx \qquad [19.47]$$

In this case the reduced apparent transport number τ_+ can be estimated from the equation:

$$\tau_+ = -\frac{F\nu_+}{RT\nu} \left(\frac{\partial \psi}{\partial \ln a_2} \right) \qquad [19.48]$$

By assuming an average value for τ_+ throughout the membrane, equation [19.47]) becomes:

$$E = -\frac{RT\nu}{\nu_+ F} \tau_+ \ln \frac{a_R}{a_L} \qquad [19.49]$$

In this equation R is the gas constant, $\nu = \nu_+ + \nu_-$, τ_+ is the average value of the apparent reduced transport number of counterions in the membrane, taking as reference frame the center of gravity of the solvent, whereas a_R and a_L represent, respectively, the activities of the electrolyte solutions on the right and left compartments of the concentration cell [COM 94].

For a binary electrolyte of monovalent ions, equation [19.49] reduces to:

$$E = -\frac{2RT}{F}\tau_{+} \ln\frac{a_R}{a_L} \qquad [19.50]$$

The advantage of the deduction presented here is that the electromotive force of concentration cells is obtained in a much more direct manner than using other alternative approaches [GAR 00].

19.6. Experimental characterization of ion exchange membranes

The characterization of ion exchange membranes for use in fuel cells involves:

1. Measurement of water sorption.
2. Determination of the ion exchange capacity.
3. Measurement of the transport number of protons.
4. Determination of membranes conductivity.
5. Measurement of the water transported by electro-osmosis.
6. Measurement of reformers (methanol) permeability in membranes.
7. Determination of the morphology of membranes.
8. Thermal stability of membranes.

19.6.1. *Water sorption*

Membranes previously dried in high vacuum are equilibrated in atmospheres of different relative humidity. The thickness of the membranes may be recorded as a function of the water content of the membrane and the value of this quantity can be further used in experiments related with the assembly membrane electrodes, impedance spectroscopy, etc. To measure the maximum water sorption of membranes, or water uptake, the following methodology is currently used. The dry membrane is weighed and then immersed in distilled water for 72 hours. Then the membrane is gently blotted with filter paper and weighed again. The weight measurement is repeated five times to obtain an average value. The water uptake is calculated by means of the following expression:

$$Water\ uptake(\%) = \frac{weight\ of\ wet\ membrane - weight\ of\ dry\ membrane}{weight\ of\ wet\ membrane} x100 \qquad [19.51]$$

Illustrative results obtained for different ion exchange membranes are shown in Table 19.1.

19.6.2. *Determination of the ion exchange capacity*

The measurement of the ion exchange capacity of membranes gives the number of moles of ionic fixed groups covalently bonded to the polyelectrolyte chains per g of dry membrane. For this purpose, dry membranes are immersed in a solution of HCl 0.5 N for 5 hours, in the case of strong ion exchange membranes, and 5 days for weak ion-exchange membranes. The system is stirred during this interval of time. Then the membranes are washed several times with distilled water and immersed again in a solution of NaCl 1 N for 5 days. During this period the system is stirred. The objective is to exchange the protons of the acid membrane for Na+ ions. The protons released by the membrane in the ion-exchange reaction are then titrated with an alkali solution. From the titration the ion-exchange capacity of the membrane is obtained. For orientative purposes, a few representative results obtained in a few membranes studied by our research group are presented in Table 19.1.

Membrane	Water uptake 100×g water/g dry membrane	mmoles H^+/g of dry membrane	moles of water /eq. fixed groups	mmoles H^+/g of water in the swollen membrane
MF4SC*	19.80	1.36	8.09	6.86
MF4SC/An*	19.24	1.28	8.35	6.65
Nafion 117*	20.09	1.35	8.27	6.72
EPDM-PP(10%)**	43.38	2.45	9.84	5.65
EPDM-PP(20%)**	23.25	1.27	10.17	5.46
PSS***	26.7	2.06	7.20	7.72

Table 19.1. *Sorption of water and ion exchange capacity for different membranes: *[COM 07], **[COM 08],***[PAR 08]*

Owing to swelling effects, membranes with high ion-exchange capacity may exhibit in numerous cases a relatively low concentration of protons in the water pores. For example, the ion exchange capacity and water uptake for the PSS membrane in Table 19.1 are, respectively, 2.06 meq H^+ and 0.267g H_2O per gram of dry membrane. The membrane has 7.72 meq. H^+/g H_2O, a quantity only somewhat higher than that of Nafion (6.72 meq. H^+/g H_2O) with IEC =1.35 meq. H^+/g dry membrane.

19.6.3. *Measurements of transport number and mobility of protons in membranes*

An alternative procedure to the Hirthof method [ATK 78] to determine the transport number of protons in cation-exchange membranes in the acid form is the use of concentration cells [SAT 02], [LAK 69] made up of two compartments separated by the membrane. The compartments are filled with HCl solutions of different concentration. Symmetric Ag/AgCl electrodes located in each compartment are used to monitor the electromotive force of the cell. The solutions are strongly stirred by means of magnetic stirrers to reduce the membrane-solution interphase to a minimum. The electrodes are connected to a potentiometer and the electromotive force of the concentration cell is recorded as a function of time.

The potentials of Ag/AgCl electrodes in the previous configuration contribute to the electromotive force (*emf*) of the concentration cell. The observable *emf* can be measured straightforwardly after pouring the solutions in the cell ($t \rightarrow 0$) (initial time method), and also when the system reaches steady state conditions ($t \rightarrow \infty$), [COM 94, 97]. The symmetry of the membranes is controlled by exchanging the concentrations of solutions in the semi-cells. If the difference of the observed *emfs* is practically zero, the membrane is considered symmetrical and quasi-homogenous [COM 97].

By assuming an average value for the transport number in reduced form $\bar{\tau}$, the integration of equation [19.47] for *HCl* solutions ($\nu+ = \nu- = 1$) becomes:

$$-E = \frac{2RT}{F}\bar{\tau}_{+}\ln\frac{a_R}{a_L} \qquad [19.52]$$

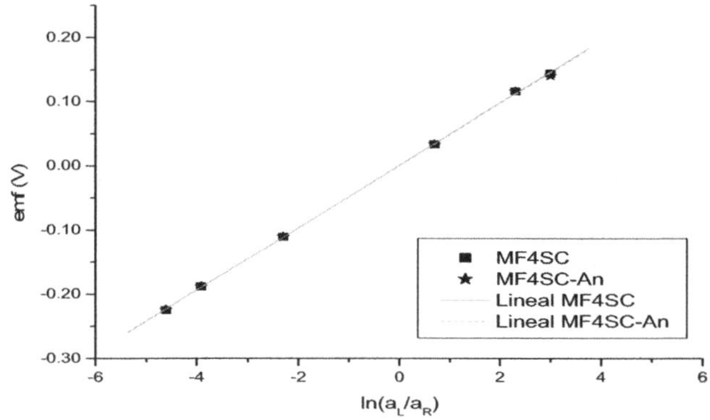

Figure 19.3. *Representation of the observed emf as a function of* $ln(c_L/c_R)$ *for the MF4SC and MF4SC-An membranes (Table 19.1), at 25°C. The measurments were performed keeping* C_R*=0.01 M. Straight lines in the figure were calculated by equation [19.52]*

Values of *emf* for different concentrations (a_R/a_L) corresponding to the MF4SC and the MF4SC-An membranes are plotted as a function of log(a_R/a_L) in Figure 19.3. The plot shows that the *emf* of these systems is practically a linear function of the logarithms of the concentration ratio. This behavior suggests a slight dependence of the apparent transport number of protons on the concentration profile of HCl in the membrane, even though this profile is associated with a large range of concentrations. The method allows the estimation of $\bar{\tau}_+$ for different concentration ratios [COM 94].

19.6.3.1. *Measurement of the mobility of protons*

The density of electric current flowing in a membrane equilibrated with an electrolyte solution, under an electrical potential driving force, is given by:

$$j = F\sum_i z_i c_i J_i \ , (i = +, -). \qquad [19.53]$$

Since the concentration of counterions across the membrane is constant, the relationship between the electrochemical potential and the electrical potential is $\frac{d\bar{\mu}_i}{dx} = z_i F \frac{d\psi}{dx}$. For the reasons outlined in section 19.5.1, the flux of the species i across the membrane in these conditions is $J_i = z_i F \frac{D_i}{RT}\frac{d\psi}{dx} = -z_i \bar{u}_i \frac{d\psi}{dx}$, where $\bar{u}_i$ is the ionic mobility.

Then the electric current density can be written as:

$$j = -F\left(\sum_i c_i z_i^2 \bar{u}_i\right)\frac{d\psi}{dx} \qquad [19.54]$$

By considering Ohm's law $j = -\bar{\kappa} d\psi / dx$, where $\bar{\kappa}$ is the average conductivity of the system, and taking into account equation [19.54], the following expression for the average conductivity is obtained:

$$\bar{\kappa} = F\sum_i z_i^2 c_i \bar{u}_i \qquad [19.55]$$

For cation-exchange membranes equilibrated with water, the concentration of counterions is equal to the concentration of fixed anions covalently bonded to the matrix of the membrane; this concentration can be obtained from the IEC of the membrane. Then the electrical current flowing through the membranes is transported by counterions and equation [19.55] can be written as:

$$\overline{\kappa} = F\overline{c}_{+}u_{+} \qquad [19.56]$$

Accordingly, the apparent mobility of counterions in cation-exchange membranes equilibrated with water can be obtained from the conductivity.

19.6.4. *Measurement of conductivity*

Impedance spectroscopy is a technique widely used to characterize materials and interphases. One of the most important characteristics of this technique is the correlation which exists between the response of a real system to an electric perturbation field and that of an ideal model represented by electric circuits composed of discrete elements (capacitances, inductances and ohmic resistances) [MAC 87], [BOU 89].

The response of a membrane to a sinusoidal voltage signal of type $V(t)=V_m \sin(\omega t)$, where ω is the angular velocity ($\omega=2\pi f$, with f the frequency in Hz), is the intensity current $I(t)= I_m \sin(\omega t+\theta)$, where θ represents the phase difference between the applied voltage and the current. Obviously when $\theta=0$ the behavior of the system is purely resistive. The complex impedance of the system will be $Z^*(\omega) = \dfrac{V(t)}{I(t)}$.

In the analytical description of conductive processes, equivalent electric circuits are used. The circuits are integrated by elements that represent the different macroscopic phenomena intervening in the transport of mass and electrical charge. Figure 19.4 shows different types of used circuits. In some cases, the relaxation processes observed in experiments can be easily interpreted by a simple equivalent circuit, as in those in Figures 19.4a and b, associated with a single relaxation time. However, in most cases the systems exhibit a distribution of relaxation times and the equivalent circuit consists of different circuits, like those described in Figure 19.4c and d. In general, it will be necessary to use a more complex analysis if complex polarization processes occurring in polyelectrolytes are accounted for. When non-linear methods are used in the analysis, the values of the parameters that describe the model can be optimized by minimising the differences between the values measured for a given experimental dielectric quantity and those calculated with the model [COM 96]. Phenomenologically, resistances and capacitors account, respectively, for energy dissipation and accumulation of charges or energy storage. A parallel R-C circuit is a dielectric polarization model that describes relaxation processes associated with a single relaxation time τ (=RC) where the migration of free charges does not prevail.

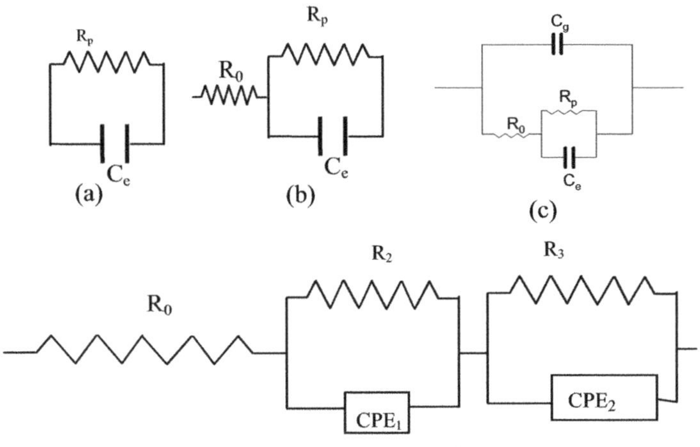

Figure 19.4. *A Debye RC model (a) and different types of circuits to describe the dielectric behavior of wet cation-exchange membranes in the acid form*

Impedance measurements are performed by placing the membrane previously equilibrated with water, between two gold circular and parallel electrodes. The response of cation exchange membranes in the acid form to an alternating electric field may be described by a circuit composed of an ohmic resistance R_0 connected in series with a parallel R-C circuit. The parameter R_0 represents the resistance to proton transport in the membrane while the RC circuit accounts for the polarization phenomena taking place in the membrane. The complex impedance Z* of the system is:

$$Z^* = R_0 + \frac{R_p}{1 + jR_p\omega C} \qquad [19.57]$$

Accordingly, the real, $Z^{'}$, and loss, $Z^{''}$, components of the complex impedance, Z^*, are

$$Z' = R_0 + \frac{R_p}{1 + (R_p\omega C)^2} \qquad [19.58]$$

$$Z"=-\frac{\omega R_p^2 C}{1+(R_p\omega C)^2} \qquad [19.59]$$

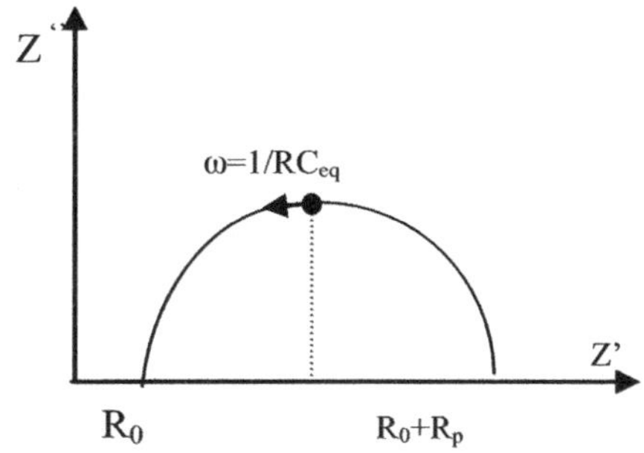

Figure 19.5. *Nyquist plot corresponding to the equivalent circuit (b) shown in Figure 19.4a. The maximum of the half circle corresponds to ωRC=1 where RC = τ, < relaxation time [BOU 89]*

As shown in Figure 19.5, The Z'' vs Z' plot, or Nyquist representation [NYQ 28], is a semicircle intersecting the abscissa axis at $Z' = R_0$ $(\omega\rightarrow\infty)$ and $Z' = R_0 + R_p$ $(\omega\rightarrow 0)$ The representation of the impedance modulus as a function of the frequency or Bode diagram [BOD 56] is an alternative method to obtain the proton resistance of cation-exchange membranes in the acid form. As in the Nyquist plots: $/Z^*/ = R_0$ $(\omega\rightarrow\infty)$ and $/Z^*/ = R_0 + R_p$ $(\omega\rightarrow 0)$. Moreover, $\phi = \tan^{-1}(Z'/Z")$ reaches a maximum ($\phi = 0$) at the plateau $(\omega\rightarrow\infty)$.

As indicated above, complex systems present a distribution of relaxation times and the Nyquist diagrams are not semicircles but arcs. In these cases, the capacitance is not ideal so that the capacitor in the circuit is replaced by a constant phase element (CPE) whose admitance is expressed by:

$$Y(\omega) = Y_0(j\omega)^{-n} \qquad [19.60]$$

where Y0 (Ω-1 s-n) and n ($0 \leq n \leq 1$) are empirical parameters that can be obtained by fitting the Nyquist diagrams to the experimental results [MAC 87]. The more n decreases, the more Nyquist plots depart from semicircles. However, as in the case of circuits with a single relaxation time, the arcs intersect the abscissa axis at ω→ ∞ (Z' = R0) and at ω→ 0 (Z'=R0 + RP).

Two particular cases appear when the dependence of the impedance on n is considered: if $n = 1$, the capacitor of the equivalent circuit is ideal but if $n = 0.5$, the element is a Warburg impedance associated with a Fick's diffusion process. In general, for complex processes, such as that shown in Figure 19.4d, the equivalent capacitance of the system C_{eq} can be determined by means of the relationship:

$$C_{eq} = \frac{(RY_0)^{1/n}}{R} \quad [19.61]$$

Accordingly, the dielectric behavior of complex systems can be studied by impedance spectroscopy (EIS) using equivalent electric circuits. From the values of the parameters that describe the models, information can be obtained concerning the the heterogenous behavior of the systems.

Finally, the proton conductivity can be obtained from R_0 using the following expression:

$$\kappa = \sigma = \frac{l}{R_0 S} \quad [19.62]$$

where S and l are, respectively, the area and thickness of the membrane in contact with the electrodes.

Figures 19.6 and 19.7 present, respectively, the Nyquist and Bode diagram for EPDM-PP20% and EPDM-PP10% membranes equlibrated with water, at different temperatures [COM 07].

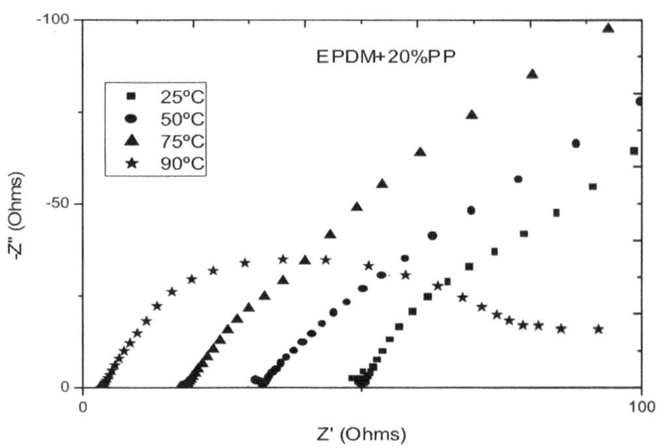

Figure 19.6. *Nyquist plot for the EPDM-PP20% membrane in equilibrium with the water for different temperatures: from left to right 90, 75, 50 and 25°C [COM 07]*

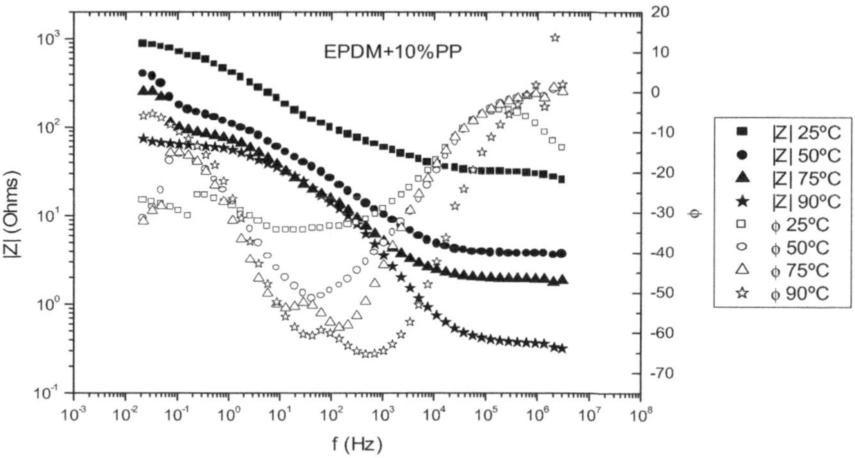

Figure 19.7. *Value of the impedance modulus, |Z*| (filled symbols) and the phase angle (open symbols) for the EPDM+10%PP membrane equilibrated with water, at different temperature: (squares) 25°C, (circles) 50°C, (triangles) 75°C and (stars) 90°C*

The points at which the arcs intersect the abscissa axis at high frequencies yield R_0, at each temperature of interest. However, because of the coupling between the pure conductivity and relaxation processes taking place in the membranes/electrodes, the curves depart from arcs as frequency increases. Figure 19.7 shows the double-logarithmic plot of the complex impedance modulus, $|Z^*|$ against log ω (Bode plot) for the EPDM-PP (10%) membrane of Table 19.1, at different temperatures.

Normally, $|Z^*|$ rapidly decreases as frequency increases reaching a plateau in the high frequencies region. On the other hand, the phase angle reaches a maximal value ($\phi = 0$) at the plateau, and the value of $|Z^*|$ at the frequency at which ϕ reaches a maximum is assumed to be R_0, the proton resistance of the membrane.

The Bode plots of Figure 19.7 show that the conductivity of the EPDM-PP(10%) membrane at 25°C is about one order of magnitude smaller than that reported for Nafion membranes, at the same temperature (1–7 S/m), [HIC 049], [MAU 00], [MAU 02].

The temperature dependence of the conductivity of EPDM-PP membranes roughly obeys Arrhenius behavior (Figure 19.8) with activation energies of 57 and 33 kJ/mol for EPDM-PP(10%) and EPDM-PP(20%) membranes, respectively (see Table 19.2 for details).

	EPDM-PP10%		EPDM-PP20%	
T(°C)	|Z| (Ohm)	κ (S/cm)	|Z| (Ohm)	κ (S/cm)
25	3,13E+01	9,84E-04	5,01E+01	7,25E-04
50	3,77E+00	8,17E-03	3,24E+01	1,12E-03
75	1,92E+00	1,60E-02	1,81E+01	2,01E-03
90	3,69E-01	8,35E-02	3,53E+00	1,03E-02

Table 19.2. *Proton resistance R, and conductivity, κ, obtained from the Bode plots for EPDMPP (10%) and EPDM-PP (20%) sulfonated membranes in acid form under 100% RH [COM 07]*

The activation energy associated with proton transport processes in these membranes is significantly lower than that reported for the transport of ionic carriers in non-polyelectrolytes or conventional polar polymers.

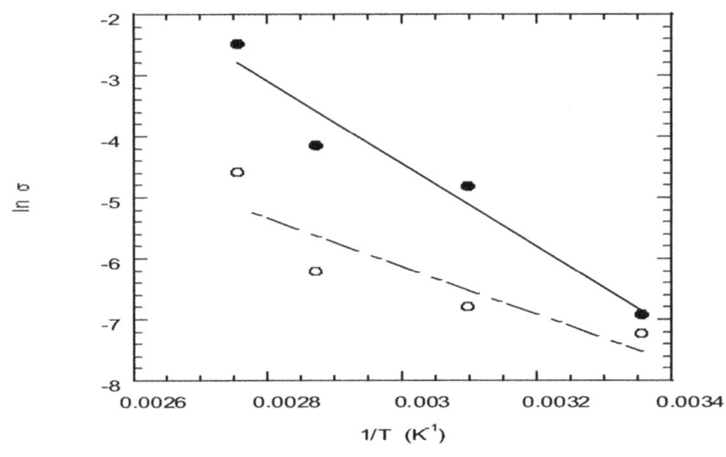

Figure 19.8. *Arrhenius plot for the conductivity in S/cm of EPDM-PP20% (filled circles) and the EPDM-PP10% (open circles) membranes at equilibrium with 100% RH humidity atmosphere [COM 07]*

19.6.5. *Electro-osmotic measurements*

The flux of solvent through ion exchange membranes dragged by proton transport is measured with electro-osmotic cells. The experimental device consists of two semi-cells separated by the membrane, according to the configuration: Ag| AgCl | solution of HCl (c)| cation exchange membrane in acid form | solution HCl (c) | Ag | AgCl.

Each semi-cell is provided with a capillary to measure the electro-osmotic flux from the displacement of liquid in the capillary. This device, in series with a coulomb-meter, is connected to a battery. The current causes the passage of water from the anode to the cathode. From the displacements of the capillary meniscus, which must be the same but in opposing direction, the change in volume ΔV by effect of the transported charge Q is measured. A balance of matter in the cathodic compartment allows the determination of the volume of water, ΔVe-o, dragged by the cations. For monovalent cations, the relevant expression is:

$$\Delta V_{e-0} = \Delta V - \frac{Q}{zF}\left(\overline{V}_{Ag} + t_+\overline{V}_{Cl_zM} - \overline{V}_{AgCl}\right) \qquad [19.63]$$

where $\overline{V}_{Ag}$, $\overline{V}_{Cl_zM}$ and $\overline{V}_{AgCl}$ are, respectively, the partial molar volumes of Ag, electrolyte and AgCl, whereas t+, z and F are, respectively, the cation transport number, the valence of the cation M and Faraday's constant. Then, the transport number of water arisen from electro-osmosis is given by:

$$t_w = \frac{F}{Q}\frac{\Delta V_{e-o}}{\overline{V}_{H_2O}} \qquad [19.64]$$

Because the conductivity of membranes increases with concentration of electrolyte at low concentrations, the electro-osmotic flux in membranes must decrease as the concentration of the solution increases, (see equation [19.34]; also Figure 19.10 in reference [SAT 02]).

19.6.6. *Measurements of the permeability of reformers in membranes: methanol permeability in vapour phase*

Though Nafion membranes present good chemical and thermal stability conditions, the large permeability of methanol in these materials decreases their performance when this reformer is used as fuel.

The permeability of reformers in ion exchange membranes can be measured by using pervaporation techniques. Pervaporation, like vapour permeability, is a separation process; the only difference between these techniques lies in the physical state of the feed which is a liquid in the former case and a vapour in the latter. However, in both processes non-porous membranes are used obtaining a gaseous product from the feed. The selective permeability of different components through a membrane is essentially a mass transfer process, the chemical potential gradient being the driving force. A theoretical analysis of the system shows that the flux of the component i of the feed in a point of the membrane is given by equation [19.1], that is:

$$J_i = -\frac{c_i D_i}{RT}\frac{d\mu_i}{dx} = -\frac{D_i}{c_i}\frac{dc_i}{dx} \quad [19.65]$$

where it is assumed that the activity coefficient of component i is the unit. Assuming that $ci=kipi$ where ki and pi are, respectively, Henry's solubility constant and the vapor pressure of the component i, equation [19.65] becomes:

$$J_i = -k_i D_i \frac{dp_i}{dx} = -P_i \frac{dp_i}{dx} \quad [19.66]$$

where $Pi=kiDi$ is the permeability coefficient. Pi is generally expressed in barrers (1 barrer = 10-10 × cm3 (STP) cm/(cm^2 s cmHg)). The integration of equation [19.66] leads to the following expression:

$$J_i = P_i \frac{p_{i1} - p_{i2}}{l} \cong P_i \frac{p_{i1}}{l} \quad [19.67]$$

Notice that vapor pressure of the component i at the side of the membrane in contact with the feed is $pi1$ whereas $pi2$ at the other side of the membranes is approximately zero because this part of the pervaporator is kept under vacuum. Moreover:

$$p_i = x_i p_i^0 \gamma_i \quad [19.68]$$

where xi, p_i^0 y γi represents the molar fraction, the vapour pressure of the pure component and the activity coefficient of the i component of the feed in the supply compartment of the pervaporatour. By measuring the flux of the component i, the permeability coefficient of this component is obtained as:

$$P_i = \frac{J_i l}{x_{i1} p_{i1}^0 \gamma_i} \quad [19.69]$$

The permselectivity of a component A of the feed with respect to other component B in the pervaporation process is given by:

$$\alpha(A/B) = \frac{y_A / y_B}{x_A / x_B} \quad [19.70]$$

where y and x represent, respectively, the molar fractions of the permeability products and the supply liquids.

To determine the methanol permeability in liquid phase we can use the method described by Yo Jin Kim *et al.* [KIM 04], or a system composed of two 50 mL glassy compartments described by Tricoli *et al.* [TRI 98, 00], [PIV 99].

19.7. Determination of membrane morphology using the SEM technique

Scanning Electron Microscopy (SEM) is an important technique that can be used for the morphological characterization of membranes, especially the topological aspects of these materials. The technique allows the imaging of the surface and/or cross-sectional areas of membranes with a high energy beam of electrons, thus providing information concerning the morphology of the surfaces, morphology of fillers dispersed in mebranes (form, size and distribution), porosity and distribution of pore sizes in membranes, etc.

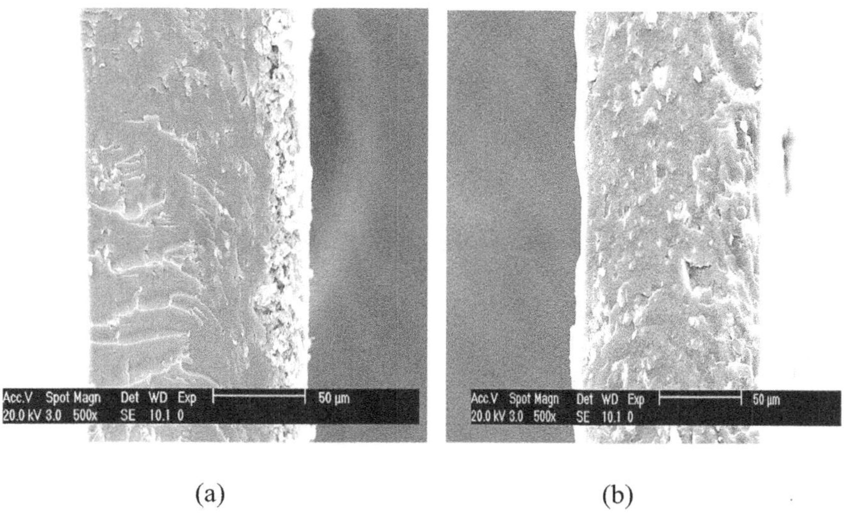

Figure 19.9. *Cross-sectional surface images obtained by SEM a) Nafion + Sepiolite membrane. b) SEBS + Sepiolite membrane*

Figures 19.9 and 19.10 present, respectively, cross-sectional and surface micrographies of Nafion and sulfonated SEBS composite membranes, both containing sepiolite as filler. Both membranes were cast at 70°C. Higher temperatures are not recommended for the casting process because a too rapid evaporation of solvent could form bubbles in the membrane. The images suggest a rather good dispersion of the fillers in the membranes. However, small sediments on one side of the membranes, somewhat smaller in the SEBS composite membrane, can be detected. The fact that the solvent evaporation rate is higher in the composite membrane based on SEBS than that based on Nafion explains the smaller sedimentation in the former membrane.

19.8. Thermal stability

Owing to the fact that fuel cell performance improves with temperature, it is fundamental to use cation-exchange membranes that are stable at temperatures near 150°C as solid polyelectrolytes in these devices. Membranes based on high thermal stability polymers, such as polysulfones, poly(aryl-ether-sulfones), polyketones, polyimides, etc., could be used for this purpose.

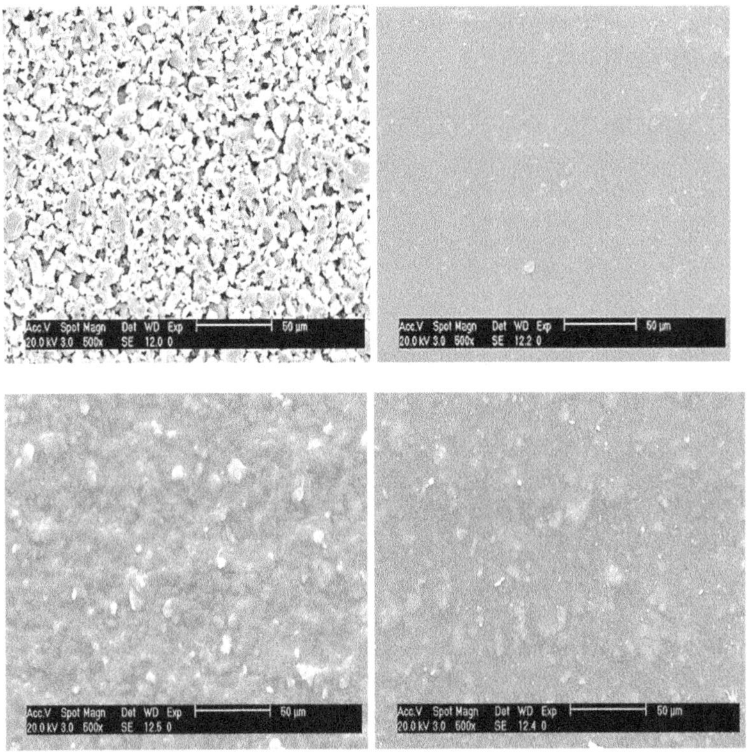

Figure 19.10. *Surface images of the Nafion+Sepiolite and the SEBS+Sepiolite membranes: a) (top left) Upper Nafion + Sepiolite side. b)(top right) Lower Nafion + Sepiolite side. c)(bottom left) Upper SEBS+Sepiolite side. d)(bottom right) Lower SEBS+Sepiolite side [FER 07]*

The thermal stability of membranes can be studied with different techniques, thermogravimetry (TG) and differential thermogravimetry (DTG) being mostly used. The first technique records the variation of mass wth temperature at different heating rates under the desired atmosphere (oxygen, nitrogen, etc.). The curve W(t) vs T, or thermogram, gives the loss of mass as a function of temperature.

DTG allows curves presenting a series of peaks (Figure 19.11), whose areas are equivalent to the changes of mass of the sample, to be obtained. This technique, together with TG experiments, are used to determin the degradation temperature of polymers, thermal stability, content of water and residual solvents, etc.

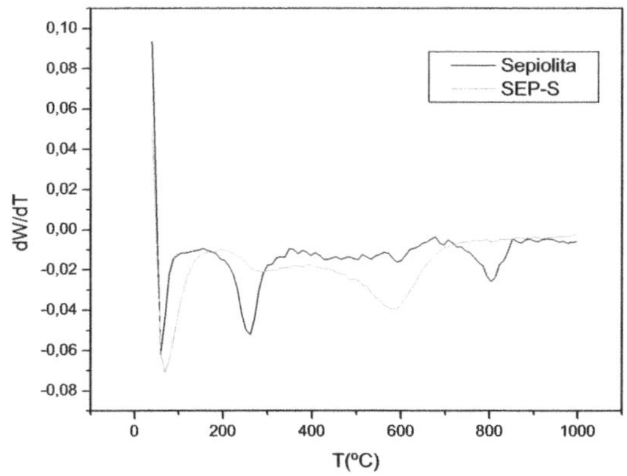

Figure 19.11. *Variation of mass as a function of the temperature for sepiolite and sulfonated sepiolite. This filler was added to Nafion SEBS membranes [FER 07]*

Figure 19.11 presents variations of mass with temperature for sepiolite and sulfonated sepiolite, fillers which have been used for the preparation of Nafion and SEBS composite membranes. The thermogram for the pristine sepiolite exhibits a mass loss slightly above 200°C which corresponds to water trapped in the canals of the filler, followed by a second mass loss in the vicinity of 800°C associated with a dehydroxylation process [FRO 03]. On the other hand, the DTG thermogram corresponding to the sulfonated filler exhibits mass losses in the vicinity of 200°C and 600°C, presumably associated with the organic-inorganic sulfonated filler [FER 07].

19.9. Acknowledgements

This work was supported by the DGICYT through Grant MAT2005-05648-C02-02 and the Generalitat Valenciana (Spain) with the IMCITA/2005/31 project between IMPIVA and the Institut de Technologie Electrique (ITE).

19.10. Bibliography

[ALB 03] ALBERTI G., CASCIOLA M. *Annu. Rev. Mater. Res.*, vol. 33,129-154, 2003.

[ALB 92] ALBERTI G., CASCIOLA M., CONSTANTINO U., PERAIO A., MONTONERI E., *Solid State Ionics*, vol. 50, p. 315-322, 1992.

[ALB 95] ALBERTI G., CASCIOLA M., CONSTANTINO U., PERAIO A., REGA T., MATER J., *J. Mater. Chem.*, vol. 5, p. 1809-1812, 1995.

[ALB 97] ALBERTI G., CASCIOLA M., *Solid State Ionics*, vol. 97, p. 17-186, 1997.

[ATK 03] ATKINS J. R., SIDES C.R, CREAGER S. E., HARRIS J.L., PENNINGTON W. T., THOMAS B.H., DESMARTEAU D.D., *J. New Mat. Electrochem. Syst.*, vol. 6, p. 9-15, 2003.

[ATK 78] ATKINS P.W., *Physical Chemistry*, W.S.H. Freeman, San Francisco, p. 840, 1978.

[BAU 62] BAUM B., HOLEY W., WHITE J.R., in *Membrane Separation Processes*, Wiley & Sons, Inc., New York, 1962.

[BOD 56] BODE W.W., *Network Analysis and Feedback Amplifier Design*, Van Nostrand, Princeton, N.J., 1956.

[BOU 89] BOUKAMP B.A., *Equivalent Circuit University of Twenty*, 1989.

[BUC 88] BUCKLEY A., STUETZ D.E., SERAD G.A., in *Encyclopedia of Polymer Science and Engineering*, 2nd. ed., John Wiley & Sons, New York, vol. 11, 1988.

[CAB 85] CABASSO J., "Materials Science of Synthetic Membranes", *ACS Symposium Series 269*, *American Chemical Society*, LLOYD D.R. (Ed.), Washington D.C., p.305–323, 1985.

[COM08] COMPAÑ V., RIANDE E.F.J., FERNANFEZ-CARRETERO F.J., BEREZINA N.P., SYTCHEVA A.A-R., *J. Membr. Sci.*, vol. 318, p. 255–263, 2008.

[COM 94] COMPAÑ V., LÓPEZ M.L., SØRENSEN T.S., GARRIDO J., *J. Phys. Chem*, vol. 98, p. 9013, 1994.

[COM 96] COMPAÑ V., SORENSEN T. S, DIAZ-CALLEJA R., RIANDE E., *J. Applied Physics*, vol. 79, no. 1, p. 1–9, 1996.

[COM 97] COMPAÑ V., SORENSEN T.S., ANDRIO A., LÓPEZ M.L., DE ABAJO J., *Jour. Membr. Sci.*, vol. 123, p. 293–302, 1997.

[FER 07] FERNANDEZ-CARRETERO F. J., COMPAÑ V., RIANDE E., *J. Power Sources*, vol. 173, p. 68–76, 2007.

[COR 03] CORNET N., DIAT O., GEBEL G., JOUSSE F., MARSACQ D., MERCIER R., PINERI M., *J. New Mat. Electrochem. Syst.*, vol. 3, no. 33, 2003.

[DIM 02] DIMITROVA P., FRIEDRICH K.A., STIMMING U., VOGT B., *Solid State Ionics*, vol. 150, p. 115–122, 2002.

[DIN 01] DING J., CHUY C., HOLDCROFT S., *Chem. Mat.*, vol. 13, p. 2231, 2001.

[DEL RIO 05] DEL RIO C., SANCHEZ F., ACOSTA J.L., *Polymer*, vol. 46, p.3975–3985, 2005.

[ENN 99] ENNARI J. *et al.*, *Macromol. Symp.*, vol 146, p. 41–45, 1999.

[FRO 03] FROST R.L., DING Z., *Thermochimica Acta*, vol. 397, no. 1–2, p. 119–128, 2003.

[GAR 00] GARRIDO J., COMPAÑ V., *J. Electrochem. Soc.*, vol. 147, no. 11, p. 4263–4267, 2000.

[COM 07] COMPAÑ V., FERNÁNDEZ-CARRETERO J., RIANDE E., LINARES A., ACOSTA J.L., *J. Electrochem. Soc.*, vol. 154, no. 2, p. B159, 2007.

[GAR 92] GARRIDO J., COMPAÑ V., *J. Phys. Chem.*, vol. 96, p. 2721, 1992.

[GAR 94] GARRIDO J., COMPAÑ V., LÓPEZ M.L.,*J. Phys. Chem.*, vol. 98, p. 6003, 1994.

[GAR 97] GARRIDO J., COMPAÑ V., LÓPEZ M.L., *Current Topics in Electrochemistry*, vol. 4, chap. 7. Research Trends, Poojapura, 1997.

[GAR 97] GARRIDO J., COMPAÑ V., LÓPEZ M.L., MILLER D.G., *J. Phys. Chem. B*, vol. 101, no. 29, p. 5740–5746, 1997.

[GEN 01] GENIES C., MERCIER R., SILLION B., PETIAUD R., CORNET N., GEBEL G., PINNERI M., *Polymer*, vol. 42, p. 359, 2001.

[GEN 01] GENOVA-DIMITROVA P., FOSCALLO D., POINSIGNON C., SANCHEZ J.Y., *J. Memb. Sci.*, vol. 185, p. 59, 2001.

[GRE 61] GREGOR H.P., *Proc. Ann. Powers Sources Conf*, vol. 15, no. 4, 1961.

[GUO 99] GUO Q., PINTAURO P.N., TANG H., O'CONNOR S., *J. Memb. Sci.*, vol. 154, p. 175, 1999.

[HAR 03] HARRISON W. *et al.*, *J. Polym. Sci.: Part A: Polym. Chem.*, vol. 41, p. 2264, 2003.

[HIC 04] HICKNER M.A., GHASEMI H., KIM YU S., EINSLA B.R., MCGRATH J.E., *Chem Rev.*, vol. 104, p. 4587–4612, 2004.

[HOG 01] HOGARTH M., GLIPA X., *High temperature membranes for solid polymer fuel cells*, TechniETSU F/02/00189/REP; DT1/Pub URN 01/893, 2001.

[HOG 05] HOGARTH W.H.J., DINIZ DA COSTA J.C., (MAX)LU G.Q., *Journal of Power Sources*, vol. 142, p. 223–237, 2005.

[JAN03] JANNASCH P., *Curr. Opin. Colloid Interface Sci.*, vol. 8, no. 96, 2003.

[JUD 50] JUDA W., MCRAE W.A.J., *Am. Chem. Soc.*, vol. 72, p. 1044, 1950.

[KIM 04] KIM YO JIN, WON CHOON CHOI, WOO SEONG.IHL, HONG WON HI., *J. Membr. Sci.*, vol. 238, p. 213–222, 2004.

[KOT 01] KOTER S., *Separ. Purif. Technol.*, vol. 22–23, p. 643, 2001.

[KOT 02] KOTER S., *J. Memb. Sci.*, vol. 206, p. 201, 2002.

[KRE 01] KREUER K.D., *J. of Membrane Sci.*, vol. 185, p. 29–39, 2001.

[KRE 98] KREUER K.D., FUCHS A., SPAETH M., MAIER J., *Electrochimica Acta*, vol. 43, p. 1281–1288, 1998.

[LAK 69] LAKSHMINARAYANAIAH N., *Transpor Phenomena in Membranes*, Academic Press, New York, 1969.

[LI 03] LI Q.F,. HE R.H., JENSEN J.O., BJERRUM N.J., *Chem. Mater.*, vol. 15, 4896, 2003.

[LÓP 01] LÓPEZ M.L., COMPAÑ V., GARRIDO J., RIANDE E., ACOSTA J.L., *J. Electrochem. Soc.*, vol. 148, E372-E377, 2001.

[MAC 87] MACDONALD J.R., *Impedance Spectroscopy: Emphasizing solid materials and systems*, Wiley-Interscience, 1987.

[MAU 00] MAURITZ K.A., PAYNE J.T., *J. Membrane Science*, vol. 168, no. 39, 2000.

[MAU 02] MAURITZ K.A., STOREY R.F., MOUNTZ D.A., REUSCHLE D.A., *Polymer*, vol. 43, p. 4315–4323, 2002.

[MAU 04] MAURTIZ K.A., MOORE R.B., *Chem. Rev.*, vol. 104, 4535, 2004.

[MIY01] MIYAKE N., WAINRIGHT J.S., SAVINELL R.F., *J. Electrochem. Soc.*, vol. 148, no. 8, A905–A909, 2001.

[NYQ 28] NYQUIST H., *Phys. Rev.*, vol. 32, p. 110, 1928.

[PAD 03] PADDISSON S.J., *Ann. Mat. Res.*, vol. 33, p.289, 2003.

[PAR 08] PARCERO E., FERNANDEZ-CARRETERO F.J., COMPAÑ V., HERRERA R., DEL CASTILLO L.F., RIANDE E., *Journal of The Electrochemical Society*, vol. 155, F245, 2008.

[PIV 99] PIVOVAR B. S., WANG Y., CUSSLER E.L., *J. Membr. Sci.*, vol. 154, p. 155–162, 1999.

[POU 92] POURCELLY G., GAVACH C., "Perfluorinated membranes", in COLOMBAN P. (ed.), *Proton Conductors*, Cambridge University Press, New York, p. 294–310, 1992.

[PUS 82] PUSCH W., *Angew. Chem.*, vol. 21, p. 660, 1982.

[R&D 03] R&D Plan for the High Temperature Membrane Working Group. Department of Energy Office of Efficiency and Renewable Energy's Hydrogen, Fuel Cells & Infrastructure Technology Program, 2003.

[REB 75] REBOIRAS M.D., RIANDE E., *An. Quím.*, vol. 71, p. 39, 1975.

[SAT02] SATA T., *Ion Exchange Membranes. Preparation, Characterization, Modification and Application*, RSC Advancing the chemical sciences, UK, 2002.

[SCH 62] SCHLÖGL R., "Ber. Bunsenges", *Physik. Chem.*, vol. 70, p. 400, 1966, HELFFERICH F., *Ion Exchange*, McGraw Hill, New York, 1962.

[TRI 00]TRICOLI V., CARRETTA N., BARTOLOZZI M., *J. Electrochim. Soc.*, vol. 147, p. 1286, 2000.

[TRI 98] TRICOLI V., *J. Electrochim Soc.*, vol. 145, p. 3798, 1998.

[TUC 97] TUCKERMAN M.E., LAASONEN K., SPRIK M., PARRINELLO M., *Science*, vol. 275, p. 817–820, 1997.

[ULB06] ULBRICHT M., "Advanced functional polymer membranes", *Polymer*, vol. 47, p. 2217–2262, 2006.

[WAI 95] WAINRIGHT J.S., WANG J-T., WENG D., SAVINELL R.F., LITT M., *J. Electrochem. Soc.*, vol. 142, L121, 1995.

[WAL 99] WALTER M., BAUMGÄRTNER K.M., KAISER M., KERRES J., ULLRICH A., RÄUCHLE E., *J. Appl. Polym. Sci.*, vol. 74, p. 67, 1999.

[WIH 02] WIHELM F.G., PUNT I.G.M., VAN DER VEGT N.F.A., STRAHMANN H., WESSLING M., *J. Memb. Sci.*, 199, p. 167, 2002.

[WIL 02] WILES K.B., BHAND V.A., WANG F., MCGRATH J.E., *Polym. Prep.*, vol. 43, p. 993, 2002.

[YAM 64] YAMABE Y., SENO M., *Ion Exchange Resin Membranas*, Gihodo, Tokyo, Japan, p.8, 1964.

[YAN 04] YANG C. SRINIVASAN S., BOCARSLY A.B., TULYANI S., BENZINGER J.B., *J. Membr. Sci.*, vol. 237, p. 145–161, 2004.

[ZHA 01] ZHANG L., CHENGSON M., MUKERJEE S., *Electrochim. Acta*, vol. 48, p. 1845–1859, 2003.

Chapter 20

Semiconducting Organic Materials for Electroluminescent Devices and Photovoltaic Conversion

20.1. Brief history

During the last 50 years, polymers have held an increasingly important place amongst dielectric materials. They are light, easy to process, and their physico-chemical properties can be modified by well-established techniques of organic chemistry. Although less present in industrial applications, semiconducting organic materials offer the same advantages as their inorganic counterparts. The strong interest aroused by organic electronics since the end of the 1980s recently led to the commercialization of organic displays. Nevertheless, more work needs to be put into research, particularly to achieve the production of flat screens and to improve the performances of organic photovoltaic cells. The organic semiconductors which constitute the active layers of these components are generally listed in two categories: polymers and materials with low molar mass, or "small molecules".

The phrase "organic semiconductor", applied today to different types of materials, was initially applied only to certain molecular crystals (naphthalene, anthracene, etc.) presenting properties similar to inorganic semiconductors. In particular, the electroluminescence of anthracene was studied from the beginning of the 1960s onwards [POP 63]. However, the functioning voltages required, higher

Chapter written by Pascale JOLINAT and Isabelle SEGUY.

than 100 V, and the weak quantum efficiency (less than 0.1%) of the devices did not allow commercial applications to be envisaged.

If the photoconductivity of organic crystals was not used either because of their weak photosensitivity or their spectral domain restricted to the ultraviolet, numerous photoconductive "small molecules" were developed from the 1970s onwards. These molecules, incorporated in a polymer insulating matrix, are commonly used in photocopiers and laser printers [LOU 88, LAW 93]. Dyes such as phthalocyanines or perylenes, widely used for the manufacture of photoconductive layers, can also form organic photovoltaic cells. Although the feasibility of organic solar cells was demonstrated from the 1970s onwards, the importance of their development only appeared recently in the more general frame of the development of renewable energies. It is by studying organic photovoltaic cells that C.W. Tang, from the research laboratory Kodak, observed the electroluminescence of thin organic layers for the first time in 1979. In 1987, the same research team reported the realization by vacuum evaporation of the first organic electroluminescent diode, whose multilayer structure permitted a net increase in the electroluminescence efficiency and a decrease in functioning voltage [TAN 87].

The development of conductive polymers was produced at the same time as that of photoconductive molecules. In intrinsic conductive polymers, the semiconductivity is related to the alternation of σ and π bonds (conjugation) and the delocalization of electrons along the polymer chain. In 1977, H. Shirakawa and his collaborators discovered that the conductivity of polyacetylene increased by twelve orders of magnitude after being exposed to gases such as chlorine, iodine or bromine. This discovery, rewarded by a Nobel Prize in 2000 [SHI 01], caused numerous investigations into what we now call "synthetic metals". When the stability issues were overcome, doped conductive polymers found diverse applications such as for sensors, electrostatic coating, electromagnetic screening, anti-corrosion coating, etc. Non-doped polymers gave rise to a strong interest as soon as the first transistors and opto-electronic components were made. Thus, the electroluminescence of poly(p-phenylene vinylene) was revealed for the first time in 1990 by J.H. Burroughes and his collaborators from the University of Cambridge [BUR 90].

We just saw that electroluminescent diodes and organic photovoltaic cells have a common history. Their development, related to that of photoconductive "small molecules" and intrinsic conductive polymers, led to two technologies of distinct manufacture. Nevertheless, the theoretical description of electroluminescence and of photovoltaic conversion is similar for both types of materials. Subsequently, we shall first explain the origin of conduction in organic semiconductors by referring to their molecular structure. We then describe the parameters which permit their electro-optical characterization. Then, the use of these materials in electroluminescent diodes or photovoltaic cells, and the current performances of these components are presented. We shall end this chapter by listing the techniques which allow the processing of organic semiconductors in the form of thin layers.

20.2. Origin of conduction in organic semiconductors

The understanding of conduction phenomena in organic semiconductors plays a very important role when using these materials in components such as electroluminescent diodes or photovoltaic cells. In the early 1980s, the nomenclature intrinsic to inorganic semiconductors was chosen to describe their properties, but a terminology giving a better account of the particularities of these materials was later developed. Strictly speaking, the energy band model assumes a perfect periodic structure identical to that of the one encountered in large monocrystals. Organic materials only very rarely possess such regularity. The nature of the atoms making up the molecule, but also their quantity therefore has an important effect on the electronic structure of the material. In general, the behavior of electrons in an atom is described by wave functions which are solutions of the Schrödinger equation. These mathematical functions, which determine the probability of a given presence of an electron around a core, are called atomic orbitals. Only the orbitals furthest away from the atom core participate in chemical bonding. By linear combination of these atomic orbitals, we define the molecular orbitals which delimit the regions of space in which the probability of electron occupation is high (95%) around a molecular system.

As already mentioned, organic semiconductors are generally listed in two categories according to their molecular structure: polymers and organic materials with low molar mass (less than 1,000 g/mol). These materials are conjugated systems made up of the alternation of single and double carbon-carbon bonds. These double bonds associate a single bond σ, corresponding to the axial covering of hybrid orbitals sp^2, with a bond π coming from the lateral covering of two non-hybrid atomic orbitals p_z. The molecular orbitals can be bonding (σ, π) or antibinding (σ^*, π^*) depending on whether the linear combination of atomic orbitals corresponds to an addition or a subtraction of associated wave functions. The difference of energy between the Highest Occupied Molecular Orbital (HOMO), and the lowest unoccupied Molecular Orbital (LUMO) corresponds to the distance between bonding π and antibonding π^* orbitals; it makes up the forbidden energy band also called the "*band gap*". Although the terms HOMO and LUMO characterize insulated molecules, it is customary to use them to describe solids. We shall therefore consider that the position of the HOMO with respect to the vacuum level corresponds with a solid to the ionization potential I_P. Similarly, the position of the LUMO with respect to the vacuum level corresponds to the electron affinity A_E of the solid (Figure 20.1).

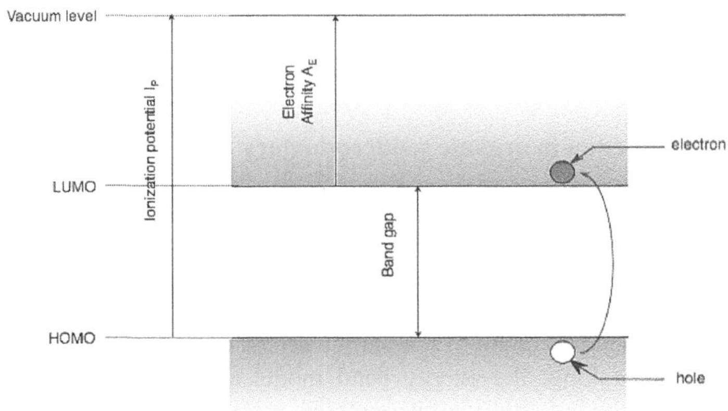

Figure 20.1. *Electronic levels in an organic semiconductor*

In organic semiconductors, the transport of charge carriers (electrons or holes) is predominantly ensured by hops between molecular sites or between different polymer chains. In strongly ordered molecular materials, such as pentacene crystals, carriers evolve between nearby molecules permitting mobilities of the order of 35 $cm^2/V.s$ to be reached at ambient temperature [JUR 04]. However, in the most disordered molecular systems and polymers, mobilities only reach 10^{-3} to 10^{-5} times this upper value [MOZ 05, CHE 00]. By comparison, they can reach $\mu_e = 1{,}500$ $cm^2/V.s$ for electrons and $\mu_t = 450 cm^2/V.s$ for holes in monocrystalline silicon. The charge transport can be improved by the conditions of preparation of thin organic layers: deposition methods, purity of materials. Weak mobility values consequently give rise to a weak conductivity of the order of 10^{-6} $S.cm^{-1}$ in organic semiconductors. However, these materials are only rarely doped, unlike their inorganic counterparts. The electron donor or acceptor character of the material is here an intrinsic property defined with respect to the electronegativity scale of the material considered. The more electronegative it is, the more its acceptor character is emphasized because it has more power to attract electrons. The material is generally called an acceptor if it possesses a large electron affinity value (of the order of 4 eV); it is considered a donor if the value of its ionization energy is weak (typically 5 eV).

20.3. Electrical and optical characteristics of organic semiconductors

The choice criteria of organic materials used in the manufacture of electroluminescent diodes and photovoltaic cells must take into account the physico-chemical parameters required for the good working order of the component. A knowledge of the electrical and optical properties of organic semiconductors therefore plays a dominating role in the understanding of the functioning and the improvement performances of these devices. Numerous measurement techniques, intrinsic or not to organic semiconductors, were implemented to characterize them. This section will only mention the most common analysis methods permitting access to the energy levels of the materials, the mobility of charges and the properties of excitons created in films.

The position of the HOMO and LUMO electronic levels of an organic material closely depend on its physical and chemical characteristics, but also on its environment. The technique chosen to determine them can perturb the measurement and distort the results. The most common experimental method to determine these values is cyclic voltammetry. This method relies on determination by electrochemistry of oxidation and reduction potentials of the material which we wish to study. It permits the characterization of thin films deposited on one of the electrodes used for measurement (a working electrode) or solution molecules in the electrolyte. Ultraviolet Photoelectron Spectroscopy (or UPS) permits a much more rigorous measurement of the ionization potential of organic semiconductors. It is made on thin films (a few hundred angströms for the thickest) placed in an ultra-high vacuum chamber and requires an intense luminous and monochromatic source, and an electron analysis system. The absorption by an insulated molecule of a UV photon of sufficient energy $h\nu$ causes the emission of an electron with kinetic energy E_K. UPS measures the quantity of electrons emitted with a given energy. The variation of the scanning voltage applied to the sample permits a spectrum to be obtained which traces the exact match between the kinetic energy distribution of these electrons and the energetic states of the molecule. This characterization technique, applied during the growing of a thin film, permits access to information concerning the following points:

– the shift of the valence bands, which corresponds to a variation of the HOMO level position during the passage from one material to another;

– the band bending caused by charge transfer corresponding to the difference in work function values of both materials put in contact [SCH 99];

– the possible formation of a dipole at the interface [HIL 00_2].

The shift of the LUMO level is only accessible by spectroscopy measurement of inverse photoelectron (IPES) made in the same conditions [HIL 00_1].

The *Time of Flight* (or TOF) is amongst the most used techniques to determine the mobility of charge carriers in a solid semiconductor. The principle of this method lies in the measurement of time τ_{tr} taken by the electrons or the holes to traverse the sample which we wish to study. Electric charges are emitted by absorption of light generated by an impulsion laser. They are led by the electric field applied to the sample. The measurement of the charge mobility can also be made by making a *Field Effect Transistor* (or FET) whose active thin layer is composed of the studied material [HOR 98]. We deduce the mobility value of the current-voltage curves. We must note that the value obtained by this calculation depends on characteristics of the structure, such as contact resistance. The PR-TRMC method (*Pulse-Radiolysis Time-Resolved Microwave Conductivity*) is based on the examination of a material sample under the irradiation by a high frequency electron beam (20 MHz). This fairly complex measurement process does not permit the mobility of holes to be distinguished from that of electrons, and constitutes a microscopic analysis of the charge's mobility [CRA 96].

All these electrical characterizations are most often completed by optical measurements (photoluminescence, UV-visible absorption, etc.) which, associated with mathematical models, allow information such as the binding energy of excitons and their diffusion length in a given material to be accessed. Finally, we should mention electro-absorption spectroscopy which, applied to a functioning multilayer component, supplies information on the internal electric field distribution.

20.4. Application to electroluminescent devices

20.4.1. *General structure of an organic electroluminescent diode (OLED)*

The basic structure of an organic electroluminescent diode is a monolayer structure in which a luminescent material is taken "in sandwich" between two electrodes, one of which must be transparent to permit the passage of photons. The use of appropriate electrode materials allows the injection of electrons and holes, which leads to the formation of excitons and their radiative recombination in the organic layer. When this structure was applied to organic crystals, the studies showed that the recombination zone was localized near one of the electrodes owing to the different injection and transport properties of electrons and holes [KAL 94]. The monolayer structures lead to weak electroluminescence efficiencies, on the one hand because the concentrations of electrons and holes are not balanced, and on the other hand because the defects near the electrodes are non-radiative recombination sources [CHO 97].

In multilayer structures, the injection and the transport of electrons and holes are attributed to distinct layers; moreover, the recombinations are localized at the level of an organic heterojunction owing to a confinement of charges. In the example of Figure 20.2, a bilayer diode consists of a material preferentially transporting holes (a Hole-Transporting Layer, or HTL) and an Electron-Transporting Layer (ETL) Emitter of Light (EML). When the diode is in direct polarization, the electrons and holes penetrate, respectively, the LUMO level of the ETL layer and the HOMO level of the HTL layer by passing or traversing energy barriers of different heights. The charge carriers are then partially blocked at the interface between both organic materials because of the difference in their HOMO and LUMO energy levels. The formation and the recombination of excitons are therefore produced at the level of this interface. We note that the representation below is very schematic. Indeed, the rigidity of energy bands supposes the absence of charges within organic layers and therefore do not correspond to the reality. Furthermore, if it is established that an effective injection of charges at the anode and the cathode, respectively, require a metal of high and low work function, the characteristics intrinsic to the materials are not enough to precisely describe the electronic structure at the interfaces. In particular, the physico-chemical interaction between the materials [NGU 01] and the space charge [TUT 03] present in the layers are likely to modify the injection and transport conditions.

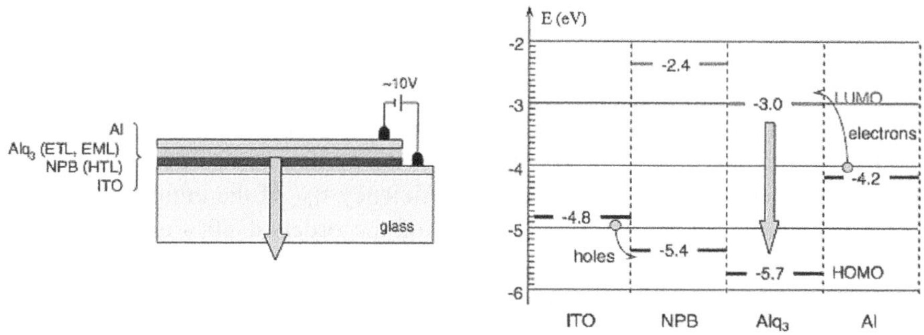

Figure 20.2. *Typical structure of an OLED and associated energy diagram*

20.4.2. *Electroluminescence efficiency*

A simplified description of the functioning of OLEDs allowss the different components of the electroluminescence efficiency to be revealed.

First, the exciton formation results from the injection and transport of electrons and holes from electrodes, and the excitation of luminescent molecules. The efficiency of this process is therefore related to the equilibrium of both currents and to the absence of defects, which would lead to non-radiative de-excitations. It can be characterized by:

$$\eta_{exc} = \frac{J_{exc}}{J_{tot}} \qquad [20.1]$$

where J_{exc} is the current density leading to the formation of excitons and J_{tot} is the total current density.

The formation of excitons can give rise to singlet or triplet excited states which are distinguished by their spin value (S = 0 or S = 1). The singlet states, which lead to fluorescence, statistically constitute 25% of the excited states. This value was experimentally confirmed in low molar mass materials, while the rates of singlet excitons greater than 50% were revealed in certain electroluminescent polymers. The phosphorescence, corresponding to a radiative de-excitation from the triplet states, is most often invisible at ambient temperature because it is associated with a forbidden transition. Although certain compounds are phosphorescent at ambient temperature and lead to a maximum theoretical efficiency of 100% [HOL 05], the fluorescence therefore leads, for "small molecules", to:

$$\eta_{\sin} = 0.25 \qquad [20.2]$$

The de-excitation from the excited singlet states can be produced in a radiative or non-radiative manner. The rate of radiative de-excitations is defined by the photoluminescence (generally fluorescence) efficiency η_{PL} of the emitting material. We note that if the fluorescence efficiencies of the order of 80% are commonly obtained in solution, the measured values for thin layers [GAR 96] are clearly reduced because of the interaction between molecules.

The last process which limits electroluminescence efficiency is the reabsorption of the light by the adjacent organic layers, and its confinement due to organic layers and the use of a glass substrate. This latter phenomenon is particularly important because we estimate that the fraction of light η_{opt} which comes out at the surface of the component is generally only of the order of 20% [GU 97].

From the above reasoning, we deduce the expression of the external quantum efficiency of electroluminescence, a ratio between the number of injected charges and the number of emitted photons:

$$\eta_{EL} = \eta_{exc} \cdot \eta_{\sin} \cdot \eta_{PL} \cdot \eta_{opt} \qquad [20.3]$$

The luminous efficiency (in lm/W) is frequently used to characterize the components. It expresses, in photometric units [SUR 00], their power efficiency:

$$\eta_P = \frac{h\nu}{qV} \cdot \eta_{EL} \qquad [20.4]$$

where $h\nu$ is the energy of the emitted photons and qV is the energy of the electrons submitted to the polarization voltage of the diode V.

20.4.3. *Advancement of the technology*

The craze created by organic electroluminescence since the late eighties allowed the understanding of its fundamental mechanisms, the development of numerous luminescent materials [MIT 00] and the development of suitable technologies (see section 20.6). Whereas the lifetime [SAT 98] of the components initially seemed to constitute a serious slowdown for their commercialization, passive matrix organic displays are today present in numerous commercial objects (cameras, telephones, etc.). Nevertheless, the interest aroused by OLEDs concerns, above all, visualization and, more recently, lighting.

Organic electroluminescence presents numerous advantages towards current visualization technologies. Indeed, it allows us to make very thin light devices; unlike liquid crystal screens, electroluminescent screens do not require any backlighting and do not present any restriction in viewing angle. As we shall see later, the processing techniques of electroluminescent materials are relatively simple. Finally, the realization of components on flexible substrates is conceivable. Numerous prototypes of organic screens have been presented, the largest with a forty inch-diagonal and a definition of 12,805,800 pixels (WXGA mode). However, the commercialization of organic screens has been slowed down by the difficulty of making large active matrices fitted to the electronic addressing of OLEDs at a reasonable cost, and for the stability problems of matrix luminescent materials.

The emission of white light by electroluminescent organic diodes (WOLEDs) can be obtained by the addition of two or three colors emitted from one [YAN 00] or several layer(s). The first advantage of these components is that their emission spectrum can be modulated on demand by modifying, for given materials, the thickness of different layers [JOL 98] or their concentration. Moreover, the possibility of making luminous surfaces extended on rigid or flexible substrates brings new perspectives in the lightning field. Finally, the interest in OLEDs appeared following a constant increase in their luminous efficiency. Owing to the recent use of phosphorescent materials, the values reached are today of the order of those measured in incandescent lamps (from 13 to 20 lm/W). Nevertheless, this technology must still be developed to be able to envisage the commercialization of electroluminescent lightning panels.

20.5. Application to photovoltaic conversion

20.5.1. *General structure of an organic photovoltaic cell*

The structure of an organic photovoltaic cell is perfectly similar to that of an OLED. We must, however, distinguish two types of structures: the bilayer structure composed of a heterojunction between a donor material and an acceptor material, and the volume heterojunction commonly called a "blend" structure. This latter category of cells consists of a mixture in volume donor and acceptor materials. The donor is most often a polymer associated with a fullerene C_{60} derivative, which plays the role of acceptor. This molecule is generally functionalized in order to present a better solubility in the matrix polymer. A rigorous control of the mixture's morphology permits an interpenetrated lattice in both materials to be obtained, which increases the dissociation rates of the created excitons. Promising efficiencies of conversion (3.1%) have thus been reached [AER 02], but the main problem still remains the transport of charges towards the electrodes. The realization of intermediate structures, where a donor–acceptor mixture is inserted between donor and acceptor layers, will help circumvent this difficulty [PEU 03].

20.5.2. *Functioning of an organic photovoltaic cell*

The functioning of such components can be described as being the inverse of that of an OLED. The different stages of organic photovoltaic conversion are detailed below.

The first stage is the absorption of photons in the organic film. It is expressed by the energy supply to the HOMO level electrons. The transition towards the LUMO level leads to the outbreak of a hole which remains connected to the electron by mutual electrostatic interaction. The electrically neutral electron-hole couple is then called an exciton. In inorganic materials, the binding energy of these excitons is very weak (14.7 meV for silicon) and the thermal energy is sufficient, at ambient temperature, to separate both connected charges. It is different in organic materials since the binding energy most often reaches values of the order of 0.2 eV. The result is a difficulty to dissociate the charge carriers which remain connected at ambient temperature.

The second stage is the diffusion of excitons over a distance, L_D, depending on the material. The separation of charges is possible if the excitons rejoin the donor-acceptor junction where the electric field is sufficiently strong.

The dissociation of the excitons which reach the interface between both organic materials represents the third stage of the process. The electrons are attracted to the material having a high electron affinity and the holes by the material having a weak ionization potential. If the dissociation is made far from the interface in a neutral region, the electron and the hole end up recombining.

Finally, the charge carriers thus created rejoin the electrodes and the circuit outside the cell. The transport of charges towards the electrodes is controlled by the mobilities of carriers in the organic layers.

The characteristic parameters of a photovoltaic cell are deduced from the recording of the current-voltage curve under illumination. The most significative parameter is the fill factor, FF, which accounts for the quality of the form of curves. It is determined from the following expression:

$$FF = \frac{P_m}{I_{cc}V_{co}} = \frac{I_m V_m}{I_{cc}V_{co}} \qquad [20.5]$$

where I_{cc} is the short circuit current, V_{co} the open circuit voltage, and I_m and V_m represent the current and the voltage of the functioning point which allows the maximum power (P_m) of the cell to be extracted.

The EQE (*External Quantum Efficiency*) is defined by the ratio of the number of electrons flowing in the external circuit connected to the cell to the number of incident photons. The EQE is deduced from the spectral response of the solar cell SR(λ):

$$EQE = \frac{J_{cc}(\lambda)}{\phi(\lambda)} \cdot \frac{hc}{\lambda q} = SR(\lambda) \cdot \frac{hc}{\lambda q} \qquad [20.6]$$

where $\phi(\lambda)$ (W/m^2) represents the luminous intensity by illuminated unit surface, $J_{cc}(\lambda)$ (W/m^2) the density of short circuit current for a given wavelength λ.

The efficiency of power conversion, η, is defined by the ratio of the maximal power delivered by the cell to the incident luminous power P_{in}:

$$\eta = \frac{P_m}{P_{in}} = \frac{I_{cc}V_{co}FF}{P_{in}} \qquad [20.7]$$

The maximal conversion efficiency of a cell is only significant for a spectral distribution and a given intensity. The most commonly used standard irradiation corresponds to a mass air number AM 1.5 [PRO 97].

20.5.3. *Advancement of the technology*

The synthesis of new molecules better fitted to photovoltaic cells and the outbreak of new architectures of devices led to the acceleration of the development of this type of structure. During the last 20 years, the main efforts concerning the manufacture of cells have been concentrated on the addition of new layers at the interface level between the electrodes and organic films. Thus, the insertion between the ITO and the donor of a poly(3,4-ethylene dioxithiophene) (PEDOT) doped with poly(styrene sulphonate) (PSS) layer allows, in certain cases, an improvement of the conversion efficiency of cells to be achieved. As for the cathode, the deposition of lithium fluoride (LiF) and/or bathocuproine (BCP) allows the acceptor material to be protected during the deposition of the metal which often causes malfunctions of the component. The role of the BCP also consists of confining excitons to avoid them coming and recombining at the cathode/acceptor interface, and thus improving the conversion efficiency of the cell [PEU 00]. The realization of piled up cells (or tandem cells) allowing the domain of absorption of the component to be widened, led to conversion efficiencies of 5% [XUE 04]. One of the parameters restricting the performances of organic photovoltaic cells is the weak mobility of charges in the materials. One of the conceived solutions for an increase in this mobility is to make anisotropic films from ordered molecules. Mobilities of the order of 10^{-1} $cm^2/V.s$ could thus be measured in columnar phases of discotic liquid crystals [ADA 94]. A few cells were produced from these materials; however, the organization of molecules remains an important challenge.

The improvement of performance of the cells composed of mixed donor-acceptor materials is strongly related to the solubility and morphology of the films. These parameters strongly depend on the chosen solvent, which can cause strong variations on the characteristics of cells. Thus, for ITO/PEDOT/MDMO-PPV:PCBM/LiF/Al type cells, the conversion efficiency increases from 0.9% to 2.5% by replacing chlorobenzene with toluene [SHA 01]. One of the routes explored to avoid the formation of aggregates during deposition is to chemically connect the donor and acceptor molecules. The weak conductivity of these components limits the efficiency of this type of cell [NIE 04]. One of the main weaknesses of these cells, however, remains the stability issue already encountered in the OLEDs [KRO 02]. It is related to the nature of molecules, sensitive to oxygen and water vapor, but also to chemical reactions which can be produced at the interface between the electrodes and organic films. It is therefore of prime importance to protect the components by a layer impermeable to oxygen and water vapor during their functioning.

Over the last five years, progress in the synthesis and finalization of organic photovoltaic materials have allowed an increase in conversion efficiency from 1 to 5%, whether for bilayer heterojunctions or in volume. However, important research efforts still need to be pursued in the improvement and understanding of the functioning of these devices in order to conceive their commercialization.

20.6. The processing of organic semiconductors

We present below the main techniques which have been proposed to process organic semiconductors. Generally, polymers are deposited in the form of a solution whereas "small molecules" are vacuum evaporated. Certain techniques only allow surfaces to be uniformly covered, whilst others, more promising, allow the formation of patterns such as pixels.

20.6.1. *Deposition of polymer solutions*

20.6.1.1. *Deposition by spreading*

The deposition of polymers by centrifugation, or "spin coating", is commonly used in microelectronics to deposit photosensitive resins. A film is obtained by spreading a polymer solution and simultaneously evaporating the solvent. Its thickness is then mainly determined by the rotation speed of the spin-coater and by the concentration, and therefore the viscosity, of the solution [FLA 84]. Although simple to implement, this technique does not permit large surfaces to be covered, does not guarantee a uniformity of the thickness and leads to large material losses. Alternative techniques such as pulverizing or spreading with the aid of a metallic strip [JOL 04] have been developed for diverse applications, but they are not very suitable for the realization of layers with a thickness smaller than a micrometer.

20.6.1.2. *"Inkjet" deposition*

"Inkjet" deposition technology [CAL 01, SHI 03] turns out to be very promising for the realization of organic components: not only does it allow the formation of polymer patterns, by it could also be applied to low molar mass dendrimers. The principle of this method consists of ejecting drops of a solution of a polymer from a micrometer sized nozzle. The red, green and blue (RGB) pixels of a screen are thus obtained by leaving drops between the polymer walls preformed on the substrate, and by realizing three successive passages. Although stemming from a very widespread printing technique, we must note that its application to organic semiconductors is not immediate. Indeed, the additives used in inks to optimize their viscosity, their surface tension and their drying speed, cannot be applied to organic semiconductors which must be pure. Furthermore, their choice lies more on their electro-optical properties and stability than on their mechanical properties.

20.6.2. *Vapor phase deposition of low molar mass materials*

20.6.2.1. *Deposition by thermal evaporation under secondary vacuum*

Vacuum thermal evaporation (or sublimation) [HOL 70] is suitable for the deposition of "small molecules" providing that they are not subjected to a thermal degradation during the operation. The materials to be evaporated are arranged in a crucible and heated by Joule effect under a secondary vacuum (~ 10^{-4} Pa); a deposition is then produced on the samples placed above, in conditions (speed, thickness) controlled *in situ* by means of a vibrating quartz microbalance. The main advantage of this technique is that it guarantees a high purity of layers, more difficult to obtain in the case of polymers. The introduction of dopants is nevertheless possible by co-evaporation of both materials. Also applied to metals, vacuum evaporation permits the manufacture of components in a unique device. The main disadvantages of this technique are the difficulty to make deposits of uniform thickness on an extended surface and the important material losses it generates.

20.6.2.2. *Low-pressure vapor phase deposition*

Low-pressure vapor phase deposition [BAL 98, SHT 01] solves the problems inherent to vacuum evaporation. This method consists of evaporating the materials, and transporting them towards a distinct deposition chamber by means of a heated inert gas. The deposition speed is then controlled by the gas flux, the pressure of the enclosure (of the order of 10^{-1} to 10^{2} Pa) and the temperature of the source and the substrate. Unlike vacuum evaporation, the organic flux can be uniformly spread on any surface by a "shower-head" at a few centimeters' distance; this also allows material losses to be minimized.

An analogous "inkjet" printing method is developed in the case of vapor phase deposition [SHT 04]. It consists of sending the organic vapor jet on the cooled substrate by a nozzle situated at very short distance (~10-100 µm). It then becomes possible to form patterns from "small molecules".

20.6.3. *Laser thermal transfer of organic materials*

Laser thermal transfer of organic materials [ELI 03, SUH 03] on a substrate was initially developed for the realization of RGB filters, used in liquid crystal screens. This method consists of transferring a polymer or a material of low molar mass from a "donor" substrate towards an "acceptor" substrate by laser ablation.

The donor substrate is composed of a photosensitive layer which converts luminous energy into thermal energy; it is covered with the organic material(s) to be transferred. This is done by assembling both substrates and by selectively irradiating certain zones with the aid of an infrared laser beam.

The advantages of this technique are its high precision (of the order of 2.5 μm), the possibility of applying it to extended surfaces and the absence of solvents. It could be applied not only to the realization of organic pixel matrices but also to the preparation of substrates (for the deposition of polymer barriers) for matrices manufactured by "inkjet".

20.7. Conclusion

This chapter has successively tackled the history of organic semiconductors, their electronic properties and the main characterization methods used to describe them. We have then presented two essential applications, electroluminescent diodes and organic photovoltaic cells, before listing the deposition methods proposed for their manufacture. We should stress that the simplicity of processing organic semiconductors is a strong point in favor of these materials. Organic electronics give rise to a lot of interest because of the new applications they shed light on (flexible luminous or photovoltaic surfaces, "roll-to-roll" production by printing techniques, etc.).

We have also seen that organic semiconductors are distinguished from their inorganic counterparts, particularly from the viewpoint of charge mobility values. The revelation of well-known problems in the field of dielectric materials (injection and transport phenomena, space charge effects, etc.) justifies the integration of this chapter in a book on dielectric materials.

20.8. Bibliography

[ADA 94] ADAM D., SHUHMACHER P., SIMMERER J., HÄUSSLING L., SIEMENSMEYER K. ETZBACH K.H., RINGSDORF H., HAARER D., "Fast photoconduction in the highly ordered columnar phase of a discotic liquid crystal", *Nature*, vol. 371, no. 6493, p. 141–143, 1994.

[AER 02] AERNOUTS T., GEENS W., POORTMANS J., HEREMANS P., BORGHS S., MERTENS R., "Extraction of bulk and contact components of the series resistance in organic bulk donor–acceptor heterojunctions", *Thin Solid Films*, vol. 403–404, p. 297–301, 2002.

[BAL 98] BALDO M., DEUTSCH M., BURROWS P., GOSSENBERGER H., GERSTENBERG M., BAN V., FORREST S., "Organic vapor phase deposition", *Adv. Mater.*, vol. 10, no. 18 p. 1505–1514, 1998.

[BUR 90] BURROUGHES J.H., BRADLEY D.D.C., BROWN A.R., MARKS R.N., MACKAY K., FRIEND R.H., BURNS P.L., HOLMES A.B., "Light-emitting diodes based on conjugated polymers", *Nature*, vol. 347, no. 6293, p. 539–541, 1990.

[CAL 01] CALVERT P., "Inkjet printing for materials and devices", *Chem. Mater.*, vol. 13, no. 10, p. 3299–3305, 2001.

[CHE 00] CHEN B., LEE C., LEE S., WEBB P., CHAN Y., GAMBLING W., TIAN H., ZHU W., "Improved time-of-flight technique for measuring carrier mobility in thin films of organic electroluminescent materials", *Jpn. J. Appl. Phys.*, vol. 39, p. 1190–1192, 2000.

[CHO 97] CHOONG V.E., PARK Y., SHIVAPARAN N., TANG C.W., GAO Y., "Deposition-induced photoluminescence quenching of tris-(8-hydroxyquinoline) aluminum", *Appl. Phys. Lett.*, vol. 71, no. 8, p. 1005–1007, 1997.

[CRA 96] VAN DE CRAATS A.M., WARMAN J.M., DE HAAS M.P., ADAM D., SIMMERER J., HAARER D., SCHUHMACHER P., "The mobility of charge carrier in all four phases of the columnar discotic material hexakis(hexylthio)triphenylene: combined TOF and PR-TRMC results", *Adv. Mater.*, vol. 8, no. 10, p. 823–826, 1996.

[ELI 03] ELIZURAND E., GELBART D., "Printing methods for flat-panel display manufacturing", *J. Graphic Technol.*, vol. 1, p. 1–5, 2003.

[FLA 84] FLACK W.W., SOONG D.S., BELL A.T., HESS D.W., "A mathematical model for spin coating of polymer resists", *J. Appl. Phys.*, vol. 56, no. 4, p. 1199–1206, 1984.

[GAR 96] GARBUZOV D.Z., BULOVIC V., BURROWS P.E., FORREST S.R., "Photoluminescence efficiency and absorption of aluminum-tris-quinolate (Alq_3) thin films", *Chem. Phys. Lett.*, vol. 249, nos 5–6, p. 433–437, 1996.

[GU 97] GU G., GARBUZOV D.Z., BURROWS P.E., VENKATESH S., FORREST S.R., "High-external-quantum-efficiency organic light-emitting devices", *Opt. Lett.*, vol. 22, no. 6, p. 396–398, 1997.

[HIL 00_1] HILL I.G., KHAN A., SOOS Z.G., PASCAL R.A. Jr., "Charge-separation energy in films of π-conjugated molecules", *Chem. Phys. Lett.*, vol. 327, nos 3–4, p. 181–188, 2000.

[HIL 00_2] HILL I.G., MILLIRON D., SCHWARTZ J., KHAN A., "Organic semiconductor interfaces: electronic structure and transport properties", *Appl. Surf. Sci.*, vol. 166, nos 1–4, p. 354–362, 2000.

[HOL 70] HOLLAND L., *Vacuum deposition of thin films*, Chapman & Hall, London, 1970.

[HOL 05] HOLDER E., LANGEVELD B.M.W., SCHUBERT U.S., "New trends in the use of transition metal-ligand complexes for applications in electroluminescent devices", *Adv. Mater.*, vol. 17, no. 9, p. 1109–1121, 2005.

[HOR 98] HOROWITZ G., "Organic field-effect transistors", *Adv. Mater.*, vol. 10, no. 5, p. 365–377, 1998.

[JOL 98] JOLINAT P., CLERGEREAUX R., FARENC J., DESTRUEL P., "The effect of the thicknesses of the various layers on the color emitted by an organic electroluminescent device", *J. Phys. D: Appl. Phys.*, vol. 31, no. 10, p. 1257–1262, 1998.

[JOL 04] JOLINAT P., "Dépôts de polymères", *Traité EGEM-Techniques de fabrication des microsystèmes*, Hermès Science Publications, vol. 1, p. 121–142, 2004.

[JUR 04] JURCHESCU O.D., BAAS J., PALSTRA T.T.M., "Effect of impurities on the mobility of single cristal pentacene", *Appl. Phys. Lett.*, vol. 84, no. 16, p. 3061–3063, 2004.

[KAL 94] KALINOWSKI J., "Space-resolved recombination electroluminescence in organic crystals", *Synth. Met.*, vol. 64, nos 2–3, p. 123–132, 1994.

[KRO 02] KROON J.M., WIENK M.M., VERHEES W.J.H., HUMMELEN J.C., "Accurate efficiency determination and stability studies of conjugated polymer/fullerene solar cells", *Thin Solid Films*, vol. 403–404, p. 223–228, 2002.

[LAW 93] LAW K.Y., "Organic photoconductive materials: recent trends and developments", *Chem. Rev.*, vol. 93, no. 1, p. 449–486, 1993.

[LOU 88] LOUTFY R.O., HOR A.M., HSIAO C.K., BARANYI G., KAZMAIER P., "Organic photoconductive materials", *Pure Appl. Chem.*, vol. 60, no. 7, p. 1047–1054, 1988.

[MIT 00] MITSCHKE U., BÄUERLE P., "The electroluminescence of organic materials", *J. Mater. Chem.*, vol. 10, no. 7, p. 1471–1507, 2000.

[MOZ 05] MOZER A.J., SARICIFTCI N.S, PIVRIKAS A. ÖSTERBACKA R., JUSKA G., BRASSAT L., BÄSSLER H., "Charge carrier mobility in regioregular poly(3-hexylthiophene) probed by transient conductivity techniques: A comparative study", *Phys. Rev. B*, vol. 71, no. 3, p. 035214_1-035214_9, 2005.

[NGU 01] NGUYEN T.P., IP J., JOLINAT P., DESTRUEL P., "XPS and sputtering study of the Alq_3/electrode interfaces in organic light emitting diodes", *Appl. Surf. Sci.*, vol. 172, nos 1–2, p. 75–83, 2001.

[NIE 04] NIERENGARTEN J.F., "Fullerene-(p-conjugated oligomer) dyads as active photovoltaic materials", *Sol. Energy Mater. Sol. Cells*, vol. 83, nos 2–3, p. 187–199, 2004.

[PEU 00] PEUMANS P., BULOVIC V., FORREST S.R., "Efficient photon harvesting at high optical intensities in ultrathin organic double-heterostructure photovoltaïc diodes", *Appl. Phys. Lett.*, vol. 76, p. 2650–2652, 2000.

[PEU 03] PEUMANS P., UCHIDA S., FORREST S.R., "Efficient bulk heterojunction photovoltaic cells using small molecular weight organic thin films", *Nature*, vol. 425, no. 6945, p. 158–162, 2003.

[POP 63] POPE M., KALLMANN H.P., MAGNANTE P., "Electroluminescence in organic crystals", *J. Chem. Phys.*, vol. 38, no. 9, p. 2042–2043, avril 1963.

[PRO 97] PROTIN L., ASTIER S., "Convertisseurs photovoltaïques", *Techniques de l'ingénieur, D3360*, 08–1997.

[SAT 98] SATO Y., ICHINOSAWA S., KANAI H., "Operation characteristics and degradation of organic electroluminescent devices", IEEE *J. Select. Top. Quant. Electron.*, vol. 4, no. 1, p. 40–48, 1998.

[SCH 99] SCHLAF R., SCHROEDER P.G., NELSON M.W., PARKINSON B.A., LEE P.A., NEBESNY K.W., ARMSTRONG N.R., "Observation of strong band bending in perylene tetracarboxylic dianhydride thin films grown on SnS_2", *J. Appl. Phys.*, vol. 86, no. 3, p. 1499–1509, 1999.

[SHA 01] SHAHEEN S.E., BRABEC C.J., SARICIFTCI N.S., PADINGER F., FROMHERZ T., HUMMELEN J.C. "2.5% efficient organic plastic solar cells", *Appl. Phys. Lett.*, vol. 78, no. 6, p. 841–843, 2001.

[SHI 01] SHIRAKAWA H., "The discovery of polyacetylene film: the dawning of an era of conducting polymers (Nobel lecture)", *Angew. Chem. Int. Ed.*, vol. 40, no. 14, p. 2574–2580, 2001.

[SHI 03] SHIMODA T., MORII K., SEKI S., KIGUCHI H., "Inkjet printing of light-emitting polymer displays", *Mater. Res. Soc. Bull.*, vol. 28, no. 11, p. 821–827, 2003.

[SHT 01] SHTEIN M., GOSSENBERGER H.F., BENZIGER J.B., FORREST S.R., "Material transport regimes and mechanisms for growth of molecular organic thin films using low-pressure organic vapor phase deposition", *J. Appl. Phys.*, vol. 89, p. 1470–1476, 2001.

[SHT 04] SHTEIN M., PEUMANS P., BENZIGER J.B., FORREST S.R. "Direct, mask- and solvent-free printing of molecular organic semiconductors", *Adv. Mater.*, vol. 16, no. 2, no. 18, p. 1615–1620, 2004.

[SUH 03] SUH M.C., CHIN B.D., KIM M.H., KANG T.M., LEE S.T., "Enhanced luminance of blue light-emitting polymers by blending with hole-transporting materials", *Adv. Mater.*, vol. 15, no. 15, p. 1254–1258, 2003.

[SUR 00] SURREL J., "Radiométrie – Photométrie", *Optique et photonique*, vol. 1, p. 51–54, 2000.

[TAN 87] TANG C.W., VANSLYKE S.A., "Organic electroluminescent diodes", *Appl. Phys. Lett.*, vol. 51, no. 12, p. 913–915, 1987.

[TUT 03] TUTIS E., BERNER D., ZUPPIROLI L., "Internal electric field and charge distribution in multilayer organic light-emitting diodes", *J. Appl. Phys.*, vol. 93, no. 8, p. 4594–4602, 2003.

[YAN 00] YANG J.P., JIN Y.D., HEREMANS P.L., HOEFNAGELS R., DIELTIENS P., BLOCKHUYS F., GEISE H.J., VAN DER AUWERAER M., BORGHS G., "White light emission from a single layer organic light emitting diode fabricated by spincoating", *Chem. Phys. Lett.*, vol. 325, nos 1–3, p. 251–256, 2000.

[XUE 04] XUE J., UCHIDA S., RAND B.P., FORREST S.R., "Assymetric tandem organic photovoltaic cells with hybrid planar-mixed molecular heterojunctions", *Appl. Phys. Lett.*, vol. 85, no. 23, p. 5757–5759, 2004.

Chapter 21

Dielectric Coatings for the Thermal Control of Geostationary Satellites: Trends and Problems

21.1. Introduction

Space is a complex and dynamic medium. It is filled with gravitational fields, made up of matter whose density becomes extremely weak, traversed by cosmic radii and by a very wide spectrum of corpuscular radiations, but also occupied by intense magnetic fields (of a greater/lesser extent). Furthermore, as they are connected by a very strong synergy, its components cannot be considered separately. Compatibility with such an environment is an essential element in the conception of any space vehicle. The constant quest for high reliability, a long lifetime and a high degree of autonomy, associated with the choice of an orbit and selection of materials, etc. means that all natural and/or artificial stresses need to be taken into account in order to best design space systems, sub-systems and equipment. On a satellite, the good functioning of any piece of equipment depends on its temperature. If temperature regulation is often made "naturally" in our terrestrial environment or in our laboratories, this is not the case in environment of space, where the obtaining of a correct temperature can only be achieved as a result of a specific science and technology called "thermal control of space vehicles". Thermal control uses passive solutions as much as possible, i.e. thermal control radiators and external insulation. The materials brought into play are essentially dielectric polymers, directly subjected to the space environment.

Chapter written by Stéphanie REMAURY.

One of the characteristics of the geostationary environment (the orbit of telecommunication satellites at 36,000 km from the Earth) is the permanent presence of electrically charged particles, presenting strong flux and associated with energies lower than 50 keV. These particles have a strong ability to charge external coatings. If these coatings accumulate the charges implanted on them, potential differences operate between diverse points in the satellite. When a configuration of electric field or of a critical potential is reached (respectively 10^7 V/m or 500 V), electrostatic phenomena such as discharges are produced and can damage the embedded electronics and the external coatings and structures of the vehicle/satellite. Electrostatic phenomena can also be produced on low orbits but the most "risky" ones remain those of the geostationary orbit.

This chapter describes the space environment and the electrostatic phenomena it induces on dielectric materials brought into play for the thermal control of satellites placed in geostationary orbit.

21.2. Space environment

21.2.1. *Orbits*

Artificial satellites placed around the Earth can circulate in different types of orbit [CNE 02]: the most common are called LEO (Low Earth Orbit) and GEO (Geostationary Orbit), at two different altitudes (Figure 21.1). In LEO orbit, also called "low orbit", the satellites are situated between 700 and 800 km of altitude, orbiting the Earth in only a few hours. A trajectory at 90° with respect to the equator is known as a "polar" orbit, if the angle is different it is known as "inclined", and at 0° it is known as "equatorial". The GEO, or geostationary orbit, is situated at about 36,000 km of the Earth on the equatorial plane.

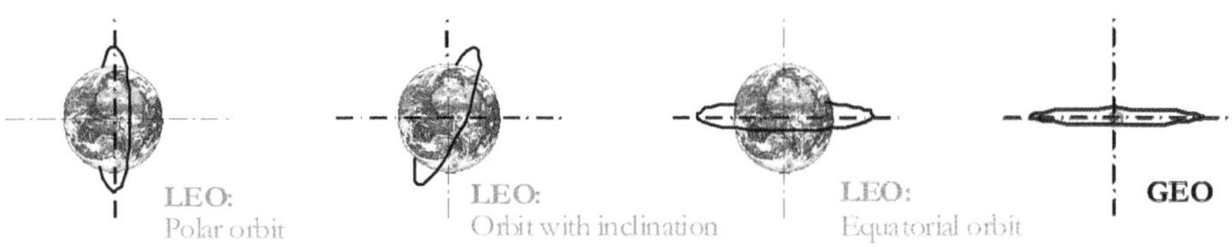

Figure 21.1. *Different orbits*

21.2.2. *Free space*

Space is empty. For an earthling, it is a difficult notion to conceive because down here, matter (in its three states, solid, liquid or gas) is everywhere: the air we breathe, water in the oceans, the rocks underground, the objects which surround us, etc. In reality, space is not absolute since we still find atoms (which make up a residual atmosphere, not very dense, up to an altitude of 1,000 km), solid debris, diverse radiations such as those coming from stars, etc. The quality of the vacuum and the nature of residual gases strongly depend on the altitude with respect to the Earth, but also on many other parameters including solar activity.

21.2.3. *Microgravity*

By definition, microgravity designates a state where the apparent gravity is very weak. We keep this term for a space environment in which the relative residual acceleration is of the order of 10^{-4} to 10^{-8} g. We meet these conditions almost perfectly in orbit above 250 km or in an Earth–Moon journey. Microgravity is the consequence of the orbital speed of the satellite. The most spectacular phenomenon ensuing from microgravity is the levitation of objects and of men during manned flights.

21.2.4. *Thermal environment*

The sun is a considerable source of energy for satellites. The energetic flux received by a satellite placed in geostationary orbit is on average 1,371 W/m² [CNE 02].

The albedo solar flux, corresponding to the reflection and diffusion of the solar flux on the terrestrial atmosphere, is also an important source of heat. Its spectral distribution differs from that of the Sun because of the absorption bands in the atmosphere, whereas its intensity is a function of the angle and of the distance of the satellite from the Earth.

Similarly, the Earth radiates an infrared thermal flux.

The direct solar flux, the albedo and the terrestrial thermal flux are, therefore, in low terrestrial orbit, the 3 sources of heat outside the satellite. In geostationary orbit, the albedo and terrestrial fluxes are negligible because of the distance from the satellite to the Earth.

Conversely, the background of the sky is the only source of cold outside the satellite.

21.2.5. *Atomic oxygen*

One of the major components of the high atmosphere is oxygen in atomic form coming from the dissociation of molecules by solar UV and X-ray radiations, mainly during phases of strong activity (with a cycle of eleven years). The density of atomic oxygen decreases exponentially with altitude but remains significant up to almost 1,000 km. The issues related to atomic oxygen are therefore only critical for these low orbits.

21.2.6. *Electromagnetic radiation*

Space vehicles are permanently subjected to radiation covering the whole band of electromagnetic wavelengths.

Infrared radiation is divided into three categories, near and medium infrared (from about 0.75 μm to 3 μm) as well as far or thermal infrared (from about 3 μm to 100 μm). Thermal infrared constitutes a large part (10%) of solar radiation. Generally, the thermal radiation of the Earth and the planets is entirely situated in the infrared spectrum.

The visible flux (from 0.3 to 0.75 μm) has a strong influence on the thermal equilibrium of satellites.

The majority of ultraviolet or UV flux (from 0.12 μm to 0.3 μm) originates from the Sun and, even though they only represent a small part (1% to 2%) of solar energy, they have particularly harmful effects such as the degradation of the surface of materials. For example, they cause the breakdown of molecular chains of polymers by weakening the material and modifying its mechanical and optical characteristics.

X-ray and gamma radiations are of solar (eruptions) or galactic origin. They are dangerous for mankind and precautions are taken in the programming of extra vehicular activities. On the other hand, the total amount of these X-ray and gamma radiations is not sufficient to notably affect the thermal equilibrium of thermal control materials.

21.2.7. *Charged particles [HID 05]*

Electrons and protons originate from the sun and make up solar winds. Their density and their energy vary a lot as a function of solar activity. The simplified representation in Figure 21.2 presents the magnetosphere. Far from being static, solar winds assign to it calm periods and more turbulent times. The terrestrial atmosphere is permanently submitted to UV and X-ray radiations, to solar particles and other cosmic radiations, which, in interaction with atoms or molecules of the upper atmosphere, produce ions and electrons. For example, at very low altitude, the most energetic radiations such as cosmic radii are capable of making up a few thousand ions per cm^3. It is only at about 50 km that the ionized part of the atmosphere really begins, called the *ionosphere*, and which extends up to around 1 R_t (1 Terrestrial Radius [R_t] ~ 6,378 km). With altitude, the ionization rate (ratio of the density of charged particles to the total density of particles) regularly increases to practically reach unity. In the ionospheric plasma, particles have an average energy of the order of 0.1 eV, and the probability to induce strong potentials on satellites is then practically zero. LEO satellites exist in this medium.

The ionosphere is dominated by the *plasmasphere* which is situated on the closed lines of the terrestrial magnetic field. It is a region of dense and cool plasma of mainly ionospheric origin, which can be split into two parts: the inner belt and the outer belt. These radiation belts are known as the *Van Allen belts*. The *inner belt* begins between 300 and 1,000 km of altitude as a function of latitude and extends up to 10,000 km. It mainly consists of electrons, whose energy can reach 1 MeV, and high energy protons (of several hundred MeV). The *outer belt*, also called the *plasmapause*, starts at about 10,000 km and extends up to about 5R_t. It is mostly composed of electrons and protons and marks the end of the plasmasphere.

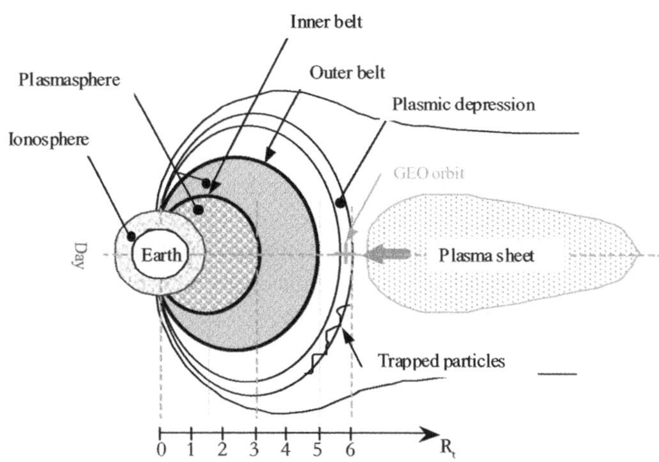

Figure 21.2. *Simplified schematic description of the magnetosphere in the noon–midnight meridian plane [HID 05]*

The *plasmic depression* zone which surrounds the plasmapause is also situated in the closed lines of the terrestrial magnetic field. It is very thin and does not move with the Earth.

Above this zone is the *plasmic layer* which is a strongly agitated zone and the main cause of issues of satellite charges in geostationary orbit. This part of the magnetosphere is a reserve of energetic plasma of over 12 keV, under flux greater than 0.5 nA/cm^2. It is composed of electrons and ions drifting in opposite directions because of the diverting action of the magnetic field. Unlike the others zones, it is only situated on the night side. In calm periods, the GEO orbit is situated outside the plasmic layer, but when the magnetosphere is agitated, during a stormy period, the plasmic layer is dilated and moves forward along the trajectory of the satellites until it reaches them.

As we have just seen, if the geostationary environment is very structured, the average energies characteristic of the particles which compose it are, however, subjected to variations. In a calm or stormy period, a zone can have its own type of particles, their density and their average energy radically changed. A satellite crossing these zones can have its functioning seriously altered if the characteristics of the medium exceed those for which it was designed.

Over the entire GEO orbit, satellites are subjected to particles trapped outside the Van Allen belt, made up of electrons whose energy varies from about 50 keV to a few MeV under an average flux of 50 pA/cm^2 and protons whose energy is of several tens of keV.

In the midnight–6am sector (noon being at the center of the day side), in a calm period (when the plasmasphere is dilated) the GEO orbit is found in an ionospheric type atmosphere with energy (~ 0.1 eV) but less dense (~ 1 cm^{-3}). A stormy period corresponds to a contraction of the plasmasphere (or forward movement of the plasmic layer) in the circulation zone of geostationary orbit satellites. In this type of environment, satellites are mainly exposed to electrons whose average energy is less than 50 keV under an average flux of 0.8 pA/cm^2.

21.2.8. *Meteoroids and cosmic debris*

Around the Earth, we find much natural debris called meteoroids, formed at the same time as the solar system. Since the launch of the first satellite in 1957, several million pieces of artificial debris have been added to it, resulting from man's activities in space. Their distribution is virtually uniformly in low orbits; the most dangerous zones are situated between 500 km and 800 km.

21.3. The thermal control of space vehicles

The functioning of any piece of equipment embedded on a satellite depends on its temperature and its variation. In space, obtaining a correct temperature is the result of the "thermal control of satellites and space vehicles".

21.3.1. *The definition of thermal control*

Thermal control must guarantee a space system (a vehicle, instrument, equipment, etc.) a thermal environment which allows it to function normally during all phases of the mission.

For this purpose, it must:

– maintain structures, equipment and components in the ranges of temperatures specified for their proper functioning;

– limit, if required, the thermal gradients in space (dT/dx) or in time (dT/dt) for a structure, a piece of equipment or a component;

– guarantee conditions to the specified thermal limits at interfaces (heat flux, interface temperatures, couplings or thermal decouplings).

These needs vary as a function of the phases of the mission and the states of activity of the equipment.

21.3.2. *Usual technologies for thermal control*

The thermal control system resorts to numerous technologies to heat or cool a space system. Readers can find further details on these technologies in the space technology lessons of the CNES [CNE 02]. Amongst them, we find external insulation and radiators which implement films and polymer coatings of a dielectric nature.

21.3.3. *Coatings for thermal control*

The global temperature of a satellite results from the good equilibrium between the dissipation of the equipment and the absorbed external flux (sun, albedo and planetary infrared) or radiated (infrared) towards the space background. The external surfaces of the satellite will, then, need to be conceived and proportioned to ensure this thermal equilibrium at its best. Surfaces which are not used (or non-usable) as radiators will be radiatively neutralized by superinsulating mattresses. The surface and coatings of radiators will be defined and proportioned to absorb and radiate the required amount of flux.

21.3.4. *Multilayer insulations (MLI)*

Multilayer insulations (MLI) are made by the superposition of radiative screens composed of films (most often in Mylar or Kapton) aluminized on both sides. Polyester tulles (a "bridal veil") are inserted between these films to reduce the conductive couplings (Figure 21.4). The external foil of these MLI will be chosen to resist the aggressive space environment (UV, particles, atomic oxygen, etc.). We usually use an external Kapton foil aluminized on one side (the inner side). It is this Kapton which gives the gilt yellow color of the external MLI of the satellites.

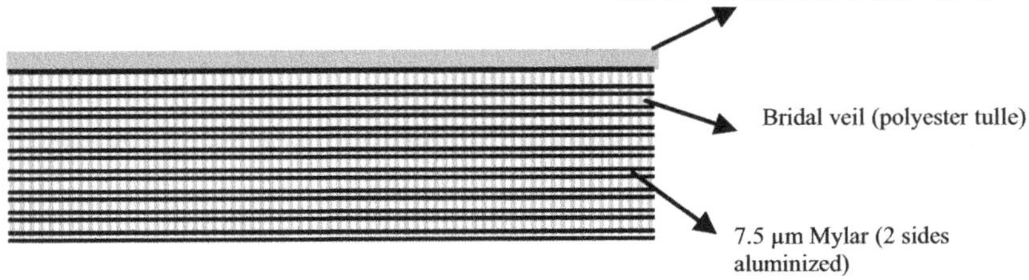

Figure 21.3. *Diagram of the principle of an MLI*

In free space, the thermal efficiency of MLI is spectacular. For example, a MLI composed of twelve layers has a thickness of about 3 mm and could have a surface conductance of 0.03 W/m².K. An equivalent polystyrene foam type insulating material should have a thickness of 1 m for the same efficiency.

21.3.5. *Radiator coatings*

In order to dose the right need of absorbed and emitted flux, these coatings are selected according to their thermo-optical properties. Let α_s be the solar absorption coefficient (solar energy absorbed by the material) and ε_{IR} the infrared emissivity (energy emitted by the material in the infrared spectrum).

We could resort to warm coatings such as black paint, or cool coatings such as white paint or quartz mirrors called Optical Solar Reflectors (OSR). We could also opt for a mixture of several coatings. Most often, cool coatings are used, having a weak solar absorptivity, to free ourselves from solar and albedo fluxes which are always variable in orbit.

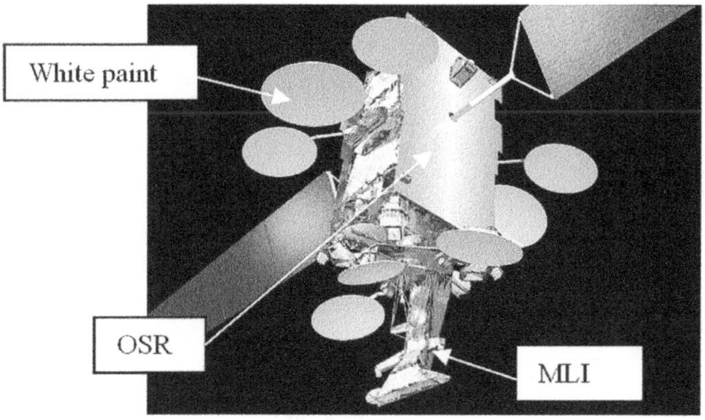

Figure 21.4. *Coatings on an Alphabus platform (EADS Astrium/Thales Alenia Space)*

21.4. Electrostatic phenomena in materials

As we have described, space vehicles placed in GEO orbit are submitted to the presence of electric charges, mainly in the form of electrons. It is now important to describe the electrostatic phenomena caused by these charges and experienced by the materials.

21.4.1. *Electrical conductivity*

Statistical data coming from flight statements during periods of strong geomagnetic activity have allowed a listing of the energetic spectra of electronic fluxes to be established. Depending on the geomagnetic activity level, the spectra are listed according to an index *Kp* [VIE 02] ranging between 0 and 9. In Figure 21.5, a spectrum designated by index $Kp>5$ is presented. The characteristics of index $Kp>5$ are described for the simulation of environmental conditions of a geostationary orbit. They come from an agreement between the rate of storms and their power. The spectrum $Kp>5$ therefore represents strong amplitude storms likely to occur.

The effects of the energetic dispersion of electrons on the external coatings are multiple and all of them are not negative [LEV 96]. The most degrading energies are those which permit the storage of electrons in the external dielectric. They are of the order of a few tens of keV and depend at the same time on the type and the thickness of the dielectric. For example, the average penetration depth of electrons of energy E = 20 keV in polyimide (Kapton) type polymeric materials, in fluorocarbon (Teflon) polymers or in silicones, is of the order of 7 to 10 μm.

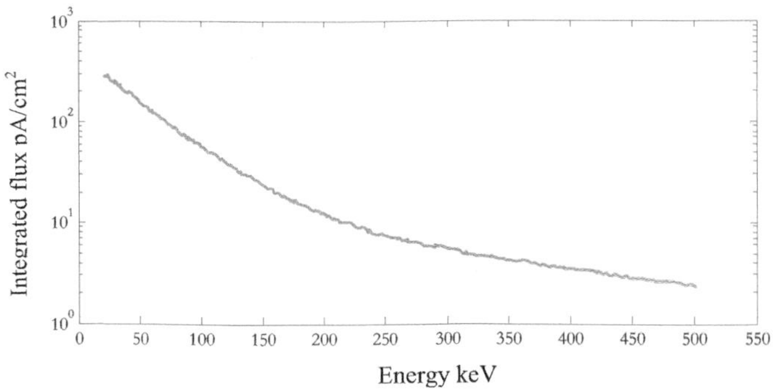

Figure 21.5. *Electron spectrum with integrated index Kp > 5 [HID 05]*

When dielectrics are irradiated by electrons, numerous phenomena come into play. The emission of secondary electrons are produced during the penetration of the electrons in the material, then, when those are stored, the intrinsic conductivity of this material permits the charges to be evacuated more or less rapidly depending on the nature of the dielectric. This effect can be accelerated if the material is irradiated by strong energy electrons which flow through the dielectric and produce induced conductivity along their path.

21.4.1.1. *The secondary electronic emission*

When a surface is irradiated by electrons of given energy E, we observe the emission of secondary electrons. The ratio of the current incident (or primary) electrons to the secondary electrons is a function of the energy E of the incident electrons. This ratio, called *secondary emission efficiency*, is presented in Figure 21.6a, and noted $\delta(E)$:

$$\delta(E) = \frac{I_S}{I_P} \qquad [21.1]$$

When the primary electrons have an energy E greater than E_2, energy of the second cross-over point (characteristic of the material), the secondary emission efficiency being less than 1, the material is charged. The surface potential increasing, the following incident electrons must overcome this barrier created by their predecessors. They then get to the surface of the material with a low energy E, thus getting closer to the second cross-over point E_2. When the incident electrons energy is equal to E_2, we find as many secondary electrons as primary electrons. In this case, the surface potential of the material no longer evolves. If a single electron gets to the surface with an energy below the second cross-over point E_2 then the secondary emission efficiency becomes greater than 1. The amount of electrons emitted by the surface therefore becomes greater than the amount of implanted electrons. The potential decreases. The barrier opposite to the primary electrons is smaller, the incident electrons energy increases. In this way, an equilibrium settles around the second cross-over point E_2.

Leung expresses the variation of $\delta(E)$ as a function of the maximum energy of the secondary emission efficiency (E_{max}) and the second cross-over point (E_2) through the equation:

$$\delta(E) = \frac{E}{E_2} \cdot e^{-\frac{2}{E_{max}^{1/2}}\left(E^{1/2} - E_2^{1/2}\right)} \qquad [21.2]$$

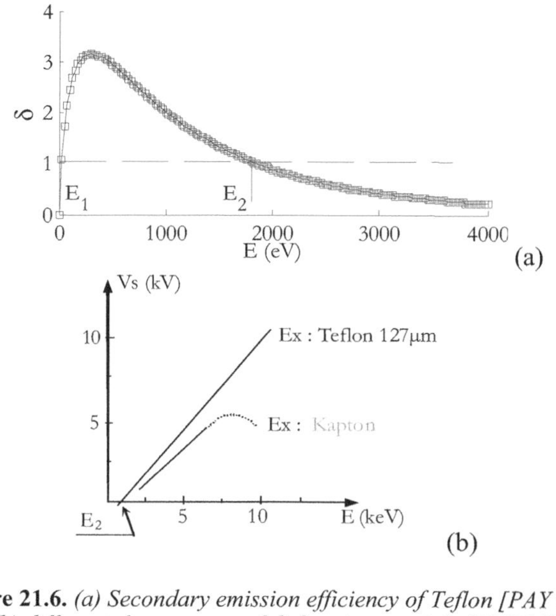

Figure 21.6. *(a) Secondary emission efficiency of Teflon [PAY 96], (b) different charge cases of dielectric materials [LEV 96]*

This secondary emission phenomenon induces a delay on the surface potential of the dielectric as a function of energy, as presented in Figure 21.6b. For example, for Teflon (Figure 21.6a) which has a very weak conductivity, the surface charge is governed by the secondary emission. For Kapton, in a thin layer which has a higher conductivity, the surface charge is controlled by the secondary emission and the intrinsic conductivity. This naturally causes a slower rate of charge accumulation.

When we increase the energy to exceed that for which the electron path becomes equivalent to the material thickness, this charge decreases. The electrons are then transmitted rather than stored in a capacitive form. This modifies the conductivity of the material irradiated over its entire thickness. This phenomenon is known as radiation-induced conductivity.

21.4.1.2. *Intrinsic conductivity*

The electrical conductivity of materials plays a crucial role in the value of the internal charge level of dielectrics. If a constant current is applied to a material, the internal electric field increases until an equilibrium is created, in which the current entering and the conduction current are both equal. For a material of thickness d, the maximum electric field at equilibrium, E_{max}, is calculated from Ohm's law:

$$V = RI \quad \text{i.e.} \quad E_{\max} = \frac{V}{d} = \frac{I}{S} \times \frac{SR}{d} = \frac{j}{\sigma} \qquad [21.3]$$

where V is the potential, I the current, R the resistance, j the current density, σ the conductivity and S the surface.

The calculation of the internal charge of dielectric materials can quickly become complex because their conductivity is not constant. It can be strongly affected by numerous factors such as the temperature, the electric field or radiations. The measurement techniques of conductivity in dielectric materials and moderately conductive materials are respectively defined by the norms: [AST 93, 87].

21.4.1.3. *Radiation-induced conductivity*

Under the effect of strong energy radiation (electromagnetic particles or waves), an insulating material is ionized and its electrical conductivity increases [SES 99] [ARK 93]. The radiation excites the electrons towards the conduction band, generating charge carriers, in direct proportion to the rate of absorbed energy (or dose) by the polymer. The conductivity is related to the density of carriers by:

$$\sigma = \sum ne\mu \qquad [21.4]$$

where σ is the conductivity, n the number of free electrons by cm^3, e the charge of the electron and μ the mobility.

Its value, for a given material, is therefore a function of the absorbed radiation dose and of the temperature. The induced conduction process is outlined in Figure 21.7.

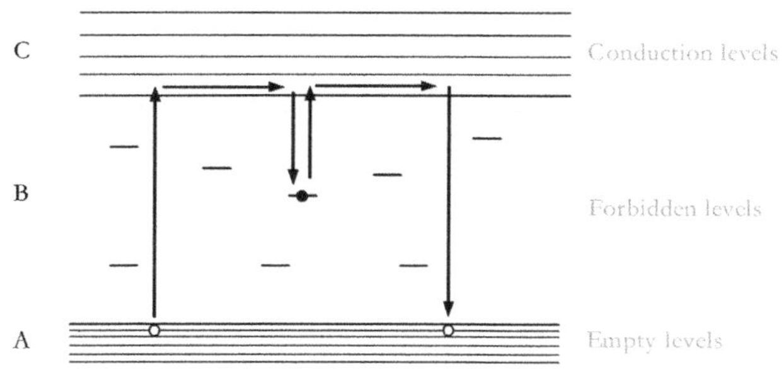

Figure 21.7. *Energy bands diagram of an insulating material. Kinetics of trapping/detrapping of carriers following an ionizing radiation. A represents the energies of the valence band, B the theoretically forbidden levels, and C the usually empty energy levels of the conduction band [HID 05]*

When an electron of band A is excited by a sufficiently energetic radiation, it moves into the conduction levels. After a certain time, it can either directly recombine with a hole, or be temporarily trapped before being thermally detrapped towards the conduction band. The trapping/detrapping process can be repeated several times before the electron falls into an energy level sufficiently low for its recombination with a free hole to be more probable than a thermal excitation.

The equation which described the conductivity σ of an irradiated dielectric material was developed by Fowler in 1956 [FOW 56]:

$$\sigma = \sigma(E) + KD^{\Delta} \qquad [21.5]$$

The parameter $\sigma(E)$ $[\Omega^{-1}.m^{-1}]$ represents the intrinsic conductivity of the material as a function of the electric field E *[V/m]*. The term KD^{Δ} represents the conduction induced under radiation. K $[\Omega^{-1}.m^{-1}rad^{-\Delta}s^{\Delta}]$ is the conductivity coefficient induced under radiation, which is a function of the irradiated material nature. D $[rad.s^{-1}]$ is the dose output absorbed by the material, i.e. the average energy quantity transmitted by the ionizing radiations to a volume element divided by the mass of this volume and by unit time. The exponent Δ is a coefficient with no unit, ranging between 0.5 and 1, depending on the irradiated material. It takes values rather close to 1 for amorphous structure materials and 0.5 for crystalline materials. It is associated with the ambient temperature and a coefficient T_1 homogenous to a temperature. This coefficient is a characteristic of the material, representative of the increased rate of the traps density. It can be described by the equation:

$$\Delta = \frac{T_1}{T + T_1} \qquad [21.6]$$

For a dielectric with surface S [cm²], density d [g.cm³], normally irradiated to its surface by electrons whose average penetration is r [cm], with acceleration voltage V [V] and absorbed current I [A] in the irradiated zone $S{\times}r$, the dose output D^o is given approximately by the formula [SIG 89]:

$$D^o = 10^3.\frac{V.I}{S.r.d} \qquad [21.7]$$

21.4.2. *Electrostatic discharges in the geostationary environment*

The internal charge of dielectric materials only becomes an issue when the accumulated charge induces a single phenomenon of electrostatic discharge or dielectric breakdown. In geostationary orbit, this phenomenon generally appears after the immersion of a satellite in a geomagnetic storm. In this case, the energetic plasma of the plasmic layer causes the differential charge of diverse parts of the satellite and, if a field or potential configuration exceeding a critical threshold appears, an electrostatic discharge is then initiated. It causes the creation of a very brief (a few hundred ns) and very intense (able to reach several amperes) current impulse. The configuration is known as critical if one of the cases presented in Figure 21.8 appears [ECS 98]. In this part, we shall designate the initiation and propagation modes of an electrostatic discharge when it is initiated in a dielectric and when it is initiated by peak effect in a metal.

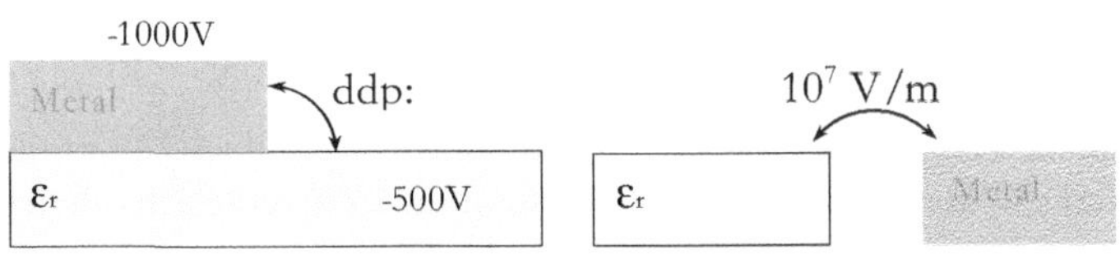

Figure 21.8. *Critical configurations of potential and electric field [ECS 98]*

21.4.2.1. *Dielectric discharge*

When a dielectric is submitted to a bombardment of electrons, some of them are trapped near the surface. Their accumulation creates a negatively charged layer, source of intense electric fields within the insulating material (Figure 21.9a). If the field becomes too intense, a breakdown (or boring) of the dielectric can result from it. In this case, the material becomes locally conductive over its entire thickness, the charges situated on both sides of the insulating material recombining on this path (Figure 21.9b). This phenomenon initiator of surface discharges (called *flash-over*) and space charge expansion (called *blow-off*) appears under the influence of an electric field and, in the presence of a desorbed gas, there can be propagation of the discharge (Figure 21.9b). During dielectric boring, the emitted energy creates a local temperature rise which leads to the creation of a conducting plasma. Under the influence of a transversal electric field, a neutral desorbed atom is split into an ion and an electron (Figure 21.9c). The electron, following the electric field, joins the closest surrounding mass and the positive ion comes to compensate the charge settling on the surface and desorbs another neutral atom. This atom is then ionized and the discharge is thus continued until total compensation of the charge.

In parallel to this discharge, a space charge expansion phenomenon appears. Unlike the surface discharge, the space charge expansion generates discernible and measurable currents in the structure of a satellite. The space charge electrons, attracted to their image charge, move along the field lines and create a current between the metallization of the dielectric and the metal structure of the satellite. It is in fact an electrostatic re-equilibrium current.

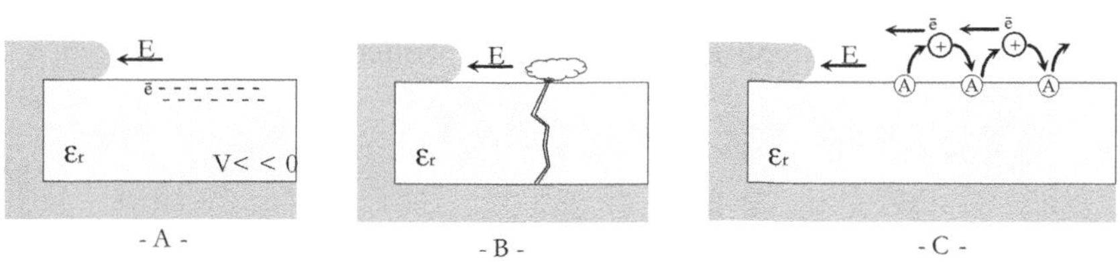

Figure 21.9. *Mechanism of the dielectric discharge [ECS 98]*

21.4.2.2. *Metal discharge*

Metal discharge is about the potential rise of metallization near a dielectric and concerns the electrons present in the metal. The release of the discharge is produced when the metal becomes more negative than a dielectric of about 500V (critical potential) [ECS 98]. By field effect, electrons are extracted from the metal and ejected towards the dielectric (Figure 21.10).

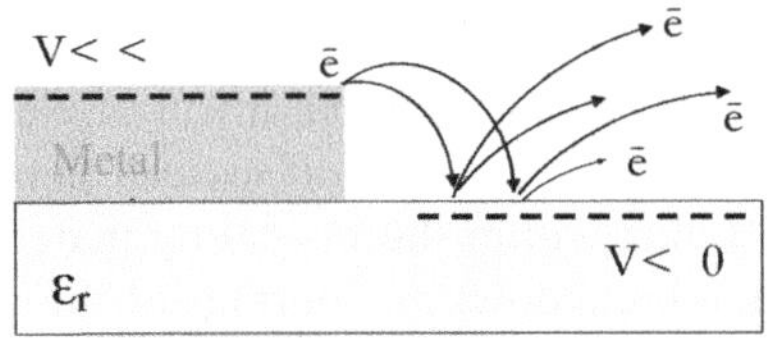

Figure 21.10. *Metal discharge [ECS 98]*

These incident electrons strike the surface of the dielectric material and by secondary emission extract several other electrons. This causes an increase in the potential difference between the metal and the insulating material and thus causes an even stronger field emission. This phenomenon is produced until the avalanche which totally discharges the material. For a better understanding of the mechanisms of this phenomenon, we shall explain in detail the phenomena of field emission and peak effect.

21.4.2.2.1. Field emission

Field emission is a phenomenon which is produced when the surface of a conducting material is submitted to an intense electric field which tends to send the electrons away. If we consider two electrodes submitted to a potential such that the created electric field is greater than 10^4 V/cm, then there is an emission of electrons from the cathode to the anode. Field emission law is in exponential form. It is governed by the tunnel effect process and its expression is that of Fowler and Nordheim. In a slightly simplified form, it can be written [PAY 96]:

$$J = \frac{1.54\times 10^{-6}.E^2}{\Phi} \times e^{-6.83\times 10^7 \frac{\Phi^{3/2}}{E} v(y)} \qquad [21.8]$$

where J [A/cm²] is the current density, E [V/cm] is the electric field, Φ [eV] is the potential barrier height to cross with the electrons and v is a function of the y parameter:

$$y = 3.79\times 10^{-4} \frac{\sqrt{E}}{\varphi} \qquad [21.9]$$

Table 21.1 gives a few values of the parameter $v(y)$.

y	0	0.2	0.4	0.6	0.8	1
V(y)	1	0.937	0.789	0.577	0.312	0

Table 21.1. *A few values of v(y) [PAY 96]*

This equation only plays a part in a strong field and was experimentally verified up to current densities of 10^7 A/cm². Figure 21.11 presents the current density as a function of the electric field for a barrier ϕ of 4 eV.

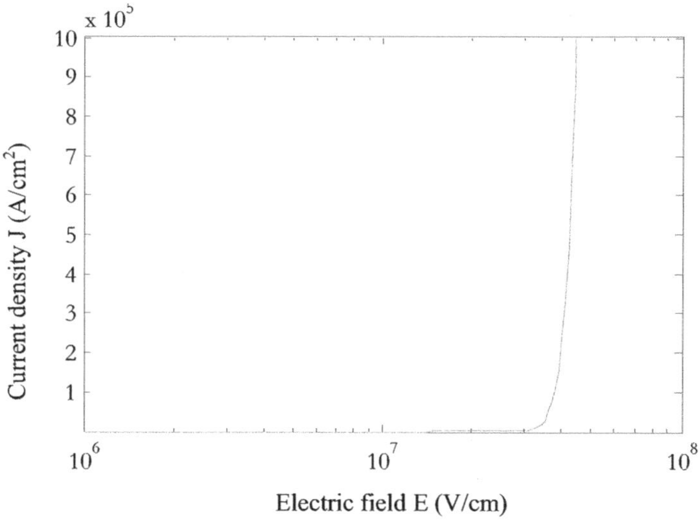

Figure 21.11. *Current density as a function of the electric field for a barrier of 4 eV [PAY 96]*

21.4.2.2.2. The peak effect

Electric fields likely to induce the field emission (E > 10^7 V/m) are generally generated by peak effect. This phenomenon takes place when a conductor, which presents an extremity of very weak diameter, is submitted to a potential. In this case, the charges tend to arrange themselves in a uniform way on the surface of the conductor. Some of them are therefore pushed towards the smallest extremity, which tends to increase very strongly the charge density at this point and therefore generates an intense electric field.

To represent this, we can idealize a point by representing it with two spheres connected by a wire, which permits them to be at the same potential (Figure 21.12).

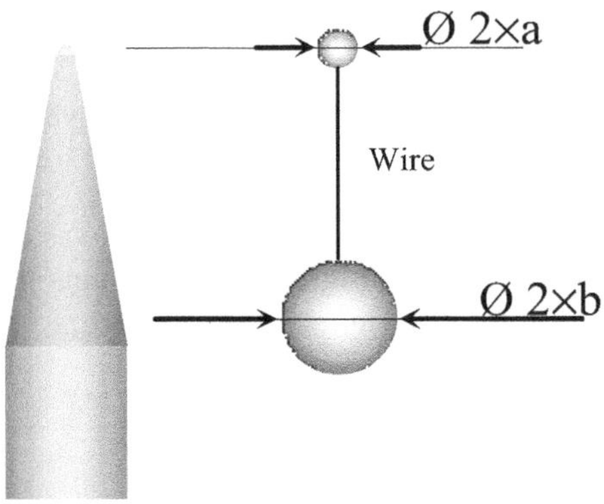

Figure 21.12. *Equivalent model of the peak [PAY 96]*

If we ignore the weak influence of the wire on the electric field and we consider that the sphere of radius a has a charge q, its potential will then be:

$$V = \frac{1}{4\pi\varepsilon_0} \times \frac{q}{a} \quad [21.10]$$

The presence of the other sphere naturally influences the distribution of charges, such that there is not really a spherical symmetry. If, despite everything, we are only interested in an estimation of fields, the electric field in the immediate neighborhood of the surface of the sphere will be:

$$E = \frac{1}{4\pi\varepsilon_0} \times \frac{q}{a^2} = \frac{V}{a} \quad [21.11]$$

The electric field is therefore a function of the inverse of the curve radius of the peak.

21.5. Conclusion

In space, satellites placed in geostationary orbit are in the flux of huge particle accelerators: the Sun and the Earth. At the end of a complex interaction process between the medium and the satellite materials, a charge potentially of several thousand volts can result from it, followed by electrostatic discharges and electromagnetic perturbations of the onboard electronics, which can lead to the loss of the vehicle. The aim of this chapter was to describe the phenomena which come into play between the charged particles of the space environment and the surface dielectric materials outside the satellites, intended for thermal control.

Today, we reduce the risk of breakdown by preventing dangerous potential increases. For this purpose, we now seek to reduce the resistance of coatings in order to send the charges towards the parts lit by the Sun, in order to take advantage of photoemission. Also, any conductive material is related to the common mass by a resistance permitting the flow of charges. Finally, very insulating dielectrics, capable of storing the charge for several hours, are prohibited.

To attempt any digital simulation, only a single simulated test, in an empty chamber, under a flux of electrons at a representative spectrum, will allow the electrostatic equilibrium of the satellite to be determined.

21.6. Bibliography

[ARK 93] ARKHIPOV V.I., RUDENKO A.I., SESSLER G.M., "Radiation-induced conductivity and charge storage in irradiated dielectrics", *J. Phys. D: Appl. Phys.*, 26, p. 1298, 1993.

[AST 87] AMERICAN STANDARD TEST METHODS, "Standard test methods for DC resistance or conductance of moderately conductive materials", ASTM D 4496–87, 1987 (Approved again in 1998).

[AST 93] AMERICAN STANDARD TEST METHODS, "Standard test methods for DC resistance or conductance of insulating materials", ASTM D 257–93, 1993 (Approved again in 1998).

[CNE 02] CNES, *Cours de technologie spatiale, Techniques et Technologies des Véhicules Spatiaux,* Conseil international de la langue française, vol. 1 and 3, September 2002.

[ECS 98] EUROPEAN CORPERATION FOR SPACE STANDARDIZATION, "Spacecraft charging: Environment-induced effects on the electrostatic behaviour of space systems", ECSS-E-20-06, Noordwijk, Holland, 1998.

[FOW 56] FOWLER J.F., "Radiation-induced conductivity in solid state, and some applications", *Proc. Royal Soc.*, London, A 236, p. 464, 1956.

[HID 05] HIDDEN G., Développement d'un nouveau revêtement de contrôle thermique pour satellite géostationnaire: Etude de l'influence de nanoparticules conductrices sur les propriétés thermo-optiques et électriques du composite, PhD thesis, Paul Sabatier University – Toulouse 3, 2005.

[LEV 96] LEVY L., "Charges des matériaux et systèmes de l'environnement spatial", *Environnement spatial: Préventions des risques liés aux phénomènes de charge*, CNES - Cours de technologie spatiale, Cépaduès Editions, France, p. 169, 1996.

[PAY 96] PAYAN D., "Les décharges électrostatiques induites par l'environnement géostationnaire", *Environnement spatial: Préventions des risques liés aux phénomènes de charge*, CNES - Cours de technologie spatiale, Cépaduès Editions, France, p. 289, 1996.

[SES 99] SESSLER G.M., YANG G.M., "Charge dynamics in electron-irradiated polymers", *Brazilian Journal of Physics*, vol. 29 no. 2, p. 233, 1999.

[SIG 89] SIGUIER J.M., "Contribution à l'étude des décharges électrostatiques sur les satellites: étude de la résistivité des matériaux de régulation thermique", *Mémoire Ingénieur CNAM*, Toulouse, 1989.

[VIE 02] VIEL V., REULET R. "Tests et moyens d'irradiation", *Environnement spatial: Préventions des risques liés aux phénomènes de charge,* CNES - Cours de technologie spatiale, Cépaduès Editions, France, p. 289, 2002.

Chapter 22

Recycling of Plastic Materials

22.1. Introduction

It is difficult to imagine our life today without plastic materials. The wide variety of plastic materials with their very large range of properties (light, heavy, flexible, rigid, thermal insulators, electric insulators, electric conductors, good optical properties) and their relatively simple transformation at a profitable cost, are the main reasons for their increasing use for the production of products and consumer goods.

The technological developments of the last decades, especially from the 1950s–1960s, have entailed a large evolution of the use and development of new plastics.

In numerous applications, plastics have replaced other traditional materials (metal, ceramics and wood) but their use has also extended owing to their application in new domains of science and technology: in microelectronics, biomedicine, telecommunications, etc.

Once their useful life is finished, plastic products and components must be disposed of. The issue of the final processing of plastic residues, as for a large part of the solid residues generated by our society, has not been completely solved. It seems logical to seek solutions other than their accumulation in rubbish tips, not only because of the impact on the environment but also because the materials in the residues could be reused; it is a waste of non-renewable resources.

Chapter written by Pilar MARTINEZ and Eva VERDEJO.

22.2. Plastic materials

22.2.1. *Introduction to plastic materials*

Plastic materials are organic polymers characterized by their plasticity, or ability to be molded under the effect of pressure and temperature. This is why, despite their diversity and their particularities, they are known under the generic name of plastics, which stems from the Greek *plastikos*: to form or prepare to mold.

The most used synthetic polymers are obtained from a fraction of petroleum. We estimate that 6% of petroleum is transformed into plastic. Its chemical structure is characterized by the repetition of small molecular units called monomers.

Polymers are macromolecules of high molecular mass, synthesized through poly-condensation or poly-addition reactions, from monomers. They contain carbon and hydrogen atoms and other elements (oxygen, nitrogen, etc.), combined altogether by chemical covalent bonds.

Plastics can be listed in three large groups according to the structure of the macromolecules and the bonds between them (this has a high influence on their thermal and mechanical behavior):

– thermoplastics: consisting of macromolecules of linear or ramified chains but without cross-linking. By increasing the temperature, they melt and can be molded in a reversible way. This characteristic permits their recycling, since owing to the application of heat, they can be remelted and remolded;

– elastomers: with an elastic and gummy aspect, their molecules are distributed by forming a tridimensional lattice, with not many inter-twinings and a low degree of cross-linking, presenting a restricted mobility. The molecular bonds between the chains break at high temperatures and do not reform when the temperature decreases;

– thermosettings: thermosetting or thermohardenable resins react to give a tridimensional lattice of macromolecules, with a high degree of cross-linking. Under the effect of heat, they conserve their shape and are maintained rigid until they reach the temperature which destroys them. The intertwining is not reversible and, consequently, the thermally stable residues and their composites cannot be recycled as easily as thermoplastics;

– plastic, thermoplastic or thermally stable materials are usually presented with other substances: strengthenings, charges and additives. These products of organic or inorganic nature are mixed up with the polymeric matrix, resulting in a plastic which can be better worked on and/or have improved properties;

– charges are organic or inorganic filling materials, which can reach relatively high percentages, and make the material more economical. They can modify and improve a few of their rheological and mechanical properties as well as their physical aspect. Silicone, silicates, calcium carbonate, clay, talc, lime and even well-pulverized synthesis polymers can be used as charges;

– strengthenings are usually fibers which are added to the plastics by forming composed or composite materials. Fiberglass, carbon, graphite, aramide, and cellulose fiber can all be used, and thus increase the solidity, the resistance to impact, the rigidity, etc. Epoxy-type resins and polyesters are much used for these thermohardenable composites;

– the term "additive" covers a wide range of chemical products which are added to plastics: antioxidants, thermal stabilizers and stabilizers exposed to light, lubricants, plasticizers, coloring agents, etc.

22.2.2. *Consumption of plastic materials*

In 2007, the consumption of plastics in Europe exceeded 52.5 million tonnes.

The consumption distribution of thermoplastics by sector is represented in Figure 22.1.

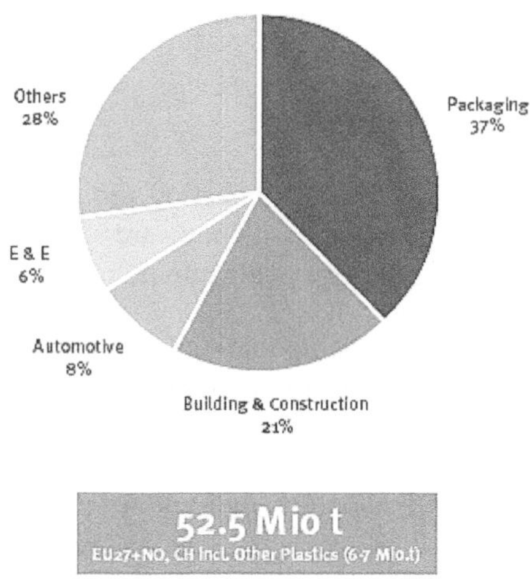

Figure 22.1. *Consumption of plastics by sector in Europe, 2007. Source: Plasticeurope Market Research Group (PEMRG)*

More than half of the total consumed plastics (Figure 22.2), correspond to four types of thermoplastics: polyethylene (PE), polypropylene (PP), polystyrene (PS) and vinyl polychloride (PVC) which are known as commercial plastics. Their good characteristics for use and their economical price allow their use for technical applications as well as for the manufacture of consumer products.

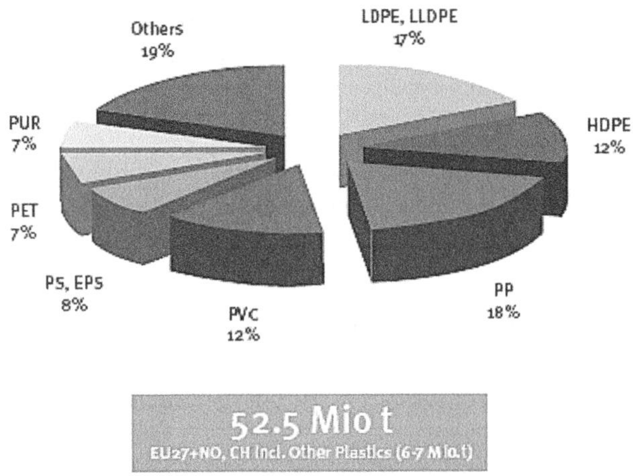

Figure 22.2. *Consumption of plastics by materials in Europe, 2007. Source: Plasticeurope Market Research Group (PEMRG)*

22.2.3. *Plastics in electrical engineering*

Polymeric materials in electrical engineering are fundamentally used for their dielectric and mechanical properties. They are used as thermoplastics and thermohardenables depending on their applications.

PE is used as an internal insulating material and as the covering for conductive cables for the transport of electric energy in high, average and low voltage lattices.

PP is used as an insulating material and as a dielectric in power capacitors, possibly impregnated by oil; it supports high voltages and is characterized by weak losses.

Epoxy resin composites with fiberglass and/or mineral charges are used in the cores of line insulators which are used in suspended air cables of the electric grid.

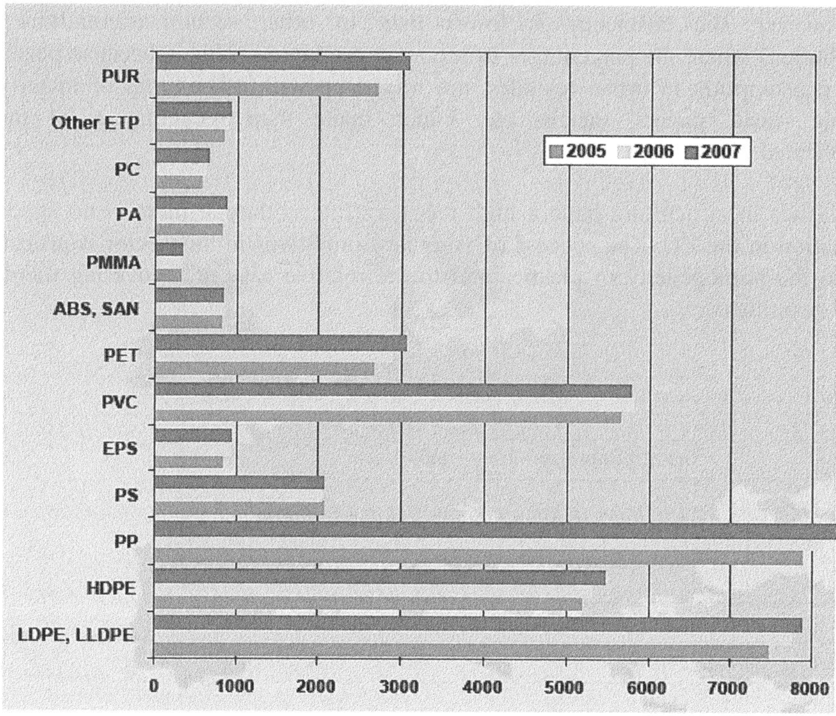

Figure 22.3. *Plastics Materials demand by type, 2005–2007. Source: Plasticeurope Market Research Group (PEMRG). Western Europe (WE) EU15 + CH + N (+ Malta + Cyprus)*

22.3. Plastic residues

22.3.1. *Generation and recovery of plastic residues*

From the estimations of the Association of Plastics Manufacturers in Europe (APME), the total amount of plastic residues after their consumption reached 20,608,000 tonnes in 2002, amongst which 38% of plastics were recovered as materials or energy. For the remaining 62%, their final destination was incineration or rubbish tips.

The highest amounts recovered, in urban residues, are due on the one hand to the high percentage of plastics in these residues (mainly as wrappings) and, on the other hand, to the strong development of selective recovery systems required to be applied by all countries of the EU, in order to reach the following recycling percentages: 15% of materials in 2001 and 22.5% of plastics in 2008.

However, the efficiency is lower than in other sectors (agriculture or distribution) where the percentages of recovery are higher. This is because plastics, after consumption in urban residues, are mixed up with other types of materials, organic waste, papers, metals, etc. which make their recycling a lot more complicated.

Plastics in agriculture have a high recovery rate, although there is no specific legislation in the EU. The success of voluntary initiatives in this sector is probably due to the homogeneity of plastics and to the relative ease of recovering them in large quantities.

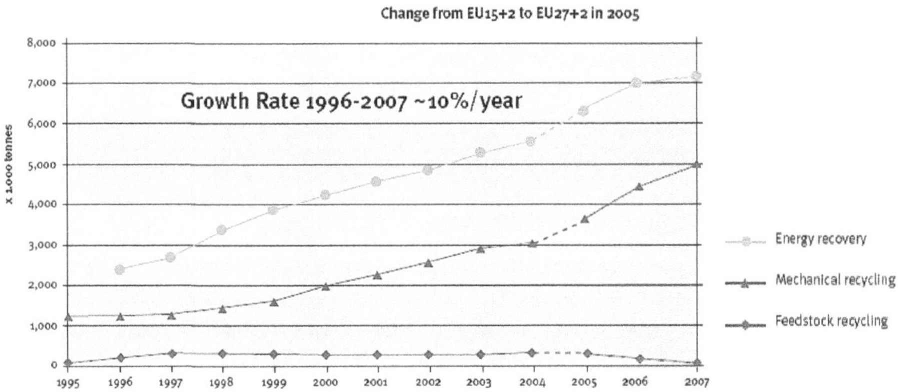

Figure 22.4. *Strong continued growth of recycling and energy recovery, 1995–2007. Source: Plasticeurope*

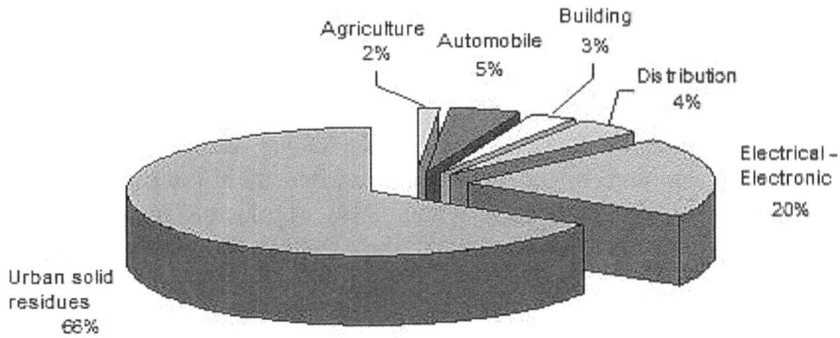

Figure 22.5. *Total plastic residues by sectors in Europe, 2002. Source: APME, Plastics in Europe 2002 & 2003*

On the other hand, in the automobile, electric and/or electronic and building equipments sectors, the recovery results are still fairly weak. These residues are generally heterogenous and often mixed up with other non-plastic materials, which makes their separation or their selection complicated. Different alternatives to recover plastics in out of service vehicles and in electrical and electronic equipment are being evaluated to reach recycling objectives required by different corresponding directives.

	Agriculture	Automobile	Building	Distribution	Electrical – Electronic	Urban solid residues	Total
Total plastic residues	311	959	628	4,190	848	13,671	20,607
Rubbish tip & Incineration (without energy recovery)	145	895	574	2,145	811	8,246	12,817
Energy recovery	1	7	0	444	3	4,222	4,678
Feedstock – chemical recycling	0	0	0	0	0	330	330
Mechanical recycling in Europe	149	58	52	1,332	32	843	2,466
Mechanical recycling exportation	16	0	2	269	0	54	341
Total recovered plastic residues	53.4%	6.7%	8.6%	48.2%	4.1%	39.7%	37.9%

Table 22.1. *Amount of recovered plastic residues in 2002, by sector and by recovery mode. Source: APME, Plastics in Europe 2002 and 2003*

22.3.2. *Processings at the end of life*

Once their useful life is finished, the final destination of plastic residues, like that of most solid residues produced by our society, is their accumulation in rubbish tips.

Outside of the use of the materials themselves, we induce a waste of non-renewable resources and, furthermore, another major issue is raised: that of the rubbish tips themselves.

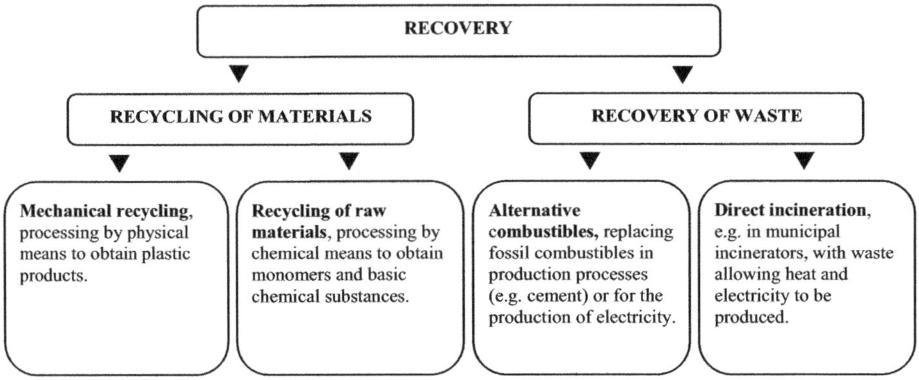

Figure 22.6. *Options for the recovery of plastic residues.*
Source: APME. http://www.plasticseurope.org

Plastic residues represent resources which can be recovered as materials or in the form of energy. The three alternatives for their recovery are:

1. mechanical recycling;
2. chemical recycling or *feedstock*;
3. energy recovery.

22.3.2.1. *Mechanical recycling of thermoplastics*

The mechanical recycling of thermoplastics is a process in several stages which requires a specific technology and equipment. Fundamentally, mechanical recycling consists of selecting and listing the plastic residues, grinding them, cleaning them, melting and extruding them in the shape of pellets or granules.

Subsequently, the recovered material can be transformed again to produce new pieces or products.

The quality of recycling depends to a great extent on the separation of different types of used polymers (particularly those which are not compatible), the degree of cleaning and the absence of impurities, such as metals, glass, paper, etc.

Ferrous contaminants are eliminated by using electromagnetic separators and/or metal detectors, and non-ferrous contaminants by using the Foucault current technique. Glass and paper are separated in the heavy or light fraction depending on their weights, by using water or air currents (via hydrocyclones, centrifuges, etc.).

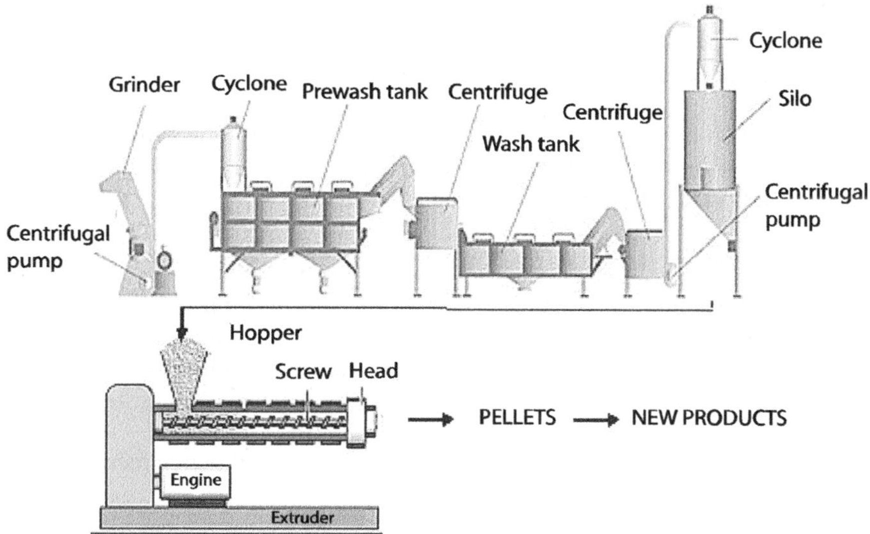

Figure 22.7. *Stages for the processing of mechanical recycling of plastics. Source: VIPLAT, recycling equipment supplier, www.viplat.com*

To separate different types of plastics from one another, we use their density differences: 0.94–0.96 g/cm^3 for high density PE, 0.89–0.91 g/cm^3 for PP and 1.16–1.38 g/cm^3 for PVC. For this purpose, we apply flotation separation in water or in water mixed with other solvents or tensioactives.

From the differences in transmission/reflection of light, and for wavelengths characteristic of the absorption spectrum, optical detectors with different emission sources (polarized light, UV, IR, X-ray, etc.), combined with air ejectors, can separate colored plastics, different types of plastics and plastics which contain certain unwanted charges or additives.

22.3.2.2. *Mechanical recycling of thermohardenables*

The recycling of thermostables and their composites with charges and strengthening fibers is essentially a division process followed by a thinner grinding until a fibrous powder with an appropriate particle size is obtained, which varies between 100 mm and 50 micrometers.

All components of the original composite are found in the recycled material, which is a mixture of polymer, fibers and charges. This recovered material possesses shorter fibers but can partially replace the charges to be used in the production of new composites. In certain cases, when the recycled material contains a larger amount of fibers, it can be used to partially replace the strengthening fiber in the molding of new pieces. However, in all cases the mechanical properties are reduced when the amounts of recycled material are large.

They can also be used as strengthening material in thermoplastic composites or mixed with tar or thermoplastics. They can also replace wood fibers for the production of posts and pylons (wood-plastics) [PIC 05].

22.3.2.3. *Chemical recycling, or feedstock*

Chemical recycling, also called feedstock, is the decomposition of polymers contained in residues to obtain starting monomers or a mixture of basic chemical products: hydrocarbons with different molecular masses, synthetic gas, etc., which could be used as combustibles or as raw material for the chemical industry.

There are two types of processes: chemolysis or chemical depolymerization, and thermolysis.

22.3.2.3.1. Chemolysis or chemical depolymerization

This process is about adding reactive chemical products which, in the presence of catalysts and under certain pressure and temperature conditions, break the macromolecules down. The reaction products are monomers or oligomers, which can be used to synthesize the same starting polymer again, or even another type of polymer.

The process can be applied to condensation polymers with functional groups combined with weak bonds likely to be dissociated by the action of certain reactives. We distinguish different types of depolymerization reactions depending on the reactive used:

– glycolysis: with polyglycols, polymeric by-products of ethylene oxide such as glycol polyethylene, glycol polypropylene and glycol diethylene, etc.;

– methanolysis: with methanol;

– alcoholysis: with other alcohols, such as butanediol;

– hydrolysis: with water or acids;

– saponification: with alcali;

– aminolysis: with amines.

22.3.2.3.2. Thermolysis

Thermolysis processes depend on the application of high temperatures to produce the breakdown of polymer chains in the presence (or not) of catalysts. This can obtain basic hydrocarbons or synthesis gases such as volatilizable hydrocarbons, naphthas, liquid and waxy hydrocarbons, bituminous components, or mixtures of combustible gases with synthesis gases.

When the process of decomposition or breakdown is undertaken in the absence of air, it is called pyrolysis or steam cracking (at 800°C without catalyst, or 400°C with a catalyst). When it takes place in a hydrogen gas atmosphere, it is called hydrogenation or hydrocracking (at 400–450°C and 50–100 bars); if it is made in the presence of a controlled amount of oxygen, the process is known as gasification (at 850–950°C).

These recovery processes are the most suitable for addition polymers such as polyolefins, acrylic and vinyl polymers. The thermally stable pyrolysis is also possible. In that case, we can obtain mixtures of organic products with weak molecular mass which, subsequently, can be separated and purified.

The thermal processing of thermally stable composites can also lead to the recovery of large value inorganic components contained in the composite, such as fiberglass for example (see Figure 22.8).

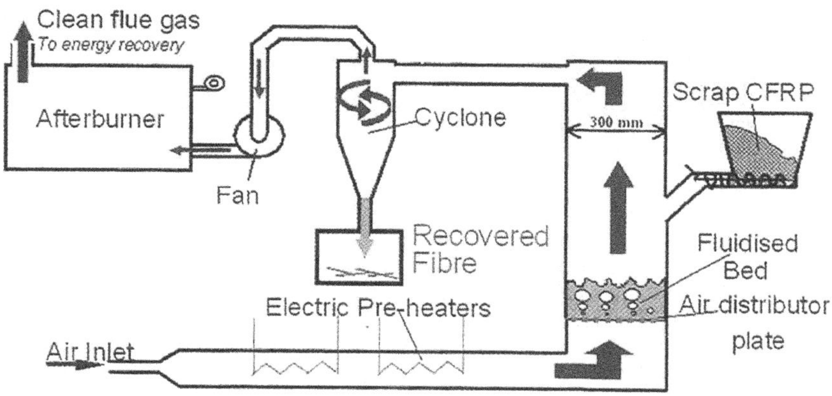

Figure 22.8. *Recovering process of composites through a fluidized bed.*
Source: "Recycling termoshet composites". JET Composites n.17 May 2005

In a bed of fluidized sand, the warm air is at temperatures between 450 and 550°C; the polymer is decomposed in its lightest volatilizable fractions. The solid mineral fibers and charges, led by the gas, are separated for their re-use. The gases coming from the decomposition of the polymer are led to a combustion enclosure at high temperature for their complete oxidation and energy recovery.

The recovered amount of fiberglass and inorganic components (classical carbonate, aluminum oxide, etc.) present in the composites can represent 70% of the weight. These recovered materials can be used as replacement raw materials in furnaces for concrete production industries (1,480°C).

22.3.2.4. *Incineration with energy recovery*

In 2003 in Occidental Europe 21,150,000 tonnes of plastic residues were produced, among which 4,750,000 tonnes were incinerated with a view to recovering energy.

The aim of incineration is the thermal destruction of residues (900–1,000°C) through their complete oxidation by bringing oxygen in greater amounts than the stereochemical amount required for their own combustion. The heat of output gases in the combustion chamber can be used as thermal energy or transformed into mechanical and/or electric energy.

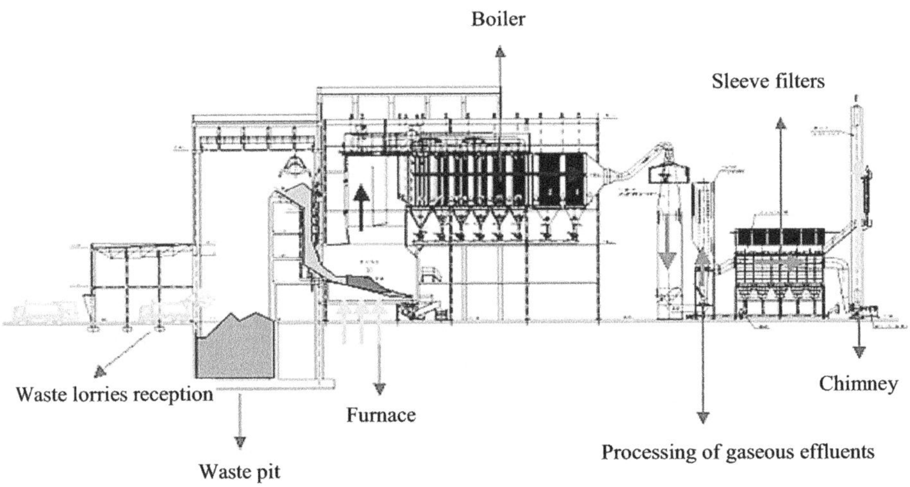

Figure 22.9. *General diagram of an incineration process with energy recovery. Source: Incinérateur RSU de Txingudi, www.txinzer.com*

The energetic recovering or incineration with energy recovery is made in incineration power stations which behave as thermal power stations. The combustibles, in this case the residues, are burnt in a furnace with appropriate characteristics where we recover the heat of combustion gases to produce water vapor, which sets in motion a turbine for electrical energy production.

During incineration, three types of pollutants are produced: gases, flying ash and cinders which must be processed to reduce their environmental impact.

Since the calorific power of plastics is relatively high, plastic residues can be used as alternative combustible or mixed with other combustibles (coal, fuel, etc.) for use in industrial boilers, industrial furnaces for concrete production, etc. This allows a portion of the combustible required to produce heat, vapor or electrical energy to be saved.

Thermally stable polymers have a calorific power near that of good quality coal. The composites contain 25–35% of thermally stable resin and can be successfully burnt to obtain energy, for example mixed with urban solid residues (USW) for incineration.

MATERIAL	MJ/Kg
Natural gas	52
Polypropylene	44
Polyethylene	43
Crude oil	42
Polystyrene	40
Polyamide	37
PET	33
Coal	29
PMMA	25
PVC	20

Table 22.2. *Energy content. Source: "Los Plásticos y la Gestión de sus Residuos" CEMAV Vicerrectorado de Metodología, Medios y Tecnología, UNED*

22.3.3. *Potential and limitations of recycling*

The solutions, known as the 3 Rs, of the environmental issue of residues are, in order: reduce, re-use and recover. But it is not always possible to use this type of solution: there are associated technical, economical and social problems, such as:

– the difficulty of controlling the generation of residues when they result from the consumption of products with a short life-cycle or unique use (disposable products);

– the difficulty of separating residues in a selective manner and in sufficient quantities, depending on their nature and composition;

– the fact of finding a combination of incompatible products in the residues, which prevents good quality recycled material being obtained.

Also, sometimes recycling is not conceivable because of difficulties or restrictions with the technique, or because costs are excessive and the recycled materials are not competitive with the raw materials they could replace.

We present below a few potentialities and limitations for the recovery of plastic residues in general and, in particular, for residues coming from electrical installations:

– today, the recovery of materials contained in all plastic residues is fundamentally made via the mechanical recycling of the main consumer thermoplastics: PE, PP, PVC, etc. Mechanical recycling is a processing which is only applied to plastic residues of one selected and homogenous material. The plastics thus recovered represent 15% of the total residues;

– there are residues which contain plastics with other types of materials (metals, ceramic-glass, etc.) that are grinded together, which prevents them from being easily dismantled or separated: insulated cables, small size components, etc. In these cases, there are effective processes to separate recyclable plastics from the rest of the materials (magnetism, extraction with solvents, gravity, flotation, etc.);

– the current processes of mechanical recycling are not efficient to recover complex plastics or multimaterials since the recovered material has no thermal and homogenous rheological properties, and the material which would be obtained could not be processed, or would give a material of weak quality;

– plastic residues are materials which might have been subjected to a degradation during their useful life, such as thermal degradation, chemical degradation, etc. However, they can still be used for less demanding products; we can combine them with the original material or improve their properties by adding corresponding additives before their reprocessing;

– the recycling of thermohardenable composites is not very developed. From a technical point of view, it has been verified that the recovery of these materials, in the form of grinded powder or by separating the strengthening fiber, is possible without any loss of mechanical properties for the new products. Others possibilities to recover the material are the combustion and use of organic scraps as subsidiary raw material in the production of concrete;

– chemical recycling, with which we claim to make residual plastic cost-effective by transforming it into a basic raw material or into a combustible, starts to appear as a real alternative owing to the improvement of separation processes and the development of new technologies. Chemical recycling is currently in an advanced experimental phase;

– the elements of electric grids: cables, insulating materials, capacitors, etc. are long life (>20 years) products. The ratio between the generation and the total volume of plastic in this type of residue is much lower than in other consumer products in the electrical–electronic sector which have a shorter life cycle. However, the residues coming from the dismantling or the renewing of large lattices and electrical installations could represent an advantage in recovering residues in large quantities;

– the thermoplastics used for the electrical insulation of cables are potentially recyclable (by mechanical recycling), except in the case of crossed-linked polyethylene (XLPE) which can only be recovered by chemical depolymerization or energetic recovery;

– in the case of unipolar cables with only one type of plastic as an insulator, recycling could be made by means of a mechanical process. For more complex cables, in which there are different types of plastics, we could use separation processes with selective solvents such as the Vinyloop process for PVC;

– the economical value of cable which conducts copper, aluminum, etc. is a decisive factor for the recovery of cable by using separation processes for the plastic which covers it. Consequently, the plastic could also be recycled;

– the high calorific power of plastics makes them potentially usable as alternative combustibles in certain industrial processes (e.g. in furnaces for concrete production). They can also be burned to recover the thermal energy or be transformed into electricity;

– the dielectric films (PP) of capacitors, impregnated with oil, would require a previous cleaning and depollution, which complicates and increases the recovery cost of the material. It is therefore more likely that the incineration of the residue is preferred in installations with systems processing pollutants present in gases and fly ashes.

22.4. Bibliography

[AGU 99] AGUADO J., SERRANO D., *Feedstock Recycling of Plastic Wastes*. RSC Clean Technology. Monographs, Ed. Royal Society of Chemistry, Cambridge, 1999.

[ANS 00] ANSEMS A.M.M., DE GROOT J.L.B., VAN DER VLUGT M., *Best Practices for the mechanical recycling of post-user plastics*. Association of Plastics Manufacturers in Europe (APME), 2000.

[APME] Association of Plastics Manufacturers in Europe, *Plastics in Europe – An analysis of plastics consumption and recovery in Europe 2002, 2003 and 2004*. www.plasticseurope.org

[PLASTIC EUROPE] The Compelling Facts About Plastics 2007 – An analysis of plastics production, demand and recovery for 2007 in Europe. www.plasticseurope.org

[EPIC] ENVIRONMENT AND PLASTICS INDUSTRY COUNCIL. *Plastics Recycling Overview. Summary Report by Environment and Plastics Industry Council – A Council of the Canadian Plastics Industry Association.* www.plastics.ca/epic

[EuPR] EUROPEAN PLASTIC RECYCLERS: http://www.eupr.org/

[MAR 04] MARTÍ-FERRER F., SANZ-BOX C., BENEDITO-BORRÁS A., TEROL-MARTÍ M.L. "Propiedades mecánicas de mezclas de HDPE con residuos de poliéster reforzado con fibra de vidrio", *VIII Congrés National de Materiaux,* Espagne, June 2004.

[PIC 05] PICKERING, S. "Recycling termoshet composites", *JET Composites,* no. 17, May 2005.

[PWMI] PLASTIC WASTE MANAGEMENT INSTITUTE, *An Introduction to Plastic Recycling in Japan 2004*. Plastic Waste Management Institute, Tokyo, Japan, May 2004. www.pwmi.or.jp/

[SCH 98] SCHEIRS J., *Polymer Recycling*, p.1, Wiley, Chichester, UK, 1998.

[VERT 05] EU Project G7RT-CT-2002-05073, "*Setting-up of European Virtual Institute for Recycling*", Technological Reference Paper on Recycling Plastics, May 2005.

Chapter 23

Piezoelectric Polymers and their Applications

23.1. Introduction

The piezoelectric effect on a polymer was revealed for the first time in Poly(vinylidene fluoride (PVDF) by Heiji Kawaï, in 1969 [KAW 69]. In general, piezoelectricity corresponds to the outbreak, or the variation, of a polarization in a material in response to the application of a mechanical stress: a phenomenon designated as the direct effect. This type of material also produces the opposite effect: the application of an electric field causes a deformation of the sample. These electrical and mechanical properties when coupled with polymers are used in several applications, such as receiving and transmitting transducers.

PVDF and its copolymer vinylidene fluoride (VDF) with trifluoroethylene (TrFE) or tetrafluoroethylene (TFE) groups are materials whose ferroelectricity has been demonstrated [LOV 83, FUR 89], and which is developed in section 23.3.1. In these semi-crystalline polymers, it is the spontaneous polarization of the crystalline phase which causes their piezoelectricity. Fluorinated polymers have good piezoelectric properties associated with intrinsic properties of softness, lightness, etc. These organic materials thus have an interesting place in applications such as transducers, for example, in comparison with a piezoelectric ceramic, the lead zirconate titanate (PZT) which is also widely used.

Chapter written by Alain BERNES.

More recently, ferroelectric properties have been shown in other semi-crystalline polymers: the odd-numbered polyamides [MEI 93]. These polyamides possess interesting piezoelectric properties, but their sensitivity to hydration has not yet permitted a commercial development [NEW 90].

Research devoted to piezoelectric amorphous polymers are much less numerous, because a piezoelectric activity is often insufficient to be developed in applications. The first studies were undertaken on polymers substituted for nitrile such as polyacrylonitrile (PAN) or vinylidene cyanide and vinyl acetate copolymers (PVDCN/VAc) [MIY 80]. Amongst these materials, the most promising ones are the vinylidene cyanide copolymers, which possess better values of piezoelectric constants and of the dielectric relaxation intensity $\Delta\varepsilon$. The origin of piezoelectricity in amorphous polymers is different to that which is observed in semi-crystalline polymers and inorganic crystals. Indeed, polarization in piezoelectric amorphous polymers does not correspond to a thermodynamic equilibrium state, but rather to a quasi-stable state of fixed molecular dipoles.

Fluorinated polymers can thus be used to make electroacoustic transducers such as loudspeakers, headphones, microphones, probes for medical echography, underwater acoustic hydrophones, acoustic emission probes for non-destructive testing [CHE 95, NAL 95]. These devices are distinguished from usual transducers by their active membrane which generally does not require any acoustic matching layers to the propagation medium. Their performances are also improved, owing to the possibility of giving the membrane a geometry calculated beforehand, by use of thermoforming processes [RAV 81].

In this chapter, we shall develop the properties of piezoelectric polymers in relation to their physical and chemical structure. We shall then present their specificities in several examples of electroacoustic and ultrasonic transducer applications in the biomedical and Non-Destructive Testing (NDT) domains.

23.2. Piezoelectric polymeric materials

23.2.1. *Poly(vinylidene fluoride)(PVDF)*

Poly(vinylidene fluoride) is a semi-crystalline thermoplastic polymer with a relatively simple chemical structure $[-CH_2-CF_2-]_n$, where n is typically larger than 10,000. Figure 23.1 shows that the constitutive motif possesses a strong dipole moment $\mu_i = 7.56\ 10^{-30}$ C.m [MAR 82] due to the presence of two fluoride atoms and the dissymmetry of the molecule. The dipole moment is defined by $\mu_i = q\ .d$, where d is the distance which separates two concentrated charges $+q$ and $-q$.

PVDF crystallizes in the shape of lamellae, called crystallites, whose thickness is generally of the order of a few tens of nanometers. The macromolecules are coiled up within these lamellae with blades perpendicular to both sides. These crystallites are arranged into essentially spherically symmetric polycrystalline aggregates designed as spherolites. The increase in crystallites occurs in a radial manner from nucleation centers which constitute the core of the spherolites. Apart from crystallites, the arrangement of macromolecules is made in a disordered manner. The volume not occupied by the crystallites is therefore amorphous. Globally, the polymer is semi-crystalline (a biphasic structure) and its crystallinity rate is of the order of 50 to 55%.

It seems that the ferroelectric properties strongly depend on the crystalline structure of the material. Indeed, numerous studies [KEP 92, SER 80] reveal that the crystalline structure of PVDF mainly present four polymorph phases. The most commonly observed structure, obtained primarily by quenching from the melt, is the non-polar α-phase. The phase which presents the most interest for electro-active properties is the ferroelectric β-phase. The other two phases are referenced as γ- and δ-phases which are also ferroelectrics. Nevertheless, the contributions of these γ- and δ-phases in the ferroelectric properties are neglected because they are constituted of cells that have one very weak dipole moment in comparison with that of the β-phase.

The α-phase of PVDF is constituted of chains in a tg+tg$^-$ formation, whereas β-phase consists of all-trans chains (ttt) (Figure 23.1). The crystalline group of phase α remains uncertain because of the conformational disorder associated with this phase, which is either orthorhombic or monoclinic, depending on the observations made. The β-phase is orthorhombic.

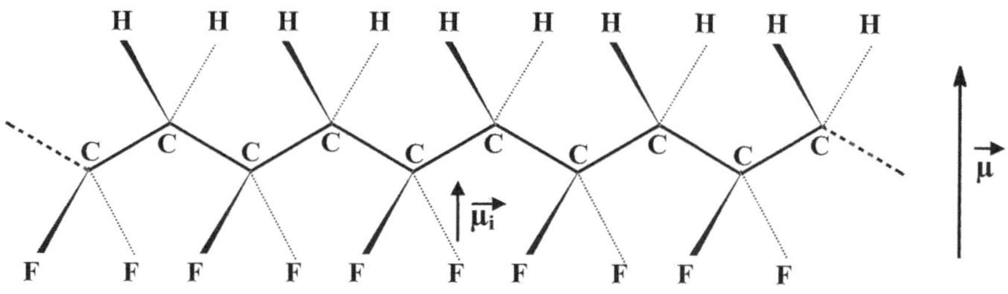

Figure 23.1. *"All-trans" conformation (β-phase) of macromolecular PVDF chains*

Figure 23.1 shows that the dipole moment $\mu = \sum \mu_i$ associated with the macromolecule is at a maximum when the configuration of the chain is in β-phase. Indeed, in this case, the carbonated chain takes all-trans conformation, thus all the CF2 dipoles are oriented in the same direction and the dipoles are added. Thus, the dipole moments oriented in a preferred direction create a macroscopic polarization P effective for each crystallite: this phenomenon causes the ferroelectric properties of the β-phase in PVDF. On the other hand, the α-phase of PVDF is apolar, although the dipole moment of the chain is not null. These dipole moments cancel each other out within a crystalline cell and therefore do not present any electro-active properties. Only PVDF in β-phase will be used as a ferroelectric material for piezoelectric applications.

The calculation of interaction energies of electrostatic and Van Der Waals forces by dissociating the inter- and intra-molecular components shows that the α-phase of PVDF is more stable than the β-phase [SER 80]. PVDF therefore crystallizes naturally in its α-phase.

The crystallization of PVDF is strongly affected by the presence of sequence defects in the macromolecular chain. These defects present themselves in the form of an inversion of one or several monomers: "head-head" (h-h) or "tail-tail" (t-t) bonds (Figure 23.2).

$$-CH_2-CF_2-CH_2-CF_2-CF_2-CH_2-CH_2-CF_2-CH_2-CF_2-$$

↑ ↑

h-h t-t

Figure 23.2. *Examples of monomer inversion with a consecutive CF2 – CF2 designed by head-head (h-h) bond and a consecutive CH2 – CH2 for a tail-tail (t-t) bond*

Generally, the rate of sequence defects is of the order of 3 to 6%. The increase in this rate allows a gradual passage of α-phase towards β-phase. A rate greater than 15.5% permits a crystallization to be obtained in the β-phase.

The crystallization in the β-phase is usually obtained by a mechanical extension of films at a temperature near to 100°C in the α-phase. This stretching of a film from 3 to 5 times its initial length causes a breakdown of the original spherulitic structure into an array of crystallites whose molecules are oriented in the direction of the force. Figure 23.6 (below) schematizes the film thus oriented, composed of crystallites scattered in an amorphous phase, where these crystallite axes are oriented in a preferred direction.

23.2.2. *The copolymers P(VDF-TrFE)*

Poly(vinylidene fluoride -trifluoroethylene) copolymers are obtained during synthesis by the introduction of a trifluoroethylene group –CF2-CFH- in PVDF chains. A copolymer possessing 25% of trifluoroetylene entities (in molar percentage) is designated by P(VDF-TrFE) 75-25.

In these copolymers, the arrangement of VDF and TrFE monomers is statistical. They behave like PVDF homopolymers in which some hydrogen atoms were replaced by fluoride atoms: they only possess, for example, one fusion temperature and only one glass transition.

The copolymer of vinylidene fluoride (VDF) with trifluoroethylene (TrFE) produces h-h type sequence defects in the PVDF chain, and thus favors an "all-trans" configuration macromolecule [FUR 89, NAL 95]. The P(VDF-TrFE) copolymers therefore crystallize naturally in the ferroelectric β-phase. This natural crystallization in the β-phase is obtained for concentrations in TrFE monomers ranging between 17 and 50%. The electro-active properties of these copolymers are very similar to those of PVDF and have the same origin. Nevertheless, the rate of trifluoroethylene units strongly influences certain physical properties of these copolymers.

The exact crystalline structure depends on the amount of trifluoroethylene entities. Matrix distorsions of the β-phase of PVDF gradually appear with an increase in the number of TrFE monomers, such that the TrFE unit rate must remain less than about 50% to conserve electro-active properties.

23.2.3. *The odd-numbered polyamides*

Polyamides are semi-crystalline thermoplastic polymers. In general, the chemical formula of polyamide x is written:

$$[—(CH_2)_{x-1}—CO—NH—]_n \qquad [23.1]$$

The number x represents the amount of carbon atoms of the repetition unit. The constituent motif is therefore composed of an apolar "paraffin" sequence (sequence of methylenes $-CH_2-$ groups), a variable length which depends on the considered polymer, and a very polar amide group. The value of the dipole moment associated with the amide group is $\mu=1.18.10^{-29}$ C.m [MEI 93]. We can distinguish even-numbered polyamides possessing an even number of carbon atoms in the repetition unit from odd-numbered polyamides possessing an odd number of carbon atoms.

The molecular cohesion and the physical properties of polyamides are dominated by the electrostatic interactions existing between the amide groups of adjacent chains. These strong interactions lead to the formation of hydrogen bonds between neighboring chains (Figures 24.2.a and 24.2.b). They play a role of first importance in the formation and stability of crystalline zones, and in the cohesion of the amorphous phase.

The arrangement of macromolecular chains in the crystalline zones has been studied in detail for several decades [SCH 92, LOV85]. In crystalline zones of even and odd polyamides, the chains adopt a planar zigzag conformation (the most favorable structure). The existence of hydrogen bonds between adjacent chains leads to the formation of a structure "in sheets" ("*Hydrogen-bonded sheets*").

2-a

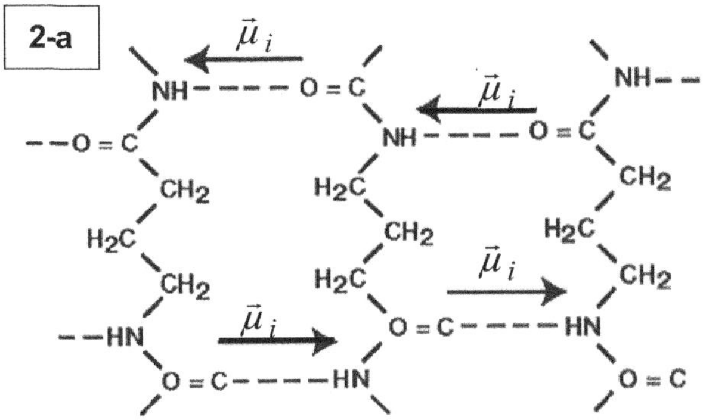

2-b

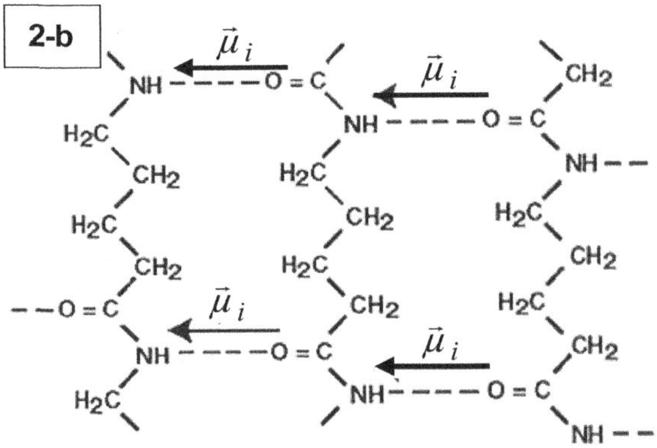

Figure 23.3. *Model of crystalline arrangement of the PA4 (2a) and the PA5 (2b)*

For each type of polyamide, either even and odd, two structures are conceivable: an antiparallel structure and a parallel structure, respectively. An example of these structures is presented for PA4, Figure 23.3.a, and for the PA5, Figure 23.3.b. For odd polyamides, the parallel structure is the most favorable. On the other hand, for even polyamides, the antiparallel structure is the most favorable because it leads to the formation of a maximal rate of formed hydrogen bonds. In the case of odd polyamides, all the dipoles are aligned in the same direction, whereas for even polyamides, they are inversed two by two. This molecular arrangement causes the ferroelectric properties of odd polyamides: the dipole moments oriented in a preferred direction creates an effective macroscopic polarization P for each crystallite.

The ferroelectric properties of odd polyamides depend on the implementation conditions of the films. The optimal properties were obtained for films quenched from the melted state and stretched in a uniaxial manner (three times the initial length), then annealed at high temperature [MAT 84]. These films present a "double-orientation": the polymeric chains are preferentially oriented in the direction of the stretching, and the sheets made up of the hydrogen bonds are oriented parallel to the film plane.

Polyamides possess an important hydrophilic character. The maximal amount of water which can be absorbed increases as a function of the density of amide groups, i.e. when the paraffin sequence size of the polymer decreases. Thus, the amount of water absorbed in bulk can reach 4.5% in PA11 and PA7, and 12% in PA5, while it is less than 0.02% in PVDF. The physical properties of polyamides are affected by the amount of absorbed water: variation in the mechanical modulus and the dielectric permittivity. Only the amorphous phase is directly affected by the hydration rate, since the water molecules are not diffused in the crystalline phase.

23.2.4. *Copolymers constituted of vinylidene cyanide monomer*

Vinylidene cyanide (VDCN) is an extremely reactive monomer which polymerizes readily with a wide variety of common monomers, such as vinyl acetate or styrene. Most of these copolymers have a high melting or softening point and high glass transition temperature (T_g). Chemical structures of a few vinylidene cyanide based copolymers are represented in Figure 23.4 [NAL 95].

In these copolymers, the nitrile group possesses a strong dipole moment of $1.12.10^{-29}$ C.m which causes the formation of hydrogen bonds and dipole interactions in the physical structure. The Nuclear Magnetic Resonance (NMR) ^{13}C spectra reveal that these copolymers are alternated with a head–tail arrangement.

This technique also shows that there is no stereoregularity of successive vinyl ester motifs, contributing to the amorphous structure of copolymers. This amorphous structure is confirmed by studies of x-ray diffraction which, however, reveal a local order in the glass state, disappearing when the temperature is raised above the glass transition.

Indeed, the dipolar interactions of nitrile groups favor the local order of the amorphous phase. Then, at the crossing of the glass transition, this weak cohesion between macromolecules is destroyed by cooperative molecular motions. This phenomenon causes an endothermal peak on the thermograms recorded by differential enthalpic analysis (DEA).

Orientation of dipole moments in the domains of local order allows electro-active properties specific to VDCN copolymers to be communicated.

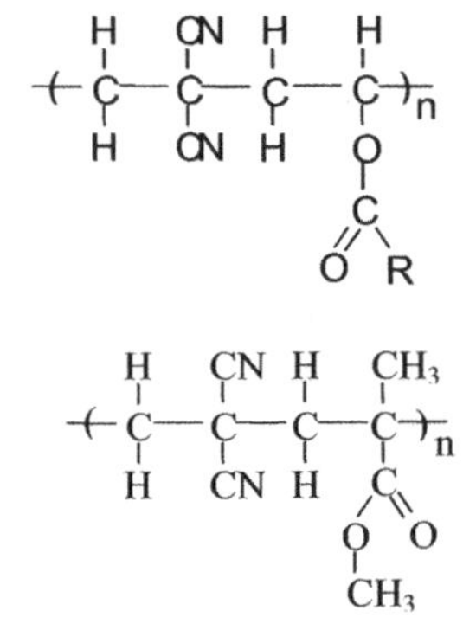

P(VDCN-VAc) where R = CH_3
P(VDCN-VPr) = CH_2CH_3
P(VDCN-VPiv) = $C(CH_3)_3$
P(VDCN-VBz) = C_6H_5

P(VDCN-MMA)

Figure 23.4. *Chemical structures of vinylidene cyanide copolymers: VAc= vinyl acetate, VBz= vinyl benzoate, VPr= vinyl propyonate, VPiv= vinyl pivarate, MMA= methyl methacrylate*

23.3. Electro-active properties of piezoelectric polymers

23.3.1. *Ferroelectricity*

By definition, a crystal, a ceramic or a polymer is ferroelectric if it can present, in the absence of an electric field, a polarization P (C.m^{-2}) which can be inversed by the application of an electric field E (V.m^{-1}) greater than a critical value E_c, called a coercive field.

Ferroelectricity is present in ten classes of crystalline symmetry which do not possess any center of symmetry. In these materials, a weak frequency alternative field with amplitude $E_m > E_c$ produces a hysteresis loop between the electric displacement vector D_i (C.m^{-2}) and the electric field E_i:

$$D_i = \varepsilon_0 E_i + P_i \qquad \text{(with } 1 \leq i \leq 3) \qquad [23.2]$$

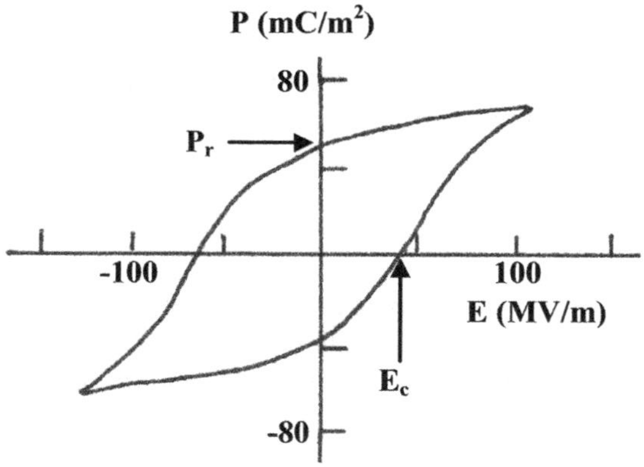

Figure 23.5. *Hysteresis loop of a ferroelectric PVDF*

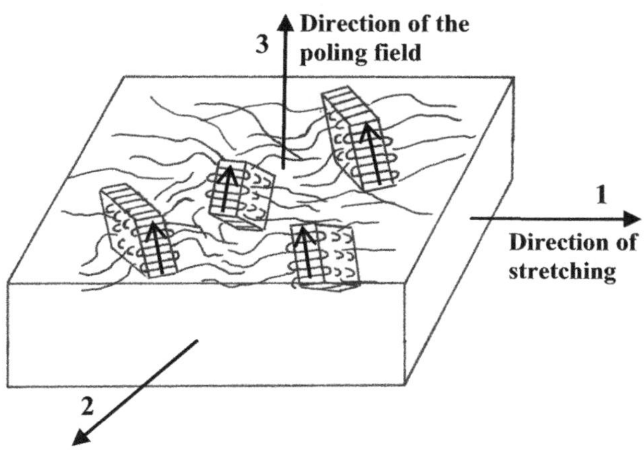

Figure 23.6. *Definition of the tensorial directions for a piezoelectric semi-crystalline polymeric film, the arrows in the crystallites designate their polarization*

The example of a hysteresis loop obtained on a ferroelectric PVDF is represented in Figure 23.5. For a null field, $D = P_r$, where P_r is called the remnant polarization of the material. The values of the index $i = 1$, 2 and 3 correspond to the numbers associated with the three axes represented for a ferroelectric polymeric film in Figure 23.6. The previous section on piezoelectric polymeric materials shows that the application of a thermomechanical orientation on the films permits the optimization of the polarization of crystallites or of the domains of local order in fluorinated polymers, odd polyamides, and Poly(vinylidene cyanide) copolymers. Thus, the axes at the scale of the film are designated by convention [LOV 83]: (1) for the direction of stretching or the preferred direction of the axis of molecular chains, (2) for the direction in the film plane and perpendicular to (1) and (3) for the direction perpendicular to the film plane corresponding to that of the poling field.

In ferroelectric materials, the polarization appears spontaneously at the microscopic scale, corresponding to the ferroelectric domains in monocrystals, to microcrystals in ceramics and crystallites in polymers. However, these polarized regions generally have a random orientation with respect to each other, such that their sum makes the material non-polarized. The application of an electric field greater than the coercive field is therefore fundamental to obtain a macroscopic polarization.

The ferroelectricity of polymers is due to the presence of permanent dipoles associated with each monomer [LOV 85]. Thus, for odd polyamides, we have described in the previous section that a uniaxial orientation produces an orientation of hydrogen bonds parallel to the film plane. The application of an electric field then causes a 90° rotation of the dipoles. The structure in sheets then passes from an orientation in the film plane to an orientation perpendicular to the film plane, by a simple twist of the chains around their main axis [SCH 92]. The film thus polarized is illustrated in Figure 23.6.

In general, the pole field E greater than E_c allows orienting the dipoles in a direction and thus creates a stable configuration under no external field: each dipole tends to orient itself in the local field created by all the other dipoles [LOV 83]. As shown in Figure 23.6, the polarization resulting from each crystallite is therefore oriented in a preferred manner according to the poling field E. In a volume element dv the polarization P is written as a function of the dipole moments μ_i:

$$\Sigma \vec{\mu}_i = \vec{P}.dv \qquad [23.3]$$

As in insulating materials, the polarization is related to the electric field by the dielectric susceptibility χij (F.m-1) (where χij is a second-rank tensor):

$$P_i = \chi_{ij} E_j \qquad [23.4]$$

Both relationships [23.2] and [23.4] allow us to obtain:

$$D_i = \varepsilon_0 E_i + \chi_{ij} E_j = \varepsilon_0 \delta_{ij} E_j + \chi_{ij} E_j = (\varepsilon_0 \delta_{ij} + \chi_{ij}) E_j = \varepsilon_{ij} E_j \qquad [23.5]$$

where $\varepsilon_{ij} = \varepsilon_0 \delta_{ij} + \chi_{ij}$ is the dielectric permittivity of the material and $\delta_{ij} = 1$ for $i = j$, $\delta_{ij} = 0$ for $i \neq j$. For most ferroelectrics, $\varepsilon_0 \delta_{ij} << \chi_{ij}$ and $\varepsilon_{ij} \approx \chi_{ij}$. In practice, the relative dielectric permittivity $\varepsilon_r = \varepsilon_{ij} / \varepsilon_0$ is commonly used.

23.3.2. *Semi-crystalline polymers: Fluorinated polymers and odd polyamides*

The origin of the polarization in fluorinated polymers and in odd polyamides was much debated for a long time. In particular, different processes were proposed in order to explain the electro-active properties of PVDF in the β-phase: the electret effect and the Maxwell–Wagner–Sillars effect notably. The revelation of the hysteresis loop and dipolar reorientation in fluorinated polymers and in odd polyamides allowed the ferroelectric character of these materials to be shown. The origin of the polarization was therefore attributed to crystalline zones [LOV 83, DAS 91]. Table 23.1 compares the characteristic parameters of PVDF in β-phase and a ferroelectric P(VDF-TrFE) 75-25 with a TGS crystal and a $BaTiO_3$ ceramic. We note that the properties of ferroelectric polymers depend on the implementation conditions of the film (uniaxial or biaxial orientation, temperature, etc.).

In comparison with crystals and ceramics, certain ferroelectric polymers, such as P(VDF-TrFE) copolymers, present a phase transition produced at the Curie temperature (T_c) [FUR 89]. Beyond this temperature, these materials lose their ferroelectric properties and become paraelectric. Depending on the transition order, the remnant polarization P_r, is cancelled in a continuous (second order transition) or discontinuous (first order transition) way. No ferro/paraelectric phase transition is observed in PVDF and odd polyamides.

Ferroelectric materials	E_c (MV/m)	P_r (mC/m²)	T_{fp} (°C)	ε_r (at 20°C)
TGS	1	28	49	50
$BaTiO_3$	0.1 to 2	260	120	1,200
PZT	2.5	140	>300	1,700
PVDF	30 to 100	30 to 60	—	10 to 15
P(VDF/TrFE) 75/25	~ 100	50 to 100	130	~ 10

Table 23.1. *Coercive field, E_c, remnant polarization, P_r, ferro / paraelectric transition temperature, T_{fp}, and relative dielectric permittivity, ε_r, of ferroelectric materials*

23.3.3. *Amorphous Poly(vinylidene cyanide) copolymers*

The electro-active properties of vinylidene cyanide (VDCN) copolymers depend on the dipole moments μ of the constituent motifs, and their concentration N. The obtained values, for example, in the P(VDCN-VAc) copolymer are respectively

$\mu = 19.10^{-30}$ C.m and N = $0.44.10^{28}$ m^{-3}, while in PVDF, they correspond to $\mu = 6.7.10^{-30}$ C.m and N = $18.40.10^{28}$ m^{-3}. The piezoelectric properties are also related to the ability of the dipoles to orient themselves during the application of an electric field E. In amorphous polymers, the orientation of dipoles gets easier as the molecular mobility is high. Thus, the poling process involved heating samples to some high temperature greater to that of the glass transition (T > Tg), applying a high electric field and then cooling the polymer to room temperature with the field applied.

The optimization of the aligning of dipoles depends on the incidence of the polarization temperature and the intensity of the electric field on the relaxation time associated with the molecular mobility. Although the intensity of the electric field must be as high as possible to obtain the best resulting polarization, the value of E is restricted to that which produces the dielectric breakdown. In general, for these amorphous copolymers, the applied electric field is less than 100 MV/m.

The polarization $P_t = N.\mu$ corresponds to the maximal theoretical value which could be obtained in a polymer. However, for the best experimental polarization conditions, the results show that the optimal orientation of the dipoles with respect to the poling field gives a remnant polarization P_r much lower than P_t. For most polymers, the measured remnant polarization corresponds to an average value of the ratio $(P_r/P_t) \approx 30\%$. This ratio can, depending on the polymer, clearly diverge from the average value; thus, for example, the ratio P_r/P_t corresponds to a value of about 30 to 42% for PVDF, while it reaches 60% for P(VDCN-VAc) with, for this amorphous copolymer, P_r = 50 mC/m^2. The remnant polarization values of VDCN copolymers are reported in Table 23.2 [NAL 95].

Copolymers	T_g (°C)	ε_r (at 20°C)	$\Delta\varepsilon_r$ (at T_g)	P_r (mC/m^2)
P(VDCN-VAc)	178	5.6	120	35
P(VDCN-VBz)	184	5.6	115	21
P(VDCN-VPr)	176	5.8	85	28
P(VDCN-VPiv)	172	5.8	100	33
P(VDCN-MMA)	146	5.4	30	12

Table 23.2. *Glass transition obtained by AED , Tg, dielectric constant at 20°C and 10 Hz, ε_r, intensity of the dielectric relaxation at T_g, $\Delta\varepsilon_r$, and remnant polarization, P_r, of vinylidene cyanide copolymers*

In semi-crystalline polymers, the polarization of crystallites is not modified for temperatures higher than glass transition. For this reason, PVDF ($T_g \approx$ -35°C) can be used up to temperatures of 80 to 100°C. In amorphous copolymers, the electro-active properties are stable until temperatures situated around the glass transition temperature zone, at which point the increase in the molecular mobility causes a loss of the orientation of dipoles. It is interesting to note that the value of T_g in amorphous copolymers reported in Table 23.2 can reach 170 to 180°C. Furthermore, the intensities of the dielectric relaxation $\Delta\varepsilon$ which are indicated in this table define the amplitude of the dielectric constant hop to the temperature zone crossing of the glass transition (T_g). Thus, the high value of $\Delta\varepsilon$ reported for vinylidene cyanide copolymers reveals cooperative molecular motions with large amplitude at the crossing of T_g.

23.3.4. *Influence of chemical composition and physical structure on the electro-active properties of polymers*

The chemical composition of polymers modifies their ferroelectric properties. Furukawa [FUR 89] revealed that in fluorinated copolymers the unit rate of trifluoroethylene has an influence on the remnant polarization. The maximum remnant polarization $P_r = 100$ mC/m^2 is reached for P(VDF-TrFE) 80-20, while it is of the order of 60 mC/m^2 in the PVDF homopolymer. We must note that the increase in the rate of TrFE motifs would tend to reduce the dipole moment of a macromolecule because the dipole moment associated with the TrFE motif is less than that of the vinylidene (VDF) motif. The results show that rates of TrFE less than 20% molar cause a reduction of the remnant polarization. This evolution is attributed to a decrease in the crystallinity rate. Moreover, it is for a TrFE rate greater than 50% molar that the ferroelectrics properties appear more clearly, because the conformational disorder leads to the outbreak of new crystalline phases.

Odd polyamides	PA 11	PA 9	PA 7	PA 5
Density of dipoles 10^{-30} C.m /100Å^3	4.29	4.93	6.59	9.41
P_r (mC/m²)	56	68	86	125
E_c (MV/m)	64	75	80	100

Table 23.3. *Density of dipoles, remnant polarization and coercive field in odd polyamides [MEI 93]*

In odd polyamides, the remnant polarization and the coercive field are both directly proportional to the density of dipoles in the material corresponding to the density of amide groups (Table 23.3). The shortening of paraffin sequences situated between amide groups increases the cohesion energy of crystalline regions. Thus, the increase in the coercive field is associated with an increase in the density of dipoles and therefore in the number of hydrogen bonds [MEI 93].

Generally, in semi-crystalline polymers, since the electro-active properties are mainly related to the nature of the crystalline phase, the improvement of the macroscopic polarization is obtained by processes of thermal processing or mechanical stretching, which either allows the crystallinity rate to increase, or to increase the relative orientation of crystallites.

In amorphous VDCN copolymers, local order plays a major role on the electro-active properties. The increase in the remnant polarization occurs if a local order exists in the amorphous polymers, or if it is produced by a thermomechanical orientation. Indeed, the orientation causes an increase in intermolecular dipolar interactions and facilitates the densification of the arrangement of macromolecules. These zones of local order thus generated will give rise to a larger polarization P_r.

23.3.5. *Protocols of polarization*

Several polarization protocols of polymeric films have been used since the discovery of the ferroelectric properties of these materials. The first method, proposed by H. Kawaï for the polarization of PVDF films, consisted of applying a continuous electric field of about 30MV/m at 90°C for a relatively long time ($\approx 10^3$s), then cooling the film under an electric field down to room temperature [KAW 69].

A method for industrial use was also developed: Corona polarization [BAU 96]. This method permits a polarization to be made at room temperature in short times ($\approx$10s), gets rid of the breakdown phenomena of films and thus polarizes very large film surfaces. The device is composed of a point situated a few centimeters from the film surface, with a very high potential (20 kV) producing air ionization. A grid placed between the point and the film surface is brought to an intermediary potential (10 kV) permitting the field to homogenize at the film surface and to fit the potential for its surface. The charges thus drained towards the film surface create a strong electric field (of a few hundred MV/m) which causes the orientation of the dipoles of the crystalline phase.

Another method, particularly developed at the Franco-German Institute of Saint Louis (ISL) [BAU 94], consists of applying a low frequency alternative electric field on an initially metallized film. The gradual increase in the poling field allows, when E is greater than the coercive field, the hysteresis loop between the polarization and the electric field to be reached. When the poling field is reduced to zero, the film possesses a permanent polarization which is very stable over time.

23.3.6. *Piezoelectricity*

Piezoelectricity is the property of certain materials to charge themselves electrically when they are submitted to a mechanical stress (a direct piezoelectric effect) or to get deformed under the action of an electric field (converse piezoelectric effect) [NYE 85]. This effect, only present in insulating materials, is equivalent to the outbreak of charges with opposite signs on both sides of the film. At the molecular level, this effect is due to the non-coincidence of centers of gravity with positive and negative charges.

The direct piezoelectric effect causes a linear relationship between the X_{ij} (N.m^{-2}) components of the second-rank tensor, representing the stress state applied to the piezoelectric material and the resulting charge density (electric displacement) D_i which is produced at constant electric field E and temperature T:

$$D_i = d_{ijk} X_{jk} \qquad \text{(with } 1 \leq i, j, \text{ and } k \leq 3) \qquad [23.6]$$

The three components of the polarization are therefore related to the nine X_{ij} components of the stress, by the 27 constant piezoelectric coefficients d_{ijk} (C.N^{-1}), a third-rank tensor.

The converse piezoelectric effect is the thermodynamic consequence of the direct effect. In the converse effect, there is a linear relationship between the components of the vector E_i, intensity of the electric field applied to the material, and the tensor components of the strain S_{ij} characterizing the change of the material dimensions (they contract or expand). Thus, at constant stress and temperature:

$$S_{ij} = d_{kij} E_k = d^t_{ijk} E_k \qquad \text{(with } 1 \leq i, j, \text{ and } k \leq 3) \qquad [23.7]$$

where the exponent t indicates the transposed matrix. The units of the converse piezoelectric coefficients are (m.V^{-1}). The piezoelectric coefficients for the direct and converse piezoelectric effects are thermodynamically identical, i.e. $d_{direct} = d_{inverse}$; they can be either positive or negative. Let us note that the sign of the charge D_i and of the strain S_{ij} resulting from the piezoelectric effect depends on the direction of the mechanical and electric fields respectively. Usually, the piezoelectric coefficient measured in the direction of applied electric field is designated as the longitudinal coefficient, whilst that measured in the direction perpendicular to the electric field is called the transverse coefficient. The coefficients having indices $i \neq j$ represent the shear components of strain and stress tensors.

The piezoelectricity of polymers is caused by the biphasic character of these materials. Thus, the specificity of the piezoelectric properties is attributed to the existence of an intrinsic polarization of crystalline domains associated with the mechanical adaptation of the amorphous phase.

According to the generalized form of Hooke's law, a solid submitted to a stress X_{jk} lower than the sufficiently weak elasticity and intensity limit, will be subjected to a strain S_{ij} in the linear approximation, such as:

$$S_{ij} = s_{ijkl} X_{kl} \qquad \text{(with } 1 \leq i, j, k \text{ and } l \leq 3) \qquad [23.8]$$

where s_{ijkl} ($m^2.N^{-1}$) represents the elastic compliance of a fourth-rank tensor.

The inverse relationship $X_{ij} = c_{ijkl} S_{kl}$ defines elastic stiffness tensor c_{ijkl} ($N.m^{-2}$) of the material. The coefficients of these two tensors are related by the relationship: $s_{ijkl} \,.\, c_{klmn} = c_{ijkl} \,.\, s_{klmn} = \delta_{im} \,.\, \delta_{jn}$.

The piezoelectric effect is present in 20 classes of crystalline symmetry which do not possess any center of symmetry. Ferroelectricity is therefore not a condition required for the existence of piezoelectricity. Indeed, piezoelectricity can exist at temperatures greater than the ferro/paraelectric transition, or in non-ferroelectric materials such as quartz.

23.3.7. *Reduction of the number of independent coefficients – Matrix notation*

In general, stress and strain are represented by symmetric tensors ($X_{ij} = X_{ji}$ and $S_{ij} = S_{ji}$), causing coefficients d_{ijk} symmetric in j and k on the piezoelectric tensor: $d_{ijk} = d_{ikj}$. The number of independent piezoelectric coefficients is thus reduced from 27 to 18. Furthermore, the symmetry of stress and strain tensors also reduces the number of elasticity and rigidity coefficients from 81 to 36. It is therefore possible to use a more concise notation to write the coefficients of piezoelectric, elasticity and rigidity tensors, known as matrix notation [NYE 85]. In this matrix notation, the second and the third index of d_{ijk}, as well as both indices of X_{jk} are reduced to only one index which varies from 1 to 6, depending on the relation: the pairs of indices ii = 11, 22, 33 are respectively replaced by m = 1, 2, 3, and the pairs of mixed indices ij = 23 or 32, 13 or 31, 12 or 21 are respectively replaced by m = 4, 5, 6. Equations [23.6], [23.7] and [23.8] are respectively written in the following matrix format:

$$D_i = d_{im} X_m \qquad [23.9]$$

$$S_m = d_{im} E_i \qquad [23.10]$$

$$S_m = s_{mn} X_n \qquad [23.11]$$

where $i = 1, 2, 3$ and $m, n = 1, 2, \ldots, 6$. It is important to note that the matrices s, d, X and S in their reduced notation do not transform as tensors when the coordinate system is changed.

23.3.8. *Piezoelectric constitutive equations*

A formalism presenting the coupling between the thermal, elastic and electric parameters of a material was introduced based on thermodynamic principles [MIT 76, MAS 50]. Thus, the choice of several thermodynamic potentials and independent variable combinations allowed a series of eight piezoelectric constitutive equations to be obtained for an isothermal process which, written in matrix notation, reduce to:

$$S_m = s_{mn}^{T,E} X_n + d_{im}^{T,X} E_i \quad [23.12] \qquad X_m = c_{mn}^{T,D} S_n - h_{im}^{T,S} D_i \quad [23.16]$$

$$D_i = d_{im}^{T,E} X_m + \varepsilon_{ij}^{T,X} E_j \quad [23.13] \qquad E_i = -h_{im}^{T,D} S_m + \beta_{ij}^{T,S} D_j \quad [23.17]$$

$$X_m = c_{mn}^{T,E} S_n - e_{im}^{T,S} E_i \quad [23.14] \qquad S_m = s_{mn}^{T,D} X_n + g_{im}^{T,X} D_i \quad [23.18]$$

$$D_i = e_{im}^{T,E} S_m + \varepsilon_{ij}^{T,S} E_j \quad [23.15] \qquad E_i = -g_{im}^{T,D} X_m + \beta_{ij}^{T,S} D_j \quad [23.19]$$

with $1 \le i$ and $j \le 3$, and $1 \le m$ and $n \le 6$. The superscripts in equations [23.12] – [23.19] indicate the variables which are maintained constant.

According to equations [23.12] and [23.13], the pure piezoelectric effect of deformation (of electric charge) is exclusively obtained under no stress conditions (no electric field); e, g and h are piezoelectric tensors; c is the calorific capacitance and β is the inverse dielectric susceptibility ($= \chi^{-1}$). The thermodynamics of piezoelectric materials also reveals that the piezoelectric coefficients of the same type are thermodynamically equivalent: $d^X = d^E$, $g^D = g^X$, $e^S = e^E$ and $h^D = h^S$, their superscripts are usually omitted. Moreover, the piezoelectric tensors' coefficients are mutually related by the following relationships:

$$d_{im} = e_{in} s_{nm}^{E} = \varepsilon_{ij}^{X} g_{jm} \qquad (\text{m.V}^{-1} \text{ or C.N}^{-1}) \qquad [23.20]$$

$$e_{im} = d_{in} c_{nm}^{E} = \varepsilon_{ij}^{S} h_{jm} \qquad (\text{C.m}^{-2} \text{ or N.V}^{-1}.\text{m}^{-1}) \qquad [23.21]$$

$$g_{im} = h_{in} s_{nm}^{D} = \beta_{ij}^{X} d_{jm} \qquad (\text{m}^{2}.\text{C}^{-1} \text{ or V.m.N}^{-1}) \qquad [23.22]$$

$$h_{im} = g_{in} c^{D}_{nm} = \beta^{X}_{ij} \, e_{jm} \qquad (\text{N.C}^{-1} \text{ or V.m}^{-1}) \qquad [23.23]$$

Thus, d_{im} is designated as the strain or the electric charge coefficients, e_{im} as the electric charge or the stress coefficients, g_{im} as the strain or the voltage coefficients and h_{im} as the stress or the voltage coefficients. Consider, as an illustration, the difference between g and d coefficients, while omitting matrix indices for simplicity. Equation $D = d.X$ gives the piezoelectric charge measured on a sample connected in short circuit, allowing the flow of charges in the external circuit. If, on the other hand, the sample is placed in an open circuit, the charges will accumulate at the surface of the sample and will produce an electric field E through the sample. This electric field depends on the capacitance (permittivity) of the sample (charge = capacitance.voltage) being related to the stress by: $E = -g \, . \, X$ with $g = d/\varepsilon$.

23.3.9. *Comparison of piezoelectric properties*

Let us take for example PVDF, one of the polymeric materials much used in applications, in which only 5 of its piezoelectric coefficients are non-zero, considering its symmetry [NYE 85]:

$$d_{ij} = \begin{bmatrix} 0 & 0 & 0 & 0 & d_{15} & 0 \\ 0 & 0 & 0 & d_{24} & 0 & 0 \\ d_{31} & d_{32} & d_{33} & 0 & 0 & 0 \end{bmatrix} \qquad [23.24]$$

d_{31}, d_{32}, d_{33} associate a longitudinal stress, depending on the axes 1, 2, 3, to a polarization in the direction (3). Both other coefficients are related to shear effects. When the film is isotropic in its plane (biaxially oriented or unoriented): $d_{31} = d_{32}$, and $d_{15} = d_{24}$.

Ferroelectric materials	d_{31} (10^{-12} C/N)	e_{31} (10^{-2} C/m²)	g_{31} (10^{-3} V.m/N)	h_{31} (10^{7} V/m)	k_{31}
PVDF	20	6.0	174	53	0.10
P(VDF/TrFE)(55/45)	25	3.0	160	19	0.07
P(VDF/TrFE) (75/25)	10	2.0	110	22	0.05
P(VDCN/VAc)	6	2.7	169	76	0.06
PZT	110	920	10	87	0.31

Table 23.4. *Comparison of piezoelectric properties of polymers and nonorganic materials [NAL 95]*

Table 23.4 compares the physical properties of the commercially available piezoelectric polymers with the properties of a typical piezoelectric ceramic, PZT [NAL 95]. We note that the piezoelectric constants d_{31} of all polymers are relatively weak in comparison to the highest values of PZT. However, this is compensated by a dielectric constant ε_r which is weaker in polymers (Tables 23.1 and 23.2). Thus, the voltage generated per unit stress, which is called g_{31}, is higher in polymers than in other materials, which gives them a better sensitivity during the application of a force (stress).

The electromechanical coupling factor k_{ij} is introduced to directly estimate in a material the conversion efficiency from one form of energy to another. The coupling factor k is therefore defined as a function of the direct or converse piezoelectric effect by the relationships:

$$k^2 = \frac{\textit{Mechanical energy converted to electrical energy}}{\textit{Total input mechanical energy}} \qquad [23.25]$$

$$k^2 = \frac{\textit{Electrical energy converted to mechanical energy}}{\textit{Total input electrical energy}} \qquad [23.26]$$

23.4. Piezoelectricity applications

23.4.1. *Transmitting transducers*

Transmitting transducers correspond to piezoelectric applications where the electrical energy provided to dielectric (free remainder to become deformed) is by unit volume (to simplify, the indices are omitted):

$$W_f = \frac{1}{2}\varepsilon^X E^2 \qquad [23.27]$$

where ε^X is the permittivity of the free dielectric and E the applied electric field. The mechanical energy converted by piezoelectric effect of the electrical energy is by unit volume:

$$W_m = \frac{1}{2}s^E S^2 \qquad [23.28]$$

where s^E is the elastic constant at given electric field and S the strain. By piezoelectric effect, the deformation is proportional to the applied field $S=d.E$. The supplied energy W_f is therefore split into two parts:

$$W_f = W_{él} + W_m = \frac{1}{2}\varepsilon^S E^2 + \frac{1}{2} s^E d^2 E^2 \qquad [23.29]$$

where ε^S is the permittivity of the clamped dielectric. So:

$$W_f = \frac{1}{2}\varepsilon^S \left(1 + \frac{s^E d^2}{\varepsilon^S}\right) E^2 = \frac{1}{2}\varepsilon^S \left(1 + k^2\right) E^2 \qquad [23.30]$$

k^2 therefore represents the fraction of supplied electrical energy which is transformed into mechanical energy [23.26]. The electromechanical coupling factor can be obtained from measured d-constant as follows:

$$k = d\sqrt{\frac{s^E}{\varepsilon^S}} \qquad [23.31]$$

For PVDF, k^2 is of the order of 5.10^{-2}, whereas for polycrystalline ferroelectrics, the coupling factor ranges between 25.10^{-2} and 40.10^{-2}. Although the coupling factor of PVDF is weaker than that of ferroelectric ceramics, the convertible mechanical energy is potentially of the same order of magnitude: indeed, the maximum electric fields applicable to PVDF are greater than those supported by ceramics (because coercive fields are higher in PVDF at room temperature, and the weak thicknesses achievable in polymers facilitate the thermal dissipation). Thus, for:

– PVDF: $E_{max} = 10^7$ V.m^{-1} and $\varepsilon^X = 10^{-10}$ F.m^{-1} gives $W_{fmax} = 5000$ J.m^{-3};

– PZT: $E_{max} = 3.10^5$ V.m^{-1} and $\varepsilon^X = 10^{-8}$ F.m^{-1} gives $W_{fmax} \approx 500$ J.m^{-3}.

The maximal energy we can supply to PVDF is therefore about 10 times higher than that admissible by a PZT ceramic, for which the electromechanical conversion efficiency is about 10 times stronger. Other polymers, P(VDF-TrFE) and P(VDNC/VAc) have ferroelectric properties comparable (Table 23.4) to that of PVDF. If we consider that the acoustic matching of the propagation media (air, water, biological tissues) is better for polymers than for PZT and certain characteristics intrinsic to polymers (soft, light, and easy to manufacture into large sheets and cut or bend into complex shapes) can be dominating, it appears that ferroelectric polymers take an interesting place in being complementary to PZT in specific applications, as in the domains of high fidelity loudspeakers and earphones, underwater acoustic hydrophones and probes for medical ultrasound.

23.4.2. *Piezoelectric sensors*

In the domain of application of piezoelectric sensors, the evaluation of the specificity of both types of polymeric and ceramic materials is made, like in the previous section, by undertaking a study based on PVDF and PZT.

As can be shown from Table 23.4, for all piezoelectric coefficients *d*, where (to simplify) the indices are omitted, a charge freed during the application of the same stress will be 10 times weaker in PVDF than in PZT: $d_{(PVDF)} \approx$ 10 to 20 $pC.N^{-1}$ and $d_{(PZT)} \approx$ 100 at 300 $pC.N^{-1}$.

The permittivity ε of PVDF is about 100 times weaker than that of PZT; the electric field generated within the dielectric, for a given stress, is also higher for PVDF. Indeed, the piezoelectric constant, $g=d/\varepsilon$, shows that $g_{(PVDF)} \approx$ 10 at 20.10^{-2} $V.m.N^{-1}$ and $g_{(PZT)} \approx$ 1 to 3.10^{-2} $V.m.N^{-1}$. Thus, for a sensitivity s_e given in voltage by unit stress, the thickness *é* of a PVDF film can be 10 times smaller than that of PZT since: $s_e = g \,.\, é$ $(V.N^{-1}.m^2)$.

The merit factor, $M = W/v$, of a low frequency sensor, particularly in the case of hydrophones, is defined as the electrical energy available by unit volume *v* for a unit stress. This merit factor is favorable to PVDF in comparison with PZT:

$$M = \frac{1}{2}\frac{s_e^2 C}{v} = \frac{1}{2}g^2\varepsilon = \frac{1}{2}g.d \qquad [23.32]$$

where *C* is the electrical capacitance of the film, thus:

$g_{(PVDF)} \,.\, d_{(PVDF)} \approx g_{(PZT)} \,.\, d_{(PZT)} \approx$ 1 at 9.10^{-12} $J.m^{-3}/N.m^{-2}$.

As in the case of transmitting transducers, PVDF cannot claim to replace PZT type ceramics, but covers applications such as light active membranes (microphones), weak dimension pressure probes (shockwave chronography), large band probes for ultrasound echography, and high dimension hydrophones required for very low frequencies.

23.5. Transducers

23.5.1. *Principle*

Electroacoustic transducers are made up of piezoelectric polymer films metallized on both sides. In the audio frequency range, the application of an alternating voltage between the metallized sides generates a surface variation by transverse piezoelectric effect, to which a vibration of the film corresponds. If the frequency imposed by the voltage is sufficiently high, the film vibrates following its thickness direction at high frequency, producing ultrasounds. Electroacoustic transducers such as headphones, speakers and ultrasonic generators all work according to this principle. Thus, for the same applied voltage, a high piezoelectric constant causes a large displacement (deflection), and for a large Young's modulus, the film gives a stronger output force during the vibration. Vice versa, a pressure difference in the front and rear sides of the piezoelectric film leads to a surface variation which causes the outbreak of an electric voltage between electrodes. This type of functioning is used in sensors devices: microphones, pressure probes, etc.

Piezoelectric polymers challenge ceramics with their large surfaces, weaker density, better acoustic matching to aqueous media and lower price. Furthermore, they present a strong advantage; they can be thermoformed according to the shapes of any surfaces whose geometries could not be made before, to constitute sensors and transducers of specific shape: spheres and segments of a sphere, pistons, surfaces adapting themselves well to that of the support, etc.

23.5.2. *Electroacoustic transducers*

The first commercial applications of piezoelectric polymeric films were electroacoustic transducers such as high-frequency loudspeakers (tweeters), stereophonic headphones and microphones, [RAV 81] which are distinguished from usual transducers by an active membrane. The piezoelectric polymeric films are well fitted for this type of acoustic transducer which possesses a wide range in frequency. Indeed, the characteristics of these polymers allow an implementation in the form of a thin film (1 – 300 μm), with low mechanical stiffness. In 1974, The Pioneer Company thus developed an omni-directional tweeter up to 20 kHz, composed of a piezoelectric PVDF film. The realization of numerous acoustic transducers possessing piezoelectric polymeric films has been described in several articles [WAN88]. The small thickness of films allows promising applications such as a transparent loudspeaker composed of a P(VDCN/VAc) film with the aspect of a sheet of paper.

23.5.3. *Characteristics of ultrasonic transducers*

The development of piezoelectric polymers applications in the domain of ultrasonic transducers is very promising. The transducers realized cover a wide range of frequency from 10 kHz to 100 Mhz, such as hydrophones, acoustic probes for non-destructive testing and probes for medical echography. Table 23.5 compares a few physical properties of these materials with those of piezoelectric ceramics much used in ultrasonic transducers. Ferroelectric polymers present several advantages in comparison with ceramics:

– these soft materials are characterized by an acoustic impedance (about 4.10^6 $kg.m^{-2}.s^{-1}$) near to that of water and the human body, allowing more complete energy transfer and a weaker reflection and distortion of the wave at the transducer interface;

– the large mechanical loss of polymers permits short pulses with a high distance resolution to be obtained;

– the implementation of polymers permits the form to be adjusted, in order to improve the ultrasound wave focus; a better resolution of the direction and the analyzed zone is thus obtained;

– the biocompatibility of these materials with human tissues is favorable to medical applications.

Ferroelectric material	Density ρ (10^3 kg/m^3)	Sound velocity (10^3 m/s)	Coupling factor k_{31}	Mechanical loss tan δe	Acoustic impedance Z (10^6 kg/m^{2}s)
PVDF	1.78	3.0	0.20	0.13	4.0
P(VDF/TrFE) 75/25	1.88	2.0	0.30	0.05	4.5
P(VDCN/VAc)	1.20	4.5	0.28	0.02	3.0
PZT	7.5	83.3	0.64	0.002	30

Table 23.5. *Electromechanical properties of polymeric and ceramic materials for ultrasonic transducers*

23.5.4. *Hydrophones*

One of the original applications of ferroelectric polymers was as an acoustic probe immersed in water. Polymeric films which make up hydrophones are active materials mainly characterized by piezoelectric constants which depend on the type of application, the structure of the transducer, the frequency band used, and the exposition conditions of the film in its environment.

Let us first consider the hydrophone in its sensor mode, when it operates as a recepting transducer, where the piezoelectric film is deformed under the effect of acoustic wave pressure. At the lowest frequencies, the piezoelectric hydrostatic constant in voltage g_{3h} is used, whilst at highest frequencies, the piezoelectric constant in voltage g_{33} is dominating. Indeed, at sufficiently low frequencies, the size of the active material can be considered small in comparison with the acoustic wavelength, so that the acoustic pressure is identical depending on the three axes of the hydrophone. In these conditions, the output signal of the hydrophone is a function of the hydrostatic mode associated with the piezoelectric constant in voltage g_{3h}. This constant is defined as the superposition of active modes according to the three axes:

$$g_{3h} = g_{31} + g_{32} + g_{33} \quad [23.33]$$

where g_{3h} corresponds to the electric excitation produced according to the three axes and h indicates a hydrostatic mode. For piezoelectric ceramics and piezoelectric copolymers, the modes (3,1) and (3,2) are identical:

$$g_{3h} = 2g_{31} + g_{33} \quad [23.34]$$

As for the hydrophone in the transmitting mode, at high frequencies the piezoelectric effects following the thickness are associated with the dominating electric charge constant d_{33}, whilst d_{3h} is used at low frequencies.

In PZT, the negative values of the constants g_{31} and g_{32} and the positive values of g_{33} give a low g_{3h} in comparison with the pure (3,3)-mode excitation, causing a weak hydrostatic-mode hydrophone. For the piezoelectric homopolymer, PVDF, the mode constants (3,1) and (3,2) are different because of the stretching and poling processes involved during its implementation. These characteristics communicate to the PVDF a relatively high and useful hydrostatic receiving sensitivity, thus favoring the development of a hydrophone in hydrostatic-mode.

Planar hydrophones were the first hydrostatic probes made with a PVDF film. In the eighties, this type of device was then improved by placing a stiffener either behind a single sheet or between two sheets of ferroelectric polymers. The stiffening layers were thick plates of epoxy-loaded graphite and aluminum. The reason for the addition of a stiffener to the hydrostatic transducer design was to improve low frequency response. However, this rigid membrane tends to block the polymeric film laterally, in such a way that the (3,1) and (3,2) modes are restricted [WAN 88].

The cylindrical hydrophone is probably the simplest device which operates in hydrostatic mode. This type of probe, made for example with a piezoelectric copolymer, is of cylindrical shape [NAL 95]. These cylinders can have a nominal external diameter ranging between 0.64 and 2.54 cm, and a length which can reach 30 cm. For a nominal wall of thickness 0.5 to 1.3 mm, the sensitivity reached is -202 to -194 dB, corresponding to 1 $V.\mu Pa^{-1}$. The performances and the characteristics thus obtained are due to the properties of active materials, but for numerous uses of these hydrophones, the cylinder must be protected by a coating.

The development of hydrophones is mainly made in the domain of excitation modes associated with the coefficients g_{33} and g_{3h}. However, more recent research has been undertaken for the conception of a probe possessing a high gain in the acoustic detection of mode (3,1) [BHA93]. This transducer is made up of a ferroelectric polymeric film tightened by springs along the axis (3,1).

23.5.5. *Probes for Non-Destructive Testing (NDT)*

Piezoelectric polymer-based transducers are also developed in a wide domain of ultrasounds for non-destructive testing (NDT) in industry. PVDF ultrasonic probes have shown that they are characterized by weak noise and a wide band, permitting measurements on metallic or non-metallic pieces to be made [CHA 88]. In the aeronautic domain, it has been shown that this type of probe is also suitable for the testing of composite materials.

In general, the effect of the devices that have been made is that, during tests in which the penetration depth of the probe is essential, piezoelectric ceramics are preferable, whereas when good resolution is preferred, ferroelectric polymers are chosen. Both of these materials can therefore be used in a complementary way in the conception of probes, where the receiving transducers are composed of PVDF and the transmitting transducers are made with PZT [FOS 81].

23.5.6. *Biomedical transducer applications*

The biomedical applications of ferroelectric polymers mainly use their electromechanical transduction properties for the measurements of dynamic pressure during the flow of fluid, heartbeats and contact pressure on soft tissue [GAL 88]. Their low acoustic impedance and their high mechanical losses compared to piezoelectric ceramics imply, for example in PVDF transducers, that the frequency responses are approximately constant in fairly large utilization bands. For thicknesses ranging between 6 and 110μm, the thickness-mode resonance frequencies between 160 kHz and 10 Mhz are achievable. Moreover, a PVDF hydrophone will not perturb the acoustic field and will register rapid dynamic behavior in pressure, as required. These ferroelectric sensors are implemented mainly for external measurements in the form of flexible bands or as implanted sensors applied to pulmonary and cardiovascular domains [MAR 82]. Systems which produce images in real time and which are composed of piezoelectric polymeric films are undergoing an important development in the domain of diagnosis by ultrasound. This type of transducer is especially developed to observe the surface of human organs.

23.6. Conclusion

The potential of piezoelectric polymer applications is great. These polymers have been introduced in numerous sensors and actuators in several domains: medical instrumentation, robotics, electroacoustics and ultrasonic transducers. One of the emerging domains is that of the biomedical world, in which polymers are studied as actuators with the idea of applying them as artificial muscles, medical robots for diagnosis and microsurgery, or implanted to stimulate the tissues.

The development of applications using piezoelectric polymers is conditioned by the improvement of their ability to get integrated into a system, and to meet more and more drastic specifications concerning piezoelectric activity and performance at temperature. The study of the relationship between structure and properties has allowed the development of several semi-crystallins or amorphous polymers owning ferroelectric properties adapted to the realization of new powerful sensors and actuators. On this basis, the investigations are pursued in the domain of the synthesis of materials possessing a high dipole concentration and good electromechanical properties. Indeed, although PVDF and P(VDF-TrFE) possess relatively high piezoelectric responses, these materials have a low relative deformation ($\approx$ 0.1%) and a sensitivity reduced to external stimulations. Recently, a new class of ferroelectric polymers was developed [BAU05], by introducing a monomer, the chlorofluoroethylene (CFE) which is copolymerised with VDF-TrFE to constitute poly(vinylidene-trifluoroethylene-1,1-chlorofluoroethylene fluoride) terpolymer P(VDF-TrFE-CFE). This terpolymer is a ferroelectric relaxor which possesses a high electrostrictive effect with a relative strain change higher than 7% (40% for the muscle) associated with a relatively high modulus of the order of 0.3 GPa (0.8 MPa for the muscle). These properties confer promising electroactive functions on this terpolymer as a sensor or an actuator.

23.7. Bibliography

[BAU 94] BAUER F., "High pressure applications of ferroelectric polymers", *Proceedings 8th International Symposium on Electrets*, IEEE 94CH3443-9, p. 617–622, 1994.

[BAU 05] BAUER F., "Ferroelectric relaxor terpolymers: properties and potential applications", *Proceedings 12th International Symposium on Electrets*, IEEE 05CH37679, p. 117–120, 2005.

[BAU 96] BAUER S., "Poled polymers for sensors and photonic applications", *J. Appl. Phys.*, vol. 80 no. 10, p. 5531–5557, 1996.

[BHA 93] BHAT J.J., THOMSON T.T., SASEENDRAN PILLAI P.R., "Development of (3,1) drive low-frequency piezofilm hydrophones with improved sensitivity", *J. Acoust. Scoc. Am.*, vol. 94, p. 3053–3056, 1993.

[CHA 88] CHANG C., SUN C.T., "A new sensor for quantitative acoustic emission measurement", *J. Acoust. Emission*, vol. 7, p. 21–29, 1988.

[CHE 95] CHEN Q.X., PAYNE P.A., "Industrial applications of piezoelectric polymer transducers", *Meas. Sci. Technol.*, vol. 6, p. 249–267, 1995.

[DAS 91] DAS-GUPTA D.K., "Pyroelectricity in polymers", *Ferroelectrics*, vol. 118, p. 165–189, 19991.

[FOS 81] FOSTER F.S., PATTERSON M.S., HUNT J.W., "The conical scanner: a two transducer ultrasound scatter imaging technique", *Ultrasonic Imaging*, vol. 3, p. 62–82, 1981.

[FUR 89] FURUKAWA T., "Ferroelectric properties of vinylidene fluoride copolymers", *Phase Transitions*, vol. 18, p. 143–211, 1989.

[GAL 88] GALLETTI P.M., DEROSSI D., DEREGGI A.S., *Medical Applications of Piezoelectric Polymers*, Gordon & Breach, New York, 1988.

[KAW 69] KAWAI H., "The piezoelectricity of poly (vinylidene fluoride)", *Jpn. J. Appl. Phys.*, vol. 8, p. 975–976, 1969.

[KEP 92] KEPLER R.G., ANDERSON R.A., "Ferroelectric polymers", *Advances in Physics*, vol. 41, p. 1–57, 1992.

[LOV 83] LOVINGER A.J., "Ferroelectric polymers", *Science*, vol. 220, no. 4602, p. 1115–1121, 10 June 1983.

[LOV 85] LOVINGER A.J., "Recent developments in the structure, properties and applications of ferroelectric polymers", *Jpn. J. Appl. Phys.*, vol. 24, Supplement 24–2, p. 18–24, 1985.

[MAR 82] MARCUS M.A., "Ferroelectric polymers and their applications", *Ferroelectrics*, vol. 40, p. 29–41, 1982.

[MAS 50] MASON W.P., *Piezoelectric Crystals and their Applications in Ultrasonics*, Van Nostrand, New York, 1950.

[MAT 84] MATHUR S.C., SCHEINBEIM J. I., NEWMAN B.A., "Piezoelectric properties and ferroelectric hysteresis effects in uniaxially stretched nylon-11 films", *J. Appl. Phys.*, vol. 56, no. 9, p. 2419–2425, 1984.

[MEI 93] MEI B.Z., SCHEINBEIM J.I., NEWMAN B.A., "The ferroelectric behavior of odd-numbered nylons", *Ferroelectrics*, vol. 144, p. 51–60, 1993.

[MIT 76] MITSUI T., TATSUZAKI I., NAKAMURA E., *An Introduction to the Physics of Ferroelectrics*, Gordon & Breach, London, 1976.

[MIY 80] MIYATA S., YOSHIKAWA M., TASAKA S., KO M., "Piezoelectricity revealed in the copolymer of vinylidene cyanide and vinylacetate", *Polymer J.*, vol. 12, no. 12, p. 857–860, 1980.

[NAL 95] NALWA H.S., *Ferroelectric Polymers*, Marcel Dekker Inc., New York, 1995.

[NEW 90] NEWMAN B.A., KIM K.G., SCHEINBEIM J.I., "Effect of water content on the piezoelectric properties of nylon 11 and nylon 7", *J. Mater. Sci.*, vol. 25, p. 1779–1783, 1990.

[NEY 85] NEY J. F., *Physical Properties of Crystals*, Oxford University Press, Oxford, 1985.

[RAV 81] RAVINET P., GUILLOU D., MICHERON F., "Transducteurs électroacoustiques à dome sphérique de polymère piézoélectrique thermoformé", *Revue Technique THOMSON-CSF*, vol. 13, no. 1, p. 5–42, March 1981.

[SCH 92] SCHEINBEIM J.I., LEE J.W., NEWMAN B.A., "Ferroelectric polarization mechanisms in nylon 11", *Macromolecules*, vol. 25, p. 3729–3732, 1992.

[SER 80] SERVET B., BROUSSOUX D., MICHERON F., BISARO R., RIES S., MERENDA P., "Phases cristallines orientées du poly(fluorure de vinylidène) et leurs propriétés piézoélectriques", *Revue Technique THOMSON-CSF*, vol. 12, no. 4, p. 761–793, December 1981.

[WAN 88] WANG T.T., HERBORT J.M., GLASS A.M., *The Applications of Ferroelectric Polymers*, Chapman & Hall, New York, 1988.

Chapter 24

Polymeric Insulators in the Electrical Engineering Industry: Examples of Applications, Constraints and Perspectives

24.1. Introduction

From the electric power station to the final user, electricity is transported through the medium of high voltage networks, i.e. 1kV to 1,000 kV, and even to 1,200 kV. To ensure the security of electricity supply, the durability of the different electrical equipment used in these networks – transformers, circuit-breakers, lines and cables, etc. – is obviously essential.

Actually, the reliability of this type of product strongly depends on insulating structures and materials. These insulating pieces and structures also play a major role in technical performance and in the cost price of this equipment.

This chapter tackles the main aspects of the engineering of insulating materials for electrical engineering applications, by presenting the typical constraints for the use of high voltage equipment through examples of applications, and then by evoking the possible evolution of the insulation world.

For engineering aspects, the case of high voltage equipment or power transformers is particularly emphasized. However, cables or rotating machines, (such as generators or motors) which are not discussed here, have the same problems

---

Chapter written by Jean-Luc BESSEDE.

and therefore very similar structures, even though thermal or environmental stresses are a bit different.

As for the evolution perspectives, we shall answer the question of optimization of electrical behavior and the prediction of ageing by means of, on the one hand, numerical simulation, and on the other hand by the control of the evolution of the electrical characteristics related to microstructural changes due to ageing.

We shall also deal with the impact of environmental questions on these materials and the possibilities of reducing their environmental impacts via eco-design.

24.2. Equipment

High voltage equipment is conceived to operate in complete safety at nominal currents in normal network conditions but also to face short circuit currents occurring under fault conditions, during the closing or opening of electrical circuits.

When they are open, the equipment must also safely insulate different parts of the network.

24.2.1. *Arc commutation*

In general, during current breaking operations (Figure 24.1), an electric arc is initiated between contacts (7) and (8) and transmits its energy to the thermal expansion volume Vt, generally filled with SF_6 gas, a very good dielectric.

During the passage of the current through zero, the created overpressure is discharged through the insulating nozzle (9) and within the mobile contact (7). This double blowing allows the arc to be cooled effectively and cuts the current.

In the arc switching chamber, the insulation nozzle is submitted to strong stresses.

It must indeed support the high dielectric gradients and the ultraviolet radiation induced by the de-excitation of gas molecules and other constituents of the arc plasma, such as metal vapors and materials coming from the erosion of the walls of the arcing chamber or the nozzle.

The nozzle must also sustain high temperatures generated by the arc, and resist projections of metal vapors and droplets [AES 02].

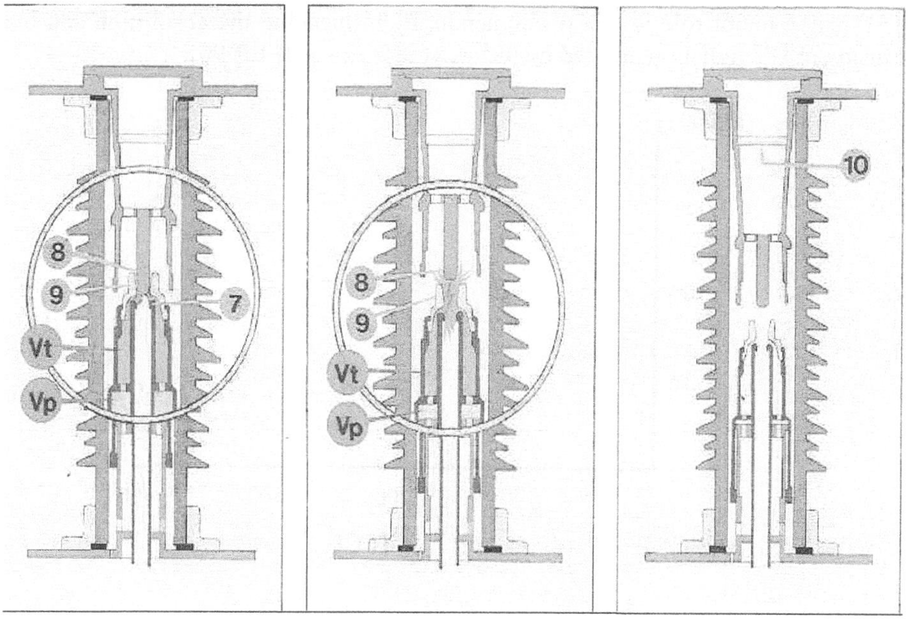

Figure 24.1. *Thermal blowing chamber*

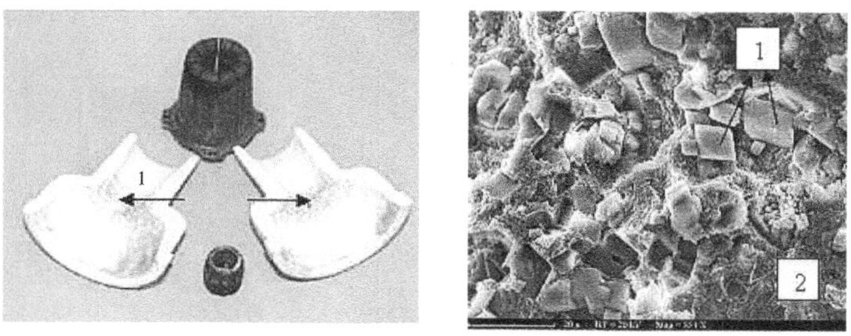

Figure 24.2. *(a) Nozzle and contact of the equipment (1: ablation zone of the nozzle)*
(b) Microstructure of the nozzle (1: mineral charge – 2: PTFE)

This leads to a degradation of the nozzle (Figure 24.2), whose material is generally composed of polytetrafluoroethylene (PTFE) filled with mineral additives.

Studies undertaken on the arc/nozzle interaction have proved that during the extinction of the arc, the erosion of the nozzle is mainly due to photo-ablation. In terms of composite structure, the mineral charges and the microstructure of the

PTFE play a major role in this phenomenon, by influencing the absorption and the diffusion of UV radiation emitted by the arc (Figure 24.3) [FER 96].

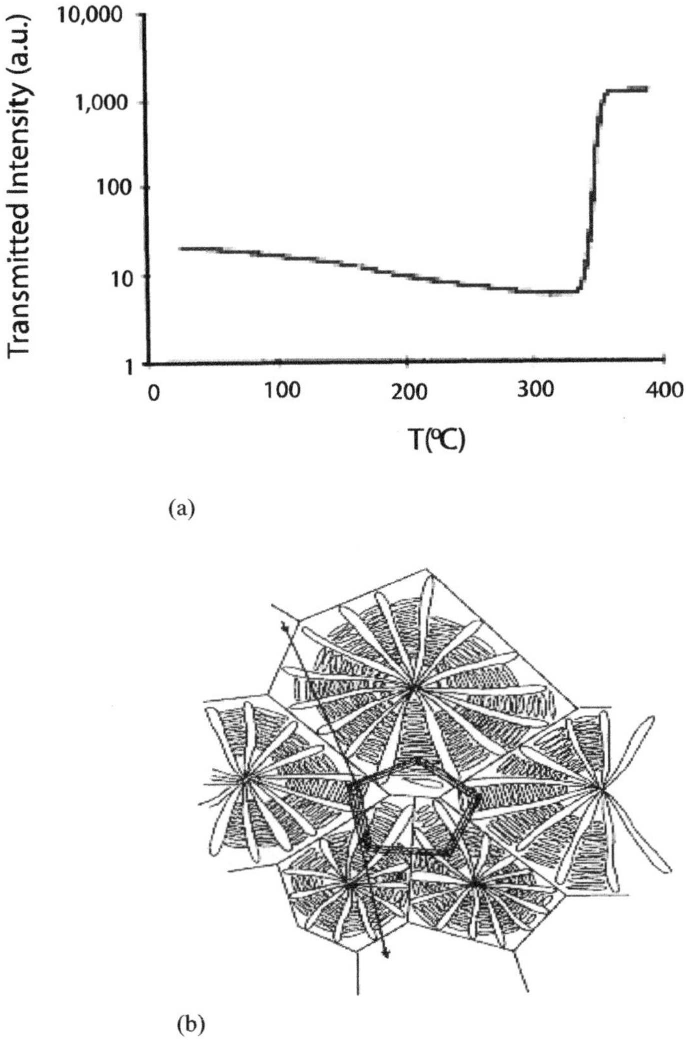

Figure 24.3. *Radiation/PTFE interactions – a) Intensity transmitted as a function of temperature; b) Schematic showing the multiple reflections between crystallites*

Indeed, the dielectric properties of the PTFE, and particularly their stability, are also essential to ensure the insulation capacity of the arcing chamber. Particular attention should be brought to the influence of humidity, as we can see in Figure 24.4 in the case of PTFE filled with CaF_2 [ETI 00].

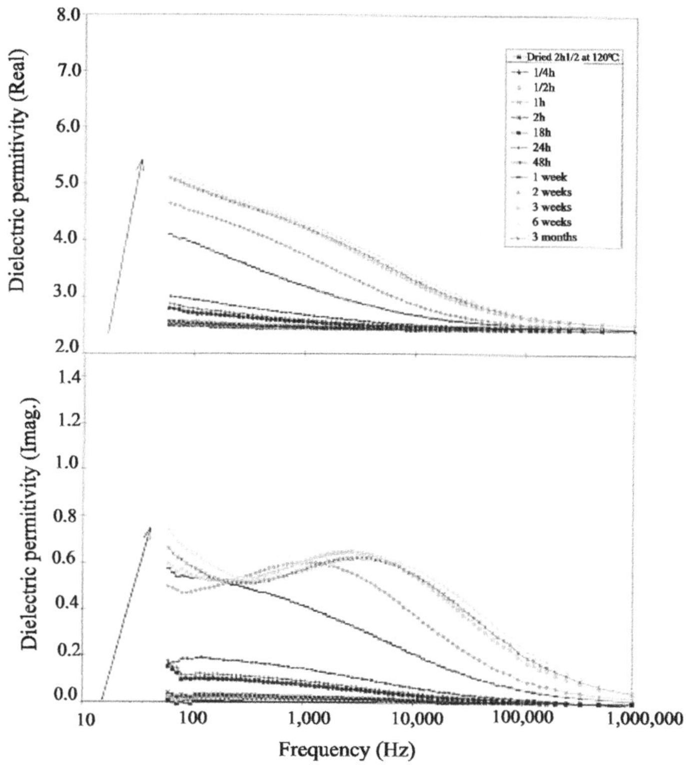

Figure 24.4. *Evolution of the real (ε') and imaginary parts (ε'') of the complex permittivity as a function of the material exposition (PTFE/CaF_2) to ambient air*

24.2.2. *Composite insulators*

Generally, the external insulators of high voltage equipment are composed of aluminous electrotechnical porcelain. This material is well-known and its ageing is very good. However, ceramic insulators are heavy, fragile and, in heavily polluted areas, must often be cleaned.

Composite insulators are therefore more and more commonly used, because they present appealing advantages compared with ceramic insulators. They do not explode in case of impact; they are flexible and light and allow a combination of functions – for example for the use of optical fiber systems.

However, the specifications for such insulators are fairly strict. Indeed, they must present a very good fit, as gas leaks at the equipment level must be lower than 0.5% per year. These insulators must also support any type of atmospheric conditions which might risk affecting their dielectric, mechanical, or watertightness characteristics.

In general, these materials and structures present a lifetime of 20 to 30 years at least, for whatever type of atmosphere (with industrial pollution, marine environment, etc.), and can be installed in conditions of extreme temperatures, from -60°C (in certain cases) up to an ambient temperature of 50°C, leading to particularly high operating temperatures (100°C at least).

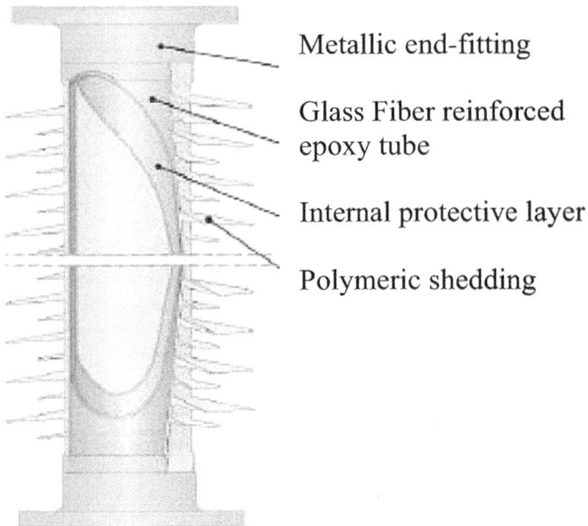

Figure 24.5. *Typical composite insulator*

In such a structure (Figure 24.5), the materials used must have a very good lifetime under voltage, with low dielectric losses as well as good resistivity and chemical resistance characteristics. The composite tube is generally made with epoxy resin reinforced with fiberglass. The winding angle is about +/- 55° in order to withstand bending and pressure stresses and an internal polyester layer is generally added to improve the stability to the arc radiation, or to acids coming from the decomposition of the SF_6 gas.

The extremities are glued to the composite tube before being coated with the shedding material – a silicone elastomer is the most commonly used – which then covers the metallic end-fittings.

These insulators are used for circuit-breakers, measurement transformers, or as bushings (on transformers or sub-stations, for example), but also as insulating enclosures for surge arresters [KIE 05]. This type of structure – with a solid core – is also used as support insulators (of conductive bars for, example), or suspension line insulators for use on overhead lines.

During the design and selection of materials, the quality of the insulation is one of the essential parameters to control, in order to ensure the durability of the insulation offered by these insulators. For example, the current leakage obtained for a given voltage, as a function of the climatic events and ageing (Figure 24.6), allows a distinction to be made between diverse materials.

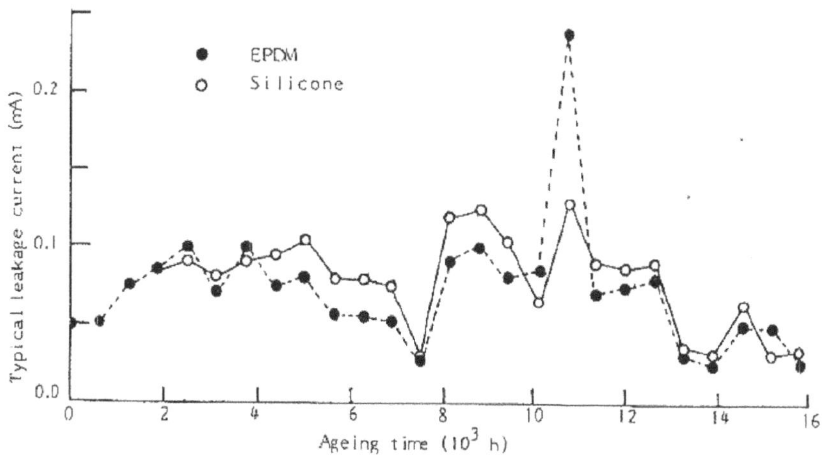

Figure 24.6. *Example of the evolution of the leakage current as a function of the blade material of an insulator*

24.3. Power transformer insulation

A power transformer is one of the most expensive strategic components of transmission and distribution electrical network, and the reliability of these networks strongly depends on the reliability of this type of equipment. Actually, the deterioration of the insulation represents the most frequent cause of failure of transformers.

The insulation of these pieces of equipment can be considered as a typical composite structure. Indeed, it is composed of a combination of insulating solid and liquid materials in which the solid material acts as a mechanical and electrical separation between the conductors (copper or aluminum) or between two loops of a coil.

For such insulation, cellulose has been used for over 100 years, mainly for economical reasons, despite numerous investigations undertaken to replace it by more effective materials. Aramid paper is also used, particularly for applications submitted to high temperatures (i.e. operating temperatures greater than 100°C).

Insulating liquid, mainly mineral oil, is extremely compatible with cellulose and presents good thermal exchange characteristics, acting as an insulator and heat transfer agent.

In this type of insulation, the control of the interactions between cellulose, water and oil, is of particular importance to ensure the durability of the insulation. Over time, a cellulose depolymerization phenomenon occurs, and therefore a degradation of the papers, cardboards and woods appears.

As can be seen in Figure 24.7, the mechanical performance of a cellulose material can thus be reduced by 50% in a relatively short time, compared with the lifetime of a transformer (over 40 years) [KIE 06].

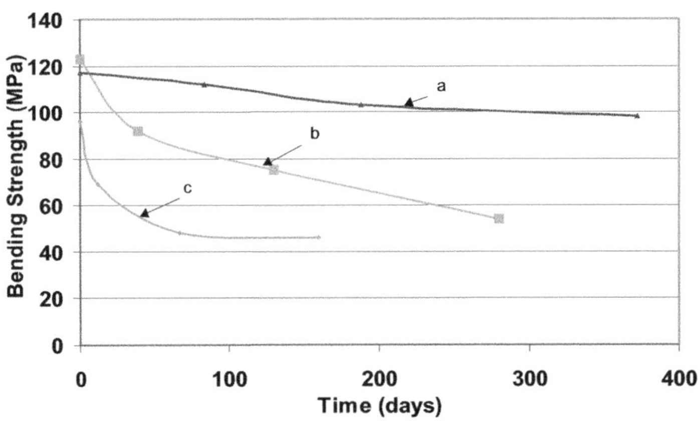

Figure 24.7. *Stress at the breakdown (4-point bending) as a function of ageing at 95°C with a humidity rate of 3% (a), 6% (b) and 10% (c) – the case of lamelled glued wood*

24.4. Perspectives

During the past twenty to thirty years, under the effect of the opening of markets and the reduction of electrical equipment programmes in numerous countries, the manufacturers of electric materials have been subjected to a heavy pressure on prices. This has led them to introduce numerous innovations towards the functional and economical optimization of products and materials, while improving their reliability.

Today, under the effect of the economic dynamism of emerging countries, and with the arrival of numerous pieces of equipment in occidental countries at the end of their lives, a resurgence in the demand for electrical products associated with a renewal of clients' expectations can be seen. This offers manufacturers with new technological innovation opportunities and allows new paths of investigations to be conceived.

In the first place, the question of the performance improvements of insulators is, and will remain, topical for a very long time.

Whether it is to increase the lifetime or reliability, to reduce the dimensions or the costs, the natural trend is to increase the breakdown voltage of insulators as much as possible, and to make this characteristic remain stable over time.

The modeling of the microscopic configuration of the electric field as a function of the electrical properties of different phases present in the material, or breakdown mechanisms, can bring interesting answers to this question. For example, the influence of the presence of different types of particles on the propagation of a discharge can be simulated (Figure 24.8) [FLA 02]. The effect of the granulometric distribution of charges on breakdown can thus be estimated, giving indications of the interest in using nano-particles in such insulating materials.

Today these tools are still at the development stage and, if 2D modeling is fairly accessible, 3D modeling still requires much research. In particular, the correlation between calculations and experiments remains to be established systematically.

The possibility of predicting the remaining lifetime of an insulating piece, from a diagnosis of its state of ageing, is also an important subject.

Indeed, numerous equipments installed from the fifties to the seventies are now reaching their theoretical lifetime limit. Network operators and electricity manufacturers therefore ask themselves, rightly, if the use of this equipment can be prolonged and, in that case, if the risk of accidental failure can be avoided.

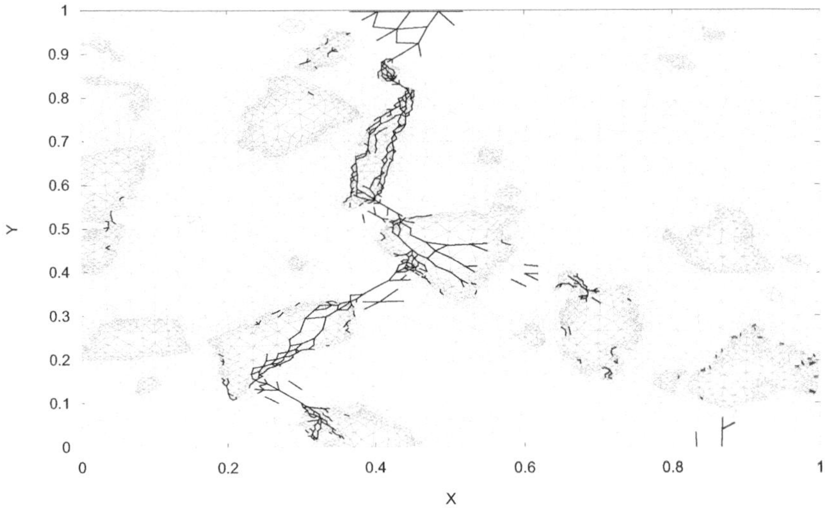

Figure 24.8. *Discharge propagation in a 2D composite structure*

This worry has already initiated many studies. For example, in the case of bulk insulating materials, such as epoxy charged with particles (silicone, calcium carbonate, etc.), the revealing of a correlation between the evolution of electrical characteristics, change in microstructure, and chemical evolution with respect to ageing is well established (Figures 24.9 and 24.10) [VOU 06].

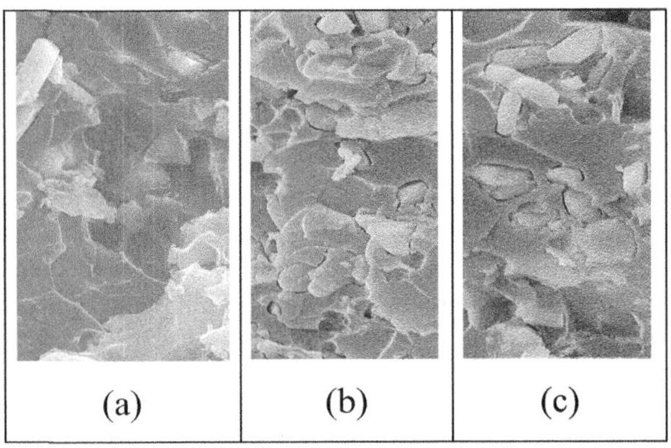

Figure 24.9. *Micro-structure of new epoxy/alumina composite (a), 30 days old (b), 6 months old (c)*

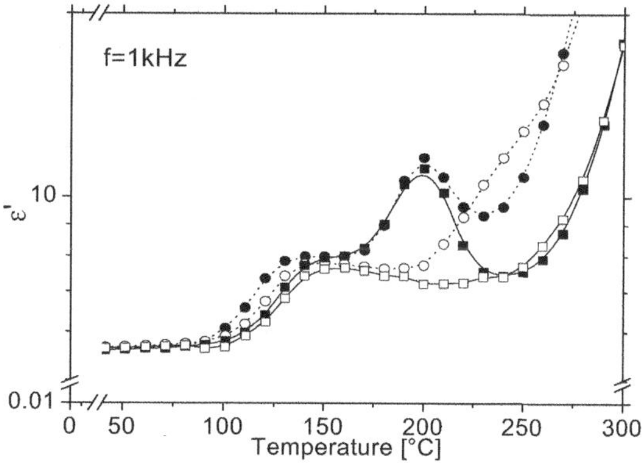

Figure 24.10. *Dielectric permittivity at 1 kHz of the new epoxy/alumina composite (■ 1st scan, □ 2nd scan) and 6 month old (● 1st scan, ○ 2nd scan)*

Other research can also be mentioned. For example the ageing state in transformers can be estimated by following the evolution of the chemical composition of oil by gas phase chromatography. The ageing of cables too, can be particularly studied by following the evolution of the distribution of space charges as a function of time.

These techniques have started to be used systematically in laboratories, and one may think that they will undergo an important development in the future.

Finally, in the domain of insulation, like in other technical domains, eco-design will represent an important means for innovation in the future. Because of the evolution of western society and its regulations, the environmental issue becomes more and more important. To face it, eco-design approaches have been developed by manufacturers, allowing, owing to a rational approach, the environmental impacts of products to be reduced, with a limited economic impact.

Thus, the improvement of recyclability (for example by using thermoplastics instead of thermosets [BES 05] or by avoiding mixtures of materials, making those inseparable at the end of life) constitutes a priority route of environmental improvement.

The use of dangerous substances (CMR: Carcinogenic, Mutagenic, Repro-toxic) in insulators is a more complex issue. Whether it is substances recording (in application of the European directive REACH), reduction of volatile organic compounds (VOC) or elimination of heavy metals or flame retardants, the evolution of authorized substances should, in the coming years, strongly impact on the formulation and the engineering of polymers, and therefore on their properties.

Furthermore, the more common use of environmental evaluations to cover the entire life cycle of products as evaluation tools, even the choice of materials, will also have an important impact on the electrical insulator sector.

24.5. Conclusion

As can be seen in this chapter, polymeric insulating materials, because of their properties and their versatilities, are likely to respond to a wide range of applications and solicitations.

They are often used in a "composite" form allowing complex parts with multiple functions to be obtained by combining different properties. Different polymers can thus be associated together, or polymers can be associated with mineral elements such as fiberglass, mineral powders, etc. Polymers can also be associated with liquids, and thus cover a whole range of insulation functions.

Depending on the application and the function to be realized, all the dielectric properties can be used to characterize and specify the materials used in electrical engineering applications. However, these are not sufficient and (as we have seen) radiation absorption characteristics, microstructural characteristics, or even mechanical characteristics are also very useful.

In the future, the technological evolutions that will prevail will be those which allow manufacturers to best meet the new expectations of their clients (essentially electrical network operators).

On this account, the new electrical properties simulation tools, the ageing studies aimed at revealing correlations between the evolution of microstructures and their properties, and eco-design methodology are the key domains in which research efforts will continue. Any progress made will then allow the designing of future insulators to better answer the needs of performance improvement, lifetime prediction, or the considerations of environmental impacts.

24.6. Bibliography

[AES 02] AESCHBACH H., VISATA O., DUFOURNET D., BESSÈDE J-L., BLATTER J., "Arcing-Contact/Insulating-Wall Interactions in High Voltage Circuit Breakers", *Proceeding of 21st International Conference on Electrical Contacts*, Zurich, Switzerland, p. 511–517, September 2002.

[BES 05] BESSÈDE J-L *et al.*, "Suitability of Polyethylene Terephtalate for the making of HV GIS Insulator", *IEEE Conference on electrical insulation and dielectric phenomena*, Nashville, USA, paper ID 306, 2005.

[ETI 00] ETIENNE S., STOCHMIL C., BESSÈDE J-L., "Dielectric properties of polymer-based Micro-heterogeneous Insulator", *12th int. conf. on internal friction and ultrasonic attenuation*, Buenos Aires, Argentina, 19–23 July 1999, *J. of Alloys and Compounds*, 310, p. 368–373, 2000.

[FER 96] FERRY L., VIGIER G., BESSÈDE J-L., "Effect of ultraviolet radiation on polytetrafluoroethylene: morphology influence", *Polymers for Advanced Technologies*, vol. 7, nos. 5 & 6, p. 493–500, 1996.

[FLA 02] FLANDIN L., VOUYOVITCH L., BESSÈDE J-L., BÉROUAL A., ALBÉROLA N.D., "Modeling of dielectric breakdown in epoxy based heterogeneous systems", *9th INSUCON International Electrical Insulation Conference*, Berlin, 18–20 June 2002.

[KIE 05] KIEFFEL Y., HERMOSILLO V., MISTIAEN B., PUYANÉ R., HASSANZADEH M., BESSÈDE J-L., "Experience with hollow core composite insulators at AREVA T&D - Research and applications", *INMR 2005 World Insulator Congress*, Hong-Kong, November 2005.

[KIE 06] KIEFFEL Y., BESSÈDE J-L., PERRIER C., DEVAUX F., SARAVOLAC M., LONG P., "Improvement of the reliability of power transformer insulation", *10th INSUCON International Electrical Insulation Conference*, Birmingham, UK, 24–26 May 2006.

[VOU 06] VOUYOVITCH L., FLANDIN L., BESSEDE J-L., ALBEROLA N.D., "Evolutions of Microstructure and Dielectric Behaviour of Epoxy Based Insulator-Insulator c composites over Long Periods of Time", *Journal of Applied Polymer Science*, vol. 100, p. 3454–3464, 2006.

List of Authors

Serge AGNEL
IES
Montpellier 2 University
France

Fulbert BAUDOIN
Laboratoire Plasma
et Conversion d'Energie
Paul Sabatier University
Toulouse
France

Alain BERNES
Laboratoire de Physique des Polymères
Paul Sabatier University
Toulouse
France

Abderrahmane BEROUAL
Laboratoire Ampère
Ecole Centrale de Lyon
France

Laurent BERQUEZ
Laboratoire Plasma
et Conversion d'Energie
Paul Sabatier University
Toulouse
France

Jean-Luc BESSEDE
AREVA T&D
ARC
Villeurbanne
France

Guy BLAISE
Laboratoire de Physique des Solides
Paris 11 University
France

Laurent BOUDOU
Laboratoire Plasma
et Conversion d'Energie
Paul Sabatier University
Toulouse
France

Jérôme CASTELLON
IES
Montpellier 2 University
France

Vicente COMPAÑ MORENO
Departamento de Termodinámica
Aplicada
Universidad Politécnica
Valence
Spain

Eric DANTRAS
Laboratoire de Physique des Polymères
Paul Sabatier University
Toulouse
France

Philippe DEMONT
Laboratoire de Physique des Polymères
Paul Sabatier University
Toulouse
France

Slim FAKHFAKH
LASSI/UTAP
Reims University
France

Jean-Numa FOULC
Laboratoire de Génie Electrique
Joseph Fourier University
Grenoble
France

Jean-Luc FRANCESCHI
Laboratoire Plasma
et Conversion d'Energie
Paul Sabatier University
Toulouse
France

Virginie GRISERI
Laboratoire Plasma
et Conversion d'Energie
Paul Sabatier University
Toulouse
France

Omar JBARA
LASSI/UTAP
Reims University
France

Pascale JOLINAT
Laboratoire Plasma
et Conversion d'Energie
Paul Sabatier University
Toulouse
France

Colette LACABANNE
Laboratoire de Physique des Polymères
Paul Sabatier University
Toulouse
France

Nadine LAHOUD
Laboratoire Plasma
et Conversion d'Energie
Paul Sabatier University
Toulouse
France

Christian LAURENT
Laboratoire Plasma
et Conversion d'Energie
Paul Sabatier University
Toulouse
France

Séverine LE ROY
Laboratoire Plasma
et Conversion d'Energie
Paul Sabatier University
Toulouse
France

David MALEC
Laboratoire Plasma
et Conversion d'Energie
Paul Sabatier University
Toulouse
France

Pilar MARTINEZ
AIMPLAS
Instituto Tecnológico del Plástico
Paterna
Spain

Juan MARTINEZ-VEGA
Laboratoire Plasma
et Conversion d'Energie
Paul Sabatier University
Toulouse
France

Didier MARTY-DESSUS
Laboratoire Plasma
et Conversion d'Energie
Paul Sabatier University
Toulouse
France

Christian MAYOUX
Laboratoire Plasma
et Conversion d'Energie
Paul Sabatier University
Toulouse
France

Christine MAYOUX
SETARAM
Caluire et Cuire
France

Jérôme MENEGOTTO
Laboratoire de Physique des Polymères
Paul Sabatier University
Toulouse
France

Philippe MOLINIÉ
Département Energie
Supélec
Gif-sur-Yvette
France

Dominique MOUZE
LASSI/UTAP
Reims University
France

Petru NOTINGHER
IES
Montpellier 2 University
France

Christophe PERRIER
AREVA T&D
Technology and Innovation Center
Massy
France

Anca PETRE
Laboratoire Plasma
et Conversion d'Energie
Paul Sabatier University
Toulouse
France

Stéphanie REMAURY
Laboratoire Matériaux-revêtements
CNES
Toulouse
France

Evaristo RIANDE GARCIA
Instituto de Tecnología de Polímeros
Madrid
Spain

Sébastien RONDOT
LASSI/UTAP
Reims University
France

Isabelle SEGUY
Laboratoire Plasma
et Conversion d'Energie
Paul Sabatier University
Toulouse
France

Gilbert TEYSSEDRE
Laboratoire Plasma
et Conversion d'Energie
Paul Sabatier University
Toulouse
France

Gérard TOUCHARD
Laboratoire d'Etudes aérodynamiques
Poitiers University
Futuroscope-Chasseneuil
France

Alain TOUREILLE
IES
Montpellier 2 University
France

Daniel TREHEUX
LTDS
Ecole Centrale de Lyon
France

Pascal VENET
Laboratoire Ampère
Claude Bernard University
Villeurbanne
France

Eva VERDEJO
AIMPLAS
Instituto Tecnológico del Plástico
Paterna
Spain

Index

R

S

T

U, V, W

Printed and bound by CPI Group (UK) Ltd, Croydon, CR0 4YY

29/09/2026

14971581-0001